# FRACTIONAL DYNAMICS
## Recent Advances

# FRACTIONAL DYNAMICS

## Recent Advances

Joseph Klafter
*Tel Aviv University, Israel*

S C Lim
*Multimedia University, Malaysia*

Ralf Metzler
*Technische Universität Munchen, Germany*

*editors*

World Scientific

NEW JERSEY • LONDON • SINGAPORE • BEIJING • SHANGHAI • HONG KONG • TAIPEI • CHENNAI

*Published by*

World Scientific Publishing Co. Pte. Ltd.
5 Toh Tuck Link, Singapore 596224
*USA office:* 27 Warren Street, Suite 401-402, Hackensack, NJ 07601
*UK office:* 57 Shelton Street, Covent Garden, London WC2H 9HE

**Library of Congress Cataloging-in-Publication Data**
Fractional dynamics : recent advances / edited by Joseph Klafter, S.C. Lim & Ralf Metzler.
p. cm.
Includes bibliographical references and index.
ISBN-13: 978-981-4340-58-8 (hardcover : alk. paper)
ISBN-10: 981-4340-58-8 (hardcover : alk. paper)
1. Fractional calculus. 2. Dynamics. 3. Diffusion--Mathematical models. I. Klafter, J. (Joseph) II. Lim, S. C., 1947– III. Metzler, Ralf.
QC20.7.F73F727 2011
530.12--dc23

2011021725

**British Library Cataloguing-in-Publication Data**
A catalogue record for this book is available from the British Library.

The cover image is based on a microscopic image from the Oddershede group (Niels Bohr Institute, Copenhagen, Denmark), artistically rendered by Mette Høst (artist-in-residence, Niels Bohr Institute).



Typeset by Stallion Press
Email: enquiries@stallionpress.com

Printed in Singapore by World Scientific Printers.

## Preface

When *The random walk's guide to anomalous diffusion* appeared some ten years ago (*Phys. Rep.* **339**, 1 (2000)), none of us anticipated how timely this publication would turn out to be. Since then the concept of fractional dynamic equations has deeply penetrated the statistical and chemical physics communities, and, in parallel, has become a major field in mathematics. In the engineering sciences, in which fractional dynamics has a long-standing history in the description of viscoelastic materials or in process control, it is popular as ever. Fractional concepts are by now also recognized in more applied domains such as biological physics or the geosciences. Visiting the scientific databases, a search for *fractional diffusion* or *fractional Fokker–Planck* results in more than 5,000 direct hits, and the related papers were cited 16,000 times in 2010 alone. For comparison, the more generic term *anomalous diffusion* currently receives 11,000 citations per year.

Fractional dynamics is going strong — time to sit back for a moment and take stock. This monograph is intended both as an introduction to the field of fractional diffusion and relaxation and as a compendium on the recent advances and novel insights in this still rapidly growing research topic. With contributions from the foremost researchers in fractional dynamics, we are confident that this book will turn out to be a useful reference for both colleagues familiar with fractional dynamics as well as those who would like to discover the strengths of this framework.

This book covers the mathematical and physical foundation of fractional dynamics both in terms of ordinary and partial fractional differential equations, as well as novel concepts such as distributed order derivatives. An important mathematical concept in the context of fractional dynamics is the principle of subordination, connecting the fractional solution to its Brownian counterpart. The book also covers a representative spectrum of applications ranging from anomalous relaxation and diffusion in classical systems, to the description of time averages of single trajectory time series, a topic coming into focus due to novel microscopic techniques.

A more recent trend is the usage of fractional concepts in quantum systems. While this field is still under development, we decided to include it in this book for completeness. It may become a useful tool, especially in the semiclassical description of complex quantum systems.

The generalization of classical concepts always leaves some space for personal styles. For instance, some authors prefer to use time-fractional derivatives of Riemann–Liouville kind, while others favor the Caputo operator. As long as the associated initial conditions are properly taken into account, both approaches, as well as the solutions of the respective equations, are identical. We did not attempt to unify the various notations, as they reflect the current diversity in literature. Specific definitions in each chapter ensure that the reader may always directly verify the meaning of the used symbols.

We would like to thank all contributors for the care they invested in the composition of their chapters. With its broad spectrum from mathematical definitions over useful concepts all the way to concrete applications, we hope that this book will be useful for all colleagues working in the field of anomalous dynamics, as well as attract newcomers to one of the most active current research fields.

None of this would have been possible without the help of World Scientific's staff. We would like to thank all of them for their professional editing of the book, as well as their patience.

*Joseph Klafter* (Tel Aviv)
*Swee Cheng Lim* (Kuala Lumpur)
*Ralf Metzler* (Munich)

## Contents


Preface v

List of Contributors ix

**Classical Systems**

1. Anomalous Diffusion and Fractional Transport Equations 3
*R. Metzler and J.-H. Jeon*

2. Stochastic Diffusion and Stable Noise-Induced Phenomena 33
*B. Dybiec and E. Gudowska-Nowak*

3. Characteristic Times of Anomalous Diffusion in a Potential 51
*W. T. Coffey, Y. P. Kalmykov and S. V. Titov*

4. Reactions in Subdiffusive Media and Associated Fractional Equations 77
*S. B. Yuste, E. Abad and K. Lindenberg*

5. Natural and Modified Forms of Distributed-Order Fractional Diffusion Equations 107
*A. Chechkin, I. M. Sokolov and J. Klafter*

6. Anomalous Transport in the Presence of Truncated Lévy Flights 129
*D. del-Castillo-Negrete*

7. Anomalous Diffusion: From Fractional Master Equations to Path Integrals 159
*R. Friedrich*

8. Fractional Feynman–Kac Equation for Anomalous Diffusion Functionals 185
*S. Carmi and E. Barkai*

9. Foundations of Fractional Dynamics: A Short Account 209
*R. Hilfer*

10. Parametric Subordination in Fractional Diffusion Processes 229
*R. Gorenflo and F. Mainardi*

11. Fractional Calculus, Anomalous Diffusion, and Probability 265
*M. M. Meerschaert*

12. Fractional Langevin Equation 285
*E. Lutz*

13. Subdiffusive Dynamics in Washboard Potentials: Two Different Approaches and Different Universality Classes 307
*I. Goychuk and P. Hänggi*

14. Identification and Validation of Fractional Subdiffusion Dynamics 331
*K. Burnecki, M. Magdziarz and A. Weron*

15. A Class of CTRWs: Compound Fractional Poisson Processes 353
*E. Scalas*

16. Origin of Allometry Hypothesis 375
*B. J. West and D. West*

**Quantum Systems**

17. Principles of Fractional Quantum Mechanics 393
*N. Laskin*

18. Two Examples of Fractional Quantum Dynamics 429
*A. Iomin*

19. Fractional Dynamics of Open Quantum Systems 449
*V. E. Tarasov*

20. Casimir Effect Associated with Fractional Klein–Gordon Field 483
*S. C. Lim and L. P. Teo*

Index 507

## List of Contributors

**E. Abad**
Departamento de Física Aplicada
Universidad de Extremadura, E-06071 Badajoz, Spain
eabad@unex.es

**Eli Barkai**
Department of Physics and Advanced Materials and
Nanotechnology Institute
Bar-Ilan University, Ramat Gan 52900, Israel
barkaie@biu.ac.il

**Krzysztof Burnecki**
Hugo Steinhaus Center, Institute of Mathematics and Computer Science
Wroclaw University of Technology
Wyspianskiego 27, 50-370 Wroclaw, Poland
krazysztof.burnecki@pwr.wroc.pl

**Shai Carmi**
Department of Physics and Advanced Materials and
Nanotechnology Institute
Bar-Ilan University, Ramat Gan 52900, Israel
scarmi@shoshi.ph.biu.ac.il

**Aleksei Chechkin**
Akhiezer Institute for Theoretical Physics, Kharkov 61108, Ukraine
and
School of Chemistry, Tel Aviv University, Tel Aviv, Israel
achechkin@kipt.kharkov.ua

**William T. Coffey**
Department of Electronic and Electrical Engineering
Trinity College, Dublin 2, Ireland
wcoffey@mee.tcd.ie

**D. del-Castillo-Negrete**
Oak Ridge National Laboratory, Oak Ridge, TN, USA
delcastillod@ornl.gov

**Bartłomiej Dybiec**
Marian Smoluchowski Institute of Physics
and
Mark Kac Center for Complex Systems Research
Jagiellonian University
ul. Reymonta 4, 30–059 Kraków, Poland
bartek@th.if.uj.edu.pl

**R. Friedrich**
Institute of Theoretical Physics
Westfälische Wilhelms-Universität
Wilhelm Klemm Str. 9
48149 Münster, Germany
fiddir@uni-muenster.de

**Rudolf Gorenflo**
Department of Mathematics and Informatics
Free University of Berlin
Arnimallee 3, D-14195 Berlin, Germany
gorenflo@mi.fu-berlin.de

**Igor Goychuk**
Institute of Physics, University of Augsburg
Universitätsstr. 1, D-86135 Augsburg, Germany
igor.goychuk@physik.uni-augsburg.de

**Ewa Gudowska-Nowak**
Marian Smoluchowski Institute of Physics
and
Mark Kac Center for Complex Systems Research
Jagiellonian University
ul. Reymonta 4, 30–059 Kraków, Poland
gudowska@th.if.uj.edu.pl

**Peter Hänggi**
Institute of Physics, University of Augsburg
Universitätsstr. 1, D-86135 Augsburg, Germany
hanggi@physik.uni-augsburg.de

**R. Hilfer**
Institut für Computerphysik, Universität Stuttgart
Pfaffenwaldring 27, 70569 Stuttgart, Germany
hilfer@icp.uni-stuttgart.de

**Alexander Iomin**
Department of Physics, Technion, Haifa, 32000, Israel
iomin@physics.technion.ac.il

**Jae-Hyung Jeon**
Physics Department, Technical University of Munich
James Franck Straße, D-85747 Garching, Germany
and
Physics Department, Tampere University of Technology
FI-33101 Tampere, Finland
jae-hyung.jeon@ph.tum.de

**Yuri P. Kalmykov**
Lab. Mathématiques et Physique
Université de Perpignan Via Domitia, 52, Avenue de Paul Alduy
66860 Perpignan Cedex, France
kalmykov@univ-perp.fr

**Joseph Klafter**
School of Chemistry, Tel Aviv University, Tel Aviv, Israel
klafter@post.tau.ac.il

**Nick Laskin**
TopQuark Inc., Toronto, ON, M6P 2P2, Canada
nlaskin@rocketmail.com

**S. C. Lim**
Faculty of Engineering, Multimedia University
Cyberjaya, 63100, Selangor, Malaysia
sclim47@gmail.com

**Katja Lindenberg**
Department of Chemistry and Biochemistry and BioCircuits Institute
University of California, San Diego
La Jolla, CA 92093-0340, USA
klindenberg@ucsd.edu

**Eric Lutz**
Department of Physics, University of Augsburg
D-86135 Augsburg, Germany
eric.lutz@physik.uni-augsburg.de

**Marcin Magdziarz**
Hugo Steinhaus Center, Institute of Mathematics and Computer Science
Wroclaw University of Technology
Wyspianskiego 27, 50-370 Wroclaw, Poland
marcin.magdziarz@pwr.wroc.pl

**Francesco Mainardi**
Department of Physics, University of Bologna
Via Irnerio 46, I-40126 Bologna, Italy
francesco.mainardi@bo.infn.it

**Mark M. Meerschaert**
Department of Statistics and Probability
Michigan State University
A430 Wells Hall
East Lansing, MI 48824-1027, USA
mcubed@stt.msu.edu

**Ralf Metzler**
Physics Department, Technical University of Munich
James Franck Straße, D-85747 Garching, Germany
and
Physics Department, Tampere University of Technology
FI-33101 Tampere, Finland
metz@ph.tum.de

**Enrico Scalas**
Dipartimento di Scienze e Tecnologie Avanzate
Università del Piemonte Orientale
viale T. Michel 11, 15121 Alessandria, Italy
and
BCAM - Basque Center for Applied Mathematics
Bizkaia Technology Park, Building 500, 48160 Derio, Spain
enrico.scalas@mfn.unipmn.it
escalas@bcamath.org

**Igor M. Sokolov**
Institute of Physics, Humboldt University, Berlin, Germany
sokolov@physik.hu-berlin.de

**Vasily E. Tarasov**
Skobeltsyn Institute of Nuclear Physics
Moscow State University, Moscow 119991, Russia
tarasov@theory.sinp.msu.ru

**L. P. Teo**
Faculty of Engineering
University of Nottingham Malaysia Campus
Semenyih, 43500, Selangor, Malaysia
LeePeng.Teo@nottingham.edu.my

**Serguey V. Titov**
Institute of Radio Engineering and Electronics
Russian Academy of Sciences, Vvedenskii Square 1, Fryazino
Moscow Region, 141190, Russian Federation
pashkin1212@yandex.ru

**Aleksander Weron**
Hugo Steinhaus Center, Institute of Mathematics and Computer Science
Wroclaw University of Technology
Wyspianskiego 27, 50-370 Wroclaw, Poland
aleksander.weron@pwr.wroc.pl

**Bruce J. West**
Information Sciences Directorate, Army Research Office
Research Triangle Park, NC, USA
bruce.j.west.civ@mail.mil

**Damien West**
Physics Department, Rensselaer Polytechnic Institute
Troy, NY 12180, USA
damien_west@rpi.edu

**Santos B. Yuste**
Departamento de Física, Universidad de Extremadura
E-06071 Badajoz, Spain
santos@unex.es

# Classical Systems

# Chapter 1

## Anomalous Diffusion and Fractional Transport Equations

Ralf Metzler and Jae-Hyung Jeon

*Physics Department, Technical University of Munich, James Franck Straße, D-85747 Garching, Germany*

*Physics Department, Tampere University of Technology, FI-33101 Tampere, Finland*

Fractional dynamic equations have assumed a central rôle in the modeling of anomalous relaxation and diffusion processes. Starting from their derivation from continuous time random walk processes, we here provide a concise account of the physics of fractional dynamic equations as well as their applications to nonequilibrium processes in complex systems. In our discussion particular emphasis is placed on subdiffusion, that is caused by long waiting times. In that case due to the lack of a characteristic time scale ergodicity is broken and even long time averages become random quantities, albeit their magnitude follows a well-defined distribution. We also consider the case of Markovian Lévy flights and their behavior in confining potentials.

1. Introduction . . . . . . . . . . . . . . . . . . . . . . . . . . . 4
2. Continuous Time Random Walk and Fractional Diffusion 7
    2.1. Physical view of CTRW . . . . . . . . . . . . . . . . 9
3. Fractional Fokker–Planck–Smoluchowski Equation . . . . 10
    3.1. Subdiffusive case . . . . . . . . . . . . . . . . . . . . 11
    3.2. Subordination scheme for the subdiffusive case . . 14
    3.3. Lévy flights in external potentials . . . . . . . . . . 15
    3.4. Boundary value problems . . . . . . . . . . . . . . . 16
        3.4.1. Subdiffusion . . . . . . . . . . . . . . . . . . 16
        3.4.2. Lévy flights . . . . . . . . . . . . . . . . . . . 18
4. Randomness of Long Time Averages in Subdiffusive CTRW Processes . . . . . . . . . . . . . . . . . . . . . . . . . 20
5. Conclusions . . . . . . . . . . . . . . . . . . . . . . . . . . . 26

## 1. Introduction

In the quantitative sciences, power-laws in fact have a long history in the description of materials properties. Thus, Wurttembergian philosopher and mathematician Georg Bernhard Bülfinger[a] already in 1729 reported the necessity to use power-laws with non-integer exponents to quantify stress-strain relations of materials with pronouncedly non-Hookean behavior [1]. His work sank into oblivion and was later revived by Bach [2]. Examples for specimen obeying power-law stress-strain relations are cast iron, stone, or concrete [3, 4]. From the early 20th century power-laws had been recognized as characteristic for the temporal relaxation of viscoelastic materials, i.e. substances combining both elastic and viscous behavior in their response to external mechanical disturbance. Indeed, following experimental evidence by Nutting [5], several authors introduced fractional constitutive equations to account for these phenomena, among others, Gemant [6], Scott Blair [7], Gerasimov [8], Rabotnov [9], and Mainardi and Caputo [10]. Physically such *fractional relaxation laws* correspond to hierarchies of springs and dashpots [11, 12], and combinations of such fractional rheological elements leads to quite complex relaxation behaviors [13, 14], describing experimental data extremely well over many orders of magnitude [15–18]. For a more complete account, see the recent book by Mainardi [19]. The development of *fractional diffusion equations* is more recent. They emerged from considerations that long-ranged correlations, temporal or spatial, of a stochastic process can be translated into time or space fractional differentials. Without attempting to be exhaustive, the earlier works contain those of Balakrishnan [20], Wyss [21], Peseckis [22], Schneider and Wyss [23], Jumarie [24], Zaslavsky [25], and Saichev and Zaslavsky [26]. Fractional diffusion equations were also proposed in connection with random walks on fractals, for instance, by Giona and Roman [27], and Metzler *et al.* [28].

A consistent derivation of the time or space fractional diffusion equation in a regular Euclidean geometry is based on the continuous time random walk (CTRW) theory [29–33], applied, for instance, in [34–41]. Alternatively, it can be derived from a Langevin equation formalism with stable noise, or based on subordination techniques [42–49]. Here we are concerned with anomalous diffusion processes characterized by a power-law mean squared displacement of the form [50, 51]

$$\langle \mathbf{r}^2(t)\rangle = \int_V \mathbf{r}^2 P(\mathbf{r},t)dV = \frac{2dK_\alpha t^\alpha}{\Gamma(1+\alpha)} \tag{1}$$

[a]Sometimes also spelled Bilfinger.

in $d$ spatial dimensions, where $\alpha$ is the anomalous diffusion exponent, and $K_\alpha$ with physical dimension $[K_\alpha] = \text{cm}^2/\text{sec}^\alpha$ the anomalous diffusion constant. In Eq. (1) we defined the second moment $\langle \mathbf{r}^2(t)\rangle$ in terms of the ensemble average, the spatial integral involving the probability density function $P(\mathbf{r}, t)$ to find the test particle at position $\mathbf{r}$ at time $t$. Depending on the magnitude of $\alpha$, we distinguish *subdiffusion* for $0 < \alpha < 1$ or *superdiffusion* for $\alpha > 1$. The case $1 < \alpha < 2$ is sometimes referred to as sub-ballistic superdiffusion, while $\alpha > 2$ is called hyperdiffusion. The two limiting cases are normal diffusion ($\alpha = 1$) and ballistic motion ($\alpha = 2$), the latter being described by the two propagation modes of the wave equation. In what follows we mainly focus on the subdiffusive regime. However, we also consider the case of Lévy flights, Markovian processes with broad jump length distributions such that the variance actually diverges. Instead, fractional order moments can be calculated, see Sec. 3.3. We discuss the fractional equations describing anomalous diffusion, paying particular attention to the motion in an external potential field $V(\mathbf{r})$. We will also consider the more recent results on the analysis of single trajectories in terms of time averages. In the ageing regime of CTRW subdiffusion ergodicity is violated, and the behavior of time averaged quantities such as the mean squared displacement consequently differs from the analogous ensemble averages. For simplicity, we will restrict the discussion to one dimension, although generalisation to higher dimensions is straightforward.

Before proceeding, a word on the mode relaxation of subdiffusive CTRW processes is in order. As we will see, each temporal eigenmode relaxes non-exponentially, following the Mittag-Leffler function

$$\Phi(t) = E_\alpha(-(t/\tau)^\alpha) = \sum_{n=0}^{\infty} \frac{(-[t/\tau]^\alpha)^n}{\Gamma(1+\alpha n)}, \tag{2}$$

which is sometimes referred to as the natural generalisation of the exponential function. At short times $t \ll \tau$ the limiting behavior is equivalent to the stretched exponential or Kohlrausch–Williams–Watts (KWW) function,

$$\Phi(t) \sim 1 - \frac{t^\alpha}{\tau^\alpha \Gamma(1+\alpha)} \sim \exp\left(-\frac{t^\alpha}{\tau^\alpha \Gamma(1+\alpha)}\right). \tag{3}$$

At long times, $t \gg \tau$, the Mittag-Leffler function has the following expansion around $t = \infty$ [52]:

$$\Phi(t) = \sum_{n=1}^{\infty} \frac{(-1)^{n-1}\,(\tau/t)^{\alpha n}}{\Gamma(1-\alpha n)}. \tag{4}$$

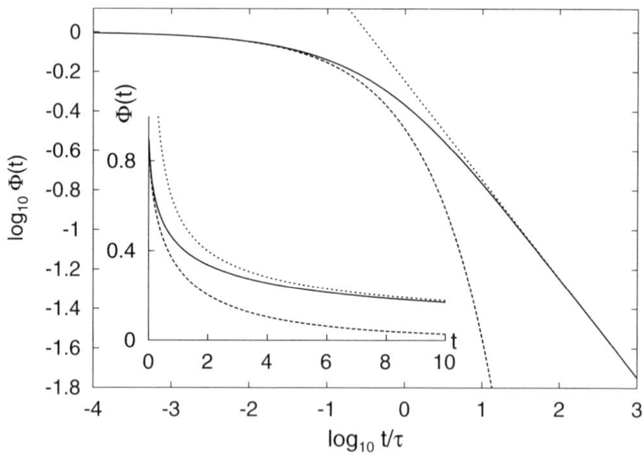


Fig. 1. Mittag-Leffler relaxation pattern. The full line (—) shows the Mittag-Leffler function (2) for $\alpha = 1/2$, while the dashed (- - -) and dotted ($\cdots$) lines represent the associated stretched exponential (3) and inverse power-law (5) behaviors, respectively. The main figure shows the behavior in $\log_{10}$–$\log_{10}$ scale, while for the inset linear scales are used.

To leading order, we thus observe the inverse power-law (Nutting) decay

$$\Phi(t) \sim \frac{1}{\Gamma(1-\alpha)} \left(\frac{\tau}{t}\right)^{\alpha}. \tag{5}$$

The interpolation property of the Mittag-Leffler function between initial stretched exponential and terminal inverse-power-law behavior is shown in Fig. 1. The Mittag-Leffler turnover has indeed been observed experimentally, from polymer rheology [15], over ligand rebinding to proteins [17] and protein conformation dynamics [53], to financial market time series [54].

Over the recent years fractional dynamic equations have found widespread applications in physics and related quantitative sciences, and are now recognized as an important tool in the description of nonequilibrium complex systems [50, 55–66]. The modeling in terms of fractional equations is bolstered up by a number of mathematical monographs on fractional calculus, see, for instance, [67–72]. The solutions of linear fractional dynamics equations of relaxation or diffusion type are often expressed in terms of special functions, in particular, in terms of Fox $H$-functions; these are discussed in [73–75]. The current book reviews the most recent advances on *Fractional Dynamics*, both from the aspect of applications and the mathematical formulation.

## 2. Continuous Time Random Walk and Fractional Diffusion

The continuous time random walk model was originally introduced by Montroll, Weiss, Scher, Lax, and Shlesinger [29–32]; for an authoritative account of CTRW theory, see the book by Hughes [33]. In a CTRW process each jump is characterized by a random jump length and waiting time, both of which are independent random variables, identically distributed according to the probability densities $\lambda(x)$ and $\psi(t)$, respectively. CTRWs are therefore renewal processes, and often called semi-Markovian.[b] The latter alludes to the fact that locally the jumps are Markovian, but the overall displacement $x = \sum_i x_i$ and process time $t = \sum_i t_i$ obtained from summation over many increments (if following a stable law) bestows long-ranging correlations in space and/or time to the overall process, respectively.

One can readily show that the probability density function of a CTRW process in Fourier–Laplace space is given by[c]

$$P(k,u) = \frac{1-\psi(u)}{u}\frac{1}{1-\lambda(k)\psi(u)}, \tag{6}$$

where we assumed that jump lengths and waiting times are not coupled [79]. For jump length and waiting time distributions with finite variance $\langle \delta x^2 \rangle = \int_{-\infty}^{\infty} x^2 \lambda(x)dx$ and mean waiting time $\langle \delta t \rangle = \int_0^{\infty} t\psi(t)dt$, independent of the details of $\lambda(x)$ and $\psi(t)$, by virtue of the central limit theorem the CTRW model reproduces the Gaussian probability density function obeying the diffusion equation. In contrast, when jump lengths and/or waiting times have distributions of the asymptotic power-law forms

$$\lambda(x) \simeq \frac{\sigma^{\mu}}{|x|^{1+\mu}}, \quad \psi(t) \simeq \frac{\tau^{\alpha}}{t^{1+\alpha}} \tag{7}$$

---

[b]The renewal property is given up in stochastic models in which waiting time and/or jump length contain a memory themselves [76–78].

[c]We denote the integral transforms of a function through explicit dependence on the image variable. The Fourier transform of a function $f(x)$ is defined through

$$f(k) = \int_{-\infty}^{\infty} f(x)\exp(ikx)dx,$$

while the Laplace transform of a function $g(t)$ is obtained from

$$g(u) = \int_0^{\infty} g(t)\exp(-ut)dt.$$

with $0 < \mu < 2$ and $0 < \alpha < 1$, anomalous diffusion occurs. The laws (7) emerge due to the generalized central limit theorem and have characteristic functions[d] with small $k$ and $u$ expansions ($0 < \mu \leq 2$ and $0 < \alpha \leq 1$)

$$\lambda(k) = \exp(-[\sigma|k|]^{\mu}) \sim 1 - (\sigma|k|)^{\mu}, \quad \psi(u) = \exp(-[u\tau]^{\alpha}) \sim 1 - (u\tau)^{\alpha}. \tag{8}$$

We distinguish the following cases: (i) normal diffusion for $\mu = 2$ and $\alpha = 1$, when the jump length variance $\sigma^2$ and the characteristic waiting time $\tau$ exist; (ii) CTRW subdiffusion for $\mu = 2$ and $0 < \alpha < 1$; (iii) Lévy flights for $0 < \mu < 2$ and $\alpha = 1$; and (iv) Lévy flights with long-tailed waiting times in the case $0 < \mu < 2$ and $0 < \alpha < 1$. In Fig. 2 we show the time series $x(t)$ for a Lévy flight and a subdiffusive CTRW process. The long jumps and waiting times are distinct. Diverging jump length variance or characteristic waiting times give rise to either subtleties in the presence of boundary conditions or to ageing and ergodicity breaking, see below.

The probability density (6) associated with the distributions (7) and (8) is governed by the time and space fractional diffusion equation

$$\frac{\partial}{\partial t}P(x,t) = {}_0D_t^{1-\alpha} K_{\alpha}^{\mu} \frac{\partial^{\mu}}{\partial|x|^{\mu}} P(x,t). \tag{9}$$

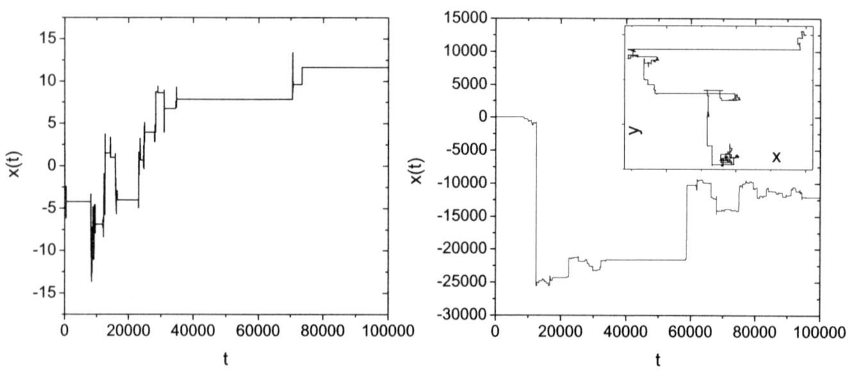


Fig. 2. Time series $x(t)$ of a subdiffusive CTRW process with $\alpha = 1/2$ (left) and a Lévy flight with $\mu = 1$ (right). Due to the scale-free nature of the waiting time and jump length distributions, sojourn times and jump lengths occur on all scales. The inset in the right graph shows the trajectory on the $(x, y)$-plane. The fractal dimensions of a Lévy flight is $d_f = \mu$.

[d]For simplicity we assume symmetric jump length distributions.

Here, $K_\alpha^\mu = \sigma^\mu/\tau^\alpha$ is the generalized diffusion coefficient of dimension $[K_\alpha^\mu] = \mathrm{cm}^\mu/\mathrm{sec}^\alpha$, and we introduce the fractional space and time derivatives, defined in terms of the Riesz–Weyl space fractional operator [71, 72]

$$\frac{\partial^\mu}{\partial|x|^\mu}P(x,t) = \frac{1}{2\Gamma(2-\mu)\,|\cos\pi\mu/2|}\frac{\partial^2}{\partial x^2}\int_{-\infty}^{\infty}\frac{P(x',t)}{|x-x'|^{\mu-1}}dx' \tag{10}$$

with the property

$$\int_{-\infty}^{\infty}\frac{\partial^\mu}{\partial|x|^\mu}P(x,t)e^{ikx}dx = -|k|^\mu P(k,t), \tag{11}$$

and the Riemann–Liouville fractional operator [67, 69–72]

$${}_0D_t^{1-\alpha}P(x,t) = \frac{1}{\Gamma(\alpha)}\frac{\partial}{\partial t}\int_0^t\frac{P(x,t')}{(t-t')^{1-\alpha}}dt'. \tag{12}$$

An equivalent form of Eq. (9) is represented by

$${}_0D_t^{\alpha}P(x,t) - \frac{t^{-\alpha}}{\Gamma(1-\alpha)}P_0(x) = K_\alpha^\mu\frac{\partial^\mu}{\partial|x|^\mu}P(x,t), \tag{13}$$

where the singular term on the left-hand side occurs due to the fact that the Riemann–Liouville fractional derivative of a constant does not vanish: ${}_0D_t^\alpha 1 = 1/[t^\alpha\Gamma(1-\alpha)]$. Therefore, the initial condition $P_0(x) = \lim_{t\to 0} P(x,t)$ explicitly occurs in the dynamic equation (13). Note that this point does not arise when one uses the fractional Caputo operator:

$${}_0^CD_t^\alpha P(x,t) = \frac{1}{\Gamma(1-\alpha)}\int_0^t(t-t')^{-\alpha}\frac{\partial}{\partial t'}P(x,t')dt'. \tag{14}$$

In that case the fractional diffusion equation (13) becomes

$${}_0^CD_t^\alpha P(x,t) = K_\alpha^\mu\frac{\partial^\mu}{\partial|x|^\mu}P(x,t). \tag{15}$$

Thus Riemann–Liouville and Caputo formulations are exchangeable, if only proper care is taken about the initial conditions.

## 2.1. *Physical view of CTRW*

Physically, long-tailed waiting time distributions and their effects have been observed in numerous systems. Examples include iterated maps [80, 81], sticking phenomena of trajectories around stable islands in chaotic systems [56, 64, 82, 83], quenched energy landscapes with exponential trap depth distribution [84, 85], or comb models [86]. In particular, in a biological context power-law waiting time distributions was observed for tracers in reconstituted actin networks [87], Mittag-Leffler and KWW dynamics was observed in the dynamics of proteins [17, 53, 95], and the associated ergodicity breaking was demonstrated for lipid granule diffusion in living cells [88] and for the lateral motion in biomembranes [89]. Finally, we mention geophysical systems [90–92] and charge carrier transport in amorphous semiconductors [32]. Subdiffusive CTRW and the associated Mittag-Leffler relaxation are also associated with ageing phenomena [80, 93, 94].

Long-tailed jump length distributions have been shown to occur naturally for motion on fast folding polymers [96, 97] and facilitated diffusion models of gene regulation including DNA looping [98]. They have been found in several studies of animal and human motion patterns, see, e.g., [99–104]. Moreover, diverging kinetic energy fluctuations were observed in optical lattices [105].

## 3. Fractional Fokker–Planck–Smoluchowski Equation

In the scenario considered above it is assumed that the jump length distribution is homogeneous and isotropic in space, i.e. that it only depends on the distance between starting and end point of the respective jump. This property is obviously violated when an external bias in the form of a potential $V(x)$ acts on the test particle, and the jump lengths depend explicitly on the starting point $x'$ and the end point $x$ of a jump. Based on the generalized master equation, it was demonstrated that the dynamic equation of such a random process is given by the time and space fractional Fokker–Planck–Smoluchowski equation [106–108]

$$\frac{\partial}{\partial t}P(x,t) = {}_0D_t^{1-\alpha}\left(\frac{\partial}{\partial x}\frac{V'(x)}{m\eta_\alpha} + \frac{\partial^\mu}{\partial|x|^\mu}K_\alpha^\mu\right)P(x,t), \tag{16}$$

for the case of diverging $\langle\delta x^2\rangle$ and $\langle\delta t\rangle$. For constant external potential, Eq. (16) reduces to the fractional diffusion equation (9). It is interesting to note that the drift term is not affected by the long-tailed jump length distribution, the associated exponent $\mu$ occurs solely in the diffusive term. The generalized friction constant $\eta_\alpha$ has dimension $[\eta_\alpha] = \sec^{\alpha-2}$.

### 3.1. *Subdiffusive case*

For pure subdiffusion ($\mu = 2$), the solution $P(x,t)$ relaxes towards the equilibrium distribution,

$$P_{\rm st}(x) = \lim_{t\to\infty} P(x,t) = \mathcal{N} \exp\left(-\frac{V(x)}{K_\alpha m\eta_\alpha}\right), \tag{17}$$

where $\mathcal{N}$ is the associated normalization factor. Comparison to the expected Boltzmann distribution $\mathcal{N}\exp\{-V(x)/[k_{\rm B}\mathcal{T}]\}$ produces the generalized Einstein–Stokes relation [106][e]

$$K_\alpha = k_{\rm B}\mathcal{T}/[m\eta_\alpha]. \tag{18}$$

Additionally, it is straightforward to show that processes governed by the time-fractional Fokker–Planck–Smoluchowski equation fulfil the second Einstein relation (linear response) between the first moment in the presence of the constant force $F_0$ and the second moment in the absence of an external bias [106]:

$$\langle x(t)\rangle_{F_0} = \frac{F_0}{2}\frac{\langle x^2(t)\rangle}{k_{\rm B}\mathcal{T}}. \tag{19}$$

The fact that the time fractional operator affects both drift and diffusion terms reflects systems in which the trapping occurs in the laboratory frame, and that the depths of the trap wells are not significantly affected by the external potential. In Fig. 3 the effect is illustrated for the case of a constant drift [109, 110]. While the test particle moves along with the drift during its mobile phases, with a certain probability $\Phi(t) = 1 - \int_0^t \psi(t')dt'$ no motion occurs up to time $t$. This corresponds to the pronounced cusp at the location of the initial condition. The result is an increasingly asymmetric shape of the probability density $P(x,t)$ originally recognized by Scher and Montroll for the drift-motion of charge carriers in amorphous semiconductors [32]. Conversely, one may think of scenarios, in which the trapping occurs in the moving frame, in which the drift is compensated. Subdiffusion in such a moving frame could correspond, for instance, to the motion of tracer particles in a flowing solution of crosslinked actin. In this case,

[e]We denote temperature by the symbol $\mathcal{T}$, contrasting the symbol $T$ for the total measurement time defined below.

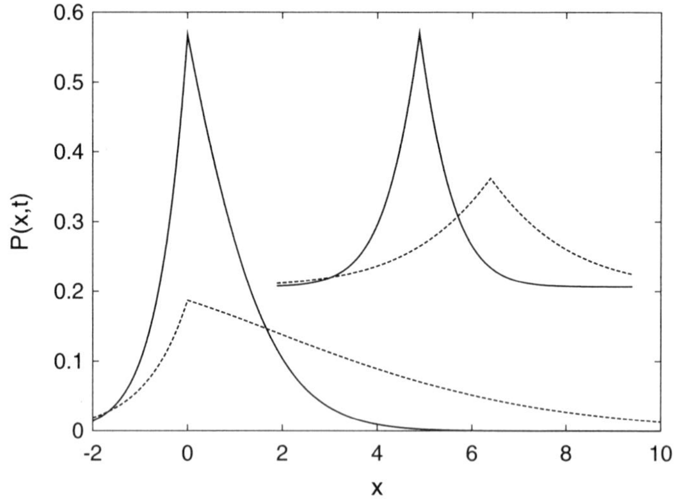


Fig. 3. Probability density function $P(x,t)$ for a subdiffusive process in the presence of a constant drift to the right. Main picture: trapping occurs in the laboratory frame, and the drift induces an increasing asymmetry of $P(x,t)$ while the maximum stays put at the origin. Inset: When the traps are located in the moving frame the subdiffusion diffusion propagator $P(x,t)$ is advected along as a whole, and remains symmetric with respect to the maximum.

the distribution remains symmetric, and is advected as a whole, see the inset in Fig. 3. It is worthwhile noting that for anomalous diffusion obeying the Galilei invariance, the root mean squared displacement and the first moment scale differently, $\sqrt{\langle(\Delta x)^2\rangle} = \sqrt{\langle x^2\rangle - \langle x\rangle^2} \simeq t^{\alpha/2}$ and $\langle x\rangle \simeq t$, while for laboratory frame trapping CTRW subdiffusion both scale identically, $\sqrt{\langle(\Delta x)^2\rangle} \simeq \langle x\rangle \simeq t^{\alpha}$. In the latter case, that is, the ratio of the width versus the distance covered by the mean of the distribution approaches a constant, in contrast to the case of Galilei invariant processes, for which the relative width continuously decreases and, compared to the distance covered by the center of the distribution, the distribution becomes increasingly sharp.

Let us further illustrate the long-ranging memory effects of subdiffusive CTRW for the example of a harmonic external potential $V(x) = \frac{1}{2}m\omega^2 x^2$, see Fig. 4. A physical example for such a Hookean force on the test particle occurs in indirect particle tracking by help of optical tweezers [88, 111, 112]. The solution of the fractional Fokker–Planck–Smoluchowski equation (18) for $\mu = 2$ is given in terms of the series over Hermite

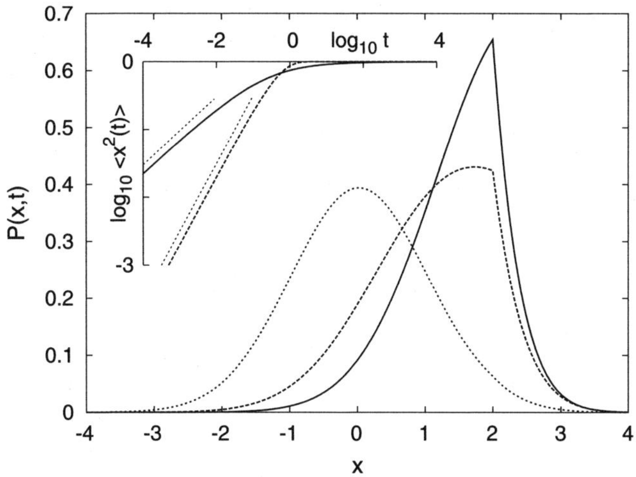


Fig. 4. Probability density function $P(x,t)$ for subdiffusive motion in an harmonic potential, $V(x) = \frac{1}{2}m\omega^2x^2$, for $\alpha = 1/2$ and at three consecutive times: initially the cusp at the location of the initial position ($x_0 = 2$) is distinct. It only vanishes in the extremely long time limit, for which the equilibrium Gaussian profile is reached. Inset: associated second moment $\langle x^2(t)\rangle$ for $\alpha = 1$ and $\alpha = 1/2$ in $\log_{10}$–$\log_{10}$ scale for $x_0 = 0$.

polynomials,

$$P(x,t) = \sqrt{\frac{m\omega^2}{2\pi k_B\mathcal{T}}}\sum_{n=0}^{\infty}\frac{1}{2^n n!}E_\alpha\left(-\frac{n\omega^2 t^\alpha}{\eta_\alpha}\right)H_n\left(\frac{\sqrt{m}\omega x_0}{\sqrt{2k_B\mathcal{T}}}\right)$$
$$\times H_n\left(\frac{\sqrt{m}\omega x}{\sqrt{2k_B\mathcal{T}}}\right)\exp\left(-\frac{m\omega^2x^2}{2k_B\mathcal{T}}\right), \tag{20}$$

and we see that individual modes relax according to the Mittag-Leffler pattern. The solution $P(x,t)$ is plotted in Fig. 4 for successive times. It can be seen that, again, the initial condition is persistent, and the associated cusp only disappears at very long times, for which $P(x,t)$ reaches the time-independent Gaussian equilibrium distribution. We also plot the associated second moment as function of time, compared to its Brownian counterpart: starting at the bottom of the potential, initially the motion is that of free (sub)diffusion, whereas at longer times a saturation to the thermal value $\langle x^2\rangle_{\rm th} = k_B\mathcal{T}/[m\omega^2]$ is reached. The relaxation is again of Mittag-Leffler type:

$$\langle x^2(t)\rangle = \langle x^2\rangle_{\rm th} + (x_0^2 - \langle x^2\rangle_{\rm th})E_\alpha\left(-\frac{2\omega^2t^\alpha}{\eta_\alpha}\right), \tag{21}$$

contrasting the exponential approach to the stationary value for normal diffusion.

### 3.2. *Subordination scheme for the subdiffusive case*

In between trapping events the motion of a subdiffusive particle can be viewed to be governed by the standard Langevin equation [113–115]. In terms of the number of steps the process therefore corresponds to Brownian motion. The mapping from the number of jumps to the *process time* including the ongoing trapping events, gives rise to the anomalies in the motion: summing over the independent, identically distributed waiting times, the temporal memory effects emerge. This property can be rephrased in terms of a subordination scheme [50, 116, 117].

Namely, it can be shown that the solution $P(x,t)$ of the purely subdiffusive fractional Fokker–Planck–Smoluchowski equation can be written in terms of the rescaled normal diffusive solution $P_M(x,t)$ in Laplace space [106]:

$$P(x,u) = u^{\alpha-1}\frac{\eta_\alpha}{\eta_1}P_M\left(x, \frac{\eta_\alpha u^\alpha}{\eta_1}\right). \tag{22}$$

This corresponds to the integral transformation [50, 117]

$$P(x,t) = \int_0^\infty \mathcal{E}_\alpha(s,t)P_M(x,s)ds. \tag{23}$$

The Laplace transform of the kernel $\mathcal{E}_\alpha(s,t)$ reads

$$\mathcal{E}_\alpha(s,u) = \frac{\eta_\alpha}{\eta_1 u^{1-\alpha}}\exp\left(-\frac{\eta_\alpha}{\eta_1}u^\alpha s\right). \tag{24}$$

As function of time, $\mathcal{E}_\alpha(s,t)$ can be represented exactly in terms of a Fox $H$-function [50]. With its series representation, for certain values of $\alpha$ simplified forms may be obtained, for instance, for $\alpha = 1/2$ one finds [50]

$$\mathcal{E}_{1/2}(s,t) = \frac{\eta_1}{\eta_{1/2}\sqrt{\pi t}}\exp\left(-\frac{1}{4t}\left[s\frac{\eta_{1/2}}{\eta_1}\right]^2\right). \tag{25}$$

This is the basis for a very efficient calculation of the solutions of the fractional Fokker–Planck–Smoluchowski equation in terms of the Markovian solution $P_M$ by numerical integration.

As originally proposed by Fogedby, a subdiffusive CTRW process can alternatively be understood in terms of the coupled Langevin equations [42] (see also [45–47])

$$\frac{dx(s)}{ds} = \frac{K}{k_{\mathrm{B}}T}F(x) + \eta(s), \tag{26a}$$

$$\frac{dt(s)}{ds} = \omega(s). \tag{26b}$$

Here the position $x$ is expressed in terms of the parameter $s$ (the internal time), and driven by the white Gaussian noise $\eta(s)$. Thus, Eq. (26a) defines standard Brownian motion $x(s)$, where $K$ is the diffusivity for the normal diffusion process in internal time $s$. Laboratory time $t$ is introduced by the subordination through the process $\omega(s)$, which is given by the probability density function [50, 117]

$$p_t(s) = \frac{1}{\alpha}\left(\frac{K_\alpha}{K}\right)^{1/\alpha}\frac{t}{s^{1+1/\alpha}}l_\alpha\left(\left[\frac{K_\alpha}{K}\right]^{1/\alpha}\frac{t}{s^{1+1/\alpha}}\right), \tag{27}$$

where $l_\alpha(z)$ is a one-sided Lévy stable probability density with Laplace transform $\int_0^\infty l_\alpha(z)\exp(-uz)dz = \exp(-u^\alpha)$. Thus, Eq. (26b) transforms the Brownian process $x(s)$ with diffusivity $K$ into the subdiffusive motion $x(t)$ with generalized diffusivity $K_\alpha$.

### 3.3. *Lévy flights in external potentials*

Lévy flights are Markovian processes ($\alpha = 1$) and can be phrased in terms of a Langevin equation with white noise, whose amplitude follows a Lévy stable distribution [43]. In the absence of an external potential, a Lévy flight can be characterized by fractional moments,

$$\langle |x(t)|^\delta \rangle^{2/\delta} = \left(\frac{2}{\mu}\frac{\Gamma(-\delta/\mu)\Gamma(1+\delta)}{\Gamma(-\delta/2)\Gamma(1+\delta/2)}\right)^{2/\delta}(K^\mu t)^{2/\mu}, \quad 0 < \delta < \mu. \tag{28}$$

This exact result can be obtained by help of the fundamental properties of Fox $H$-functions [50]. From expression (28) we see that the nature of Lévy flights is superdiffusive, $x^2$ scaling like $t^{2/\mu}$.

Under the influence of a confining external potential $V(x)$ Lévy flights attain a stationary distribution. Thus, in a harmonic potential $V(x) = \frac{1}{2}m\omega^2x^2$ the solution of the fractional Fokker–Planck–Smoluchowski equation (16) with $\alpha = 1$ reads [44]

$$P(k,t) = \exp\left\{-\frac{\eta_1 K^\mu |k|^\mu}{\mu\omega^2}\left[1 - \exp\left(-\frac{\mu\omega^2 t}{\eta_1}\right)\right]\right\} \tag{29}$$

in Fourier space. Thus, the Lévy stable index, and therefore the asymptotic spatial power-law behavior $\simeq |x|^{-1-\mu}$ is preserved throughout the process. In particular, the associated variance $\langle x^2(t)\rangle$ diverges at all times.

As soon as the external potential becomes steeper than harmonic, i.e. $V(x) \simeq |x|^c$ with $c > 2$, the probability density leaves the basin of attraction of Lévy stable distributions. $P(x,t)$ still has a power-law tail, however, now with an exponent larger than two. Let us illustrate this for a Cauchy flight ($\mu = 1$) in a quartic potential $V(x) \propto x^4$. In dimensionless coordinates, the dynamic equation for this process reads

$$\frac{\partial}{\partial t}P(x,t) = \left(\frac{\partial}{\partial x}x^3 + \frac{\partial}{\partial |x|}\right)P(x,t). \tag{30}$$

At stationarity the solution

$$P_{\rm st}(x) = \frac{1}{\pi}\frac{1}{1 - x^2 + x^4} \tag{31}$$

readily yields. Apart from the asymptotic behavior $P_{\rm st}(x) \simeq x^{-4}$ leading to a converging variance of the process, the probability density function is also characterized by two maxima, at $x_{\rm max} = \pm 1/\sqrt{2}$. This bimodality is indeed typical for Lévy flights in steeper than harmonic potentials. These behaviors are studied in detail in a series of papers by Chechkin *et al.* [118–120]. Lévy flights in softer-than-harmonic potentials of the type $V(x) \simeq |x|^c$ with $0 < c < 2$ are investigated in [121].

### 3.4. *Boundary value problems*

While CTRW subdiffusion shares the same trajectory with its Brownian analog as the subordination scheme only affects the temporal evolution of the stochastic process and leaves the (decoupled) spatial properties unchanged, it is less trivial to impose boundary conditions on Lévy flights due to their long-ranged spatial correlations.

#### 3.4.1. *Subdiffusion*

In the case of CTRW subdiffusion all kinds of boundary conditions are included in complete analogy to the case of Brownian diffusion.

In particular, the method of images [122] can be applied. For instance, for unbiased subdiffusion on a semi-infinite domain with absorbing (reflecting) boundary condition at $x = 0$ and initial condition $P_0(x) = \delta(x - x_0)$ $(x_0 > 0)$, the solution to the boundary value problem is given by

$$Q(x,t) = P(x - x_0, t) \mp P(x + x_0, t), \tag{32}$$

where $P(x,t)$ is the solution of the fractional diffusion equation (9) with $P_0(x) = \delta(x)$; the minus (plus) sign corresponds to the case of an absorbing (reflecting) boundary condition. In the presence of an absorbing boundary condition, the spatial integral over Eq. (32) renders the survival probability

$$\mathcal{S}(t) = \int_0^\infty Q(x,t)dx = \int_0^\infty \mathcal{E}_\alpha(s,t)\mathcal{S}_M(s)ds, \tag{33}$$

that is connected to its Brownian counterpart $\mathcal{S}_M(t)$ through subordination, in analogy to the propagator itself [compare Eq. (23)]. For the probability density of first passage, $\wp(t) = -d\mathcal{S}(t)/dt$, one can show that it is related to its Brownian analogue by the following subordination,

$$\wp(t) = \int_0^\infty \mathcal{E}_\alpha^\wp(s,t)\wp_M(s)ds \quad \therefore \quad \mathcal{E}_\alpha^\wp(s,u) = \exp\left(-\frac{\eta_\alpha}{\eta_1}u^\alpha s\right). \tag{34}$$

The difference from Eq. (23) is due to the additional time derivative with respect to the survival probability. The subordination kernel $\mathcal{E}_\alpha^\wp(s,t)$ for the first passage density is a completely one-sided Lévy stable law with Fox function representation

$$\mathcal{E}_\alpha^\wp(s,t) = \frac{1}{t}H_{1,1}^{1,0}\left[\frac{\eta_\alpha s}{\eta_1 t^\alpha}\left|\begin{matrix}(0,\alpha)\\(0,1)\end{matrix}\right.\right] = \frac{1}{t}\sum_{n=0}^\infty \frac{(-1)^n}{n!\Gamma(-\alpha n)}\left(\frac{\eta_\alpha s}{\eta_1 t^\alpha}\right)^n. \tag{35}$$

In the limit $\alpha = 1$, we see that $\mathcal{E}_\alpha^\wp(s,u) = \exp(-us)$ such that $\mathcal{E}_\alpha^\wp(s,t) = \delta(s-t)$, as it should.

Furthermore, one can show [55, 123] that the first passage density on a semi-infinite domain changes from the asymptotic Markovian scaling $\wp_M(t) \simeq t^{-1-1/2}$ to $\wp(t) \simeq t^{-1-\alpha/2}$, in the absence of a drift. When a constant bias $V$ is acting towards the absorbing boundary, the fast exponential decay $\wp_M(t) \simeq t^{-1-1/2}\exp(-V^2t/[4K_1])$ in the Brownian case is converted to a power-law decay $\wp(t) \simeq t^{-1-\alpha}$ [55, 124]. If initially the probability is spread out over an interval, the famed Scher–Montroll result yields, that follows a turnover from $\wp(t) \simeq t^{\alpha-1}$ to $\wp(t) \simeq t^{-1-\alpha}$, such that the sum of the two scaling exponents is −2 [32, 90, 125]. Finally, in a finite domain with one or two absorbing boundary condition(s) the Markovian exponential decay $\wp_M(t) \simeq \exp(-t/\tau^*)$, where $\tau^*$ is related to the lowest eigenvalue of the boundary value problem, also changes to the inverse power-law form $\wp(t) \simeq t^{-1-\alpha}$ [55, 123]. That is, in all cases the mean first passage time diverges for CTRW subdiffusion, as it should by definition of the process with a diverging characteristic waiting time.

### 3.4.2. *Lévy flights*

The dynamic formulation of Lévy flights in terms of the fractional diffusion equation (9) might suggest to solve boundary value problems for Lévy flights by similar methods as for compact diffusion processes such as Gaussian diffusion or CTRW subdiffusion. However, naive application of the method of images produces wrong results. From the point of view of the space-fractional operator (10), this complication can be qualitatively appreciated: the fractional derivative includes the values of $P(x,t)$ over the entire interval $(-\infty, \infty)$, weighted by the slowly decaying inverse power-law kernel. For the case of a semi-infinite domain with a reflecting boundary condition at $x = 0$ and initial condition at $x_0 > 0$, all values of $P(x,t)$ at $x \leq 0$ necessarily need to vanish. That is, the governing dynamic equation is changed when non-natural boundary conditions come into play [126–128]. Let us highlight a few important results, schematically illustrated in Fig. 5.

The *first passage* of a Lévy flight with symmetric jump length distribution (distances for jumping left or right are distributed equally) across an absorbing boundary on a semi-infinite domain is governed by the Sparre

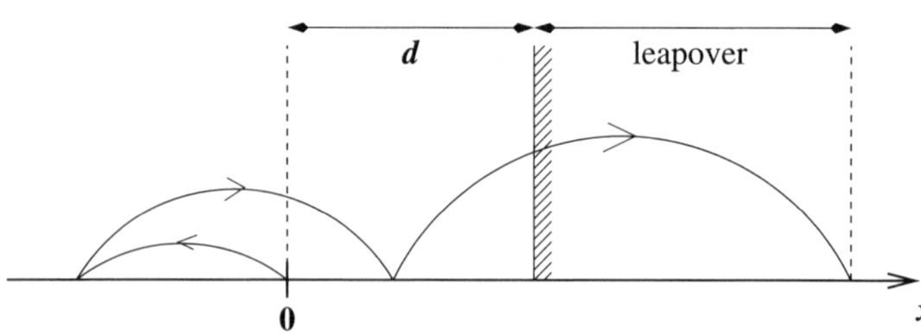


Fig. 5. First passage and leapover of a Lévy flight. The random walk starts at $x = 0$ at time $t = 0$. After a number of steps the walker crosses the absorbing boundary (shaded) at distance $d$ from the origin and is removed. A Lévy flight thereby overshoots the boundary by the leapover length $\ell$.

Andersen universality $\wp(t) \simeq t^{-3/2}$ [122, 129, 130]. Indeed, the exact asymptotic result

$$\wp(t) \sim \frac{d^{\mu/2}}{\mu\sqrt{\pi K^{(\mu)}}\Gamma(\mu/2)}t^{-3/2}, \tag{36}$$

follows the $\wp(t) \simeq t^{-3/2}$ law, and the Lévy index $\mu$ only enters the prefactor. In contrast, for a one-sided Lévy flight the first passage time density decays exponentially fast [127, 128]. For the special case $\mu = 1/2$, the involved Mainardi function reduces to a simple Gaussian, and we find [127]

$$\wp(t) \sim \sqrt{\frac{2}{\pi d}}K^{(\mu)} \exp\left(-\frac{[K^{(\mu)}]^2 t^2}{2d}\right). \tag{37}$$

The case of a finite domain is discussed in [131].

A related quantity is the *first passage leapover* of a Lévy flight, as sketched in Fig. 5. It quantifies how far the random walker overshoots a point $x = d$ on its first crossing. For symmetric Lévy flights the exact result reads [127]

$$\wp_l(\ell) = \frac{\sin(\pi\mu/2)}{\pi}\frac{d^{\mu/2}}{\ell^{\mu/2}(d+\ell)}. \tag{38}$$

Interestingly, the asymptotic behavior follows $\wp_l(\ell) \simeq \ell^{-1-\mu/2}$, despite the scaling $\lambda(x) \simeq |x|^{-1-\mu}$ of the underlying jump length distribution. In the Gaussian limit $\mu \to 2$ no leapovers exist, and the process is completely local. For completely one-sided Lévy stable jump length distribution the Lévy index $\mu$ is preserved for the leapovers [127, 128]

$$\wp_l(\ell) = \frac{\sin(\pi\mu)}{\pi}\frac{d^{\mu}}{\ell^{\mu}(d+\ell)}. \tag{39}$$

For regular Brownian motion, due to the incremental steps resulting in the continuum limit, the first passage across an absorbing boundary and the first arrival to a point are identical. For the above reasons this is no longer true for Lévy flights. The smaller the Lévy index $\mu$, the less likely it is for the random walker to hit a point (or, more practically, a small interval) on a line. In the range $1<\mu<2$, for which the first absolute moment $\langle|x|\rangle$ still exists, the probability density of first arrival times to a point reflects this reduced arrival efficiency compared to the Brownian case [123],

$$\wp_{\rm fa}(t) \sim C(\mu)\frac{x_0^{\mu-1}}{[K^{(\mu)}]^{1-1/\mu}t^{2-1/\mu}}, \tag{40}$$

where $C(\mu) = \mu\Gamma(2-\mu)\Gamma(2-1/\mu)/\sin(\pi[2-\mu]/2)\sin^2(\pi/\mu)/[\pi^2(\mu-1)]$.

We note that Lévy flights can also be constructed by a subordination method (different from the above); in particular, one can show that the space fractional diffusion equation corresponds to two completely different processes [49]. To overcome the Sparre Andersen universality the Markovian time dependence of the process needs to be changed [122, 132]. We note in passing that Lévy flights have an exponential decay of the probability density of escape times from a potential well [133–135].

## 4. Randomness of Long Time Averages in Subdiffusive CTRW Processes

So far we have been dealing with ensemble averages of the type (1). These are purely spatial averages, in which time assumes the role of a parameter. Often, however, the behavior of diffusing particles is inferred from the time series $x(t)$ of individual trajectories. Single particle tracking is a technique dating back to Nordlund [136]. With modern microscopy tools single trajectories of particles down to the nanoscale can be obtained, even in complex systems such as biological cells [88, 111, 112, 137–142]. The time series of such measurements is then analyzed in terms of single trajectory time averages, for instance, the time averaged mean squared displacement

$$\overline{\delta^2(\Delta,T)} = \frac{1}{T-\Delta}\int_0^{T-\Delta}(x(t+\Delta)-x(t))^2dt, \tag{41}$$

or its discrete counterpart. In Eq. (41) $T$ is the total measurement time setting the length of the obtained time series. $\Delta$, often referred to as the lag time, is the width of the time window sliding across the time series in the evaluation. In an ergodic system the long-time average will provide the same information as the ensemble average. As we will see, interesting effects arise in subdiffusive, ageing systems, as exemplified in Fig. 6.

Thus, for Brownian systems, the time averaged second moment of a free diffusion process scales linearly with time [line number (1) in Fig. 6], while under confinement a plateau is reached (2). Active processes show ballistic scaling $\simeq \Delta^2$ (3). Under subdiffusion conditions, different scenarios

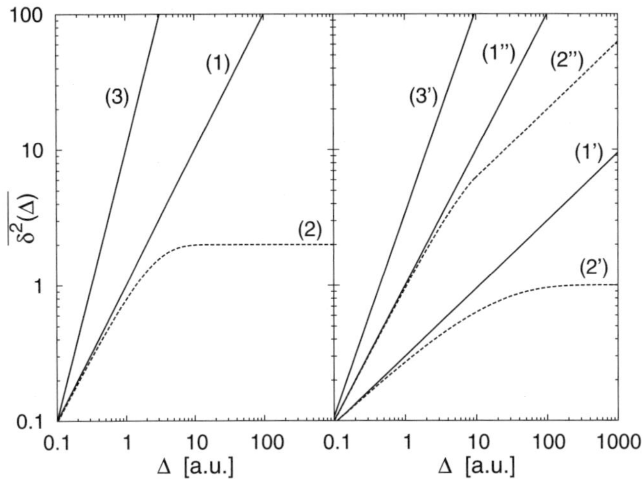


Fig. 6. Diffusion modes of the time averaged mean squared displacement (41) as function of the lag time $\Delta$. Left: Normal diffusion growing like $\overline{\delta^2} \simeq \Delta$ (1), restricted (confined) diffusion with a turnover from $\simeq \Delta$ to $\simeq \Delta^0$ (2), drift diffusion $\simeq \Delta^2$ (3). Right: Ergodic subdiffusion $\simeq \Delta^\alpha$ (1′), restricted ergodic subdiffusion turning over from $\simeq \Delta^\alpha$ to $\simeq \Delta^0$ (2′), non-ergodic subdiffusion $\simeq \Delta$ (1″), restricted non-ergodic subdiffusion turning over from $\simeq \Delta$ to $\simeq \Delta^{1-\alpha}$ (2″), superdiffusion $\Delta^{1+\alpha}$ (3′). Here, $0 < \alpha < 1$. Note the double-logarithmic scale.

may occur. Thus, ergodic subdiffusion will lead to a sublinear slope $\simeq \Delta^\alpha$ (1′), and under confinement a plateau will be reached (2′). Sub-ballistic superdiffusion corresponds to a slope between one and two (3′). However, for ageing, non-ergodic subdiffusion processes the time averaged mean squared displacement scales linearly $\simeq \Delta$, and confinement leads to a power-law $\simeq \Delta^{1-\alpha}$.

For a Brownian particle with a typical waiting time $\tau$ between successive jumps the number of jumps will on average increase like $\langle n(t)\rangle = t/\tau$. The root mean square of the step size of the jumps is $\sqrt{\langle \delta x^2\rangle}$. The ensemble average of the quantity (41) on an infinite domain therefore behaves like

$$\left\langle \overline{\delta^2(\Delta, T)} \right\rangle = 2K_1\Delta \quad \therefore \quad K_1 \equiv \frac{\langle \delta x^2\rangle}{2\tau}, \tag{42}$$

where we defined the diffusion constant $K_1$ in the typical random walk sense. For a Brownian particle we conclude that the time averaged mean squared displacement $\overline{\delta^2(t,T)}$ is completely equivalent to the ensemble average $\langle x^2(t)\rangle = 2K_1 t$. Fluctuations around the ensemble average (42) will be small for a Brownian system.

Ergodicity is no longer expected to hold for processes with diverging time scales [143–145].[f] For CTRW subdiffusion with long-tailed waiting time distribution $\psi(t) \simeq t^{-1-\alpha}$ and $0 < \alpha < 1$, the average number of steps grows like $\langle n(t)\rangle \simeq t^{\alpha}$ [33]. Consequently, it can be shown that the time averaged mean squared displacement behaves like [147, 148]

$$\left\langle\overline{\delta^2(\Delta,T)}\right\rangle \sim \frac{2K_\alpha}{\Gamma(1+\alpha)}\frac{\Delta}{T^{1-\alpha}}, \tag{43}$$

for $\Delta \ll T$. That means, that by measuring the lag time dependence of the time averaged mean squared displacement, no anomaly would be detected, instead, the behavior would point at normal diffusion. Only the dependence on the total measurement time $T$ reveals the anomalous diffusion characteristics. In fact, the effective diffusivity $K_\alpha/T^{1-\alpha}$ decreases in the course of time, a reminiscence of the ageing property of CTRW subdiffusion.

We note that in the presence of a drift, $\langle\delta x\rangle \neq 0$, the time averaged first moment,

$$\overline{\delta(\Delta,T)} = \frac{1}{T-\Delta}\int_0^{T-\Delta}(x(t+\Delta)-x(t))\,dt, \tag{44}$$

assumes the form

$$\left\langle\overline{\delta(\Delta,T)}\right\rangle \sim \frac{\langle\delta x\rangle}{\Gamma(1+\alpha)}\frac{\Delta}{T^{1-\alpha}}, \tag{45}$$

and the distribution of the long-time average $\overline{\delta(\Delta,T)}$ follows the same function $\phi_\alpha(\xi)$ defined below, but with $\xi = \overline{\delta(\Delta,T)}/\left\langle\overline{\delta(\Delta,T)}\right\rangle$ [148]. Moreover, one can show that the analogue in the time-averaged sense of the second Einstein relation (19) becomes [148]

$$\left\langle\overline{\delta(\Delta,T)}\right\rangle_{F_0} = \frac{F_0}{2}\frac{\left\langle\overline{\delta^2(\Delta,T)}\right\rangle_0}{k_B T}, \tag{46}$$

connecting the drift effected by the constant force $F_0$ with the second moment in the absence of the force. What happens when the motion is confined by an external potential $V(x)$? The motion is then described by the fractional Fokker–Planck–Smoluchowski equation (16). From a dual subordination scheme it is possible to derive the behavior of the time averaged mean squared displacement. Remarkably, it does not saturate to a plateau

[f] Ergodicity is also violated in the presence of power-law jump length distributions with diverging variance [146].

value characteristic of the ensemble average, but the confinement induces a power-law behavior of the form [149, 150]

$$\left\langle\overline{\delta^2(\Delta,T)}\right\rangle \sim \left(\langle x^2\rangle_{\rm B} - \langle x\rangle_{\rm B}^2\right)\frac{2\sin(\pi\alpha)}{(1-\alpha)\pi\alpha}\left(\frac{\Delta}{T}\right)^{1-\alpha}, \tag{47}$$

valid for $\Delta \ll T$ and $\Delta \gg (1/K_\alpha\gamma_1)^{1/\alpha}$, where $\gamma_1$ is the lowest temporal eigenvalue of the corresponding Fokker–Planck–Smoluchowski operator. Result (47) is universal in the sense that the sole dependence on the specific form of confinement, given by $V(x)$, enters in the form of the first two Boltzmann moments

$$\langle x^n\rangle_{\rm B} = \mathcal{Z}^{-1}\int_{-\infty}^{\infty} x^n e^{-V(x)/[k_{\rm B}T]}dx, \quad \mathcal{Z} = \int_{\infty}^{\infty} e^{-V(x)/[k_{\rm B}T]}dx. \tag{48}$$

The result $\left\langle\overline{\delta^2(\Delta,T)}\right\rangle \simeq (\Delta/T)^{1-\alpha}$ was also reported in [149]. A subdiffusing particle that starts its motion at the vertex of the potential will initially move freely. After some time, corresponding to the eigenvalue $\gamma_1$, a restoring force starts to act on the particle. In terms of the time averaged mean squared displacement this corresponds to the turnover between the initial linear scaling $\langle\overline{\delta^2(\Delta,T)}\rangle \simeq \Delta$ to the power-law form $\langle\overline{\delta^2(\Delta,T)}\rangle \sim \Delta^{1-\alpha}$. In Fig. 7 we show simulation results for the time

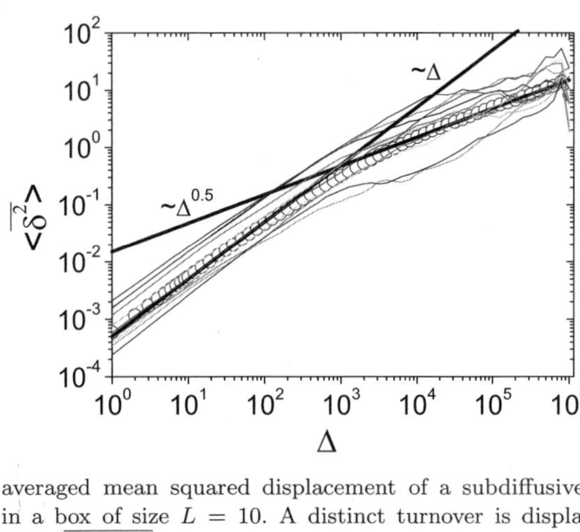


Fig. 7. Time averaged mean squared displacement of a subdiffusive CTRW process with $\alpha = 1/2$ in a box of size $L = 10$. A distinct turnover is displayed between the initial linear scaling $\overline{\delta^2(\Delta,T)} \simeq \Delta$ to the confinement-dominated regime $\overline{\delta^2(\Delta,T)} \simeq \Delta^{1-\alpha}$. The circles (○) correspond to the average over the individual trajectories, which show pronounced fluctuations. Note that the nature of these fluctuations is preserved throughout the duration. The straight lines show the predicted scaling regimes.

averaged mean squared displacement of CTRW subdiffusion in a box, showing the distinct turnover predicted by our theory. Such a turnover behavior is indeed observed experimentally for lipid granule particles in fission yeast cells [88]. In that experiment the granules are indirectly tracked by optical tweezers, exerting a harmonic trapping potential on the particle.

When the additional ensemble average is not applied to the time averaged mean squared displacement, individual realizations of the process display effects of few or single long waiting time events, as can be seen in Fig. 7. Within one given trajectory this produces variations in the slope, while between individual trajectories pronouncedly different amplitudes are obtained. This is a signature for the randomness of time averages in CTRW subdiffusion systems. As the underlying waiting time distribution is scale-free, this means that this randomness persists even when the averaging time is taken to infinity. Still, the amplitude of the time averages follow a well-defined distribution. Thus, for the dimensionless quantity $\xi = \overline{\delta^2(\Delta,T)}/\left\langle\overline{\delta^2(\Delta,T)}\right\rangle$, we find the scatter distribution [148]

$$\lim_{T\to\infty}\phi_\alpha(\xi) = \frac{\Gamma(1+\alpha)^{1/\alpha}}{\alpha\xi^{1+1/\alpha}} l_\alpha\left(\frac{\Gamma(1+\alpha)^{1/\alpha}}{\xi^{1+1/\alpha}}\right), \tag{49}$$

with the one-sided Lévy stable law $l_\alpha$. The distribution $\phi_\alpha(\xi)$ can be expressed analytically in terms of a Fox $H$-function [150], and it exhibits an exponential decay, such that $\phi_\alpha(\xi)$ has a finite width. For $\alpha = 1/2$, we find the Gaussian form $\phi_{1/2}(\xi) = (2/\pi)\exp(-\xi^2/\pi)$ [150]. Note that this distribution $\phi_\alpha$ was also found in tilted washboard potentials [152]. The non-ergodic nature of CTRW subdiffusion leads to non-uniform exploration of the otherwise completely accessible phase space in a given trajectory: extremely long waiting times lead to the preference of occupation of one subdomain $\mathbb{S}$ of the system in terms of the time averaged occupation probability, $\overline{p}_{\mathbb{S}} = \lim_{t\to\infty} t_{\mathbb{S}}/t$ [153–157]. In contrast, stationary processes such as fractional Brownian motion or fractional Langevin equations [63, 158–160] are ergodic [161–163], and time averages narrowly distributed around the ergodic value $\xi = 1$ by an approximately Gaussian distribution [150, 164]. Similar behavior is followed in the diffusion on fractal supports [86, 165, 166]. Apart from the lipid granule motion of [88] and subdiffusion in membranes [89], signatures of CTRW subdiffusion-induced weak ergodicity breaking have been reported from the blinking dynamics of quantum dots [167–169].

In this context, let us address another subtle point concerning CTRW subdiffusion processes. The velocity–velocity correlation function

$$C_v^{(\epsilon)}(\tau) = \frac{1}{\epsilon^2}\langle (x(\tau+\epsilon) - x(\tau))(x(\epsilon) - x(0))\rangle \tag{50}$$

for CTRW subdiffusion is non-negative [150],

$$\frac{C_v^{(\epsilon)}(\tau)}{C_v^{(\epsilon)}(0)} = \begin{cases} \dfrac{\epsilon^\alpha - \tau^\alpha}{\epsilon^\alpha}, & \tau \leq \epsilon, \\ 0, & \tau \geq \epsilon, \end{cases} \tag{51}$$

due to the absence of correlations for subsequent jumps. However, under confinement, it can be shown analytically [150] that indeed a pronounced negative regime for the velocity–velocity correlations appears. Similarly to the behavior of subdiffusive fractional Brownian motion, CTRW subdiffusion has a power-law recovery: For long $\tau$ the velocity autocorrelation function decays to zero (from the negative side) as the power-law

$$\frac{C_v^{(\epsilon)}(\tau)}{C_v^{(\epsilon)}(0)} \sim -\frac{1}{\Gamma(\alpha)\Gamma(1-\alpha)}\left(\frac{\epsilon}{\tau}\right)^{1+\alpha} \tag{52}$$

from its minimum, valid for $\tau \gg \epsilon$.

In Fig. 8 we display the behavior for free and confined CTRW subdiffusion. As can be seen, the behavior of $C_v^{(\epsilon)}(\tau)$ is under confinement hardly distinguishable from the corresponding behavior of subdiffusive fractional Brownian motion, despite the fundamental difference between both

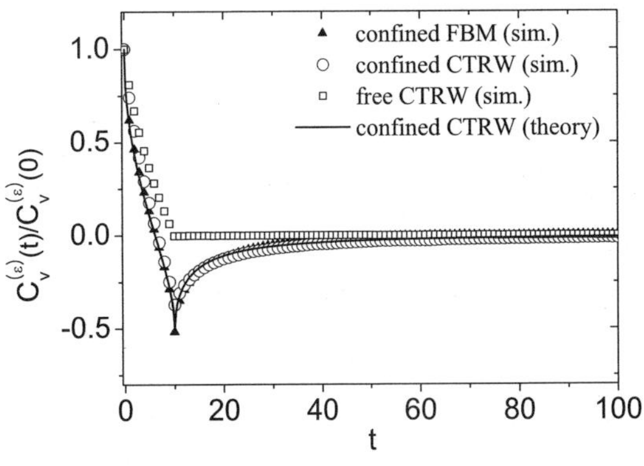


Fig. 8. Velocity–velocity correlation function for subdiffusive CTRW and FBM processes for $\alpha = \frac{1}{2}$. While the velocity autocorrelation function for unconfined CTRW is negative nowhere, the behavior of confined CTRW (motion in a box of size $L = 4$), free FBM, and confined FBM are hardly distinguishable.

processes. More efficient methods to distinguish different stochastic mechanisms are suggested in [170], compare also the discussions in [171–174].

## 5. Conclusions

Fractional dynamics has a long history in mathematics, the engineering sciences, and in control theory. During the recent years, it has also taken root in the natural sciences, as description for relaxation and diffusion processes in complex systems. An important impetus towards that development is the intimate connection of fractional dynamic equations with random walk concepts. Thus, time fractional relaxation and diffusion equations arise from long-tailed CTRW-waiting time distributions with diverging characteristic time scale. From this renewal process the long-ranged memory, symbolized by the fractional derivative, emerges due to the generalized central limit theorem. Similarly, long-ranged jump length distributions transfer to space fractional derivatives corresponding to long-ranging correlations. The particular advantage of the formulations in terms of fractional dynamic equations thereby lies in the direct generalization of their classical counterparts, the relaxation, diffusion, and, in particular, Fokker–Planck–Smoluchowski equations. In the presence of external potentials, the formulation in terms of fractional Fokker–Planck–Smoluchowski equations is practically superior: in analogy to the classical Fokker–Planck–Smoluchowski equation the external potential directly enters the equation. Similar advantages pertain to the formulation of subdiffusion reactions [175–180]. Here we summarized the results of decoupled CTRW processes in external potentials, described by space- or time-fractional Fokker–Planck–Smoluchowski equations.

A further aspect treated in this chapter are time-averaged moments. Due to the non-ergodic, ageing properties of subdiffusive CTRW processes even long-time averages become random variables, and their scaling differs from the ensemble averaged counterparts. We here addressed some of the emerging properties of time averages both without and in confinement, including the distributions of the random time averages.

## Acknowledgments

We would like to acknowledge helpful discussions and fruitful collaborations with Eli Barkai, Kirstine Berg-Sørensen, Dirk Brockmann, Stas Burov, Aleksei Chechkin, Albert Compte, Hans Fogedby, Walter Glöckle, Vsevolod

Gonchar, Yossi Klafter, Tal Koren, Katja Lindenberg, Michael Lomholt, Theo Nonnenmacher, Lene Oddershede, Helmut Schiessel, Christine Selhuber, Igor Sokolov, Vincent Tejedor, Matthias Weiss, and Irwin Zaid. We also acknowledge funding from the Deutsche Forschungsgemeinschaft and from the Academy of Finland within the FiDiPro scheme.

## References

1. G. B. Bülfinger, *Comm. Acad. Petrop.* **4**, 164 (1729).
2. C. Bach, *Z. d. Ver. deutsch. Ing.* **32**, 193 (1888); *ibid.* **48**, 1040 (1904).
3. M. Reiner, *Deformation, Strain and Flow*, 3rd Edition (H. K. Lewis & Co., 1969).
4. M. Reiner and G. W. Scott Blair, *Agricultural Rheology* (Routledge and Paul, 1957).
5. P. G. Nutting, *J. Franklin Inst.* **191**, 679 (1921); *ibid.* **235**, 513 (1943); *ibid.* **242**, 449 (1946).
6. A. Gemant, *Physics* **7**, 311 (1936); *Phil. Mag. (Ser. 7)* **25**, 540 (1938).
7. G. W. Scott Blair, *Rev. Scient. Instruments* **21**, 80 (1944); *J. Colloid Sci.* **2**, 21 (1947).
8. A. Gerasimov, *Prikl. Matem. i Mekh. (PMM)* **12**, 251 (1948).
9. Yu. N. Rabotnov, *Prikl. Matem. i Mekh. (PMM)* **12**, 81 (1948).
10. M. Caputo and F. Mainardi, *Riv. Nuovo Cimento* **1**, 161 (1971).
11. H. Schiessel and A. Blumen, *J. Phys. A* **26**, 5057 (1993).
12. N. Heymans and J.-C. Bauwens, *Rheol. Acta* **33**, 210 (1994).
13. M. Caputo and F. Mainardi, *Riv. Nuovo Cimento* **1**, 161 (1971).
14. H. Schiessel, R. Metzler, A. Blumen and T. F. Nonnenmacher, *J. Phys. A* **28**, 6567 (1995).
15. W. G. Glöckle and T. F. Nonnenmacher, *Macromol.* **24**, 6426 (1991).
16. R. Metzler, W. Schick, H.-G. Kilian and T. F. Nonnenmacher, *J. Chem. Phys.* **103**, 7180 (1995).
17. W. G. Glöckle and T. F. Nonnenmacher, *Biophys. J.* **68**, 46 (1995).
18. R. Hilfer, *Chem. Phys.* **284**, 399 (2002); *J. Phys. Cond. Mat.* **14**, 2297 (2002).
19. F. Mainardi, *Fractional Calculus and Waves in Linear Viscoelasticity* (Imperial College Press, 2010).
20. V. Balakrishnan, *Physica A* **132**, 569 (1985).
21. W. Wyss, *J. Math. Phys.* **27**, 2782 (1986).
22. F. E. Peseckis, *Phys. Rev. E* **36**, 892 (1987).
23. W. R. Schneider and W. Wyss, *J. Math. Phys.* **30**, 134 (1989).
24. G. Jumarie, *J. Math. Phys.* **33**, 3536 (1992).
25. G. M. Zaslavsky, *Physica D* **76**, 110 (1994).
26. A. I. Saichev and G. M. Zaslavsky, *Chaos* **7**, 753 (1997).
27. M. Giona and H. E. Roman, *Physica A* **185**, 87 (1992).
28. R. Metzler, W. G. Glöckle and T. F. Nonnenmacher, *Physica A* **211**, 13 (1994).

29. E. W. Montroll and G. H. Weiss, *J. Math. Phys.* **6**, 167 (1965).
30. H. Scher and M. Lax, *Phys. Rev. B* **7**, 4502 (1973).
31. M. F. Shlesinger, *J. Stat. Phys.* **10**, 421 (1974).
32. H. Scher and E. Montroll, *Phys. Rev. B* **12**, 2455 (1975).
33. B. D. Hughes, *Random Walks and Random Environments*, Vol. 1: *Random Walks* (Oxford Univ. Press, 1995).
34. R. Hilfer and L. Anton, *Phys. Rev. E* **51**, R848 (1995).
35. R. Hilfer, *Fractals* **3**, 211 (1995).
36. A. Compte, *Phys. Rev. E* **53**, 4191 (1996).
37. B. J. West, P. Grigolini, R. Metzler and T. F. Nonnenmacher, *Phys. Rev. E* **55**, 99 (1997).
38. R. Metzler and T. F. Nonnenmacher, *Phys. Rev. E* **57**, 6409 (1998).
39. R. Gorenflo, F. Mainardi, D. Moretti and P. Paradisi, *Nonlin. Dyn.* **29**, 129 (2002).
40. R. Gorenflo, F. Mainardi, D. Moretti, G. Pagnini and P. Paradisi, *Chem. Phys.* **284**, 521 (2002).
41. E. Scalas, R. Gorenflo and F. Mainardi, *Phys. Rev. E* **69**, 011107 (2004).
42. H. C. Fogedby, *Phys. Rev. E* **50**, 1657 (1994).
43. H. C. Fogedby, *Phys. Rev. Lett.* **73**, 2517 (1994).
44. S. Jespersen, R. Metzler and H. C. Fogedby, *Phys. Rev. E* **59**, 2736 (1999).
45. A. Baule and R. Friedrich, *Phys. Rev. E* **71**, 026101 (2005).
46. M. Magdziarz, A. Weron and K. Weron, *Phys. Rev. E* **75**, 016708 (2007).
47. M. Magdziarz, A. Weron and J. Klafter, *Phys. Rev. Lett.* **101**, 210601 (2008).
48. A. Piryatinska, A. I. Saichev and W. A. Woyczynski, *Physica A* **349**, 375 (2005).
49. I. M. Sokolov and R. Metzler, *J. Phys. A* **37**, L609 (2004).
50. R. Metzler and J. Klafter, *Phys. Rep.* **339**, 1 (2000).
51. J.-P. Bouchaud and A. Georges, *Phys. Rep.* **195**, 127 (1990).
52. A. Erdélyi, *Higher Transcendental Functions*, Vol. 1, *Bateman Manuscript Project* (McGraw-Hill, 1954).
53. H. Yang, G. Luo, P. Karnchanaphanurach, T.-M. Louie, I. Rech, S. Cova, L. Xun and X. S. Xie, *Science* **302**, 262 (2003).
54. F. Mainardi, M. Raberto, R. Gorenflo and E. Scalas, *Physica A* **287**, 468 (2000).
55. R. Metzler and J. Klafter, *J. Phys. A* **37**, R161 (2004).
56. G. M. Zaslavsky, *Phys. Rep.* **371**, 461 (2002).
57. R. Metzler and T. F. Nonnenmacher, *Chem. Phys.* **284**, 67 (2002).
58. R. Metzler and J. Klafter, *Adv. Chem. Phys.* **116**, 223 (2001).
59. A. V. Chechkin, V. Y. Gonchar, J. Klafter and R. Metzler, *Adv. Chem. Phys.* **133**, 439 (2006).
60. I. M. Sokolov, J. Klafter and A. Blumen, *Phys. Today* **55**, 48 (2002).
61. R. Hilfer, *Applications of Fractional Calculus in Physics* (World Scientific, 2000).
62. B. J. West, M. Bologna and P. Grigolini, *Physics of Fractal Operators* (Springer, 2003).

63. W. T. Coffey, Yu. P. Kalmykov and J. T. Waldron, *The Langevin Equation* (World Scientific, 2004).
64. G. M. Zaslavsky, *Hamiltonian Chaos and Fractional Dynamics* (Oxford Univ. Press, 2005).
65. Eds. Yu. P. Kalmykov and W. T. Coffey, *Fractals, Diffusion, and Relaxation in Disordered Complex Systems*, Advances in Chemical Physics, Vol. **133**, Parts A and B (Wiley, 2006).
66. R. Klages, G. Radons and I. M. Sokolov, *Anomalous Transport: Foundations and Applications* (Wiley-VCH, 2008).
67. K. B. Oldham and J. Spanier, *The Fractional Calculus: Theory and Applications of Differentiation and Integration to Arbitrary Order* (Academic Press, 1974).
68. K. Diethelm, *The Analysis of Fractional Differential Equations: An Application-Oriented Exposition Using Differential Operators of Caputo Type* (Springer, 2010).
69. A. A. Kilbas, H. M. Srivastava and J. J. Trujillo, *Theory and Applications of Fractional Differential Equations* (Elsevier, 2006).
70. K. S. Miller and B. Ross, *An Introduction to the Fractional Calculus and Fractional Differential Equations* (Wiley, 1993).
71. S. G. Samko, A. A. Kilbas and O. I. Marichev, *Fractional Integrals and Derivatives* (Taylor & Francis, 2002).
72. I. Podlubny, *Fractional Differential Equations* (Academic Press, 1999).
73. A. M. Mathai and R. K. Saxena, *The H-function with Applications in Statistics and Other Disciplines* (Wiley Eastern, 1978).
74. H. M. Srivastava, K. C. Gupta and S. P. Goyal, *The H-functions of One and Two Variables with Applications* (South Asian Publishers, 1982).
75. A. M. Mathai, R. K. Saxena and H. J. Haubold, *The H-function: Theory and Applications* (Springer, 2009).
76. V. Tejedor and R. Metzler, *J. Phys. A* **43**, 082002 (2010).
77. A. V. Chechkin, M. Hofmann and I. M. Sokolov, *Phys. Rev. E* **80**, 031112 (2009).
78. N. Kumar, U. Harbola and K. Lindenberg, *Phys. Rev. E* **82**, 021101 (2010).
79. J. Klafter, A. Blumen and M. F. Shlesinger, *Phys. Rev. A* **35**, 3081 (1987).
80. E. Barkai, *Phys. Rev. Lett.* **90**, 104101 (2003).
81. G. Radons, *Adv. Solid State Phys.* **38**, 439 (1999).
82. T. H. Solomon, E. R. Weeks and H. L. Swinney, *Phys. Rev. Lett.* **71**, 3975 (1993).
83. J. Klafter, M. F. Shlesinger and G. Zumofen, *Phys. Today* **49**, 33 (1996).
84. J.-P. Bouchaud, *J. Phys. I* (*Paris*) **2**, 1705 (1992).
85. S. Burov and E. Barkai, *Phys. Rev. Lett.* **98**, 250601 (2007).
86. S. Havlin and D. Ben-Avraham, *Adv. Phys.* **36**, 695 (1987).
87. I. Y. Wong, M. L. Gardel, D. R. Reichman, E. R. Weeks, M. T. Valentine, A. R. Bausch and D. A. Weitz, *Phys. Rev. Lett.* **92**, 178101 (2004).
88. J.-H. Jeon, V. Tejedor, S. Burov, E. Barkai, C. Selhuber-Unkel, K. Berg-Sørensen, L. Oddershede and R. Metzler, *Phys. Rev. Lett.* **106**, 048103 (2011).

89. A. V. Weigel, B. Simon, M. M. Tamkun and D. Krapf, *Proc. Natl. Acad. Sci.* USA **108**, 6438 (2011).
90. H. Scher, G. Margolin, R. Metzler, J. Klafter and B. Berkowitz, *Geophys. Rev. Lett.* **29**, 1061 (2002).
91. B. Berkowitz, A. Cortis, M. Dentz and H. Scher, *Rev. Geophys.* **44**, RG2003 (2006).
92. B. Berkowitz, J. Klafter, R. Metzler and H. Scher, *Wat. Res. Res.* **38**, 1191 (2002).
93. E. Barkai and Y. C. Cheng, *J. Chem. Phys.* **118**, 6167 (2003).
94. C. Monthus and J.-P. Bouchaud, *J. Phys. A* **29**, 3847 (1996).
95. R. Metzler, J. Klafter, J. Jortner and M. Volk, *Chem. Phys. Lett.* **293**, 477 (1998).
96. J. Mai, I. M. Sokolov and A. Blumen, *Phys. Rev. Lett.* **77**, 4462 (1996).
97. D. Brockmann and T. Geisel, *Phys. Rev. Lett.* **91**, 048303 (2003).
98. M. A. Lomholt, T. Ambjörnsson and R. Metzler, *Phys. Rev. Lett.* **95**, 260603 (2005).
99. D. Brockmann, L. Hufnagel and T. Geisel, *Nature* **439**, 462 (2006).
100. N. E. Humphries *et al.*, *Nature* **465**, 1066 (2010).
101. F. Bartumeus, *Phys. Rev. Lett.* **88**, 097901 (2002).
102. M. A. Lomholt, T. Koren, R. Metzler and J. Klafter, *Proc. Natl. Acad. Sci.* **105**, 11055 (2008).
103. C. M. Song, T. Koren, P. Wang and A. L. Barabasi, *Nature Phys.* **6**, 818 (2010).
104. M. C. González, C. A. Hidalgo and A.-L. Barabási, *Nature* **453**, 779 (2008).
105. H. Katori, S. Schlipf and H. Walther, *Phys. Rev. Lett.* **79**, 2221 (1997).
106. R. Metzler, E. Barkai and J. Klafter, *Phys. Rev. Lett.* **82**, 3563 (1999).
107. R. Metzler, E. Barkai and J. Klafter, *Europhys. Lett.* **46**, 431 (1999).
108. E. Barkai, R. Metzler and J. Klafter, *Phys. Rev. E* **61**, 132 (2000).
109. R. Metzler and A. Compte, *J. Phys. Chem. B* **104**, 3858 (2000).
110. R. Schumer, M. M. Meerschaert and B. Baeumer, *J. Geophys. Res.* **114**, F00A07 (2009).
111. I. M. Tolić-Nørrelykke, E. L. Munteanu, G. Thon, L. Oddershede and K. Berg-Sorensen, *Phys. Rev. Lett.* **93**, 078102 (2004).
112. C. Selhuber-Unkel, P. Yde, K. Berg-Sørensen and L. B. Oddershede, *Phys. Biol.* **6**, 025015 (2009).
113. R. Metzler and J. Klafter, *Phys. Rev. E* **61**, 6308 (2000).
114. R. Metzler and J. Klafter, *J. Phys. Chem. B* **104**, 3851 (2000).
115. R. Metzler, *Phys. Rev. E* **62**, 6233 (2000).
116. W. Feller, *An Introduction to Probability Theory and Its Applications*, Vol. 2 (John Wiley, 1971).
117. E. Barkai and R. J. Silbey, *J. Chem. Phys. B* **104**, 3866 (2000).
118. A. Chechkin, V. Gonchar, J. Klafter, R. Metzler and L. Tanatarov, *Chem. Phys.* **284**, 233 (2002).
119. A. V. Chechkin, J. Klafter, V. Yu. Gonchar, R. Metzler and L. V. Tanatarov, *Phys. Rev. E* **67**, 010102(R) (2003).

120. A. V. Chechkin, V. Yu. Gonchar, J. Klafter, R. Metzler and L. V. Tanatarov, *J. Stat. Phys.* **115**, 1505 (2004).
121. B. Dybiec, I. M. Sokolov and A. V. Chechkin, *J. Stat. Mech.* P07008 (2010).
122. S. Redner, *A Guide to First-Passage Processes* (Cambridge Univ. Press, 2001).
123. R. Metzler and J. Klafter, *Physica A* **278**, 107 (2000).
124. R. Metzler and J. Klafter, *Biophys. J.* **85**, 2776 (2003).
125. E. Barkai, *Phys. Rev. E* **63**, 046118 (2001).
126. A. V. Chechkin, R. Metzler, J. Klafter, V. Yu. Gonchar and L. V. Tanatarov, *J. Phys. A* **36**, L537 (2003).
127. T. Koren, M. A. Lomholt, A. V. Chechkin, J. Klafter and R. Metzler, *Phys. Rev. Lett.* **99**, 160602 (2007).
128. T. Koren, A. V. Chechkin and J. Klafter, *Physica A* **379**, 10 (2007).
129. E. Sparre, *Andersen, Math. Scand.* **1**, 263 (1953).
130. E. Sparre, *Andersen, Math. Scand.* **2**, 195 (1954).
131. B. Dybiec, E. Gudowska-Novak and P. Hänggi, *Phys. Rev. E* **73**, 046104 (2006).
132. B. Dybiec and E. Gudowska-Novak, *Europhys. Lett.* **88**, 10003 (2009).
133. A. V. Chechkin, V. Yu. Gonchar, J. Klafter and R. Metzler, *Europhys. Lett.* **72**, 348 (2005).
134. A. V. Chechkin, O. Yu. Sliusarenko, J. Klafter and R. Metzler, *Phys. Rev. E* **75**, 041101 (2007).
135. B. Dybiec, E. Gudowska-Novak and P. Hänggi, *Phys. Rev. E* **75**, 021109 (2007).
136. I. Nordlund, *Z. Phys. Chem.* **87**, 40 (1914).
137. I. Golding and E. C. Cox, *Phys. Rev. Lett.* **99**, 098102 (2006).
138. G. Seisenberger, M. U. Ried, T. Endreß, H. Büning, M. Hallek and C. Bräuchle, *Science* **294**, 1929 (2001).
139. A. Caspi, R. Granek and M. Elbaum, *Phys. Rev. Lett.* **85**, 5655 (2000).
140. I. Bronstein, Y. Israel, E. Kepten, S. Mai, Y. Shav-Tal, E. Barkai and Y. Garini, *Phys. Rev. Lett.* **103**, 018102 (2009).
141. S. C. Weber, A. J. Spakowitz and J. A. Theriot, *Phys. Rev. Lett.* **104**, 238102 (2010).
142. N. Gal and D. Weihs, *Phys. Rev. E* **81**, 020903(R) (2010).
143. J.-P. Bouchaud, *J. Phys. I (Paris)* **2**, 1705 (1992).
144. A. Rebenshtok and E. Barkai, *J. Stat Phys.* **133**, 565 (2008).
145. G. Bel and E. Barkai, *Phys. Rev. Lett.* **94**, 240602 (2005).
146. E. Lutz, *Phys. Rev. Lett.* **93**, 190602 (2004).
147. A. Lubelski, I. M. Sokolov and J. Klafter, *Phys. Rev. Lett.* **100**, 250602 (2008).
148. Y. He, S. Burov, R. Metzler and E. Barkai, *Phys. Rev. Lett.* **101**, 058101 (2008).
149. S. Burov, R. Metzler and E. Barkai, *Proc. Natl. Acad. Sci. USA* **107**, 13228 (2010).
150. S. Burov, J.-H. Jeon, R. Metzler and E. Barkai, *Phys. Chem. Chem. Phys.* **13**, 1800 (2011).

151. T. Neusius, I. M. Sokolov and J. C. Smith, *Phys. Rev. E* **80**, 011109 (2009).
152. I. M. Sokolov, E. Heinsalu, P. Hänggi and I. Goychuk, *Europhys. Lett.* **86**, 30009 (2009).
153. A. Rebenshtok and E. Barkai, *Phys. Rev. Lett.* **99**, 210601 (2007).
154. E. Barkai, *J. Stat. Phys.* **123**, 883 (2006).
155. S. Burov and E. Barkai, *Phys. Rev. Lett.* **98**, 250601 (2007).
156. M. A. Lomholt, I. M. Zaid and R. Metzler, *Phys. Rev. Lett.* **98**, 200603 (2007).
157. I. M. Zaid, M. A. Lomholt and R. Metzler, *Biophys. J.* **97**, 710 (2009).
158. A. N. Kolmogorov, *Dokl. Acad. Sci. USSR* **26**, 115 (1940).
159. B. B. Mandelbrot and J. W. van Ness, *SIAM Rev.* **10**, 422 (1968).
160. E. Lutz, *Phys. Rev. E* **64**, 051106 (2001).
161. W. H. Deng and E. Barkai, *Phys. Rev. E* **79**, 011112 (2009).
162. R. Metzler, V. Tejedor, J.-H. Jeon, Y. He, W. Deng, S. Burov and E. Barkai, *Acta Phys. Pol. B* **40**, 1315 (2009).
163. J.-H. Jeon and R. Metzler, *Phys. Rev. E* **81**, 021103 (2009).
164. J.-H. Jeon and R. Metzler, *J. Phys. A* **43**, 252001 (2010).
165. A. Klemm, R. Metzler and R. Kimmich, *Phys. Rev. E* **65**, 021112 (2002).
166. Y. Moroz, I. M. Sokolov and J. Klafter, *Phys. Rev. E* **81**, 010101(R) (2010).
167. F. D. Stefani, J. P. Hoogenboom and E. Barkai, *Phys. Today* **62**, 34 (2009).
168. G. Margolin and E. Barkai, *Phys. Rev. Lett.* **94**, 080601 (2005).
169. X. Brokmann, J.-P. Hermier, G. Messin, P. Desbiolles, J.-P. Bouchaud and M. Dahan, *Phys. Rev. Lett.* **90**, 120601 (2003).
170. V. Tejedor, O. Bénichou, R. Voituriez, R. Jungmann, F. Simmel, C. Selhuber, L. Oddershede and R. Metzler, *Biophys. J.* **98**, 1364 (2010).
171. S. Condamin, V. Tejedor, R. Voituriez, O. Bénichou and J. Klafter, *Proc. Natl. Acad. Sci. USA* **105**, 5675 (2008).
172. M. Magdziarz, A. Weron, K. Burnecki and J. Klafter, *Phys. Rev. Lett.* **103**, 180602 (2009).
173. M. Magdziarz and J. Klafter, *Phys. Rev. E* **82**, 011129 (2010).
174. K. Burnecki and A. Weron, *Phys. Rev. E* **82**, 021130 (2010).
175. S. B. Yuste, G. Oshanin, K. Lindenberg, O. Bénichou and J. Klafter, *Phys. Rev. E* **78**, 021105 (2008).
176. E. Abad, S. B. Yuste and K. Lindenberg, *Phys. Rev. E* **81**, 031115 (2010).
177. I. M. Sokolov, S. B. Yuste, J. J. Ruiz-Lorenzo and K. Lindenberg, *Phys. Rev. E* **79**, 051113 (2009).
178. D. Froemberg and I. M. Sokolov, *Phys. Rev. Lett.* **100**, 108304 (2008).
179. I. M. Sokolov, M. G. W. Schmidt and F. Sagues, *Phys. Rev. E* **73**, 031102 (2006).
180. B. I. Henry, T. A. M. Langlands and P. Straka, *Phys. Rev. Lett.* **105**, 170602 (2010).

# Chapter 2

## Stochastic Diffusion and Stable Noise-Induced Phenomena

Bartłomiej Dybiec and Ewa Gudowska-Nowak
*Marian Smoluchowski Institute of Physics*
*and*
*Mark Kac Center for Complex Systems Research*
*Jagiellonian University*
*ul. Reymonta 4, 30–059 Kraków, Poland*

We discuss ubiquity of noise effects generated in non-equilibrium systems driven by "stable" random forces interpreted as limit cases of pure jump stochastic processes. In particular, such an approach leads to a generalization of the common Brownian motion by a Lévy diffusion process. This chapter briefly analyzes common properties of the anomalous transport and investigates the asymptotic relations between properly scaled continuous time random walks (CTRW) and fractional Smoluchowski–Fokker–Planck equations.

1. Introduction . . . . . . . . . . . . . . . . . . . . . . . . . 33
2. White Non-Gaussian Noises . . . . . . . . . . . . . . . . . . . 36
   2.1. Escape from finite intervals . . . . . . . . . . . . . 37
   2.2. Markovian non-Gaussian Kramers problem . . . . 41
3. Bi-Fractional Kinetics . . . . . . . . . . . . . . . . . . . . 43
   3.1. Escape from finite intervals . . . . . . . . . . . . . 43
   3.2. Non-Markovian Kramers problem . . . . . . . . . . 46
4. Summary . . . . . . . . . . . . . . . . . . . . . . . . . . . 47

### 1. Introduction

Any stochastic processes can be defined as a sequence of random variables parametrized by time $\{X(t_i), i = 1, \ldots\}$. The joint probability density function $p(x_1, t_1; x_2, t_2, \ldots x_{n-1}, t_{n-1}; x_n, t_n)$ characterizing such a sequence incorporates information about the value the observable $X$ has taken at times $t_1 < t_2 < \cdots < t_n$ and can be determined from the distribution of lower order by using the Bayes rule $p(x_1, t_1; x_2, t_2, \ldots; x_{n-1}, t_{n-1}; x_n, t_n) = p(x_n, t_n | x_{n-1}, t_{n-1}; \ldots; x_1, t_1) \cdots p(x_2, t_2 | x_1, t_1) p(x_1, t_1)$. If the conditional probability density satisfies the relation

$$p(x_n, t_n | x_{n-1}, t_{n-1}; \ldots ; x_1, t_1) = p(x_n, t_n | x_{n-1}, t_{n-1}), \tag{1}$$

known as a Markov property, the process can be fully specified by a conditional probability $p(x_n, t_n | x_{n-1}, t_{n-1})$ and a one-point probability function $p(x, t)$. This idealization (termed sometimes the lack of memory) makes Markov processes a powerful and an appealing tool in modeling of various physical, biological, economic or even social phenomena [1–3]. A continuous time parameter process with a Markov property, for which sample paths (the process realizations) $x(t)$ are continuous functions of time is called a stochastic diffusion. For many Markov processes a proper transformation of time and renormalization of the state variable results in a stochastic diffusion approximation. Any process of this kind satisfies the condition

$$\lim_{\Delta t \to 0} \frac{1}{\Delta t} \mathrm{Prob}\left\{ |X(t + \Delta t) - x| > \varepsilon \middle| X(t) = x \right\} = 0 \tag{2}$$

for every $\varepsilon > 0$. This relation clearly suggests that for the diffusion process, any large displacements exceeding a given value of $\varepsilon$ are unlikely to happen in sufficiently short time intervals $\Delta t$. In contrast, in the forthcoming sections we will turn to processes termed "anomalous diffusion", which in general violate the above property.

An alternative approach to derive the evolution equation for the transition probability densities of a Markov diffusion process was presented by Itô. Itô's version of stochastic calculus starts with the stochastic differential equation of the form

$$dx(t) = \mu(x, t)dt + \sigma(x, t)dW(t), \tag{3}$$

where $W(t)$ is a unit Wiener process describing Brownian motion (i.e. a normal Gaussian process with stationary independent increments fulfilling the relations $E[W(t)] = 0$ and $E[(W(t) - W(s))^2] = |t - s|$). Accordingly, given that $X(t) = x$, the increment $\Delta x(t)$ in a small time interval $\Delta t$ has the mean value $\mu(x, t)\Delta t$, the variance $\sigma^2(x, t)\Delta t$ and is independent of all previous increments [4].

With the results summarized above, the stochastic diffusion process may be viewed as "locally" Gaussian. Further generalizations of Itô's approach are possible by e.g., introducing increments following the stable law (self-similar but non-Gaussian) or, by analyzing properties of $\Delta W(t)$ incorporated in Eq. (3) as a limit of a random sum of random elements

$$\Delta W(t) = \Delta x(t) = \sum_{i=1}^{N(t)} X_i, \tag{4}$$

where the number of summands $N(t)$ is statistically independent of $X_i$ and governed by a renewal process $\sum_{i=1}^{N(t)} T_i \leq t < \sum_{i=1}^{N(t)+1} T_i$ with $t > 0$. Let us assume further that $T_i$, $X_i$ belong to the domain of attraction of stable distributions, $T_i \sim L_{\lambda,1}$ and $X_i \sim L_{\alpha,\beta}$, whose corresponding characteristic functions $\phi(k) = \langle \exp(ikL_{\alpha,\beta}) \rangle = \int_{-\infty}^{\infty} e^{ikx} l_{\alpha,\beta}(x;\sigma,\mu = 0)dx$, with the probability density function $l_{\alpha,\beta}(x;\sigma,\mu = 0)$, are given by

$$\phi(k) = \exp\left[-\sigma^\alpha |k|^\alpha \left(1 - i\beta \, \mathrm{sign}k \tan\frac{\pi\alpha}{2}\right)\right] \quad \text{for } \alpha \neq 1, \tag{5}$$

and

$$\phi(k) = \exp\left[-\sigma|k| \left(1 + i\beta\frac{2}{\pi} \mathrm{sign}k \log|k|\right)\right] \quad \text{for } \alpha = 1. \tag{6}$$

Here the parameter $\alpha \in (0, 2]$ denotes the stability index, yielding the asymptotic long tail power law for the $x$-distribution, which for $\alpha < 2$ is of the $|x|^{-(1+\alpha)}$ type. The parameter $\sigma$ ($\sigma \in (0,\infty)$) characterizes the scale, $\beta$ ($\beta \in [-1, 1]$) defines an asymmetry (skewness) of the distribution and $\mu$ represents the shift, which for the strictly stable distributions [5] is set to 0. For $0 < \nu < 1$, $\beta = 1$, the stable variable[a] $L_{\nu,1}$ is defined on a positive semi-axis only. Within the above formulation, the counting process $N(t)$ satisfies

$$\begin{aligned}
\lim_{t\to\infty} \mathrm{Prob}\left\{\frac{N(t)}{(t/c)^\nu} < x\right\} &= \lim_{t\to\infty} \mathrm{Prob}\left\{\sum_{i=1}^{[(t/c)^\nu x]} T_i > t\right\} \\
&= \lim_{n\to\infty} \mathrm{Prob}\left\{\sum_{i=1}^{[n]} T_i > \frac{cn^{1/\nu}}{x^{1/\nu}}\right\}
\end{aligned}$$

[a] Within the text we use the notation $L_{\alpha,\beta}$ for a stable random variable, $L_{\alpha,\beta}(x)$ for its cumulative distribution function and similarly $l_{\alpha,\beta}(x)$ for its probability density. Finally, $L_{\alpha,\beta}(t)$ denotes $\alpha$-stable Lévy type process.

$$= \lim_{n\to\infty} \text{Prob}\left\{ \frac{1}{cn^{1/\nu}} \sum_{i=1}^{[n]} T_i > \frac{1}{x^{1/\nu}} \right\}$$
$$= 1 - L_{\nu,1}(x^{-1/\nu}), \tag{7}$$

where $[(t/c)^\nu x]$ denotes the integer part of the number $(t/c)^\nu x$. Since $\lim_{n\to\infty} \text{Prob}\left\{ \frac{1}{c_1 n^{1/\alpha}} \sum_{i=1}^n X_i < x \right\} \rightarrow L_{\alpha,\beta}(x)$ and $p(x,t) = \sum_n p(x|n)p_n(n(t))$, asymptotically one gets $(dL_{\alpha,\beta}(x)/dx = l_{\alpha,\beta}(x))$

$$p(x,t) \sim (c_2 t)^{-\nu/\alpha} \int_0^\infty l_{\alpha,\beta}((c_2 t)^{-\nu/\alpha} x \tau^{\nu/\alpha}) l_{\nu,1}(\tau) \tau^{\nu/\alpha} d\tau, \tag{8}$$

where $c_1$ and $c_2$ are constants. The resulting (non-Markov) process becomes $\nu/\alpha$-self-similar Lévy random walk [3, 6, 7] whose probability density function (PDF) has a scaling property $p(x,t) = t^{-\nu/\alpha} p(xt^{-\nu/\alpha}, 1)$.

In general, Lévy random walks are neither Gaussian nor Markov [8–10]. An interesting class of these processes can be built up by allowing coupling between $N(t)$ and $X_i$ and introducing the hierarchical clustering transformation. In consequence, the asymptotic distribution of the resulting diffusion front $W(t)$ depends on the way the jumps are correlated within the clusters [11]. Time relaxation patterns of such fronts are interesting models of transport and relaxation phenomena in complex, inhomogeneous systems [12–17].

Following the above formulation, see Eqs. (4)–(7), it is clear that in a broader context, the noise term entering the Langevin equation Eq. (3) can be viewed as a time-derivative of a generalized Wiener process. In the next two paragraphs we will restrict ourselves to one-dimensional systems driven by Lévy white noises, i.e. time derivatives of the Lévy motion characterized by independent, stationary increments distributed according to a Lévy stable law [5, 18, 19], see Eqs. (5) and (6). Finally, we extend inspection of noise induced diffusion to situations when Lévy flights coexist with long residence times, see Sec. 3.

## 2. White Non-Gaussian Noises

In systems operating under far from equilibrium conditions, the random force acting on a particle can be of the more general type than white Gaussian noise. One of the possible extensions is to assume that the noise is still of the white type, however the noise pulses are distributed according to some more general distributions. By virtue of the generalized central limit theorem, it is natural to consider $\alpha$-stable distributions as limiting probability distribution for sums of properly normalized independent identically distributed random variables sampled from heavy tailed distributions.

As two paradigmatic scenarios we discuss an escape from finite and semi-infinite intervals and the escape from a potential well in far from equilibrium situations, when motion of the particle is subject to anomalously large random forces. The escape from finite intervals is analyzed with the special attention to the problem of boundary conditions.

### 2.1. *Escape from finite intervals*

In order to inspect properties of the anomalous diffusion we first consider a free motion of a particle in the finite interval restricted by two absorbing boundaries located at $\pm L$. In the continuous limit, the position of a randomly moving particle is described by the Itô stochastic differential equation

$$dx(t) = dL_{\alpha,0}(t), \tag{9}$$

where $L_{\alpha,0}(t)$ stands for a symmetric Lévy motion [19, 20] being the generalization of the Wiener process, i.e. a process with stationary, independent increments distributed according to a symmetric $\alpha$-stable density [5, 18], see Eqs. (5) and (6). Without loss of generality we assume for a starting position $x(0) = x_0$. Equation (9) can then be associated with the following (space-fractional) Smoluchowski–Fokker–Planck equation [12, 15, 16, 19, 21]

$$\frac{\partial p(x,t|x_0,0)}{\partial t} = \sigma^\alpha \frac{\partial^\alpha p(x,t|x_0,0)}{\partial |x|^\alpha}. \tag{10}$$

In the above equation the Riesz–Weyl fractional derivative is defined by using the Fourier transform $\mathcal{F}\left[\frac{\partial^\alpha f(x)}{\partial |x|^\alpha}\right] = -|k|^\alpha \mathcal{F}[f(x)]$. For $\alpha = 2$, any Lévy stable noise is equivalent to the Gaussian white noise (with the intensity $\sqrt{2}\sigma$, see [5]) and the fractional Smoluchowski–Fokker–Planck equation (10) takes its standard form, i.e. $\partial^\alpha/\partial|x|^\alpha \to \partial^2/\partial x^2$.

From Eq. (10) it is possible to obtain the first passage time density $f(t)$ by using the relation [1, 22]

$$f(t) = f(t|x_0,0) = -\frac{d}{dt}\int_{-L}^{L} p(x,t|x_0,0)dx = -\frac{d}{dt}S(t), \tag{11}$$

where $p(x,t|x_0,0)$ is the solution to Eq. (10) subjected to the appropriate boundary and initial conditions [23, 24]. More precisely, for a regular diffusion process ($\alpha = 2$) assumption of absorbing boundaries located at $x = \pm L$ implies $p(x = \pm L, t) = 0$. This is not, however, the case for $\alpha < 2$ when the nonlocal character of the random process (possible large excursions over the point barriers) requires rephrasing of the absorption condition to $p(|x| \geq L, t) = 0$, see [23]. The survival probability $S(t) = \int_{-L}^{L} p(x, t|x_0, 0)dx$ is the probability that at time $t$ a random walker is still in the $[-L, L]$ interval, i.e. it is the complementary cumulative distribution function $S(t) = 1 - F(t) = 1 - \int_0^t f(u)du = \int_t^\infty f(u)du$ of the first passage time density.

For the Gaussian case ($\alpha = 2$), the solution to Eq. (10) can be constructed using Fourier series [25, Eq. (81)] and for $x_0 = 0$, it reads

$$p(x, t|0, 0) = \sum_{n=1}^{\infty} \frac{1}{L} \sin\left[\frac{n\pi}{2}\right] \sin\left[\frac{n\pi(x + L)}{2L}\right] \exp\left[-\frac{n^2\pi^2\sigma^2}{4L^2}t\right]. \tag{12}$$

From Eq. (12) one gets

$$S(t) = \sum_{n=1}^{\infty} \frac{4}{n\pi} \sin^3\left[\frac{n\pi}{2}\right] \exp\left[-\frac{n^2\pi^2\sigma^2}{4L^2}t\right] \tag{13}$$

and

$$f(t) = \sum_{n=1}^{\infty} \frac{n\pi\sigma^2}{L^2} \sin^3\left[\frac{n\pi}{2}\right] \exp\left[-\frac{n^2\pi^2\sigma^2}{4L^2}t\right]. \tag{14}$$

Accordingly, the first passage time density has the exponential asymptotics determined by the lowest eigenvalue, see Eq. (14), and the mean first passage time $\langle\tau\rangle$ from a bounded interval is finite

$$\langle\tau\rangle = \int_0^\infty f(t)t\,dt = \frac{L^2}{2\sigma^2}. \tag{15}$$

Alternatively, the above formula can be otherwise obtained by solving the backward Kolmogorov equation [1], i.e. the equation which is formally adjoint to the Smoluchowski–Fokker–Planck Eq. (10) with $\alpha = 2$.

For $\alpha < 2$, the exponential asymptotics of the survival probability can also be deduced from Eq. (10). In fact, by using of the separation of variables, the full solution $p(x, t)$ to Eq. (10) can be expressed [1, 2, 26] conveniently as a sum of eigensolutions

$$p(x, t) = \sum_i c_i p_i(x, t) = \sum_i c_i T_i(t)\varphi_i(x), \tag{16}$$

where $T_i(t)$ and $\varphi_i(x)$ fulfill

$$\frac{dT_i(t)}{dt} = -\lambda_i^{(\alpha)} T_i(t) \tag{17}$$

and

$$\frac{\partial^\alpha \varphi_i(x)}{\partial |x|^\alpha} = -\lambda_i^{(\alpha)} \varphi_i(x). \tag{18}$$

Consequently, the eigensolutions $T_i(t)$ are given by the exponential functions

$$T_i(t) = \exp[-\lambda_i^{(\alpha)} t] \tag{19}$$

and eigenvalues $\lambda_i^{(\alpha)}$ and constants $c_i$ can be determined by boundary and initial conditions, respectively. Due to a nonlocal character of the boundary conditions for Lévy flights, the major problem in solving Eq. (10) is construction of eigenfunctions to the fractional Laplacian on bounded domains [24]. Nevertheless, the knowledge of eigenvalues $\lambda_i^{(\alpha)}$ and eigenfunctions $\varphi_i(x)$ is not necessary to demonstrate exponential asymptotics of the first passage time density $f(t)$ and the survival probability $S(t)$ which, due to the exponential form of $T_i(t)$, are determined by the lowest eigenvalue $\lambda_i^{(\alpha)}$, cf. Fig. 1.

In the discussed case the mean first passage time from a finite interval (restricted by at least one absorbing boundary, the other one can be reflecting) is finite for every value of the stability index $\alpha$, see [23, 24]. The corresponding expression for the mean first passage time reads [24]

$$\langle \tau \rangle_T = \frac{L^\alpha}{\sigma^\alpha \Gamma(1+\alpha)} \tag{20}$$

and converts to Eq. (15) for $\alpha = 2$. Figure 1 presents survival probabilities (top panel) and the mean first passage time (bottom panel) as a function of the stability index $\alpha$ for the interval half-width $L = 120$. The survival probabilities have exponential asymptotics, see top panel. The bottom panel of Fig. 1 demonstrates the perfect agreement between simulated $\langle \tau \rangle_S$ and theoretical $\langle \tau \rangle_T$ values of the mean first passage times. The inset in the bottom panel of Fig. 1 displays $\langle \tau \rangle_S / \langle \tau \rangle_T$ ratio. In more general situations of continuous time random walks (CTRWs) with jump lengths and waiting times between subsequent jumps both governed by heavy-tailed distributions, deviations from the exponential type of the survival probabilities are induced by trapping events (long waiting times) [27], as discussed in Sec. 3.

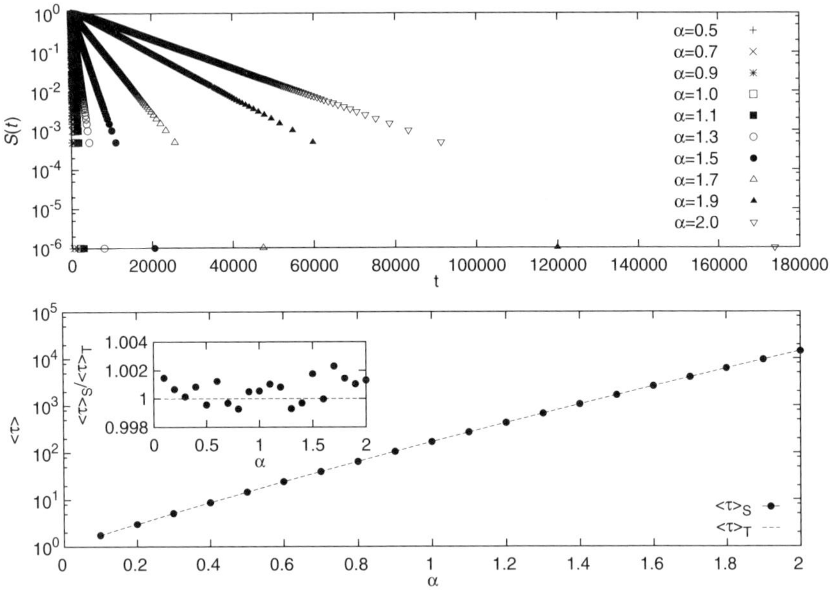


Fig. 1. The survival probability (top panel) and the mean first passage time (bottom panel) in a free Lévy motion on a finite interval. The inset in the bottom panel presents $\langle\tau\rangle_S/\langle\tau\rangle_T$, i.e. the ratio of simulated $\langle\tau\rangle_S$ and theoretical $\langle\tau\rangle_T$ values of the mean first passage times. Note the log-lin scale. For simulation of Eq. (9), the time step $\Delta t = 10^{-2}$ has been used with the number of realizations $N = 10^6$. Other simulation parameters: $L = 120$ and $\sigma = 1/\sqrt{2}$.

The escape from the finite interval (restricted by two absorbing boundaries) should be contrasted with the escape from a positive semi-line. For a Brownian particle starting its motion at $x_0 > 0$, the first time density from the positive semi-axis is [22]

$$f_{\mathrm{s-a}}(t) = \frac{t^{-3/2}x_0}{\sqrt{2\pi\sigma^2}} \exp\left[-\frac{x_0^2}{2\sigma^2 t}\right] \tag{21}$$

and the mean first passage time diverges. Nevertheless, the particle moving along the real line will certainly cross the origin ($\int_0^\infty f_{\mathrm{s-a}}(t)dt = 1$). The observed asymptotics, i.e. $t^{-3/2}$, is in the line with the Sparre Andersen theorem which states that for a free stochastic process driven by symmetric white noise, the PDF of the first exit time from the positive half line asymptotically behave like $t^{-3/2}$. Left panel of Fig. 4 presents survival probability for the escape from positive semi-line in the case of Markovian, symmetric Lévy flights. Again, deviations from the $t^{-3/2}$ asymptotics can be observed in more general cases of anomalous diffusion, due to the presence of long waiting times [28–31], see right panel of Fig. 4 and Sec. 3.

### 2.2. *Markovian non-Gaussian Kramers problem*

The problem of the escape of a massive particle over the potential barrier in the presence of an assisting random noise had been analyzed in a seminal paper by Kramers [32]. A mathematical description of the Kramers problem in the position and momentum space is provided by the Klein–Kramers equation [1, 2] which in the limit of large damping reduces to the common Smoluchowski–Fokker–Planck equation [2, 33] representing solely the diffusion process in a position coordinate.

In order to investigate an analogous problem for barrier crossing triggered by Lévy flights, let us assume that initially ($t = 0$) a test particle is located in the left potential well of the generic, fourth-order, double-well potential

$$V(x) = -\frac{a}{2}x^2 + \frac{b}{4}x^4. \tag{22}$$

The minima of the potential, see Eq. (22), are located at $\pm x_{\mathrm{w}} = \pm\sqrt{a/b}$. The height of the potential barrier is $\Delta V = V(0) - V(x_{\mathrm{w}}) = a^4/4b$. For the purpose of modeling, we have chosen the same set of parameters as in [34], i.e. $a = 128$ and $b = 512$. As a result, the potential wells are located at $x_{\mathrm{w}} = \pm 1/2$ while the barrier height is $\Delta V = 8$. For small noise intensities, it guarantees that (in the Gaussian case) the potential barrier is high enough to properly induce the separation of time scales [35] between the inter-well relaxation and the barrier crossing events.

For a particle moving in an external potential, driven by an $\alpha$-stable white noise the evolution of the probability density is described by the following fractional Smoluchowski–Fokker–Planck equation [15, 16]

$$\frac{\partial p(x,t|x_0,0)}{\partial t} = \frac{\partial}{\partial x}[V'(x)p(x,t|x_0,0)] + \sigma^\alpha \frac{\partial^\alpha p(x,t|x_0,0)}{\partial |x|^\alpha}. \tag{23}$$

In the limit of $\alpha = 2$, the standard Kramers problem [32] is reconstructed, i.e. the mean first passage time from the initial point (e.g., $x_0$) to the final point (e.g., $x$) can be calculated exactly and reads [1]

$$T_{\mathrm{MFPT}}(x_0 \to x) = \frac{1}{\sigma^2}\int_{-x_0}^{x} \exp\left[\frac{V(z)}{\sigma^2}\right]\int_{-\infty}^{z} \exp\left[-\frac{V(y)}{\sigma^2}\right] dy\, dz, \tag{24}$$

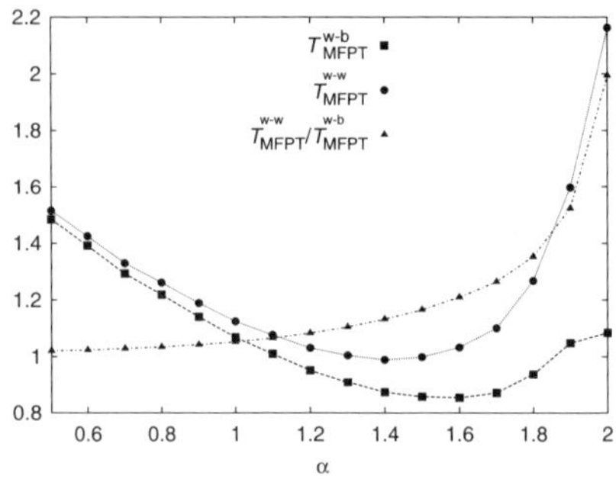


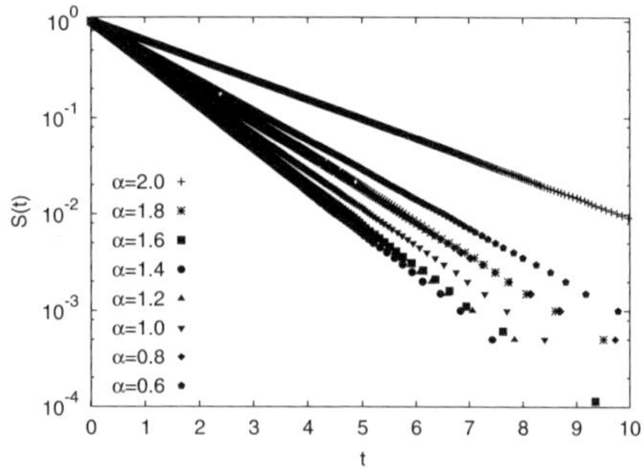


Fig. 2. Left panel presents: the mean first passage time from the potential well to the barrier top ($T^{\mathrm{w-b}}_{\mathrm{MFPT}}$), the mean first passage time from the potential well to the other minima of the potential ($T^{\mathrm{w-w}}_{\mathrm{MFPT}}$) and the ratio between them ($R = T^{\mathrm{w-w}}_{\mathrm{MFPT}}/T^{\mathrm{w-b}}_{\mathrm{MFPT}}$). Right panel presents survival probability from the potential well to the other potential well for different values of the stability index $\alpha$. Numerical results were constructed with the time step of integration $\Delta t = 10^{-5}$ and averaged over $N = 3 \times 10^5$ realizations. In the limit of $\alpha = 2$, values of MFPTs agree with their exact values, see Eq. (24). The noise intensity is set to $\sigma = \sqrt{2}$.

where $\sigma$ stands for the noise intensity. For $\alpha < 2$, the mean first passage time can be estimated numerically [34, 36, 37].

In the standard Kramers problem, for the sufficiently high potential barrier the mean first passage time from the potential well to the (other) potential well is twice the mean first passage time from the potential well to the barrier top ($2T^{\mathrm{w-b}}_{\mathrm{MFPT}} = T^{\mathrm{w-w}}_{\mathrm{MFPT}}$), see [1, 34]. The presence of Lévy flights violates this property, see [34] and Fig. 2. Consequently, due to long jumps, for small values of the stability index $\alpha$ both mean first passage times are the same ($T^{\mathrm{w-b}}_{\mathrm{MFPT}} = T^{\mathrm{w-w}}_{\mathrm{MFPT}}$). Left panel of Fig. 2 shows values of the mean first passage time from the potential well to the barrier top ($T^{\mathrm{w-b}}_{\mathrm{MFPT}}$), values of the mean first passage time from the potential well to the potential well ($T^{\mathrm{w-w}}_{\mathrm{MFPT}}$) and their ratio ($R = T^{\mathrm{w-w}}_{\mathrm{MFPT}}/T^{\mathrm{w-b}}_{\mathrm{MFPT}}$).

Numerically estimated survival probability $S(t)$ displays exponential long-time relaxation, see right panel of Fig. 2. In fact, this type of the asymptotic behavior can be deduced from a general structure of the solution to Eq. (23) describing motion subject to symmetric Lévy flights. In this case single modes of the fractional Smoluchowski–Fokker–Planck equation decay exponentially in time, see Eq. (17) and right panel of Fig. 2. Analogously, like in the case of escape from finite intervals, the lowest eigenvalue $\lambda_i^{(\alpha)}$ which determines the asymptotics of the survival probability can be estimated numerically by fitting the exponential decay to the survival probability.

## 3. Bi-Fractional Kinetics

Transport in porous, fractal-like media or relaxation kinetics in inhomogeneous materials lead to long trapping events. Consequently, such processes are usually ultraslow, i.e. subdiffusive [26, 38]. The most interesting situations take place, however, when both effects — the presence of long waiting times for the next step and long jumps — are combined in the same scenario [39]. Such realms can be described by the bi-fractional Smoluchowski–Fokker–Planck equation which in the force free case takes the form

$$\frac{\partial p(x,t)}{\partial t} = {}_0D_t^{1-\nu}\left[\sigma^\alpha \frac{\partial^\alpha}{\partial |x|^\alpha} p(x,t)\right], \tag{25}$$

with the initial condition $p(x,0|0,0) = p(x,0) = \delta(x)$. In the above equation ${}_0D_t^{1-\nu}$ denotes the Riemann–Liouville fractional derivative ${}_0D_t^{1-\nu} = \frac{\partial}{\partial t}{}_0D_t^{-\nu}$ defined by the relation

$${}_0D_t^{1-\nu} f(x,t) = \frac{1}{\Gamma(\nu)}\frac{d}{dt}\int_0^t dt' \frac{f(x,t')}{(t-t')^{1-\nu}}. \tag{26}$$

Stochastic representation of solutions to Eq. (25) is provided by the subordination method [8–10, 12, 40] which allows for an easy treatment of boundary conditions [23, 26].

The competition between long waiting times and long jumps, which is well visible in the case of non-Markovian Lévy flights [12, 17, 26] ($0 < \nu < 1$ with $0 < \alpha < 2$), can mask some temporal properties of the process providing a stochastic representation of solutions to Eq. (25). Consequently, such a process may display features typical for Markovian diffusion (e.g. linear time-dependence of the mean-square deviation), cf. [41]. First, we demonstrate how the aforementioned competition affects the escape from finite and semi-infinite intervals. Finally, we comment on the Kramers problem in the region of non-Markovian Lévy flights [42].

### 3.1. *Escape from finite intervals*

The solution $p(x,t)$ to Eq. (25) can be written [26, 43, 44] as a sum of eigenfunctions

$$p(x,t) = \sum_{i=1}^{\infty} c_i p_i(x,t) = \sum_{i=1}^{\infty} c_i T_i(t)\varphi_i(x) \tag{27}$$

with $T_i(t)$ fulfilling

$$\frac{dT_i(t)}{dt} = -\lambda_i^{(\alpha,\nu)} \, {}_0D_t^{1-\nu} T_i(t) \tag{28}$$

and $\varphi_i(x)$ satisfying

$$\frac{\partial^\alpha}{\partial |x|^\alpha} \varphi_i(x) = -\lambda_i^{(\alpha,\nu)} \varphi_i(x). \tag{29}$$

The solution to Eq. (28) is given in terms of the Mittag-Leffler function [45]

$$T_i(t) = E_\nu(-\lambda_i^{(\alpha,\nu)} t^\nu) \equiv \sum_{j=0}^{\infty} \frac{(-\lambda_i^{(\alpha,\nu)} t^\nu)^j}{\Gamma(1+\nu j)} \tag{30}$$

which is a generalization of the standard exponential function ($\nu = 1$). For small values of its arguments the Mittag-Leffler function becomes a stretched exponential [26, 46]

$$E_\nu(-\lambda_i^{(\alpha,\nu)} t^\nu) \propto \exp\left[-\frac{\lambda_i^{(\alpha,\nu)} t^\nu}{\Gamma(1+\nu)}\right] \tag{31}$$

whereas in the limit of large $t$, it displays a power-law asymptotics [26, 46]

$$E_\nu(-\lambda_i^{(\alpha,\nu)} t^\nu) \propto \frac{t^{-\nu}}{\lambda_i^{(\alpha,\nu)} \Gamma(1-\nu)}. \tag{32}$$

Analogously to Sec. 2.1, we adhere here to absorbing boundaries at the borders of the interval $[-L, L]$. Consequently, for $\alpha = 2$ one has $p(x = \pm L, t) = 0$ while for $\alpha < 2$ (due to discontinuity of trajectories) we restrict to absorbing semi-lines $p(|x| \geq L, t) = 0$, see [23, 24].

In the case of subdiffusion ($\alpha = 2$ with $\nu < 1$) the space eigenfunctions can be found exactly, see [25, 27] and Eq. (12), and the survival probability can be calculated analytically [27], see left panel of Fig. 3. In contrast, for non-Markovian Lévy flights (see right panel of Fig. 3) exact forms of the space eigenfunctions cannot be found analytically. One can however estimate the long time asymptotics of the survival probability [47, 48] which is determined by properties of the Mittag-Leffler function and displays a characteristic power-law decay with the exponent given by the subdiffusion parameter $\nu$, see [27] and right panel of Fig. 3.

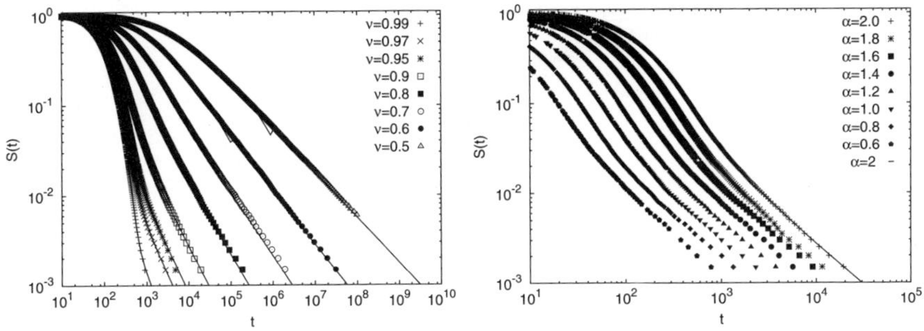


Fig. 3. The survival probability for the escape from finite intervals. Left panel presents the survival probability for a subdiffusion ($\nu < 1$ with $\alpha = 2$) for different values of the subdiffusion parameter $\nu$. Solid lines in left panel correspond to analytical solutions [27]. Right panel presents results for non-Markovian Lévy flights ($\nu < 1$ with $\alpha < 2$) with $\nu = 0.9$ for different values of the stability index $\alpha$. The interval half width is $L = 10$, the time step of integration $\Delta t = 10^{-4}$ and the number of realizations $N = 10^5$.

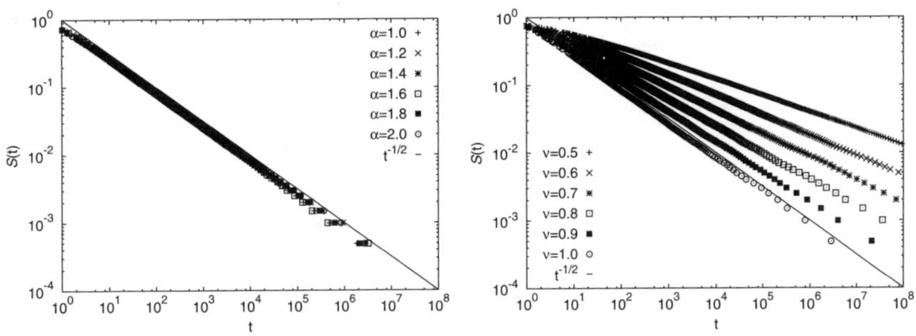


Fig. 4. The survival probability for the escape from the positive semi-line. Left panel presents the survival probability for Markovian Lévy flights ($\nu = 1$ with $\alpha < 2$) for different values of the stability index $\alpha$. Right panel presents results for non-Markovian Lévy flights ($\nu < 1$ with $\alpha < 2$) with $\alpha = 1.5$ for different values of the subdiffusion parameter $\nu$. The results were constructed numerically with the time step of integration $\Delta t = 10^{-4}$ and the number of realizations $N = 10^5$.

The non-Markovian character of Lévy flights coexisting with long waiting times can be further confirmed by referring to the Sparre Andersen scaling: For $\nu < 1$ the survival probability estimated for the positive semi-line $S(t) = \int_0^\infty p(x, t|x_0, 0)dx$ deviates distinctly from the Markovian $t^{-1/2}$ scaling (see right panel of Fig. 4) and assigns the $t^{-\nu/2}$ asymptotics [28–31].

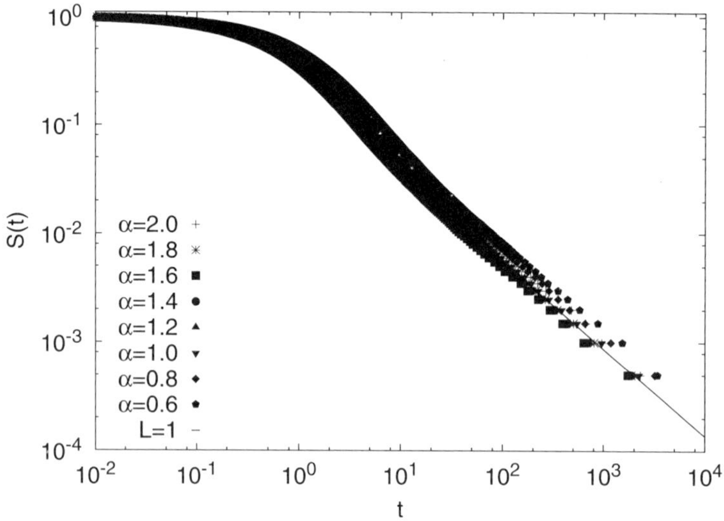


Fig. 5. The survival probability $S(t)$ for the Kramers problem in the case of non-Markovian Lévy flights. The absorbing boundary is located at the barrier top, i.e. $x_b = 0$. The value of the subdiffusion parameter is $\nu = 0.8$. Various symbols correspond to various values of the stability index $\alpha$. The solid lines present the survival probability for the escape from the finite interval of the length $L = 1$, see [47].

### 3.2. *Non-Markovian Kramers problem*

The basic fractional Kramers problem for sub- and super-diffusive motion has been investigated in a series of papers [19, 26, 36]. An interesting counterpart to those investigations is an analysis of bi-fractional (in time and space) Smoluchowski–Fokker–Planck equation describing motion affected by non-Markovian Lévy flights [8, 17, 49]. In this case, long waiting times are responsible for a heavy tailed characteristics of the survival probability which is solely determined by the subdiffusion parameter $\nu$, cf. Fig. 5. More precisely, single modes of the bi-fractional Smoluchowski–Fokker–Planck equation relax according to the Mittag-Leffler function. Therefore, asymptotically they are determined by the power law decay of $t^{-\nu}$ type [42, 50]. This slow decay of the first time distributions is in turn responsible for divergence of the mean first passage time. Consequently, in such case the mean first passage time ceases to be a proper measure of a kinetic rate. Nevertheless, one can still characterize the Kramers problem by using the first passage time density and its properties [42].

## 4. Summary

Anomalous diffusion processes are observed and anticipated in various systems operating under far from equilibrium conditions. Heavy-tailed disturbances have been reported in turbulent fluid flows, magnetized plasma, random walks along conformational looping of polymer chains and optical materials, to name but few [51–58]. On the other hand, fractional time derivatives are known to describe motion of particles immobilized for extended period of time. Interplay of both factors – long waiting times between jumps and long excursions during the motion may change profoundly typical characteristics of kinetics.

Different types of fractional generalizations of the Smoluchowski–Fokker–Planck and Klein–Kramers equations describing anomalous diffusion of inertial particle can be found in the literature [49, 59, 60]. Their derivation usually assumes various forms of subordination at the level of the Langevin approach [8, 12, 61]. Within the subordination scheme the process of primary interest $X(t)$ is obtained by randomizing the time of the original process using a different "clock" which links the physical time $t$ with some operational time $s$. In an intuitive way, the technique has been described by Fogedby [12], Sokolov [59] and Eule [62]. Their formulation of the procedure for a generalized Wiener process has been based on the analysis of the set of coupled Langevin equations

$$\begin{cases} \dot{x}(s) \equiv \dfrac{dx}{ds} = \xi(s), \\ \dot{t}(s) = \eta(s), \end{cases} \tag{33}$$

where the random walk $x(t)$ becomes parametrized by variable $s$. In the above equations $\xi(s)$ and $\eta(s)$ are assumed to be independent, random noises and the pair process $(x(s), t(s))$ preserves the Markov property. The requirement of causality ($t(s)$ is a physical time) limits choice of $\eta(s)$ to functions returning positive values only. The combined process in physical time $t$ is described by the trajectories $x(t) = x(s(t))$ and is subordinated to the parent process with corresponding realizations $x(s)$. Moreover, the time transformation implies

$$s(\tau) = T \Leftrightarrow t(T) = \tau \tag{34}$$

or

$$\text{Prob}\,\{s(\tau) < T\} = \text{Prob}\,\{t(T) \geq \tau\}. \tag{35}$$

If $\xi(t)$ stands for a white, symmetric Lévy flight [19, 23], then $x(s) = \int_0^s ds'\xi(s') \approx \sum_{i=0}^{N-1}(\Delta s')^{1/\alpha}\xi_i$ becomes Lévy motion $L_{\alpha,0}(s)$. In addition, by assuming $\eta(s)$ to be a white $\nu$-stable Lévy noise taking positive values only, the integrated process $t(s) = \int_0^s ds'\eta(s')$ converges to a $\nu$-stable totally skewed Lévy motion with the index $0 < \nu < 1$ and the probability density of the random variable $s$ at time $t$, $p(s,t)$ is given by the inverse stable distribution $l_{\nu,1}(s,t) = \frac{d}{ds}(1 - L_{\nu,1}(t/s^{1/\nu}))$. Consequently, the process $s(t)$ becomes an asymptotic (continuous) analog of the number of steps $N(t)$ in Eqs. (4) and (7). Accordingly, the probability density $p(x,t)$ of the subordinated process $X(t)$ fully coincides [10, 12] with the solution to the fractional Smoluchowski–Fokker–Planck Eq. (25). It should also be stressed that by virtue of limit theorems for stable random variables [63] the asymptotic behavior of the resulting $p(x,t)$ coincides with a probability density for an underlying continuous time random walk scheme, in which long Lévy excursions are interrupted by residence times sampled from a Mittag-Leffler distribution [60].

The non-Markovian Lévy flights are natural generalization of white Gaussian noise. Therefore, understanding their action and effects induced in dynamics of non-equilibrium systems necessitates further research on a mixture of superdiffusive and subdiffusive motion.

## Acknowledgments

This project has been supported in part (B.D.) by the Foundation for Polish Science and Polish Ministry of Science and Higher Education.

## References

1. C. W. Gardiner, *Handbook of Stochastic Methods for Physics, Chemistry and Natural Sciences* (Springer-Verlag, 2009).
2. H. Risken, *The Fokker–Planck Equation. Methods of Solution and Application* (Springer-Verlag, 1984).
3. O. E. Barndorff-Nielsen, T. Mikosch and S. I. Resnick, Eds., *Lévy Processes: Theory and Applications* (Birkhäuser, 2001).
4. N. G. van Kampen, *Stochastic Processes in Physics and Chemistry* (North–Holland, 1981).
5. A. Janicki and A. Weron, *Simulation and Chaotic Behavior of α-stable Stochastic Processes* (Marcel Dekker, 1994).
6. A. I. Saichev and G. M. Zaslavsky, *Chaos.* **7**, 753 (1997).
7. M. F. Shlesinger, G. M. Zaslavsky, and J. Frisch, *Lévy Flights and Related Topics in Physics* (Springer-Verlag, 1995).

8. M. Magdziarz and A. Weron, *Phys. Rev. E* **75**, 056702 (2007).
9. M. Magdziarz, A. Weron and J. Klafter, *Phys. Rev. Lett.* **101**, 210601 (2008).
10. M. Magdziarz, A. Weron and K. Weron, *Phys. Rev. E* **75**, 016708 (2007).
11. M. Teuerle and A. Jurlewicz, *Acta Phys. Pol. B* **40**, 1333 (2009).
12. H. C. Fogedby, *Phys. Rev. E* **50**, 1657 (1994).
13. H. C. Fogedby, *Phys. Rev. Lett.* **73**, 2517 (1994).
14. H. C. Fogedby, *Phys. Rev. E.* **58**, 1690 (1998).
15. V. V. Yanovsky, A. V. Chechkin, D. Schertzer and A. V. Tur, *Physica A* **282**, 13 (2000).
16. D. Schertzer, M. Larchevêque, J. Duan, V. V. Yanowsky and S. Lovejoy, *J. Math. Phys.* **42**, 200 (2001).
17. R. Klages, G. Radons and I. M. Sokolov, *Anomalous Transport Foundations and Applications* (Wiley-VCH, 2008).
18. A. Weron and R. Weron, *Lect. Not. Phys.* **457**, 379 (1995).
19. A. A. Dubkov, B. Spagnolo and V. V. Uchaikin, *Int. J. Bifurcat. Chaos. Appl. Sci. Eng.* **18**, 2649 (2008).
20. A. V. Chechkin, V. Y. Gonchar, J. Klafter and R. Metzler, *Adv. Chem. Phys.* **133**, 439 (2006).
21. R. Metzler, E. Barkai and J. Klafter, *EPL* **46**, 431 (1999).
22. S. Redner, *A Guide to First Passage Time Processes* (Cambridge Univ. Press, 2001).
23. B. Dybiec, E. Gudowska-Nowak and P. Hänggi, *Phys. Rev. E* **73**, 046104 (2006).
24. A. Zoia, A. Rosso and M. Kardar, *Phys. Rev. E* **76**, 021116 (2007).
25. D. R. Cox and H. D. Miller, *The Theory of Stochastic Processes* (Chapman and Hall, 1965).
26. R. Metzler and J. Klafter, *Phys. Rep.* **339**, 1 (2000).
27. B. Dybiec, *J. Stat. Mech.* P01011 (2010).
28. V. Balakrishnan, *Physica A* **132**, 569 (1985).
29. E. Barkai, *Phys. Rev. E* **63**, 046118 (2001).
30. T. Koren, J. Klafter and M. Magdziarz, *Phys. Rev. E* **76**, 031129 (2007).
31. B. Dybiec and E. Gudowska-Nowak, *EPL* **88**, 10003 (2009).
32. H. A. Kramers, *Physica Utrecht.* **7**, 284 (1940).
33. P. Hänggi, P. Talkner and M. Borkovec, *Rev. Mod. Phys.* **62**, 251 (1990).
34. B. Dybiec, E. Gudowska-Nowak and P. Hänggi, *Phys. Rev. E* **75**, 021109 (2007).
35. L. Gammaitoni, P. Hänggi, P. Jung and F. Marchesoni, *Rev. Mod. Phys.* **70**, 223 (1998).
36. A. V. Chechkin, V. Y. Gonchar, J. Klafter and R. Metzler, *EPL* **72**, 348 (2005).
37. A. V. Chechkin, O. Y. Sliusarenko, R. Metzler and J. Klafter, *Phys. Rev. E* **75**, 041101 (2007).
38. R. Metzler and J. Klafter, *J. Phys. A: Math. Gen.* **37**, R161 (2004).
39. G. Zumofen and J. Klafter, *Phys. Rev. E* **51**, 1818, (1995).
40. I. M. Sokolov, *Phys. Rev. E* **66**, 041101 (2002).
41. B. Dybiec and E. Gudowska-Nowak, *Phys. Rev. E* **80**, 061122 (2009).

42. B. Dybiec, *J. Chem. Phys.* **133**, 244114 (2010).
43. E. Scalas, R. Gorenflo and F. Mainardi, *Phys. Rev. E* **69**, 011107 (2004).
44. E. Scalas, *Physica A* **362**, 225 (2006).
45. I. Podlubny, *Fractional Differential Equations* (Academic Press, 1999).
46. W. G. Glöckle and T. F. Nonnenmacher, *Rheol. Acta.* **33**, 337 (1994).
47. B. Dybiec, *J. Stat. Mech.* P01011 (2010).
48. B. Dybiec, *Acta. Phys. Pol. B* **41**, 1127 (2010).
49. S. Eule, R. Friedrich, F. Jenko, and D. Kleinhans, *J. Phys. Chem. B* **111**, 11474 (2007).
50. R. Metzler and J. Klafter, *Chem. Phys. Lett.* **321**, 238 (2000).
51. M. F. Shlesinger, G. M. Zaslavski and J. Klafter, *Nature London.* **363**, 31 (1993).
52. J. Klafter, M. F. Shlesinger and G. Zumofen, *Phys. Today.* **49**, 33 (1996).
53. T. H. Solomon, E. R. Weeks, and H. L. Swinney, *Phys. Rev. Lett.* **71**, 3975 (1993).
54. D. del Castillo-Negrete, *Phys. Fluids* **10**, 576 (1998).
55. A. V. Chechkin, V. Y. Gonchar and M. Szydłowski, *Phys. Plasmas* **9**, 78 (2002).
56. D. del Castillo-Negrete, B. A. Carreras and V. E. Lynch, *Phys. Rev. Lett.* **94**, 065003 (2005).
57. H. Katori, S. Schlipf and H. Walther, *Phys. Rev. Lett.* **79**, 2221 (1997).
58. M. A. Lomholt, T. Ambjörnsson and R. Metzler, *Phys. Rev. Lett.* **95**, 260603 (2005).
59. I. M. Sokolov, J. Klafter and A. Blumen, *Phys. Today.* **55**, 48 (2002).
60. B. Dybiec and E. Gudowska-Nowak, *Chaos* **20**, 043129 (2010).
61. S. Bochner, *Proc. Natl. Acad. Sci. U.S.A.* **48**, 19 (1962).
62. S. Eule and R. Friedrich, *EPL* **86**, 30008 (2009).
63. A. Piryatinska, A. Saichev and W. Woyczynski, *Physica A* **349**, 375 (2005).

# Chapter 3

## Characteristic Times of Anomalous Diffusion in a Potential

William T. Coffey
*Department of Electronic and Electrical Engineering, Trinity College, Dublin 2, Ireland*

Yuri P. Kalmykov
*Laboratoire de Mathématiques et Physique, Université de Perpignan Via Domitia, 52, Avenue de Paul Alduy, 66860 Perpignan Cedex, France*

Serguey V. Titov
*Institute of Radio Engineering and Electronics, Russian Academy of Sciences, Vvedenskii Square 1, Fryazino, Moscow Region, 141190, Russian Federation*

Characteristic times of the noninertial *fractional diffusion* of a particle in a potential are defined in terms of three time constants, viz., the integral, effective and longest relaxation times. These times are described using the eigenvalues of the corresponding Fokker–Planck operator for the *normal* diffusion. Knowledge of them is sufficient to accurately predict the *anomalous* relaxation behavior for all time scales of interest. As a particular example, we consider the subdiffusion of a particle in a bistable potential.

| | | |
|---|---|---|
| 1. | Introduction | 51 |
| 2. | Normal Diffusion | 55 |
| 3. | Anomalous Diffusion | 62 |
| 4. | Fractional Diffusion of a Particle in a Double Well Potential | 66 |
| 5. | Concluding Remarks | 72 |

### 1. Introduction

Relaxation and diffusion processes in complex disordered systems such as amorphous polymers, glass forming liquids, etc. exhibit temporal nonlocal behavior arising from energetic disorder causing obstacles or traps both slowing down the motion of the particle and introducing memory effects [1, 2]. The memory effects can be described by a fractional diffusion equation incorporating a waiting time probability density function governing the random time intervals between single microscopic jumps of the particles. The fractional diffusion equation stems from the integral equation for a continuous time random walk (CTRW) [3, 4]. The situation is thus unlike that encountered in a conventional random walk which is characterized by a microscopic time scale small compared to the observation time. The microscopic time in the conventional random walk is the time the random walker takes to make a single microscopic jump. In this context one should recall that the Einstein theory of the *normal* Brownian motion relies on the diffusion limit of a *discrete* time random walk. Here the random walker makes a jump of a *fixed mean square length* in a *fixed time*. Thus the only random variable is the direction of the walker, leading automatically via the central limit theorem (in the limit of a large sequence of jumps) to the Wiener process describing the Brownian motion. The CTRW, on the other hand, was introduced by Montroll and Weiss [4] as a way of rendering time continuous in a random walk without necessarily appealing to the diffusion limit. In the most general case of the CTRW, the random walker may jump an arbitrary length in arbitrary time. However, the jump length and jump time random variables are not statistically independent [1, 5, 6]. In other words, a given jump length is penalized by a time cost, and vice versa. A simple case of the CTRW arises when one assumes that the jump length and jump time random variables are decoupled. Thus the jump length variances are always finite; however, the jump times may be arbitrarily long so that they obey a Lévy distribution with its characteristic long tail [5, 6]. Thus the jump length distribution ultimately becomes Gaussian with finite jump length variance, while the mean waiting time between jumps diverges due to the underlying Lévy waiting time distribution. Such walks, which possess a discrete hierarchy of time scales, not all of which have the same probability of occurrence, are known as *fractal time random walks* [5]. In the limit of a large sequence of jump times, they yield a fractional Fokker–Planck equation in configuration space [1, 2, 5–7].

Now the relevant fractional diffusion (Fokker–Planck) equation for the distribution function $W(x,t)$ of the one-dimensional noninertial translational motion of a particle in a potential $V(x)$ may be written as [1, 2]

$$\frac{\partial W}{\partial t} = {}_0D_t^{1-\sigma} K_\sigma \frac{\partial}{\partial x}\left[\frac{\partial W}{\partial x} + \frac{W}{kT}\frac{\partial V}{\partial x}\right]. \tag{1}$$

Here $x$ specifies the position of the particle at time $t$, $-\infty < x < \infty$, $kT$ is the thermal energy, $K_\sigma = \zeta_\sigma/kT$ is a generalized diffusion coefficient, and $\zeta_\sigma$ is a generalized viscous drag coefficient arising from the heat bath. The operator ${}_0D_t^{1-\sigma} \equiv \partial_t\, {}_0D_t^{-\sigma}$ in Eq. (1) is given by the convolution (the Riemann–Liouville fractional integral definition) [1]

$${}_0D_t^{-\sigma} W(x,t) = \frac{1}{\Gamma(\sigma)} \int_0^t \frac{W(x,t')dt'}{(t-t')^{1-\sigma}}, \tag{2}$$

where $\Gamma(z)$ is the gamma function [8]. The physical meaning of the parameter $\sigma$ is the order of the fractional derivative in the fractional differential equation describing the continuum limit of a random walk with a chaotic set of waiting times (often known as a fractal time random walk). Values of $\sigma$ in the range $0 < \sigma < 1$ correspond to subdiffusion phenomena ($\sigma = 1$ corresponds to *normal diffusion*). However, a more physically useful definition of $\sigma$ is as the fractal dimension of the set of waiting times. The fractal dimension is the scaling of the waiting time segments in the random walk with magnification of the walk. Thus, $\sigma$ measures the statistical self-similarity (or how the whole resembles its individual constituent parts [5]) of the waiting time segments. In order to construct such an entity in practice a whole discrete hierarchy of time scales, e.g. arising from energetic disorder, is needed. For example a fractal time Poisson process [5] with a waiting time distribution assumes the typical form of a Lévy stable distribution in the limit of large $\tau$. This is explicitly discussed in [5] where a formula for $\sigma$ is given and is also discussed in [9]. The fractal time process is essentially generated by the energetic disorder treated as far as the ensuing temporal behavior is concerned by considering jumps over the wells of a chaotic potential barrier landscape.

The fractional diffusion equation (1) can in principle be solved by the same methods as the normal Fokker–Planck equation (Sturm–Liouville problem, matrix continued fractions, etc.). Various explicit solutions for the fractional diffusion in a potential have been presented (see, e.g., [1, 2, 6, 7] and references therein). In particular, the solution for the harmonic potential has been given by Metzler *et al.* [10] in terms of an eigenfunction expansion with Mittag-Leffler temporal behavior. This approach has been extended to the analogous fractional translational and rotational diffusion models in a periodic potential by Coffey *et al.* [11–14]. There the authors have developed effective methods of solution of fractional diffusion equations based on both ordinary and matrix continued fractions (which comprise an extremely powerful tool for the solution of normal diffusion equations [7, 15]). However, despite the similarity in the solutions of normal and fractional diffusion equations, an essential difference appears insofar as the characteristic times associated with normal diffusion such as mean first passage time, integral relaxation time, etc. [7, 15] do not exist in anomalous diffusion. This behavior is only to be expected in the fractal time random walk model due to the absence of a characteristic waiting time. We illustrate the divergence of characteristic times in anomalous diffusion by considering the integral relaxation time $\tau_{\text{int}}$ which is defined as the area under the normalized relaxation function $f(t)$, i.e. $f(0) = 1$. If we consider the simplest possible case of relaxation for the normal diffusion, namely, the exponential relaxation function $f(t) = e^{-t/\tau}$, for which the Laplace transform is

$$\tilde{f}(s) = L\{f(t)\} = \int_0^\infty f(t)e^{-st}dt = \frac{\tau}{s\tau + 1} \tag{3}$$

and set $s = 0$, we have the integral relaxation time $\tau_{\text{int}} = \tilde{f}(0) = \tau$. Now in anomalous relaxation with $f(t) = E_\sigma\left[-(t/\tau)^\sigma\right]$, where $E_\sigma(z)$ is the Mittag-Leffler function defined as [1]

$$E_\sigma(z) = \sum_{n=0}^{\infty} \frac{z^n}{\Gamma(1 + n\sigma)}$$

with Laplace transform

$$\tilde{f}(s) = L\{E_\sigma[-(t/\tau)^\sigma]\} = \frac{1}{s + \tau^{-\sigma}s^{1-\sigma}}, \tag{4}$$

the integral relaxation time tends to infinity. This may be seen by evaluating Eq. (4) as $s \to 0$, which is singular except in the special case of normal diffusion, where $\sigma = 1$. The singular behavior, in retrospect, is obvious because of the long tailed character of the Mittag-Leffler function, and emphasizes the *nonlocal* temporal character of the fractal time relaxation process, which may arise from obstacles or traps which delay the motion of the Brownian particle and introduce memory effects into the motion.

The divergence of the characteristic times for anomalous diffusion, as defined in their conventional sense (which is a natural consequence of the underlying Lévy distribution) rendering them useless as a measure of the relaxation behavior, signifies the importance of seeking characteristic times for such processes in terms of their frequency domain representation. Here we shall show, in particular, that the *anomalous diffusion process* in a potential will have several associated characteristic times. These can be expressed in terms of the characteristic relaxation times of the *normal diffusion process* which appear as parameters in the definitions of the characteristic times of the anomalous diffusion process. Thus, the difficulties associated with divergent characteristic times in anomalous diffusion may be avoided.

## 2. Normal Diffusion

A concept which is of central importance in the calculation of the time behavior of statistical averages from the Fokker–Planck equation, is the *linear response* [7, 15] of a system to an applied stimulus. Linear response theory is quite general and has a wide range of applicability in physics. Here, we consider as the simplest example the noninertial one-dimensional Brownian motion of a particle in a potential $V(x)$ and we assume that the dynamics of the particle obey the Fokker–Planck equation for the distribution function $W(x,t)$ of the realizations of a random variable $x$, such that

$$\frac{\partial}{\partial t} W = L_{\mathrm{FP}} W. \tag{5}$$

Thus let us consider the Fokker–Planck operator $L_{\mathrm{FP}}$ of a system subjected to a *small* perturbing force $F(t)$. On account of this, $L_{\mathrm{FP}}$ may be represented as [15]

$$L_{\mathrm{FP}} = \frac{\partial}{\partial x}\left[K(x)e^{-V(x)+B(x)F(t)}\frac{\partial}{\partial x}e^{V(x)-B(x)F(t)}\right] = L_{\mathrm{FP}}^{0} + L_{\mathrm{ext}}F(t) \tag{6}$$

with

$$L_{\mathrm{FP}}^{0} = \frac{\partial}{\partial x}\left[K(x)e^{-V(x)}\frac{\partial}{\partial x}e^{V(x)}\right],$$

$$L_{\mathrm{ext}} = -\frac{\partial}{\partial x}\left(K(x)\frac{\partial}{\partial x}B\right), \quad L_{\mathrm{FP}}^{0}W_0(x) = 0,$$

where $L_{\mathrm{FP}}^{0}$ is the Fokker–Planck operator in the *absence* of the perturbation, $W_0$ is the equilibrium (stationary) distribution function, $V$ is called a generalized (effective) potential (in Secs. 2 and 3, we shall assume that

$V$ is normalized by $kT$), $K(x)$ is the diffusion coefficient, and $B(x)$ denotes a dynamical quantity.

The time dependence of the average of a dynamical variable $A(x)$ can be expressed using the principle of superposition as

$$\langle A\rangle(t) = \int_{-\infty}^{t} \Phi(t-t')F(t')dt', \tag{7}$$

where $\Phi(t)$ is the pulse response function (inverse Fourier transform of the transfer function) defined by

$$\Phi(t) = -\frac{d}{dt}C_{AB}(t). \tag{8}$$

Here $C_{AB}(t)$ is the *equilibrium* (stationary) correlation function defined by

$$\begin{aligned} C_{AB}(t) &= \langle A\left[x(0)\right] B\left[x(t)\right]\rangle_0 - \langle A\rangle_0 \langle B\rangle_0 \\ &= \int_{x_1}^{x_2} \left[A(x) - \langle A\rangle_0\right] e^{L^0_{\mathrm{FP}}t}[B(x) - \langle B\rangle_0]W_0(x)dx, \end{aligned} \tag{9}$$

where the symbols $\langle\ \rangle$ and $\langle\ \rangle_0$ denote the statistical averages over $W$ and $W_0$, respectively, with $x$ defined in the range $x_1 \le x \le x_2$. The step-off and step-on relaxation functions (obtained when either a small constant force $F_1$ is suddenly switched off or on at time $t = 0$, with statistical equilibrium having been achieved prior to the imposition of the stimulus in both instances) for a dynamical variable $A(x)$ are then

$$\langle A\rangle^{\mathrm{off}}(t) - \langle A\rangle_0 = F_1 C_{AB}(t), \tag{10}$$

$$\langle A\rangle^{\mathrm{on}}(t) - \langle A\rangle_0 = F_1\left[C_{AB}(0) - C_{AB}(t)\right], \quad t > 0. \tag{11}$$

Furthermore, the spectrum of $\langle A\rangle(t)$ (ac response) is

$$\langle A\rangle_\omega = F_\omega\left[C_{AB}(0) - i\omega\int_0^\infty C_{AB}(t)e^{-i\omega t}dt\right], \tag{12}$$

where $\langle A\rangle_\omega$ and $F_\omega$ are the Fourier components of $\langle A\rangle(t)$ and $F(t)$, respectively. The expression in parenthesis denotes the dynamical suseptibility $\chi_{AB}(\omega)$, i.e.

$$\chi_{AB}(\omega) = C_{AB}(0) - i\omega\int_0^\infty C_{AB}(t)e^{-i\omega t}dt. \tag{13}$$

In order to proceed, let us first discuss the formal step-off solution of the Fokker–Planck equation. The solution of Eq. (5) can be obtained from the Sturm–Liouville representation [15]

$$W(x,t) = W_0(x) + \sum_{k=1}^{\infty} \Phi_k(x) F_k(t). \tag{14}$$

Here, $\Phi_k(x)$ are the eigenfunctions of the Fokker–Planck operator $L_{\text{FP}}$, i.e.

$$L_{\text{FP}} \Phi_k(x) = -\lambda_k \Phi_k(x), \tag{15}$$

$\lambda_1 < \lambda_2 < \lambda_3 < \ldots$, and the decay modes $F_k(t)$ obey the equation

$$\frac{d}{dt} F_k(t) = -\lambda_k F_k(t). \tag{16}$$

The solution of Eq. (16) is

$$F_k(t) = \exp(-\lambda_k t). \tag{17}$$

Thus Eq. (14) becomes

$$W(x,t) = W_0(x) + \sum_{k=1}^{\infty} \Phi_k(x)\; e^{-\lambda_k t}. \tag{18}$$

Noting Eqs. (9) and (18), the *equilibrium* correlation function can be given as

$$C_{AB}(t) = \sum_k c_k e^{-\lambda_k t}, \tag{19}$$

i.e. in the form of a set of decaying exponentials characterized by a set of distinct eigenvalues $\{\lambda_k\}$ and their corresponding amplitudes $\{c_k\}$.

Various methods of calculating the eigenvalues have been discussed in detail in [15]. For example, they can be evaluated in terms of continued fractions [7, 15]. The continued fraction method for calculating the eigenvalues has been discussed in detail by Risken [15]. This approach was further developed in [19]. In contrast to the previously available solution [15], the method developed in [19] does not require one to solve numerically a high order polynomial equation since $\lambda_1$ may be represented as a sum of products of infinite continued fractions. Besides its usefulness in numerical calculations, the equation so obtained is very suitable for analytical purposes, e.g., for certain problems it may be expressed in terms of known mathematical (special) functions (examples are given in [7]).

For the *normal* diffusion in a potential, one can formally introduce the longest relaxation time $\tau_{\text{long}} = 1/\lambda_1$ associated with the *long time behavior* of the relaxation function $C_{AB}(t)$ which is determined solely by the slowest low-frequency relaxation mode. This mode governs transitions of the Brownian particle over the barriers from one potential well into another. The characteristic frequency of this overbarrier relaxation mode is determined by the smallest nonvanishing eigenvalue $\lambda_1$ of the Fokker–Planck operator. A knowledge of $\lambda_1$ is of importance because in many cases $\lambda_1$ provides complete information about the low-frequency dynamics of the system.

Furthermore, $\lambda_1$ of the Fokker–Planck equation, written for the Brownian motion in a potential, yields in the high barrier limit, the Kramers escape rate for normal diffusion $\Gamma \sim \lambda_1 \sim e^{-\Delta V}$ ($\Delta V$ is the barrier height) [7, 15, 20]. In formulating his escape rate theory Kramers [21] obtained an escape rate formula by assuming that apart from a small nonequilibrium region near the top of a potential barrier the distribution function of the positions and momenta of a particle is the equilibrium Maxwell–Boltzmann distribution prevailing in the depths of the well. Thus it is tacitly assumed that the nonequilibrium region lies *far inside* the region near the top of the well where the potential may be represented by an inverted oscillator potential. He was then able to obtain, postulating a steady current representing the slow leaking of particles over the barrier top, the quasi-stationary solution of the Fokker–Planck equation (Klein–Kramers equation in this instance) for the distribution function of positions and momenta in the vicinity of the barrier. Hence he derived a formula for the escape rate in the so-called intermediate-to-high damping (IHD) limit. If the dissipation constant tends to zero, this formula yields the transition state theory (TST) result where the dissipation to the bath is ignored. In the opposite limit, when the dissipation constant becomes very large, the IHD Kramers escape rate formula yields his so-called very high damping (IHD) equation

$$\Gamma = \frac{\omega_c \omega_a}{2\pi\gamma} \Xi e^{-\Delta V} \tag{20}$$

which is the case of interest here. The quantities $\omega_c$ and $\omega_a$ are the barrier and well frequencies associated with the potential $V(x)$, and $\gamma = \zeta/m$ is a damping coefficient.

We shall now introduce the concept of the integral relaxation (or correlation) time. The integral relaxation time $\tau_{\text{int}}$ of a correlation (relaxation) function $C_{AB}(t)$ is defined as the area under the curve of $C_{AB}(t)$ and is a global characteristic of a decay process [7, 15]. By definition, the correlation time $\tau_{\text{int}}$ is the area under the curve of the normalized correlation function

$C_{AB}(t)$, viz.,

$$\tau_{\text{int}} = \frac{1}{C_{AB}(0)} \int_0^\infty C_{AB}(t)dt = \frac{\sum_k c_k \lambda_k^{-1}}{\sum_k c_k}. \tag{21}$$

Thus, the relaxation times $\tau_{\text{int}}$ contain contributions from *all* the eigenvalues. Hence in order to evaluate the integral relaxation time, a knowledge of the eigenvalues $\lambda_k$ and their amplitudes $c_k$ is required. Nevertheless, the behavior of the integral relaxation time and the longest relaxation time $\tau_{\text{long}} = 1/\lambda_1$ is often similar. However, in certain cases (e.g., in a potential with non-equivalent wells in the presence of an external dc bias field [7]), the behavior of these times can be quite different and they may diverge exponentially at certain values of the bias field. Now for one-dimensional systems, there exists an exact integral formula for the correlation time $\tau_{\text{int}}$ [7, 15], namely,

$$\tau_{\text{int}} = \frac{1}{\langle AB\rangle_0 - \langle A\rangle_0 \langle B\rangle_0} \int_{x_1}^{x_2} \frac{f_A(x) f_B(x) dx}{K(x) W_0(x)}, \tag{22}$$

where $f_Z(x) = \int_{x_1}^{x} [Z(y) - \langle Z\rangle_0] W_0(y)dy$. For $A = B$, Eq. (22) reduces to Eq. (S9.14) of Risken [15], namely,

$$\tau_{\text{int}} = \frac{1}{\langle A^2\rangle_0 - \langle A\rangle_0^2} \int_{x_1}^{x_2} \frac{f_A^2(x) dx}{K(x) W_0(x)}. \tag{23}$$

We remark that the relaxation time in integral form, Eq. (23), was first given by Szabo [16], and later reproduced by other authors in various guises (see, e.g., [7, 17, 18]). The detailed derivation of Eqs. (22) and (23) and their applications to particular systems is given in [7, 15].

However, there is yet another possibility to introduce a characteristic time $\tau_{ef}$ associated with the short time behavior of the relaxation function. This is the effective eigenvalue method [22] which may be illustrated as follows. We suppose that in Eq. (19) the correlation function $C_{AB}(t)$ may be approximated by a *single* exponential $C_{AB}(t) \approx C_{AB}(0)e^{-t/\tau_{ef}}$ for all $t$ so that $C_{AB}(t)$ obeys

$$\dot{C}_{AB}(t) + \frac{C_{AB}(t)}{\tau_{ef}} = 0. \tag{24}$$

Since these equations are valid for all $t$ it then follows that

$$\tau_{ef} = -\frac{C_{AB}(0)}{\dot{C}_{AB}(0)}. \tag{25}$$

Thus, the effective relaxation time $\tau_{ef}$ is expressed entirely in terms of $C_{AB}(0)$ and $\dot{C}_{AB}(0)$, which may be evaluated from Eq. (19) in terms of the eigenvalues $\{\lambda_k\}$ and the corresponding amplitudes $\{c_k\}$ as

$$\tau_{ef} = \frac{\sum_k c_k}{\sum_k \lambda_k c_k}. \tag{26}$$

Just as the integral relaxation time, it is again difficult to evaluate $\tau_{ef}$ from this formula directly because knowledge of $\{c_k\}$ and $\{\lambda_k\}$ is required. However, using Eq. (25), $\tau_{ef}$ may also be expressed in terms of appropriate *equilibrium averages* [7, 22]. Clearly $\tau_{ef}$ is the time constant associated with the *initial slope* of the correlation function $C_{AB}(t)$ and it contains contributions from *all* the eigenvalues as does the integral relaxation time $\tau_{\text{int}}$. However, the behavior of $\tau_{\text{int}}$ and $\tau_{ef}$ may again be quite different. An example of this is the Brownian motion in a bistable potential with equivalent wells for high barriers, where $\tau_{\text{int}}$ and $\tau_{ef}$ always *diverge exponentially* [7].

Thus we have formally introduced using a time domain representation *three* time constants characterizing the time behavior of $C_{AB}(t)$. These are the *integral relaxation* (or, in linear response, correlation) time $\tau_{\text{int}}$, which characterizes the overall behavior of $C_{AB}(t)$ in the time domain, the *effective relaxation time* $\tau_{ef}$, which gives precise information on the initial decay of $C_{AB}(t)$, and the *inverse of the smallest nonvanishing eigenvalue* $\lambda_1$ of the Fokker–Planck operator $L_{FP}$, which is usually associated with the long time behavior of $C_{AB}(t)$ (slowest relaxation mode) and is proportional to the Kramers escape rate. These three time constants are important in the interpretation of the dynamics of a Brownian particle in a potential.

Now we show that they also characterize the spectrum of the generalized complex susceptibility $\chi_{AB}(\omega)$. According to Eq. (13), in order to evaluate the susceptibility $\chi_{AB}(\omega)$, one must calculate the spectrum (one-sided Fourier transform) of the *equilibrium* correlation function $C_{AB}(t)$. Furthermore, the behavior of $\chi_{AB}(\omega)$ in the frequency domain is completely determined by the time behavior of $C_{AB}(t)$. According to Eqs. (13) and (19), the susceptibility $\chi_{AB}(\omega)$ is given as an infinite sum of Lorentzians

$$\chi_{AB}(\omega) = \sum_k \frac{c_k}{1 + i\omega/\lambda_k}. \tag{27}$$

In the low- ($\omega \to 0$) and high-frequency ($\omega \to \infty$) limits, we have from Eq. (27)

$$\frac{\chi_{AB}(\omega)}{\chi_{AB}} = 1 - i\omega\tau_{\text{int}} + \cdots, \tag{28}$$

$$\frac{\chi_{AB}(\omega)}{\chi_{AB}} \sim -i(\omega\tau_{\mathit{eff}})^{-1} + \cdots, \tag{29}$$

where $\chi_{AB} = C_{AB}(0) = \sum_k c_k$ is the static susceptibility. Thus both the low- and high-frequency behavior of $\chi_{AB}(\omega)$ are completely determined by $\tau_{\text{int}}$ and $\tau_{ef}$, respectively. We remark that the asymptotic Eqs. (28) and (29) are quite general and hold for *any* system. Noting Eqs. (28) and (29), we can also give entirely equivalent definitions of $\tau_{\text{int}}$ and $\tau_{ef}$ in terms of $\chi_{AB}(\omega)$, viz.,

$$\tau_{\text{int}} = \lim_{\omega\to 0} \frac{\chi_{AB} - \chi_{AB}(\omega)}{i\omega\chi_{AB}}, \quad \tau_{ef} = \lim_{\omega\to\infty} \frac{\chi_{AB}}{i\omega\chi_{AB}(\omega)} + \cdots. \tag{30}$$

For the Brownian motion of a particle in a potential, the correlation function $C_{AB}(t)$ from Eq. (19), which in general comprises an infinite number of decaying exponentials, may be approximated by two exponentials only (see [7, Chap. 1, Sec. 1.20])

$$C_{AB}(t) \approx \Delta_1 e^{-\lambda_1 t} + (1 - \Delta_1)e^{-t/\tau_W}, \tag{31}$$

where

$$\Delta_1 = \frac{\tau_{\text{int}}/\tau_{ef} - 1}{\lambda_1\tau_{\text{int}} - 2 + 1/(\lambda_1\tau_{ef})}, \quad \tau_W = \frac{\lambda_1\tau_{\text{int}} - 1}{\lambda_1 - 1/\tau_{\mathit{eff}}}. \tag{32}$$

Thus, the spectrum $\chi_{AB}(\omega)$ corresponding to Eq. (31) is a sum of two Lorentzians with characteristic frequencies $\lambda_1$ and $\tau_W^{-1}$, viz.,

$$\frac{\chi_{AB}(\omega)}{\chi_{AB}} = \frac{\Delta_1}{1 + i\omega/\lambda_1} + \frac{1 - \Delta_1}{1 + i\omega\tau_W}. \tag{33}$$

Here, we implicitly assume that the contribution of high-frequency "intra-well" modes to $\chi_{AB}(\omega)$ may be approximated as a single Lorentzian with characteristic frequency and half-width given by the inverse of the characteristic time $\tau_W$. We remark that Eq. (33) also obeys the *exact* asymptotic Eqs. (28) and (29). We also remark that the parameters $\Delta_1$ and $\tau_W$ can be estimated by the method described in [7, Chap. 1, Sec. 1.20]. Thus having evaluated $\tau_{\text{int}}$, $\tau_{ef}$ and $\lambda_1$, we may calculate $\Delta_1$ and $\tau_W$ so that we may *predict* $\chi_{AB}(\omega)$ in all frequency ranges of interest as Eqs. (32) and (33) will describe $\chi_{AB}(\omega)$ in the entire $\omega$ range $0 \leq \omega < \infty$. A variety of examples of applications of the bimodal approximation to normal diffusion in a potential is given in [7].

## 3. Anomalous Diffusion

For anomalous diffusion the solution of the fractional diffusion equation (1) may again be converted to a Sturm–Liouville problem with the time behavior described by an infinite set of Mittag-Leffler functions rather than the decaying exponentials characteristic of the normal diffusion. Indeed the step-off solution of Eq. (1) is obtained from the Sturm–Liouville representation Eq. (14). Now the decay modes $F_k(t)$ obey the equation

$$\frac{d}{dt}F_k(t) = -\lambda_{k,\sigma 0}D_t^{1-\sigma}F_k(t). \tag{34}$$

The eigenvalues $\lambda_{k,\sigma}$ may be expressed in terms of the eigenvalues $\lambda_k$ of the Fokker–Planck operator $L_{FP}$ for *normal diffusion* Eq. (15) so that

$$\lambda_{k,\sigma} = \lambda_k \tau^{1-\sigma}. \tag{35}$$

The solution of Eq. (34) may be expressed in terms of the Mittag-Leffler functions $E_\sigma(z)$ [1, 6], viz.,

$$F_k(t) = E_\sigma\left[-\lambda_k \tau\,(t/\tau)^\sigma\right]. \tag{36}$$

Equation (35) exemplifies how the eigenvalues of the normal distribution process are altered, in this case reduced, by the nonlocal character of the anomalous diffusion process. In general, the Mittag-Leffler functions all have a long time tail characteristic of anomalous diffusion. Moreover, their arguments depend on the eigenvalues of the normal diffusion process. The presence of the long time tail means that the Mittag-Leffler functions describe the *nonlocal* behavior of the system. The advantage of posing the solution of Eq. (1) as a Sturm–Liouville problem now becomes apparent. Namely by formulating the problem in this way it is very easy to determine how the eigenvalues of the normal diffusion process are scaled (in the case of subdiffusion $\sigma < 1$ considered here reduced) by the *nonlocal character resulting from the microscopic disorder* of the anomalous diffusion process. The eigenvalues of the local process $\lambda_p$ are related to their Brownian counterparts by the prefactor $\tau^{1-\sigma}$. This scaling effect is significant in the context of escape of particles over potential barriers. There, we have seen that the smallest nonvanishing eigenvalue $\lambda_1$ of the Fokker–Planck equation, written for the Brownian motion in a potential, yields in the high barrier limit, the Kramers escape rate $\Gamma \sim \lambda_1 \sim e^{-\Delta V/(kT)}$. Therefore, in the context of anomalous diffusion, the Kramers escape rate $\Gamma$ can be best understood as playing the role of a decay parameter in the Mittag-Leffler functions governing the highly *non-exponential* relaxation behavior of the system. In particular, an important consequence of this is that one may determine the slowing-down of the Kramers escape rate, in essence the lowering due to nonlocal effects of the frequency of maximum loss (and consequent slowing-down of the relaxation process) of the low frequency absorption. We note that the characteristic times of the normal diffusion $\tau_{\text{int}}$ and $\tau_{ef}$ do not exist in anomalous diffusion ($\sigma \neq 1$). This is obvious from the properties of the Mittag-Leffler function, which has short time stretched exponential (Kohlrausch) form [3, 4] $E_\sigma(-t^\sigma) \sim \exp[-t^\sigma/\Gamma(1+\sigma)]$ and long time inverse power law behavior $E_\sigma(-t^\sigma) \sim t^{-\sigma}/\Gamma(1-\sigma)$ and so describes nonexponential relaxation. Nevertheless, we can formally define the corresponding relaxation times for anomalous diffusion in the frequency domain by generalizing Eq. (30).

In order to understand how the anomalous relaxation behavior influences the spectral properties, we first recall that according to linear response theory [7], the dynamic susceptibility $\chi_{AB}(\omega)$ is defined as

$$\chi_{AB}(\omega) = C^\sigma_{AB}(0) - i\omega \int_0^\infty e^{-i\omega t} C^\sigma_{AB}(t)dt, \tag{37}$$

where now

$$C^\sigma_{AB}(t) = \sum_{k=1}^\infty c_k E_k\left[-\lambda_{k,\sigma} t^\sigma\right] = \sum_{k=1}^\infty c_k E_k\left[-\tau\lambda_k (t/\tau)^\sigma\right]. \tag{38}$$

Equation (38) again exemplifies how the eigenvalues of the normal distribution process are altered, in this case reduced, by the nonlocal character of the anomalous diffusion process and how the Mittag-Leffler behavior replaces the exponential relaxation patterns of normal Brownian motion in a potential [for normal diffusion, Eq. (38) takes the familiar form of Eq. (19)]. Noting that the Laplace transform of the Mittag-Leffler function is given by Eq.(4), we have from Eqs. (37) and (38)

$$\chi_{AB}(\omega) = \sum_{k=1}^\infty \frac{c_k}{1+(i\omega)^\sigma/\lambda_{k,\sigma}} = \sum_{k=1}^\infty \frac{c_k}{1+(i\omega\tau)^\sigma/(\tau\lambda_k)}. \tag{39}$$

The susceptibility $\chi_{AB}(\omega)$ may now be simply evaluated in the low- and high-frequency limits. As before, we obtain from Eq. (39)

$$\frac{\chi_{AB}(\omega)}{\chi_{AB}} \approx 1 - (i\omega T_{\text{int}})^{\sigma} + \cdots, \quad \text{for } \omega \to 0, \tag{40}$$

and

$$\frac{\chi_{AB}(\omega)}{\chi_{AB}} \sim (i\omega T_{\mathit{eff}})^{-\sigma} + \cdots \quad \text{for } \omega \to \infty, \tag{41}$$

where

$$T_{\text{int}} = \left(\frac{\sum_{k=1}^{\infty} c_k \lambda_{k,\sigma}^{-1}}{\sum_{k=1}^{\infty} c_k}\right)^{1/\sigma} \quad \text{and} \quad T_{\mathit{eff}} = \left(\frac{\sum_{k=1}^{\infty} c_k}{\sum_{k=1}^{\infty} c_k \lambda_{k,\sigma}}\right)^{1/\sigma}. \tag{42}$$

Noting Eq. (35), Eq. (42) can also be rewritten in terms of the integral and effective relaxation times $\tau_{\text{int}}$ and $\tau_{\mathit{eff}}$ for normal diffusion defined, respectively, by Eqs. (21) and (26), viz.,

$$T_{\text{int}} = \tau\left(\frac{\tau_{\text{int}}}{\tau}\right)^{1/\sigma} \quad \text{and} \quad T_{\mathit{eff}} = \tau\left(\frac{\tau_{\mathit{eff}}}{\tau}\right)^{1/\sigma} \tag{43}$$

Furthermore, noting Eqs. (40) and (41), we can also give equivalent definitions of $T_{\text{int}}^{\sigma}$ and $T_{\mathit{eff}}^{\sigma}$ in terms of $\chi_{AB}(\omega)$, viz.,

$$T_{\text{int}} = \lim_{\omega\to 0} \frac{1}{i\omega}\left[1 - \frac{\chi_{AB}(\omega)}{\chi_{AB}}\right]^{1/\sigma}, \tag{44}$$

$$T_{\mathit{eff}} = \lim_{\omega\to\infty} \frac{1}{i\omega}\left[\frac{\chi_{AB}}{\chi_{AB}(\omega)}\right]^{1/\sigma}, \tag{45}$$

Thus just in the normal diffusion by using the *frequency domain representation*, we have formally introduced *three* time constants characterizing the anomalous behavior of $C_{AB}^{\sigma}(t)$ and $\chi_{AB}(\omega)$. These are the *integral relaxation* time $T_{\text{int}}$ characterizing the anomalous low-frequency behavior of $\chi_{AB}(\omega)$ and the overall time behavior of $C_{AB}(t)$ in the time domain, the *effective relaxation time* $T_{\mathit{eff}}$ giving precise information on the high-frequency behavior of $\chi_{AB}(\omega)$ in the frequency domain and the short time behavior of $C_{AB}(t)$ in the time domain, and the longest relaxation time $T_{long}$ defined as

$$T_{long} = \tau\left(\frac{\tau_{long}}{\tau}\right)^{1/\sigma} = \frac{\tau}{(\tau\lambda_1)^{1/\sigma}} = \frac{\tau}{(\tau\Gamma)^{1/\sigma}}. \tag{46}$$

The latter may be associated with the anomalous long-time behavior of $C^{\sigma}_{AB}(t)$ (slowest relaxation mode) and is inversely proportional to the scaled Kramers escape rate $\Gamma^{1/\sigma}$. Here the characteristic times $\lambda_1^{-1}$, $\tau_{\text{int}}$, and $\tau_{\mathit{eff}}$ of the normal diffusion appear simply as parameters of the anomalous diffusion. In normal diffusion, they always exist because an underlying microscopic time scale exists, namely, the duration of an elementary jump, which is characteristic of the discrete time random walk as used by Einstein.

Just in the normal diffusion, these three time constants are important in the interpretation of the dynamics of a system governed by the fractional Fokker–Planck equation (1). As already known [7, 11–14], two bands appear in the loss spectrum of $\chi''_{AB}(\omega)$. The characteristic frequency $\omega_c$ and the half-width of this low-frequency band are determined by $(\tau\lambda_1)^{1/\sigma}/\tau$. Thus, the anomalous low frequency behavior is dominated by the barrier crossing mode as in the normal diffusion. The high-frequency band is due to "intrawell" modes corresponding to $(\tau\lambda_k)^{1/\sigma}/\tau$ $(k \neq 1)$. These near degenerate "intrawell" modes are indistinguishable in the frequency spectrum of $\chi''_{AB}(\omega)$ appearing merely as a single high-frequency band with characteristic frequency $\omega_W$. Thus, the spectrum of the longitudinal susceptibility $\chi_{AB}(\omega)$ may essentially be approximated by a sum of two Cole–Cole functions, viz.,

$$\frac{\chi_{AB}(\omega)}{\chi_{AB}} = \frac{\Delta_1}{1+(i\omega/\omega_c)^{\sigma}} + \frac{1-\Delta_1}{1+(i\omega/\omega_W)^{\sigma}}, \tag{47}$$

where $\omega_c = \tau^{-1}(\tau\lambda_1)^{1/\sigma}$ and $\omega_W = \tau^{-1}(\tau/\tau_W)^{1/\sigma}$ are the characteristic frequencies and $\Delta_1$ and $\omega_W$ are given by Eq. (32). In the time domain, such a bimodal approximation is equivalent to assuming that the relaxation function $C_{AB}(t)$ as determined by the exact Eq. (38) (which in general comprises an *infinite number* of Mittag-Leffler functions) may be approximated by *two* Mittag-Leffler functions only

$$C_{AB}(t) \approx \Delta_1 E_{\sigma}\left[-(t/\tau)^{\sigma}\tau\lambda_1\right] + (1-\Delta_1)E_{\sigma}\left[-(t/\tau)^{\sigma}\tau/\tau_W\right].$$

Equation (47) correctly predicts $\chi_{AB}(\omega)$ both at low $(\omega \to 0)$ and high $(\omega \to \infty)$ frequencies. Moreover, $\chi_{AB}(\omega)$ may be determined in the entire frequency range $0 \leq \omega < \infty$ as we shall presently see.

It should be mentioned that if one is interested only in the low-frequency $(\omega \leq \omega_c)$ part of the dielectric spectrum, one may use a simpler single mode

(Cole–Cole) equation for the normalized complex susceptibility, namely,

$$\frac{\chi_{AB}(\omega)}{\chi_{AB}} = \frac{\Delta_1}{1+(i\omega/\omega_c)^{\sigma}} + 1 - \Delta_1. \tag{48}$$

In passing, we must remark that just as in the conventional normal relaxation ($\sigma = 1$), the Cole–Cole like Eq. (48) may be derived from a number of very different models (see, e.g., [9, 23, 24]). However, the advantage of using an approach based on a kinetic equation (such as the fractional Fokker–Planck equation) over all others is that one may explicitly include an external potential and so exactly calculate its effect on the relaxation process.

## 4. Fractional Diffusion of a Particle in a Double Well Potential

As a particular example, we shall now treat the anomalous diffusion of a particle in a double-well potential, viz.,

$$V(x) = \frac{ax^2}{2} + \frac{bx^4}{4}, \tag{49}$$

where $a$ ($a < 0$) and $b$ are constants. The model of normal diffusion in the bistable potential Eq. (49) is almost invariably used to describe the noise driven motion in bistable physical and chemical systems. Examples of such diverse subjects are simple isometrization processes [25–29], chemical reaction rate theory [30–38], bistable nonlinear oscillators [39–41], second-order phase transitions [42], nuclear fission and fusion [43, 44], stochastic resonance [45, 46], etc.

The *normal* diffusion in the bistable potential Eq. (49) in the very high damping limit, where the inertia of the particle may be neglected, has been extensively studied either by using the Kramers escape rate theory or by solution of the appropriate Fokker–Planck (Smoluchowski) equation (see, e.g., [7, 32, 35, 46–48] and references therein). In the noninertial limit, the conventional analysis of the problem proceeds from the Smoluchowski equation by either rendering that equation as a Sturm–Liouville problem (e.g., [35, 49]) or by the solution of an infinite hierarchy of linear differential-recurrence relations for statistical moments (e.g., [50, 51]). In particular, in [51] such a hierarchy has been derived directly from the corresponding Langevin equation and the exact solution is obtained via matrix continued fractions. This is accomplished by using a representation which symmetrizes the recurrence relations for the observables generated by the averaging procedure leading to convergence of these recurrence relations unlike the previous approaches to the problem [50], where the recurrence equations diverge. A reliable approximate solution based on the exponential separation of the time scales of the fast intrawell and low overbarrier relaxation processes associated with the bistable potential is also given. It is shown that a knowledge of the three characteristic relaxation times (the integral, effective and the longest relaxation times) of the position correlation function allows one to accurately predict the relaxation behavior of the system in the overdamped limit for all time scales of interest.

The above results of course can be generalized to anomalous diffusion. The fractional diffusion equation (1) can in principle be treated in like manner. The subdiffusion in the double-well potential Eq. (49) has been considered, e.g., in [14, 52–54]. Here in order to ascertain how the anomalous diffusion in a bistable potential, $b > 0$ and $a < 0$, modifies the behavior of the position correlation function

$$C_\sigma(t) = \frac{\langle x(0)x(t)\rangle_0}{\langle x^2(0)\rangle_0}$$

and its spectra (which characterize the anomalous relaxation), we shall give exact and approximate solutions for these quantities. Furthermore, we shall demonstrate following [14] that the characteristic times of the normal diffusion process, namely, the inverse of the smallest nonvanishing eigenvalue of the Fokker–Planck operator, the integral and effective relaxation times allow us to accurately describe the anomalous relaxation behavior.

Now it is useful to normalize variables and parameters as [7, 51]

$$y = \frac{x}{\langle x^2\rangle_0^{1/2}}, \quad V(y) = \frac{V(x)}{kT} = Ay^2 + By^4,$$
$$A = \frac{a\langle x^2\rangle_0}{2kT}, \quad B = \frac{b\langle x^2\rangle_0^2}{4kT},$$

where the angular brackets $\langle(\bullet)\rangle_0 = Z^{-1}\int_{-\infty}^{\infty}(\bullet)e^{-V(y)}dy$ mean equilibrium ensemble averages and $Z$ is the partition function given by for $A < 0$

$$Z = \int_{-\infty}^{\infty} e^{-V(y)}dy = \sqrt{\pi}(2B)^{-1/4}e^{Q/2}D_{-1/2}\left(-\sqrt{2Q}\right),$$

where $D_v(z)$ are Whitaker's parabolic cylinder functions of order $v$ [8] and $Q = A^2/4B$ is the barrier height for the potential $V(y) = Ay^2 + By^4$

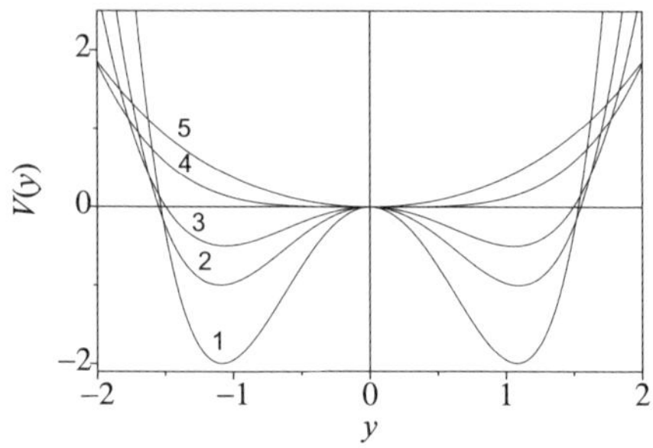


Fig. 1. Potential $V(y) = Ay^2 + By^4$ for various values of $A/(2B^{1/2}) = -\sqrt{2}$ (curve 1), $-1$ (2), $-1/\sqrt{2}$ (3), 0 (4), and 1 (5).

(see Fig. 1). The normalization condition $\langle y^2 \rangle_0 = 1$ implies that the constants $A$ and $B$ are no longer independent [7], viz.,

$$B = B(Q) = \frac{D^2_{-3/2}(\mathrm{sgn}(A)\sqrt{2Q})}{8D^2_{-1/2}\left(\mathrm{sgn}(A)\sqrt{2Q}\right)}.$$

In dimensionless form the fractional Fokker–Planck Eq. (1) becomes [14]

$$\frac{\partial W}{\partial t} = \tau^{-\sigma}{}_0D_t^{1-\sigma}\frac{\partial}{\partial y}\left[\frac{\partial W}{\partial y} + W\frac{\partial V}{\partial y}\right], \tag{50}$$

where $\tau = \langle x^2 \rangle_0 / K_1$ has the meaning of the characteristic time (waiting time between jumps) and $K_1$ is the diffusion coefficient for normal diffusion. One may seek a solution of Eq. (50) for $t \geq 0$ as [14]

$$W(y,t) = \kappa e^{-[\kappa^2 y^2 + V(y)]/2}\sum_{q=1}^{\infty}\frac{c_q(t)H_q(\kappa y)}{\sqrt{\pi 2^q q!}}, \tag{51}$$

where $H_q(z)$ are the Hermite polynomials [8] and $\kappa$ is a scaling factor chosen so as to ensure optimum convergence of the continued fractions involved (all results for the observables are independent of $\kappa$). By substituting Eq. (14) into Eq. (1) and noting the recurrence relations $dH_n(z)/dz = 2nH_{n-1}(z)$ and $H_{n+1} = 2zH_n - 2nH_{n-1}$ [8], we have a seven-term fractional differential recurrence relation for $c_q(t)$, viz., [14]

$$\begin{aligned}\dot{c}_q(t) = \tau^{-\sigma}{}_0D_t^{1-\sigma}\big[&g_q^+ c_{q+6}(t) + f_q^+ c_{q+4}(t) + e_q^+ c_{q+2}(t)\\ &+ d_q c_q(t) + e_q^- c_{q-2}(t) + f_q^- c_{q-4}(t) + g_q^- c_{q-6}(t)\big],\end{aligned} \tag{52}$$

where

$$d_n = -\sqrt{B}\Big[4\sqrt{Q}\alpha^6 + (\alpha^4 + 4Q - 6)(2n+1)\alpha^4$$
$$- 12\sqrt{Q}\alpha^2(2n^2+2n+1) + 5(2n+1)(2n^2+2n+3)\Big]/(2\alpha^6),$$
$$e_n^+ = e_{n+2}^- = (1/2)\alpha^{-6}\sqrt{B(n+1)(n+2)}$$
$$\times\Big[(\alpha^4 - 4Q + 6)\alpha^4 + 8\sqrt{Q}\alpha^2(2n+3) - 15(n^2+3n+3)\Big],$$
$$f_n^+ = f_{n+2}^- = -(1/2)\alpha^{-6}\sqrt{B(n+1)(n+2)(n+3)(n+4)}$$
$$\times\Big[-4\sqrt{Q}\alpha^2 + 3(2n+5)\Big],$$
$$g_n^+ = g_{n+2}^- = -(1/2)\alpha^{-6}\sqrt{B(n+1)(n+2)(n+3)(n+4)(n+5)(n+6)},$$

and $\alpha = \kappa/B^{1/4}$. Equation (52) can be solved exactly using matrix continued fractions as described in [14]. Having determined $c_{2q-1}(t)$, one can calculate the distribution function $W(y,t)$, the position correlation function

$$C_\sigma(t) = \left(\alpha Z B^{1/4}/\sqrt{\pi}\right)\sum_{q=1}^{\infty} c_{2q-1}(0)c_{2q-1}(t), \tag{53}$$

its spectrum $\tilde{C}_\sigma(\omega) = \int_0^\infty C_\sigma(t)e^{-i\omega t}dt$, and the dynamic susceptibility $\chi(\omega) = \chi'(\omega) - i\chi''(\omega)$ defined as [7]

$$\chi(\omega) = 1 - i\omega\tilde{C}_\sigma(\omega). \tag{54}$$

We remark that the dynamic susceptibility $\chi(\omega)$ characterizes the ac response of the system to a small perturbation. A knowledge of $\chi(\omega)$ allows one, in particular, to evaluate the signal-to-noise ratio (SNR) of the stochastic resonance at small signal amplitudes, see, e.g., [14, 55].

As we shall see two distinct bands again appear in the spectrum of $\chi''(\omega)$. The low-frequency band is due to the slowest (overbarrier) relaxation mode; the characteristic frequency $\omega_c$ and the half-width of this band are determined by the smallest nonvanishing eigenvalue $\lambda_1$. Thus, the anomalous low frequency behavior is dominated by the barrier crossing mode as in the normal diffusion. The high-frequency band is due to "intrawell" modes corresponding to the remaining eigenvalues $\lambda_k$ $(k \neq 1)$. These near degenerate modes are indistinguishable in the frequency spectrum of $\chi''(\omega)$ appearing merely as a single high-frequency band. Hence the susceptibility $\chi(\omega)$ can be effectively described via the two-mode approximation Eq. (47) with the time constants characterizing anomalous diffusion given in terms of the characteristic relaxation times for normal diffusion Eqs. (43) and (46). Thus the known equations for $\lambda_1$, $\tau_{\text{int}}$, and $\tau_{\text{eff}}$ for *normal* diffusion may also be used in order to predict the anomalous relaxation behavior. First we note that $\tau_{\text{int}}$ and $\tau_{\text{eff}}$ for *normal* diffusion may be expressed using their respective definitions in exact closed form, viz. ([7, Chap. 6]),

$$\tau_{\text{eff}} = -\frac{1}{\dot{C}_1(0)} = \tau, \tag{55}$$

$$\begin{aligned}\tau_{\text{int}} &= \tau \int\limits_{-\infty}^{\infty} e^{V(y)} \left[\int_{-\infty}^{y} z e^{-V(z)} dz\right]^2 dy \\ &= \tau \frac{\sqrt{\pi} e^{Q/2} D_{-1/2}(-\sqrt{2Q})}{2^{3/4} D_{-3/2}^2(-\sqrt{2Q})} \\ &\quad \int_0^\infty e^{\left(s-\sqrt{Q}\right)^2} \left[1 - \text{erf}\left(s - \sqrt{Q}\right)\right]^2 \frac{ds}{\sqrt{s}},\end{aligned} \tag{56}$$

where $\text{erf}(z) = (2/\sqrt{\pi}) \int_0^z e^{-z^2} dz$ is the error function [8]. Next the smallest nonvanishing eigenvalue $\lambda_1$ can be estimated in terms of matrix continued fractions as described in [14]. Moreover, for all values of $Q$, $\lambda_1$ can be evaluated from the approximate equation [7]

$$\begin{aligned}\lambda_1 = \frac{D_{-3/2}(-\sqrt{2Q})}{\tau D_{-1/2}(-\sqrt{2Q})} &\left[\frac{e^Q}{1 + \text{erf}(\sqrt{Q})}\right. \\ &\left.\times \int_0^\infty \int_0^\infty e^{-(s-\sqrt{Q})^2 - (t-\sqrt{Q})^2} \frac{\text{erf}(\sqrt{2st})}{\sqrt{st}} ds\, dt\right]^{-1}.\end{aligned} \tag{57}$$

Furthermore, in the low temperature limit, $Q \gg 1$, $\lambda_1^{-1}$ and $\tau_{\text{int}}$ have the simple asymptotic behavior [7, 25]

$$1/\lambda_1 \sim \frac{\tau \pi e^Q}{4\sqrt{2}Q}\left(1 + \frac{5}{8Q} + \cdots\right), \quad \tau_{\text{int}} \sim \frac{\tau \pi e^Q}{4\sqrt{2}Q}\left(1 + \frac{1}{2Q} + \cdots\right). \tag{58}$$

The simple asymptotic Eq. (58) allows one to easily evaluate the characteristic frequencies of the spectrum in terms of the physical model parameters such as barrier height, fractional exponent, etc. In particular, Eqs. (32) and (58) yield simple asymptotic equations for the amplitude $\Delta_1$ and characteristic frequencies $\omega_c$ and $\omega_W$, viz.,

$$\Delta_1 \sim 1 - 1/(8Q), \quad \omega_c \sim (4\sqrt{2}Q/\pi)^{1/\sigma} e^{-Q/\sigma}/\tau, \quad \text{and} \quad \omega_W \sim (8Q)^{1/\sigma}/\tau. \tag{59}$$

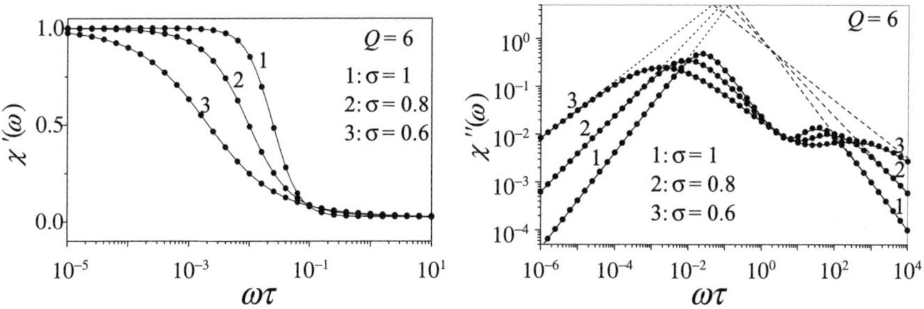


Fig. 2. The real $\chi'$ and imaginary $\chi''$ parts of the dynamic susceptibility vs. $\omega\tau$ for the barrier height $Q= 6$ and various values of the fractional exponent $\sigma = 1$ (normal diffusion), 0.8, and 0.6. Solid lines are the matrix continued fraction solution. Filled circles: the bimodal approximation, Eq. (47). Dashed and dotted lines: the low and high frequency asymptotes, Eqs. (40) and (41).

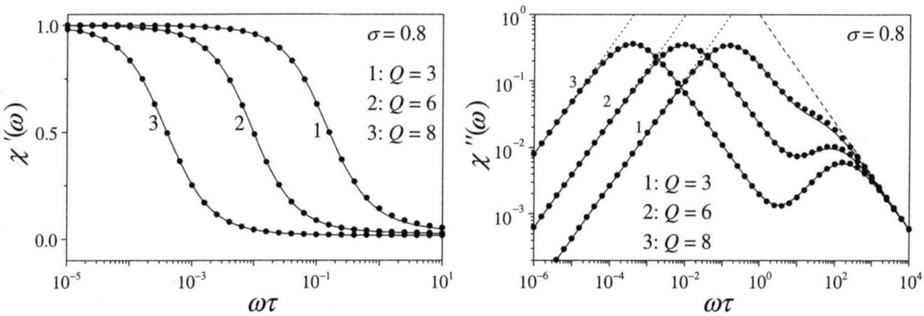


Fig. 3. $\chi'$ and $\chi''$ vs. $\omega\tau$ for $\sigma = 0.8$ and various values of $Q = 3$, 6, and 8. Key as in Fig. 2.

The above expressions allow one to readily evaluate $\chi(\omega)$ and $C_\sigma(t)$ at high barriers, $Q \gg 1$.

Calculations of the susceptibility $\chi(\omega)$ from the exact matrix continued fraction solution [14] and the approximate Eq. (47) are shown in Figs. 2 and 3. Here the low- and high-frequency asymptotes, Eqs. (40) and (41), are also shown for comparison. The two distinct bands arising in the spectrum of $\chi''(\omega)$ exhibit maxima at characteristic frequencies $\omega_c$ and $\omega_W$. Apparently, the agreement between the exact continued fraction calculations and the approximate Eq. (47) is very good [the maximum relative deviation between the corresponding curves does not exceed a few percent]. Similar (or even better) agreement exists for *all* values of $Q$ and $\sigma$. Such a high degree of accuracy of the bimodal approximation arises because the infinite number of high-frequency "intrawell" modes (these individual near degenerate modes are indistinguishable appearing merely as a single high-frequency band in the spectrum) may again be approximated effectively by a single mode. Thus, one may conclude that Eq. (47) accurately describes the behavior of $\chi(\omega)$ for all frequencies of interest and for all values of the barrier height ($Q$) and anomalous exponent ($\sigma$) parameters. The results obtained may be regarded as a generalization of the solution for the normal Brownian motion in a double-well potential [51] to fractional dynamics (giving rise to anomalous diffusion). Furthermore, one may conjecture that this generalized model can explain the anomalous relaxation of complex systems where the anomalous exponent $\sigma$ differs from unity, i.e. the relaxation process is characterized by a broad distribution of relaxation times. We remark that the bimodal approximation works extremely well in many other examples both for the noninertial anomalous and normal diffusion in a potential (various applications are given in [7, 11–13]).

## 5. Concluding Remarks

We have demonstrated how conventional Brownian motion solutions for normal diffusion of a Brownian particle in a potential may be generalized to fractional dynamics so providing one with a reasonably well grounded framework for treating the diffusion in disordered fractal systems. This was accomplished by formally introducing *three* time constants characterizing the anomalous behavior of the relaxation function $C^{\sigma}_{AB}(t)$ and the dynamic susceptibility $\chi_{AB}(\omega)$. These are the *integral relaxation* time $T_{\rm int} = \tau(\tau_{\rm int}/\tau)^{1/\sigma}$ characterizing the anomalous low-frequency behavior of $\chi_{AB}(\omega)$, the *effective relaxation time* $T_{\rm eff} = \tau(\tau_{\rm eff}/\tau)^{1/\sigma}$ yielding precise information on the high-frequency behavior of $\chi_{AB}(\omega)$ in the frequency domain and the short-time behavior of $C^{\sigma}_{AB}(t)$ in the time domain, and the longest relaxation time $T_{\rm long} = \tau/(\tau\lambda_1)^{1/\sigma} = \tau/(\tau\Gamma)^{1/\sigma}$. Thus the *anomalous* diffusion in a potential is also effectively determined by the characteristic times of the *normal* diffusion process, because the inverse of the smallest nonvanishing eigenvalue $\lambda_1$, the integral $\tau_{\rm int}$ and effective $\tau_{eff}$ relaxation times for *normal diffusion* naturally appear as time parameters. For normal diffusion, the *integral relaxation* (or, in linear response, correlation) time $\tau_{\rm int}$ is defined as the area under the appropriate relaxation function $C_{AB}(t)$; the *effective relaxation time* $\tau_{eff}$ yields precise information on the initial decay of $C_{AB}(t)$ in the time domain while the *inverse of the smallest nonvanishing eigenvalue* $\lambda_1$ of the Fokker–Planck operator $L_{FP}$ (which is usually associated with the long time behavior of $C_{AB}(t)$ and is proportional to the Kramers escape rate) is the longest relaxation time and concerns the barrier crossing process. The advantage of such a formulation is that one may treat anomalous relaxation processes knowing the characteristic times of the normal diffusion. Furthermore, because of the simple scaling relation which exists between the eigenvalues of the fractional probability density diffusion equation and those of the normal diffusion equation, the effect of anomalous diffusion on important parameters of the relaxing system such as the Kramers escape rate, etc. is readily apparent. Finally, just as in normal diffusion one may derive simple formulas for the dynamic susceptibility, etc. using a bimodal approximation. These accurately predict the effect of anomalous diffusion on the interwell (overbarrier) and intrawell relaxation processes. The overall conclusion is that one may analyze systems governed by such fractal time random walks in a manner almost as simple as the exponential relaxation, which arises from normal diffusion.

Here we have considered as the simplest example the noninertial one-dimensional anomalous diffusion of a particle in a potential $V(x)$ and we assume that the dynamics of the particle obeys the fractional Fokker–Planck equation (1) for the distribution function $W(x,t)$ of a random variable $x$. However, our principal definitions for the characteristic times of the anomalous diffusion are also valid for multidimensional fractional Fokker–Planck equations. We remark that the fractional diffusion theory can be used only at low frequencies ($\omega\tau \leq 1$) just as its normal diffusion counterpart since inertial effects are completely ignored in that theory. For normal diffusion, these effects can be included via the inertial Fokker–Planck (Klein–Kramers) equation [7, 15]. One may also include inertial effects in fractional relaxation processes. This is accomplished by utilizing a fractional version of the Klein–Kramers equation for diffusion in phase space [7, 13, 56–60].

## References

1. R. Metzler and J. Klafter, *Phys. Rep.* **339**, 1 (2000).
2. R. Metzler and J. Klafter, *Adv. Chem. Phys.* **116**, 223 (2001).
3. E. W. Montroll and M. F. Shlesinger, On the wonderful world of random walks, eds. J. L. Lebowitz and E. W. Montroll, in *Non Equilibrium Phenomena II from Stochastics to Hydrodynamics*, (Elsevier, 1984).
4. E. W. Montroll and G. H. Weiss, *J. Math. Phys.* **6**, 167 (1965).
5. W. Paul and J. Baschnagel, *Stochastic Processes from Physics to Finance* (Springer, 1999).

6. B. J. West, M. Bologna and P. Grigolini, *Physics of Fractal Operators* (Springer, New York, 2003).
7. W. T. Coffey, Yu. P. Kalmykov, and J. T. Waldron, *The Langevin Equation*, 2nd edn. (World Scientific, 2004).
8. *Handbook of Mathematical Functions*, eds. M. Abramowitz and I. Stegun (Dover, 1964).
9. V. V. Novikov and V. P. Privalko, *Phys. Rev. E* **64**, 031504 (2001); V. V. Novikov, K. W. Wojciechowski and V. P. Privalko, *J. Phys. Condens. Matter* **12**, 4869 (2000).
10. R. Metzler, E. Barkai and J. Klafter, *Phys. Rev. Lett.* **82**, 3563 (1999).
11. Yu. P. Kalmykov, W. T. Coffey, and S. V. Titov, *Phys. Rev. E* **69**, 021105 (2004).
12. W. T. Coffey, Yu. P. Kalmykov, S. V. Titov and J. K. Vij, *Phys. Rev. E* **72**, 011103 (2005).
13. W. T. Coffey, Yu. P. Kalmykov and S. V. Titov, *Adv. Chem. Phys.* **133(B)**, 285 (2006).
14. Yu. P. Kalmykov, W. T. Coffey and S. V. Titov, *Phys. Rev. E* **74**, 011105 (2006).
15. H. Risken, *The Fokker–Planck Equation*, 2nd edn. (Springer, 1989).
16. A. Szabo, *J. Chem. Phys.* **72**, 4620 (1980).
17. G. Moro and P. L. Nordio, *Mol. Phys.* **56**, 255 (1985).
18. D. A. Garanin, V. V. Ischenko and L. V. Panina, *Teor. Mat. Fiz.* **82**, 242 (1990).
19. Yu. P. Kalmykov, *Phys. Rev. E* **61**, 6205 (2000); *ibid.* **62**, 227 (2000).
20. P. Hänggi, P. Talkner, and M. Borcovec, *Rev. Mod. Phys.* **62**, 251 (1990).
21. H. A. Kramers, *Physica (Amsterdam)* **7**, 284 (1940).
22. W. T. Coffey, Yu. P. Kalmykov and E. S. Massawe, *Adv. Chem. Phys.* **85(2)**, 667 (1993).
23. W. T. Coffey, Yu. P. Kalmykov and S. V. Titov, *J. Chem. Phys.* **116**, 6422 (2002).
24. W. G. Glöckle and T. F. Nonnenmacher, *Macromolecules* **24**, 6426 (1991); K. Weron and M. Kotulski, *Physica A* **232**, 180 (1996).
25. D. Chandler, *J. Chem. Phys.* **68**, 2959 (1978).
26. B. J. Berne, J. L. Skinner, and P. G. Wolynes, *J. Chem. Phys.* **73**, 4314 (1980).
27. D. L. Hasha, T. Eguchi, and J. Jonas, *J. Chem. Phys.* **73**, 1571 (1981); *J. Am. Chem. Soc.* **104**, 2290 (1982).
28. D. K. Garrity and J. L. Skinner, *Chem. Phys. Lett.* **95**, 46 (1983).
29. B. Carmeli and A. Nitzan, *J. Chem. Phys.* **80**, 3596 (1984).
30. H.C. Brinkman, *Physica* **22**, 29 (1956).
31. H.C. Brinkman, *Physica* **22**, 149 (1956).
32. C. Blomberg, *Physica A***86**, 49 (1977).
33. P. B. Visscher, *Phys. Rev. B* **14**, 347 (1976).
34. J. L. Skinner and P. G. Wolynes, *J. Chem. Phys.* **69**, 2143 (1978); *J. Chem. Phys.* **72**, 4913 (1980).
35. R. S. Larson and M. D. Kostin, *J. Chem. Phys.* **69**, 4821 (1978).

36. S. C. Northrup and J. T. Hynes, *J. Chem. Phys.* **69**, 5246 (1978); *ibid.* **69**, 5261 (1978); *ibid.* **73**, 2700 (1980); R. F. Grote and J. T. Hynes, *ibid.* **73**, 2715 (1980).
37. M. Mangel, *J. Chem. Phys.* **72**, 6606 (1980).
38. K. Schulten, Z. Schulten and A. Szabo, *J. Chem. Phys.* **74**, 4426 (1981).
39. M. Bixon and R. Zwanzig, *J. Stat. Phys.* **3**, 245 (1971).
40. M. I. Dykman, S. M. Soskin and M. A. Krivoglaz, *Physica A* **133**, 53 (1985).
41. P. Hänggi, *Phys. Lett. A* **78**, 304 (1980).
42. J. A. Krumhansl and J. R. Schrieffer, *Phys. Rev. B* **11**, 3535 (1975).
43. J. D. Bao and Y. Z. Zhuo, *Phys. Rev. C* **67**, 064606 (2003).
44. V. M. Kolomietz, S. V. Radionov and S. Shlomo, *Phys. Rev. C* **64**, 054302 (2001).
45. M. I. Dykman, G.P. Golubev, D.G. Luchinsky, P. V. E. McClintock, N. D. Stein and N. G. Stocks, *Phys. Rev. E* **49**, 1935 (1994).
46. L. Gammaitoni, P. Hänggi, P. Jung and F. Marchesoni, *Rev. Mod. Phys.* **70**, 223 (1998).
47. W. T. Coffey, M. W. Evans and P. Grigolini, *Molecular Diffusion and Spectra* (Wiley, 1984); Russian edition (Mir, 1987).
48. A. Perico, R. Pratolongo, K. F. Freed, R. W. Pastor and A. Szabo, *J. Chem. Phys.* **98**, 564 (1993).
49. A. Schenzle and H. Brand, *Phys. Rev. A* **20**, 1628 (1979).
50. Yu. P. Kalmykov, W. T. Coffey and J. T. Waldron, *J. Chem. Phys.* **105**, 2112 (1996).
51. Yu. P. Kalmykov, W. T. Coffey and S. V. Titov, *Physica A* **377**, 412 (2007).
52. F. So and K. L. Liu, *Physica A* **331**, 378 (2004).
53. C. W. Chow and K. L. Liu, *Physica A* **341**, 87 (2004).
54. Y. M. Kang and Y. L. Jiang, *Phys. Rev. E* **81**, 021109 (2010).
55. Y. J. Li and Y. M. Kang, *Commun. Theor. Phys. (China)* **54**, 29 (2010).
56. E. Barkai and R. S. Silbey, *J. Phys. Chem. B* **104**, 3866 (2000).
57. R. Metzler and J. Klafter, *J. Phys. Chem. B* **104**, 3851 (2000).
58. R. Friedrich, F. Jenko, A. Baule and S. Eule, *Phys. Rev. E* **74**, 041103 (2006).
59. W. T. Coffey, Yu. P. Kalmykov and S. V. Titov, *Phys. Rev. E* **65**, 032102 (2002); *ibid.* **65**, 051105 (2002).
60. Yu. P. Kalmykov, W. T. Coffey and S. V. Titov, *Phys. Rev. E* **75**, 031101 (2007).

# Chapter 4

## Reactions in Subdiffusive Media and Associated Fractional Equations

Santos B. Yuste
*Departamento de Física, Universidad de Extremadura, E-06071 Badajoz, Spain*

E. Abad
*Departamento de Física Aplicada, Universidad de Extremadura, E-06071 Badajoz, Spain*

Katja Lindenberg
*Department of Chemistry and Biochemistry and BioCircuits Institute, University of California, San Diego, La Jolla, CA 92093-0340, USA*

We construct and solve fractional equations for the description of reactions in subdiffusive media starting from a mesoscopic continuous time random walk model. Our goal is to understand the spatial and temporal evolution of the reactant concentrations. Our discussion is presented in two parts. In the first part, the reactions occur at fixed locations such as immobile traps or hypersurfaces enclosing a volume. Here the reactions appear as boundary conditions. Since the boundary conditions are identical for diffusive and subdiffusive reactants, there is a close connection between the solutions in both environments. In the second part we deal with the more complex problem of reactions that occur at random locations, either because all reactants are mobile or because the reaction is a degradation process that may occur anywhere in space. Now in general the reaction has to be built into the equations themselves, and each situation demands a particular derivation. In one case in this class it is possible to find asymptotic solutions using a bounding procedure that requires only the solution of boundary value problems, but this fortuitous solution is applicable only to this particular case. More generally, this chapter serves to point to the difficulties inherent in the reaction-subdiffusion problem.

1. Introduction . . . . . . . . . . . . . . . . . . . . . . . . 78
2. Subdiffusion and Fractional Calculus . . . . . . . . . . . . 80
3. Reactions Occurring at Spatially Fixed Locations . . . . . 81
   3.1. Single-particle target problem . . . . . . . . . . . 82

3.2. Many-particle target problem . . . . . . . . . . . . 85
3.3. Escape problems . . . . . . . . . . . . . . . . . . . 87
4. Reactions Occurring at Random Locations . . . . . . . . 89
4.1. Mobile particles and traps . . . . . . . . . . . . . . 90
4.2. Fractional diffusion-reaction equations . . . . . . . 94
4.3. Single-particle target problem with a reactivity field 97
4.4. Reaction-subdiffusion equations and morphogen gradient formation . . . . . . . . . . . . . . . . . . . 98
4.4.1. Constant reactivity . . . . . . . . . . . . 99
4.4.2. Piecewise constant reactivity . . . . . . . 99
4.4.3. Exponentially decaying reactivity . . . . 101
5. Final Remarks . . . . . . . . . . . . . . . . . . . . . . 103

## 1. Introduction

Diffusive motion of particles in dense media is usually understood to be random motion characterized by a mean square displacement that grows linearly with time, $\langle r^2 \rangle \sim Dt$ ($D$ is the diffusion coefficient). Reactions among particles that undergo diffusive motion are traditionally described by well-established reaction-diffusion equations in which a diffusion equation for the concentrations is simply augmented by a local reaction term typically constructed as a local version of the laws of mass action [1, 2]. However, in recent years it has become clear that motion in complex environments [3], especially in biological environments [4], is often hindered by the presence of traps or obstacles or other impediments, leading to subdiffusive motion, that is, motion characterized by a mean square displacement that grows sublinearly with time, $\langle r^2 \rangle \sim D_\gamma t^\gamma$ with $0 < \gamma < 1$ ($D_\gamma$ is a generalized subdiffusion coefficient). The equations describing the spatio-temporal evolution of reactions among subdiffusive species are considerably more complicated than the familiar reaction-diffusion equations. Approaches based on heuristic macroscopic equations are often insufficient and can lead to incorrect conclusions. The derivation of these equations requires the formulation of a microscopic or mesoscopic model as a starting point. In this chapter we start from a particular mesoscopic point of view, namely, a continuous time random walk (CTRW) description of the motion in which the waiting time distributions between successive steps have a long or heavy tail. Starting from such a model, in Sec. 2 we discuss the derivation of a well-known generic fractional subdiffusion equation for the concentrations of a subdiffusive species as it evolves in space and time. The resulting equation has to be appropriately modified when the species in question can undergo reactions. There is no universal way to do this, contrary to the case of ordinary diffusion where reaction terms are simply added. In subsequent sections we deal with this problem in a variety of scenarios.

Next we discuss a variety of ways to include reactive processes for our subdiffusive particles. We separate our discussion into two parts, one presented in Sec. 3 and the other in Sec. 4. First, in Sec. 3 we focus on systems in which reactions occur at fixed locations. Starting from a CTRW model, we arrive at a pure fractional diffusion equation in which the reactions are manifest as spatially fixed boundary conditions (we do not consider systems subject to external potentials). These systems adhere to a description parallel to that of the corresponding normally diffusive system except for the fact that the time evolution for long times is slower. Typically many of the results for the subdiffusive system can be found from those associated with the diffusive system if in the results for the latter we make the substitution $t \to \alpha t^\gamma$, where $\alpha$ is a parameter that depends on $\gamma$ and on $D_\gamma$ (see pp. 375 in Ref. 3). In Sec. 3.1 we focus on the survival probability of a fixed target that reacts with a subdiffusive particle. Section 3.2 extends the calculation of the target survival probability to the case where the target is surrounded by a sea of subdiffusive particles rather than by a single particle. In Sec. 3.3 we present the survival probability of particles that subdiffuse in a finite region surrounded by a boundary with which the particles can react and consequently vanish.

The second part of our discussion, in Sec. 4, deals with reactions that occur at random locations either because all reactants are mobile, or because the reaction is a degradation process that may occur anywhere in space. The reactions can thus no longer be handled as boundary value problems but are instead far more complex many-body problems. The results for subdiffusive particles are in general very different from those found for particles that undergo normal diffusion. Furthermore, the solution of these problems is typically far more complicated than the corresponding solution for the diffusive problem. In Sec. 4.1 we deal with a generalization of the problem of Sec. 3.2 in which all reactants are mobile. The reaction is of the form $A + B \to B$, where a particle ($A$) survives only if it does not touch any traps ($B$). The calculation of the survival probability of a particle even when both particle and traps move according to the normal laws of diffusion is complicated and has only been solved asymptotically relatively recently. The generalization to the situation where one or the other or both species are subdiffusive presents interesting challenges, especially when the particle is diffusive and the traps are subdiffusive. In Sec. 4.2 we discuss some of the difficulties that are encountered when the fractional subdiffusion equation is to be modified by reactive contributions. It is now no longer possible to generalize reaction-diffusion equations in any simple manner. In particular, it is in general no longer appropriate to simply add a reaction term to the subdiffusion equation, and it is no longer possible to devise a general scheme to include the effects of reactions. We derive a reaction-subdiffusion equation for a class of problems that serves to illustrate the complicated nature of the task. In Sec. 4.3 we use this equation to find the survival probability of a particle in the presence of a fixed target surrounded by a "reactivity field." In Sec. 4.4 we again solve this equation, but now for a scenario that may be appropriate in the context of the formation of morphogen gradients in an embryonic environment.

We conclude this chapter with a few final remarks in Sec. 5.

## 2. Subdiffusion and Fractional Calculus

A well-known model of anomalous diffusion is based on a CTRW in which the particles are described as random walkers whose step lengths $\mathbf{r}$ and waiting times $t$ are characterized by probability distributions $w(\mathbf{r})$ and $\psi(t)$, respectively. Let $\hat{w}(\mathbf{q})$ be the Fourier transform of $w(\mathbf{r})$ and $\tilde{\psi}(u)$ the Laplace transform of $\psi(t)$. We assume a (symmetric) walk with

$$\hat{w}(\mathbf{q}) \sim 1 - (\sigma q)^{\mu} \tag{1}$$

for small $q \equiv |\mathbf{q}|$, and

$$\tilde{\psi}(u) \sim 1 - (\tau u)^{\gamma}, \quad 0 < \alpha \leq 1 \tag{2}$$

for small $u$. The parameters $\sigma^{\mu}$ and $\tau^{\gamma}$ appear in combined form in the generalized diffusion coefficient $D_{\gamma} = \sigma^{\mu}/\tau^{\gamma}$. If $\mu = 2$, then $\sigma^2$ is half the variance of the jump length. If additionally $\gamma < 1$, then the mean square displacement of the walkers grows sublinearly with time, $\langle \mathbf{r}^2 \rangle \sim 2dD_{\gamma}t^{\gamma}/\Gamma(1+\gamma)$ [5], where $d$ denotes dimensionality. The limit of long time scales and large displacement scales yields the fractional diffusion equation [6, 7]

$$\frac{\partial c(\mathbf{r},t)}{\partial t} = D_{\gamma}\, {}_0\mathcal{D}_t^{1-\gamma} \nabla^{\mu} c(\mathbf{r},t), \quad 0 < \alpha < 1 \tag{3}$$

where $c(\mathbf{r}, t)$ is the concentration of walkers, ${}_0\mathcal{D}_t^{1-\gamma}$ is the operator whose Laplace transform $\mathcal{L}$ is

$$\mathcal{L}\ {}_0\mathcal{D}_t^{1-\gamma} f(t) = u^{1-\gamma} \tilde{f}(u), \tag{4}$$

and $\nabla^\mu$ is the operator whose Fourier transform $\mathcal{F}$ is

$$\mathcal{F} \nabla^\mu g(\mathbf{r}) = -q^\mu \hat{g}(\mathbf{q}) \tag{5}$$

for sufficiently smooth functions $f$ and $g$ [8, 9]. Equation (3) is a particular case of the reaction-subdiffusion equation derived later in Sec. 4.2. The operator ${}_0\mathcal{D}_t^{1-\gamma}$ is the Grünwald–Letnikov fractional derivative [8]. When operating on well-behaved functions [functions $f(t)$ for which $\lim_{t\to 0} \int_0^t d\tau (t-\tau)^{\gamma-1} f(\tau) = 0$], this operator is equivalent to the Riemann–Liouville fractional derivative [8],

$${}_0D_t^{1-\gamma} f(t) = \frac{1}{\Gamma(\gamma)} \frac{\partial}{\partial t} \int_0^t dt' \frac{f(t')}{(t-t')^{1-\gamma}}. \tag{6}$$

The fractional spatial derivative $\nabla^\mu$ is the negative of the Riesz fractional derivative of order $\mu$ [9]. For $\mu = 2$ this operator is simply the Laplacian operator $\nabla^2$ and Eq. (3) becomes the fractional subdiffusion equation

$$\frac{\partial c(\mathbf{r}, t)}{\partial t} = D_\gamma\ {}_0\mathcal{D}_t^{1-\gamma} \nabla^2 c(\mathbf{r}, t), \quad 0 < \alpha < 1. \tag{7}$$

From here on we focus on this case.

## 3. Reactions Occurring at Spatially Fixed Locations

If a reaction occurs when a moving particle reaches a fixed target, the problem can be formulated as a boundary value problem. The boundary value formulation is essentially the same whether the moving particle undergoes normal diffusion or subdiffusion: the (sub)diffusion equation describes the motion of the particle in a region $\Omega$ in a region $\Omega$ external to the target, and this equation is augmented by the reactive boundary condition $F[c(\mathbf{r}, t), \nabla c(\mathbf{r}, t)] = 0$ at the reactive boundary $\mathbf{r} \in \partial\Omega$. Two types of boundary conditions are often considered in diffusion problems. One is the totally absorbing (or perfectly reactive) boundary, for which

$$c(\mathbf{r}, t) = 0, \quad \mathbf{r} \in \partial\Omega, \tag{8}$$

indicating immediate reaction (death) upon encounter with the boundary. The other is the radiation (or Robin, or partially absorbing) boundary condition

$$c(\mathbf{r},t) = -\Lambda \frac{\partial}{\partial n} c(\mathbf{r},t), \quad \mathbf{r} \in \partial\Omega. \tag{9}$$

Here $\partial c/\partial n$ is the component of the gradient of $c$ perpendicular to the boundary and pointing away from the region $\Omega$, and $\Lambda$ is a reaction rate parameter. This latter condition describes partially reactive boundaries, that is, situations where it is possible for a particle to come in contact with the boundary and yet not react. That this is in fact the correct mathematical description of a partially reactive boundary is not immediately obvious [10–12].

Our starting point is the Laplace transform of Eq. (7),

$$u\tilde{c}(\mathbf{r},u) - c(\mathbf{r},0) = u^{1-\gamma} D_\gamma \nabla^2 \tilde{c}(\mathbf{r},u). \tag{10}$$

The Laplace transformed equation makes it clear that the subdiffusion equation can formally be obtained from the normal diffusion equation if in the latter we implement the change $D \to u^{1-\gamma} D_\gamma$. Since the boundary conditions (8) and (9) are the same for the diffusion and subdiffusion problem, it would seem reasonable to expect that the solution for the subdiffusive problem can be obtained from that of the diffusive problem with the simple replacement $D \to u^{1-\gamma} D_\gamma$. However, this must be implemented with care because there are exceptions to this assumption, as we will see in Sec. 3.3.

### 3.1. *Single-particle target problem*

In this section we consider a specific application of the boundary value problem described by Eq. (7) with (8) or (9), namely, the computation of the survival probability of a fixed target in the presence of a subdiffusive random walker (hereafter called the "particle") in an unbounded $d$-dimensional domain. When the particle reaches the boundary $\partial\Omega$ of the target, reaction takes place with a given probability, as a result of which the target disappears. Since the ultimate fate of the particle after the reaction is irrelevant for the computation of the survival probability of the target, we can assume without loss of generality that both the targets *and* the particle vanish upon reaction. The survival probability of the target and the survival probability of the particle thus become the same. The latter can then be computed from

Eq. (7) complemented by the boundary condition (8) in the fully absorbing case (reaction with unit probability upon encounter) or Eq. (9) in the partially absorbing case (reaction with probability $< 1$ upon encounter). The fully absorbing boundary condition described by Eq. (8) (formally obtained from (9) in the limit $\Lambda \to 0$) corresponds to a so-called first passage problem since in this case the probability of first arrival of the particle at the target is precisely the probability of the target being absorbed by the particle. First passage statistics for the subdiffusive case are comprehensively dealt with in Sec. 5.1 of [13].

The target problem is relevant in a number of fields, including the optimization of target search strategies [14, 15], recombination kinetics [16], scavenging reactions [17], and photoluminescence quenching of excited states [18]. It is also the starting point for the case of many (sub)diffusing particles (see next subsection) and may be used as a reference to assess the effect of finite sized domains or confining boundaries [19, 20].

We begin by defining the survival probability $Q_1(\mathbf{r}, t)$ as the probability that the target and a particle that initially started at location $\mathbf{r}$ have not reacted up to time $t$. The survival probability of the target (or of the particle) can be written as

$$Q_1(\mathbf{r}, t) = \int_\Omega d\mathbf{r}' c(\mathbf{r}', t; \mathbf{r}), \tag{11}$$

where $c(\mathbf{r}', t; \mathbf{r})$ is now interpreted as the probability density that the particle that initially started at $\mathbf{r}$ is found at $\mathbf{r}'$ at time $t$, that is, it is the solution of the boundary value problem (7) with (9) or with (8), with the initial condition $c(\mathbf{r}', 0; \mathbf{r}) = \delta(\mathbf{r}' - \mathbf{r})$ [21]. The survival probability $Q_1(r, t)$ of a hyperspherical target of radius $R$ centered at the origin is especially simple to deal with. The calculation for the case of a subdiffusive particle can be straightforwardly derived from the corresponding calculation for a diffusive particle [17, 21, 22]. In the Laplace representation the boundary value problem becomes

$$u\widetilde{Q}_1(r, u) - 1 = u^{1-\gamma} D_\gamma \nabla^2 \widetilde{Q}_1(r, u), \tag{12}$$

$$\widetilde{Q}_1(R, u) = -\Lambda \frac{\partial}{\partial n} \widetilde{Q}_1(r, u)|_{r=R}, \tag{13}$$

$$\lim_{r\to\infty} \widetilde{Q}_1(r, u) = \frac{1}{u}. \tag{14}$$

The equations for the subdiffusive case are obtained from the normal diffusive case with the replacement $D \to u^{1-\gamma} D_\gamma$. The subdiffusive solution can thus be found by performing the same replacement in the normal diffusion solution [21–23]. The result is

$$u\widetilde{Q}_1(r,u) = 1 - \left(\frac{r}{R}\right)^{1-d/2} \frac{K_{d/2-1}(rz)}{K_{d/2-1}(Rz) + \Lambda z K_{d/2}(Rz)}, \tag{15}$$

where we have set $z = (u^\gamma / D_\gamma)^{1/2}$. In the above equation the $K$'s stand for modified spherical Bessel functions of the third kind [24]. In Laplace space, the solution for $d = 1$ and $d = 3$ in the fully absorbing case $\Lambda = 0$ takes a rather simple form, namely,

$$\widetilde{Q}_1(r,u) = \frac{1}{u} - \left(\frac{R}{r}\right)^\alpha \frac{e^{(R-r)\sqrt{u^\gamma/D_\gamma}}}{u}. \tag{16}$$

Here $\alpha = 0$ for $d = 1$ and $\alpha = 1$ for $d = 3$. Its inverse Laplace transform can be expressed in terms of Fox's $H$ function as [6, 25]

$$Q_1(r,t) = 1 - \left(\frac{R}{r}\right)^\alpha H_{11}^{10}\left[\frac{r-R}{\sqrt{D_\gamma t^\gamma}} \middle| \begin{matrix}(1,\gamma/2)\\(0,1)\end{matrix}\right]. \tag{17}$$

The Fox function reduces to a complementary error function when $\gamma = 1$, and in this limit we recover the correct result for classical diffusion, see, for example, [22].

The asymptotic long-time behavior of the survival probability depends strongly on dimension. In $d = 1$, one has a power-law decay, $Q_1(r,t) \sim (r-R)/t^{\gamma/2}$; in $d = 2$, one has a slower inverse logarithmic decay, $Q_1(r,t) \sim \ln(r/R)/\ln(D_\gamma t^\gamma)$. For $d = 3$, the target has a finite survival probability, $Q_1(r, t \to \infty) \sim 1 - R/r$. Asymptotically, the subdiffusive character of the particle is thus relevant only in the one-dimensional case and has only a marginal effect in $d = 2$.

As an aside, we note that the result (17) can also be obtained from the analogous result for normal diffusion via the relation $\widetilde{Q}_1(u|\gamma) = u^{\gamma-1}$ $\widetilde{Q}_1(u^\gamma|\gamma = 1)$ between the Laplace transform of the survival probability $Q_1(t|\gamma)$ of the target when the particle is subdiffusive and of the corresponding quantity $Q_1(t|\gamma = 1)$ when it is diffusive [5, 13, 21]. This is a consequence of the integral transformation relating the propagators for fractional and normal diffusion [5, 6, 13, 26–28].

### 3.2. *Many-particle target problem*

The generalization of the one-particle problem treated in the previous subsection to the many-body case is of special interest in the framework of Smoluchowski's theory of diffusion-controlled reactions [1]. This theory has been applied to numerous problems such as electron trapping and recombination [29], free radical and ion-scavenging [17], geminate recombination [30], luminescence quenching [31], exciton trapping [32–34], and site location in DNA [35]. The theory deals with two-species binary reactions $A+B \rightarrow products$, where both species may undergo classical diffusion with different diffusion coefficients $D_A$ and $D_B$ until they meet and react at a rate which increases monotonically with increasing reaction rate parameter $\Lambda$. The theory allows one to compute the survival probability of the $A$ particles as well as the associated decay of the concentration of $A$ under the assumptions that the $B$ species is in vast excess with respect to the $A$ species, and that each $A$ particle is surrounded by a sea of uncorrelated $B$ particles. The Smoluchowski approach becomes exact in the limit $D_A \rightarrow 0$, a situation in which each $A$ particle may be regarded as an immobile target immersed in a sea of diffusing $B$ particles. This case is usually called the "target problem" in the literature. While extensive results for the survival probability and other related quantities (e.g., the distribution of reaction times) are available for the normal diffusive case, the subdiffusive version of the problem has remained unexplored until recently. Here we briefly outline the method and main findings for this problem in continuous space. Our results for $\Lambda = 0$ are the continuous counterparts to those of Section 13.3.1 of [3] for the case of anomalous diffusion on a discrete spatial support.

Our starting point is a setting similar to that of the previous subsection, the difference being that instead of a target and a single subdiffusive particle, we now consider a target surrounded by $N$ independent subdiffusive point particles. The target is a hypersphere of radius $R$. Of interest here is the survival probability $Q(t)$ of the target. This survival probability is simply the probability that none of the $N$ particles has been absorbed upon contact with the surface of the sphere. Under the assumption that the independently moving particles are randomly distributed in the volume $V$ of $\Omega$, one obtains the factorized form

$$Q(t) = \left[\frac{1}{V}\int_{r>R} d\mathbf{r}\, Q_1(r,t)\right]^N . \tag{18}$$

Equation (18) provides the connection between the solution for the many-body case and the one-particle problem treated in the previous subsection. In the thermodynamic limit $N \to \infty, V \to \infty$ at fixed density $\rho = N/V$, one finds

$$Q(t) = \exp\{-\rho R^d f(t)\}, \tag{19}$$

with

$$f(t) = \frac{1}{R^d} \int_{r>R} d\mathbf{r}\, [1 - Q_1(r,t)]. \tag{20}$$

The problem is readily solved in Laplace space. From Eq. (15), one obtains

$$\widetilde{f}(u) = \frac{S_d}{Ruz} \frac{K_{d/2-1}(Rz)}{K_{d/2-1}(Rz) + \Lambda z K_{d/2}(Rz)}, \tag{21}$$

where $S_d = 2\pi^{d/2}/\Gamma(d/2)$ is the surface of a sphere of unit radius. The large- and small-$u$ behavior of the above expression respectively provide the short and long time behaviors of $f(t)$ and the associated survival probability via the pertinent Tauberian theorems. In the special cases $d = 1$ and $d = 3$ the quantity $\widetilde{f}(u)$ can be inverted exactly to obtain results valid for arbitrary $t$. For example, the one-dimensional case yields [23]

$$f(t) = \frac{\omega^{1/2}}{\Gamma(1+\alpha/2)} - 2\frac{\Lambda}{R}[1 - E_{\gamma/2}(-\omega^{1/2}R/\Lambda)], \tag{22}$$

where $\omega = 4D_\gamma t^\gamma/R^2$, and

$$E_\gamma(z) = \sum_{n=0}^{\infty} \frac{z^n}{\Gamma(1+n\gamma)} \tag{23}$$

stands for the Mittag-Leffler function [36, 37]. It is interesting to note that for long times $t$ the partially reflecting nature of the surface is not relevant for $d \le 2$ as $f(t) \sim \omega^{1/2}/\Gamma(1+\gamma/2)$ for $d = 1$ and $f(t) \sim \pi\omega/[\Gamma(1+\gamma)\ln(\omega)]$ for $d = 2$, both independent of $\Lambda$. However, for $d = 3$, the value of $\Lambda$ appears explicitly in the characteristic decay function, i.e. $f(t) \sim \pi\omega/[\Gamma(1+\gamma)(1+\Lambda/R)]$. For later comparison, it is useful to exhibit these asymptotic survival probability results explicitly [21, 23]:

$$Q(t) \sim \begin{cases} \exp\left[-\dfrac{2\rho}{\Gamma(1+\gamma/2)}\left(D_\gamma t^\gamma\right)^{1/2}\right], & d=1, \\ \exp\left[-\dfrac{4\pi\rho}{\Gamma(1+\gamma)}\dfrac{D_\gamma t^\gamma}{\ln\left(4D_\gamma t^\gamma/R^2\right)+2\Lambda/R}\right], & d=2, \\ \exp\left[-\dfrac{2(d-2)\pi^{d/2}\rho R^{d-2}}{\Gamma(d/2)\Gamma(1+\gamma)(1+(d-2)\Lambda/R)}D_\gamma t^\gamma\right], & d\geq 3. \end{cases} \quad (24)$$

The above results show that the behavior of the survival probability changes drastically with respect to the one-particle case. Focusing on the fully absorbing case $\Lambda = 0$, one sees that the asymptotic long-time survival probability is given by a stretched negative exponential whose argument $f(t)$ grows more rapidly with time with increasing dimension $d$ (the typical inverse logarithmic correction appears for $d = 2$). This implies that $Q(t)$ decays more rapidly with increasing dimension $d$, as opposed to the increase observed for $Q_1(r,t)$ with increasing $d$.

### 3.3. *Escape problems*

Escape problems are relevant for the study of processes that are triggered when a critical threshold value is attained. Numerous examples are found in physics, chemistry, biology, economics and social sciences [38]. One of the main goals in escape problems is to compute the distributions of the times (and the moments of these distributions) needed by one or more particles to irreversibly escape a given domain $\Omega$. Familiar questions include those associated with first passage processes [38, 39], where one asks how long it takes the $j$th particle to reach an absorbing boundary $\partial\Omega$ (order statistics problems) [40–44]. Conversely, one can enquire about the number of particles that have escaped a domain $\Omega$ up to time $t$ and the associated decay of the probability that a given particle has not escaped (survival probability) [6, 38, 39, 45].

In order to find the survival probability for this latter problem, Eq. (7) must be solved subject to the condition $c(\mathbf{r},t) = 0$ at the boundary. In our route to the solution, we use the method of separation of variables [6, 13, 46] by making the superposition ansatz $c(\mathbf{r},t) = \sum_{j=1}^{\infty} a_j\, T_j(t)\psi_j(\mathbf{r})$. We henceforth assume that the domain $\Omega$ that initially contains all the particles is a hypersphere of radius $R$. Our ansatz leads to

$$\frac{T_j'(t)}{D_\gamma\ {}_0\mathcal{D}_t^{1-\gamma}T(t)} = \frac{\psi_j''(r) + (d-1)\psi_j'(r)/r}{\psi_j(r)} = \lambda_j, \tag{25}$$

where $\lambda_j$ is the $j$th eigenvalue. The temporal part of the solution satisfies the equation $T_j'(t) = \lambda\, D_\gamma\ {}_0\mathcal{D}_t^{1-\gamma}T_j(t)$ [6, 36], i.e. $T_j(t) = E_\gamma(\lambda_j D_\gamma t^\gamma)$. This Mittag-Leffler function becomes equal to an exponential when $\gamma = 1$ (normal diffusion case), but displays rather different properties for *any* $\gamma < 1$. In contrast, the eigenfunctions $\psi_j(r)$ describing the spatial part of the solution and the set of acceptable eigenvalues $\lambda_j$ which satisfy the equation $\psi_j''(r) + (d-1)\psi_j'(r)/r = \lambda_j\psi_j(r)$ are independent of $\gamma$ and are thus *exactly* the same for the subdiffusion and normal diffusion problems [47–49]. The full solution has the form

$$c(r,t) = \sum_{j=1}^{\infty} a_j\psi_j(r)E_\gamma\left(-\lambda_j t^\gamma\right), \tag{26}$$

where the coefficients $a_j$ are determined from the initial condition $c(r,0)$ via the orthogonality properties of the eigenfunctions $\psi_j(r)$. The so-called propagator solution $c(r,t) \equiv G(r,t;r_0)$ corresponding to the initial condition $c(r,0) = \delta(r - r_0)$ (representing a single particle initially located at $r_0$) has a special relevance, since it can be used to construct other types of more complex solutions involving, for example, localized particle sources. Recently, Grebenkov [39] has studied subdiffusive propagator solutions for other confining domains and boundary conditions.

The behavior of (26) at long times may be very different from that describing normal diffusion. For instance, if one takes a homogeneous initial concentration $c_0$ in the above example of a $d$-dimensional absorbing hyperspherical surface, one gets

$$c(r,t)/c_0 = 2(r/R)^{1-d/2}\sum_{j=1}^{\infty}\frac{J_{d/2-1}(z_j r/R)}{z_j J_{d/2}(z_j)}E_\gamma[-(z_j/R)^2 D_\gamma t^\gamma], \tag{27}$$

where $z_j$ denotes the $j$th zero of the Bessel function $J_{d/2}$. For long times, one can use the asymptotic expansion

$$E_\gamma(z) \approx \sum_{m=1}^{\infty}\frac{(-z)^{m+1}}{\Gamma(1-m\gamma)}z^{-m}, \quad z \to \infty \tag{28}$$

and one can rearrange terms to get

$$c(r,t)/c_0 = \sum_{m=1}^{\infty} \frac{(-1)^{m+1}}{\Gamma(1-m\gamma)} \left(\frac{R^2}{D_\gamma t^\gamma}\right)^m \Psi_m(r/R). \tag{29}$$

It can be proved that the $\Psi_m(x)$'s are *polynomials* of degree $2m$ with a zero at $x = 1$ [50, 51]. Neglecting subdominant terms with $m > 1$ in the above long-time solution, one finds $c(r,t)/\varphi(t) \propto \Psi_1(r/R)$ where $\varphi(t) = 1/(D_\gamma t^\gamma/R^2)$ and $\Psi_1(r/R) \propto 1 - r^2/R^2$. In contrast, a similar analysis for the normal diffusion case not only leads to a faster (exponential) time decay $\varphi(t) = \exp(-z_1 D_\gamma t^\gamma/R^2)$, but also to a different radial concentration profile at late times. One finds $c(r,t)/\varphi(t) \propto (r/R)^{1-d/2} J_{d/2-1}(z_1 r/R)$, which depends on dimensionality. In the subdiffusive case, the spatial dependence is universal in the sense that it depends *neither* on the dimension $d$ *nor* on the anomalous diffusion exponent $\gamma$ or the diffusion coefficient $D_\gamma$.

In conclusion, a minute amount of subdiffusivity in the particle motion is seen to destroy the form $(r/R)^{1-d/2} J_{d/2-1}(z_1 r/R)$ of the long-time normal diffusion mode and leads to the subdiffusive form $1-r^2/R^2$ which holds for *any* $\gamma < 1$. The aforementioned singular long-time behavior of the solution with respect to $\gamma$ is not an exclusive feature of the escape problem. A similar singular behavior is seen in the survival probability of a diffusing particle in a sea of subdiffusive traps, treated later in Sec. 4.1.

## 4. Reactions Occurring at Random Locations

In this section we address reactions that occur at random locations as well as random times. When all the reactants in a subdiffusive medium are mobile, the problem of describing any quantity used as an indicator of the progression of the kinetics is much more complex than it is in a diffusive medium, and also much more complex than it is if one of the reactants is immobile. These indicators include the quantities that we have introduced already: survival probabilities, escape probabilities, and local concentrations.

Reactions may occur at random locations for two reasons: either reactants that are all mobile must meet for a reaction to take place, or the reaction is a degradation process that is not restricted to a fixed boundary as in the previous section. Such reactions can in general no longer be treated as boundary value problems but instead become much more complicated many-body problems that require the introduction of new ideas. Furthermore, the subdiffusive character of the reactant motion leads to memory effects that introduce yet another level of complexity.

In Sec. 4.1 we deal with a problem similar to that of Sec. 3.2, but substantially complicated by the fact that the previously immobile target is now also mobile (diffusive or subdiffusive). As we did there, we focus on the survival probability of the target. Then, in Sec. 4.2 we derive a reaction-subdiffusion equation starting from a CTRW model. We subsequently use this equation in Sec. 4.3 to study the reaction between a set of mobile particles and a fixed target surrounded by a "reactivity field" $k(\mathbf{r})$. This reactivity field allows the reaction to take place at random distances between a mobile particle and the target with a probability determined by the field. Finally, in Sec. 4.4 we use the reaction-subdiffusion equation to study the reaction of a constant flux of particles that emerge from a given location (source), move subdiffusively, and disappear with a probability that in general depends on their location. This scenario is particularly interesting for the study of the timely problem of morphogen gradients. In these last two sections our indicator of the reaction is the concentration of surviving reactant.

### 4.1. *Mobile particles and traps*

We begin by considering the survival probability of a mobile particle in a sea of mobile traps. The characteristic exponent for the particle is $\mu$. The exponent for the traps is $\gamma$. In earlier sections we called the exponent $\gamma$ in all cases because only one of the species was mobile, but here it is important to differentiate between them and so we introduce a distinct notation for them. This problem has in fact not been solved for any of the indicators listed above. However, it has been possible to establish the asymptotic behavior of the survival probability of the particle in most if not all situations, as we will now see. Note that these results do not include an estimate of how long it takes the system to arrive at asymptotia (a numerical study of this can be found in Ref. [53]).

We suppose that a particle $A$ moves diffusively ($\mu = 1$) or subdiffusively ($0 < \mu < 1$) in a sea of traps $B$ that move diffusively ($\gamma = 1$) or subdiffusively ($0 < \gamma < 1$), with $\gamma$ and $\mu$ not necessarily equal. One can imagine, for instance, a large difference in the sizes or in the interactions of $A$s and $B$s with the surrounding medium that may lead to large differences in the way they move. In chemical kinetics parlance we are looking at the reaction $A + B \to B$ when all components are mobile. Our goal is to calculate the survival probability $Q(t)$ of the $A$ particle.

At the end of Sec. 3.3 we mentioned the fact that it makes a major difference in the asymptotic behavior of the survival probability of $A$ whether it moves diffusively or subdiffusively. In addition, it makes an essential difference whether either of the species move at all or do not. The collection of results is thus somewhat confusing and perhaps best captured by writing the asymptotic survival probability in the form

$$Q(t) \sim \begin{cases} \exp(-\theta t^z), & d \neq 2, \\ \exp(-\theta t^z / \ln t), & d = 2, \end{cases} \tag{30}$$

where the constants $\theta$ and $z$ depend on the nature of the motion of the particles of each species. Until quite recently, results (some exact, some approximate) were only available for situations where mobility necessarily implied diffusion. The entry of subdiffusion into the discussion is a rather modern event [53–56].

First, we differentiate the cases of mobile and immobile $B$ particles (traps). When the traps are immobile, the survival probability of a diffusive particle $A$ is given by the stretched exponential form [57]

$$Q(t) \sim \exp\left(-k_d \rho^{2/(d+2)} (D_A t)^{d/(d+2)}\right), \tag{31}$$

where $\rho$ is the mean density of traps, $D_A$ is the particle diffusion coefficient, and $k_d$ is the $d$-dependent constant

$$k_d = \left(\frac{d+2}{2d}\right) (d v_d)^{2/(d+2)} \left(2\nu_d^2\right)^{d/(d+2)}. \tag{32}$$

Here $\nu_d$ is the first zero of the Bessel function $J_{(d-2)/2}(z)$ and $v_d = 2\pi^{d/2}/[d\,\Gamma(d/2)]$ denotes the volume of a $d$-dimensional sphere of unit radius. Note that there is no logarithmic correction for $d = 2$ in (31). The constants $\theta$ and $z$ can immediately be read from these equations; specifically, $z = d/(d+2)$.

On the other hand, when all the reactants move diffusively, we immediately enter more complex territory. Standard chemical kinetics based on a reaction-diffusion equation leads to [58]

$$Q(t) \sim \begin{cases} \exp[-\rho(4Dt/\pi)^{1/2}], & d = 1, \\ \exp[-4\pi\rho Dt/\ln(Dt/(a+b)^2)], & d = 2, \\ \exp(-d(d-2)v_d(a+b)^{d-2}\rho Dt), & d \geq 3, \end{cases} \tag{33}$$

where $D$ is the sum of the diffusion coefficients of the particle and the traps, and $a$ and $b$ are the radii of the $A$ and $B$ particles, respectively. This is a two-body approximation to a many-body problem and is in general not exact. The result is in fact exact only in the limit of an immobile $A$ particle. Note that these decays are entirely different from the stretched exponential of Eq. (31).

It was only many years after there results were established that Bramson and Lebowitz [59] were able to show that the time dependences ($z = 1/2$ when $d = 1$, $z = 1$ for $d \geq 2$) in Eq. (33) are in fact asymptotically correct even when $A$ does move, but with different constant prefactors than given above. They were not able to determine these prefactors. Here the field stood for many years, until Bray and Blythe [60, 61] managed to determine the prefactors for $d = 1$ and 2. Their methodology consisted in constructing upper and lower bounds to the survival probability that meet asymptotically. These bounds could be calculated exactly because they only involve boundary value problems. The great insight of their solution was this: the prefactors were found to be independent of the diffusion coefficient $D_A$ of the particle! This immediately points to a startling conclusion: the asymptotic survival probability of the diffusing $A$ is the same as it would be if the $A$ remained immobile. In other words, the many-body problem is reduced to the boundary value problem discussed in Sec. 3.2, with the asymptotic results of Eq. (24) with $\gamma = 1$.

Of interest here is what happens when the particle and/or the traps move subdiffusively. If both move subdiffusively, then a fairly straightforward extension of the diffusive approach is possible and one finds once again that the survival probability of the subdiffusive $A$ is the same as that of an immobile $A$ particle in a sea of subdifussive traps [52], that is, as given in (24). We immediately see that the exponent $z$ can be read off of those earlier results: $z = \gamma/2$ in one dimension and $z = \gamma$ in two and three dimensions and above. The prefactors $\theta$ can also be read off of the earlier results. In summary so far, we see that if $A$ is subdiffusive, no matter how the $B$s move, the asymptotic survival probability of $A$ is the same as it would be if $A$ stood still.

The case that presents the most serious difficulties occurs when the traps move subdiffusively and the $A$ particle is diffusive. To solve this problem, the upper and lower bound approach can again be invoked, but it turns out that additional bounds to those used in the earlier Bray–Blythe construction are needed because these earlier bounds do not always meet.

A lower bound is constructed as originally devised for diffusive particles and traps and later extended to the case of (sub)diffusive particles and traps. This bound involves only boundary value problems. An upper bound is also constructed as originally devised for the diffusive case, based on the (proven) assertion that the $A$ particle lives at least as long as it does in the original problem if the $A$ stands still. This bound has been called the "Pascal principle" [56, 62]. However, this upper bound is not always the tightest bound, and another upper bound was devised on the (proven) basis that the worst possible strategy for traps in their search for a target is to remain immobile. The associated survival probability is again the solution to a boundary value problem and is precisely as given in Eq. (31). This bound has been called the "anti-Pascal principle" [56]. The interesting point is that depending on the value of the subdiffusive exponent $\gamma$ of the traps, one or the other of these bounds is tighter. In other words, depending on how fast the traps move in the face of a diffusive $A$, the better upper bound is provided by the problem where the $A$ stands still or by the problem where the $B$s stand still. Fortunately, in most cases (albeit not in all, see below) the tighter upper bound asymptotically approaches the lower bound, thus providing a solution to our problem. The outcome depends on $\gamma$ and also on dimensionality.

Table 1 collects the results for the exponents $z$ in Eq. (30) for the survival probability. A number of points are of special note. First, *when the particle is diffusive* ($\mu = 1$), there is a singular change in behavior of the exponent $z$ as the exponent $\gamma$ of the traps sweeps the range from immobile to diffusive. When the traps move very slowly, the survival probability of the particle is the same as it would be if the traps simply stood still,

Table 1. Collected results for the asymptotic survival probability exponent $z$.

| Dimension | Trap exponent | Particle exponent | Survival probability exponent $z$ |
|---|---|---|---|
| $d = 1$ | $0 \le \gamma \le 2/3$ | $\mu = 1$ | $1/3$ |
| | $2/3 \le \gamma \le 1$ | $\mu = 1$ | $\gamma/2$ |
| | $0 < \gamma \le 1$ | $0 \le \mu < 1$ | $\gamma/2$ |
| $d = 2$ | $0 \le \gamma \le 1/2$ | $\mu = 1$ | $1/2$ |
| | $1/2 < \gamma \le 1$ | $\mu = 1$ | $\gamma$ |
| | $0 < \gamma \le 1$ | $0 \le \mu < 1$ | $\gamma$ |
| $d = 3$ | $0 \le \gamma \le 2/5$ | $\mu = 1$ | $3/5$ |
| | $2/5 \le \gamma \le 3/5$ | $\mu = 1$ | $3/5 \le z \le 1/3 + 2\gamma/3$ |
| | $3/5 \le \gamma \le 1$ | $\mu = 1$ | $\gamma \le z \le 1/3 + 2\gamma/3$ |
| | $0 < \gamma \le 1$ | $0 \le \mu < 1$ | $\gamma$ |

that is, $z = d/(d+2)$ as in the Donsker–Varadhan formula (31). If, on the other hand, the traps move sufficiently quickly, the survival probability in $d = 1$ and $d = 2$ is that of the Bramson–Lebowitz–Bray–Blythe problem associated with an immobile particle, that is, $z = d\gamma/2$. In three or more dimensions we also see the Donsker–Varadhan behavior if the traps move sufficiently slowly, but we are not able to determine the exponent $z$ if the traps move too quickly. In this case we can only determine that there is, again, a singular change in behavior and we can also determine a range for the value of the exponent, but the precise determination is still an open problem. Second, *when the particle is subdiffusive* ($0 \leq \mu < 1$), then $z = \gamma/2$ for $d = 1$ and $z = \gamma$ for $d \geq 2$. These results are all given in the table. The values of the prefactor $\theta$ are also available for most of these cases [52, 56] but have not been collected here.

### 4.2. *Fractional diffusion-reaction equations*

When normal diffusive particles react, the standard reaction-diffusion equation that describes the spacetime evolution of their concentrations $c(\mathbf{r}, t)$ is given by the normal diffusion equation plus a reaction term, say $F(c)$, that takes into account the rate of change of $c(\mathbf{r}, t)$ due to reactions,

$$\frac{\partial}{\partial t} c(\mathbf{r}, t) = D\nabla^2 c(\mathbf{r}, t) + F(c). \tag{34}$$

However, when particles that diffuse anomalously react, the deduction of the corresponding subdiffusion-reaction equation for $c(\mathbf{r}, t)$ is more elusive [10, 16, 63–69]. An extensive and recent discussion of this topic can be found in [70]. Here we focus on the deduction of the subdiffusion-reaction equation starting from the CTRW model for the case in which the *loss* of particles at location $\mathbf{r}$ due exclusively to reactions is given by

$$\left. \frac{\partial}{\partial t} c(\mathbf{r}, t) \right|_{\text{reaction}} = -k(\mathbf{r}, t) c(\mathbf{r}, t), \tag{35}$$

where $k(\mathbf{r}, t)$ is the reaction rate coefficient (also hereafter called the reactivity). Integrating this equation leads to $c(\mathbf{r}, t')/c(\mathbf{r}, t) = A(\mathbf{r}, t, t')$, with

$$A(\mathbf{r}, t, t') = \exp\left( - \int_{t'}^{t} k(\mathbf{r}, t'') dt'' \right), \tag{36}$$

which describes the time evolution of the ratio of the concentration of the particles at $\mathbf{r}$ at time $t$ and the concentration at $\mathbf{r}$ at the initial time $t'$ when the number of particles at $\mathbf{r}$ changes due *exclusively* to reactions, that is, when the changes in concentration at $\mathbf{r}$ caused by particle jumps into or out of a small volume element centered at $\mathbf{r}$ are not considered.

Next we introduce $j(\mathbf{r},t)$ and $i(\mathbf{r},t)$, the incoming and outgoing fluxes of particles at location $\mathbf{r}$ at time $t$, which are related by the equation

$$j(\mathbf{r},t) = \int i(\mathbf{r}-\mathbf{r}',t)w(\mathbf{r}')d\mathbf{r}', \tag{37}$$

where $w(\mathbf{r})$ is the probability distribution introduced just before Eq. (1).

The above equation simply states that the incoming flux at $\mathbf{r}$ at time $t$ arises from outgoing fluxes at all other locations $\mathbf{r}-\mathbf{r}'$ at that time. Including all of the contributions to the change in the concentration then yields

$$\frac{\partial}{\partial t}c(\mathbf{r},t) = j(\mathbf{r},t) - i(\mathbf{r},t) - k(\mathbf{r},t)c(\mathbf{r},t) \tag{38}$$

because the changes in the concentration at $\mathbf{r}$ are due to the incoming and outgoing fluxes and to the reaction process at that location. An additional relation connecting the fluxes and concentration is

$$i(\mathbf{r},t) = \psi(t)A(\mathbf{r},t,0)c(\mathbf{r},0) + \int_0^t \psi(t-\tau)A(\mathbf{r},t,\tau)j(\mathbf{r},\tau)d\tau, \tag{39}$$

which states that the outgoing flux from $\mathbf{r}$ at time $t$ arises from two sources. One is the contribution of the particles that started out at $\mathbf{r}$ at time $t=0$, did not react or move anywhere up to time $t$, and then took a step away from $\mathbf{r}$ at time $t$. The other is from those particles that arrived at $\mathbf{r}$ at some earlier time $\tau$, waited there up to time $t$ without degradation, and then stepped away.

Defining $c^*(\mathbf{r},t) \equiv c(\mathbf{r},t)A(\mathbf{r},0,t)$, we can write Eq. (38) as

$$A(\mathbf{r},t,0)\frac{\partial}{\partial t}c^*(\mathbf{r},t) = j(\mathbf{r},t) - i(\mathbf{r},t). \tag{40}$$

Equivalently, from Eq. (37),

$$A(\mathbf{r},t,0)\frac{\partial}{\partial t}c^*(\mathbf{r},t) = \mathcal{F}^{-1}\left\{[\hat{w}(\mathbf{q})-1]\,\hat{i}(\mathbf{q},t)\right\} \tag{41}$$

or, equivalently, from Eq. (5),

$$A(\mathbf{r},t,0)\frac{\partial}{\partial t}c^*(\mathbf{r},t) = \sigma^\mu\nabla^\mu i(\mathbf{r},t). \tag{42}$$

Upon Laplace transforming Eq. (39) with respect to time one gets

$$\mathcal{L}\left[A(\mathbf{r},0,t)i(\mathbf{r},t)\right] = \tilde{\psi}(u)c^*(\mathbf{r},0) + \tilde{\psi}(u)\mathcal{L}\left[A(\mathbf{r},0,t)j(\mathbf{r},t)\right]. \tag{43}$$

Taking into account that, from Eq. (38),

$$\mathcal{L}\left[A(\mathbf{r},0,t)j(\mathbf{r},t)\right] = \mathcal{L}\left[\frac{dc^*}{dt}\right] + \mathcal{L}\left[A(\mathbf{r},0,t)i(\mathbf{r},t)\right], \tag{44}$$

Eq. (43) can be rewritten as

$$i(\mathbf{r},t) = A(\mathbf{r},t,0)\mathcal{L}^{-1}\left\{\frac{u\tilde{\psi}(u)}{1-\tilde{\psi}(u)}\tilde{c}^*(\mathbf{r},u)\right\}. \tag{45}$$

Using the Laplace transform expression (4) of the fractional derivative operator along with the explicit form Eq. (2), this expression can be rewritten as

$$i(\mathbf{r},t) = A(\mathbf{r},t,0)\tau^{-\gamma}\, {}_0\mathcal{D}_t^{1-\gamma} c^*(\mathbf{r},t). \tag{46}$$

Finally, inserting Eq. (46) into Eq. (42) and expanding the abbreviated notation, we arrive at the general reaction-subdiffusion equation that is the starting point of our analysis in the next two sections:

$$\frac{\partial}{\partial t}c(\mathbf{r},t) = D_\gamma \nabla^\mu \left\{e^{-\int_0^t k(\mathbf{r},t')dt'}\, {}_0\mathcal{D}_t^{1-\gamma}\left[e^{\int_0^t k(\mathbf{r},t')dt'} c(\mathbf{r},t)\right]\right\} - k(\mathbf{r},t)c(\mathbf{r},t). \tag{47}$$

Note that in the absence of reactions, $k(\mathbf{r},t) = 0$, Eq. (3) is recovered. More specifically, we focus on the case $\mu = 2$ so that the spatial operator is the Laplacian, as mentioned earlier, and we also do not consider a time-dependent reactivity here. Our starting equation therefore is

$$\frac{\partial}{\partial t}c(\mathbf{r},t) = D_\gamma \nabla^2 \left\{e^{-k(\mathbf{r})t}\, {}_0\mathcal{D}_t^{1-\gamma}\left[e^{k(\mathbf{r})t} c(\mathbf{r},t)\right]\right\} - k(\mathbf{r})c(\mathbf{r},t). \tag{48}$$

In order to solve (48), it is convenient to work with a new function $v(\mathbf{r},t)$:

$$\tilde{v}(\mathbf{r},u) = [u + k(\mathbf{r})]^{1-\gamma}\tilde{c}(\mathbf{r},u). \tag{49}$$

Taking into account that

$$\mathcal{L}\left[e^{-k(\mathbf{r})t}\, {}_0\mathcal{D}_t^{1-\gamma}\left(e^{k(\mathbf{r})t}c(\mathbf{r},t)\right)\right] = [u + k(\mathbf{r})]^{1-\gamma}\tilde{c}(\mathbf{r},u), \tag{50}$$

Eq. (48) becomes

$$u\tilde{c}(\mathbf{r},u) - c(\mathbf{r},0) = D_\gamma \nabla^2 \left\{[u + k(\mathbf{r})]^{1-\gamma}\tilde{c}(\mathbf{r},u)\right\} - k(\mathbf{r})\tilde{c}(\mathbf{r},u) \tag{51}$$

or, equivalently,

$$[u + k(\mathbf{r})]^\gamma \tilde{v}(\mathbf{r}, u) - c(\mathbf{r}, 0) = D_\gamma \nabla^2 \tilde{v}(\mathbf{r}, u). \quad (52)$$

This is then a variant of our starting equation (48).

One final remark is in order. It should be noted that Eq. (35) does not necessarily imply a first-order reaction, since the reaction rate $k(\mathbf{r}, t)$ is quite arbitrary and might itself depend on $c(\mathbf{r}, t)$ or on the concentration of some other species. For example, suppose that we have two species with concentrations $c_1(\mathbf{r}, t)$ and $c_2(\mathbf{r}, t)$ that disappear due to bimolecular reactions at the rate $\dot{c}_1 = \dot{c}_2 = -k_0 c_1 c_2$. The fractional diffusion equation for the species $i$ is then Eq. (47) with a reaction rate given by $k(\mathbf{r}, t) = k_0 c_j(\mathbf{r}, t)$.

### 4.3. *Single-particle target problem with a reactivity field*

The problem we consider here is that of Sec. 3.1, i.e. the single-particle target problem, but now when the reaction between the target and the particle does not necessarily take place only at one fixed distance $R$ between them. Specifically, we now assume that the probability that a reaction can take place at the distance $r$ is determined by an isotropic and time-independent reactivity $k(r)$, that is, $k(\mathbf{r}, t) = k(r)$. In this case, with $c(\mathbf{r}, 0) = \delta(\mathbf{r} - \mathbf{r}_0)$, the equation for $\mathbf{r} \neq \mathbf{r}_0$ is

$$[u + k(\mathbf{r})]^\gamma \tilde{v}(\mathbf{r}, u) = D_\gamma \nabla^2 \tilde{v}(\mathbf{r}, u). \quad (53)$$

This problem has been studied by Seki *et al.* in [63] in their analysis of the kinetics of geminate reactions in three-dimensional systems. They considered three different spatial dependences for $k(r)$ (square box or step function, exponential, and inverse power), paying special attention to the escape probability $Q_1(r_0, t \to \infty)$. For the Heaviside step function reactivity $k(r) = k_0 \Theta(R - r)$, the reaction occurs with rate coefficient $k_0$ at any distance between particle and target centers that is smaller than $R$. This problem can be solved exactly. For $r_0 < R$, the survival probability is

$$Q_1(r, u) = \frac{1}{u + k_0} \left[ 1 + \frac{k_0}{u} \frac{(1 + Rq)\sinh(r_0 p)}{r_0 p \cosh(Rp) + r_0 q \sinh(Rp)} \right], \quad (54)$$

with $q = \sqrt{u^\gamma / D_\gamma}$ and $p = \sqrt{(u + k_0)^\gamma / D_\gamma}$. No exact solutions are known for the exponential and inverse power reactivity cases. For these unsolved cases one can try to apply the approximate method, proposed by Seki *et al.* [63], in which the term $[u + k(\mathbf{r})]^\gamma$ in Eq. (53) is replaced by $u^\gamma + k^\gamma(x)$.

Note that the equation resulting from this replacement can also be obtained from the corresponding normal diffusion equation (the equation with $\gamma = 1$) by the replacements $D \to u^{1-\gamma} D_\gamma$ and $k(\mathbf{r}) \to u^{1-\gamma} k(\mathbf{r})$, so that one can then arrive at the approximate solution from the solution of the normal diffusion problem. Since the approximation $[u + k(\mathbf{r})]^\gamma \to u^\gamma + k(\mathbf{r})^\gamma$ improves as $u \gg k(\mathbf{r})$ and $u \ll k(\mathbf{r})$, it is in fact quite accurate for short and long times [63].

### 4.4. *Reaction-subdiffusion equations and morphogen gradient formation*

In Sec. 4.3, Eq. (48) was taken as a starting point to study the escape problem of a single subdiffusive particle subject to a degradation process. Beyond this specific application, there are many other classical reaction-transport problems of interest whose solution can be extended to the anomalous diffusion case via Eq. (48). In certain cases the interplay between the reaction kinetics and memory effects associated with anomalous diffusion may drastically change the phenomenology of the system at hand. One such case concerns a problem of utmost importance in developmental biology, namely, the formation of morphogen gradients leading to patterning in embryonic tissues [72].

The location, differentiation and fate of many embryonic cells is governed by the spatial distribution of special signaling molecules called morphogens. These molecules are secreted at localized sites of the embryo at a fairly constant rate. The secreted morphogens then undergo degradation as they disseminate through the tissue, and a concentration gradient forms as a result of the balance between the secretion and degradation processes. Different target genes in embryonic cells are activated when the morphogen gradient exceeds appropriate concentration thresholds, implying that the cell response to the local environment depends on the local concentration. As a result of this differential response, cells are able to interpret the morphogen gradient and translate it into specific "code" for their further development via the expression of relevant genes.

Traditional models of morphogen gradient formation are based on classical diffusion equations with an added linear degradation term. The typical decay profile of the gradient is modulated by both the value of the diffusion constant and the degradation rate [72]. Recently there has been a recognition that subdiffusive rather than diffusive transport might more appropriately describe temporary morphogen trapping upon binding to proteins. Such trapping events may be mimicked in the framework of a CTRW model by considering the morphogens as random walkers subject to a long-tailed waiting time distribution [73]. Degradation may occur either as a death process caused by enzymes inside the cell, or as irreversible escape from the surface of the cell.

Neglecting finite size effects, the concentration profile for the morphogen problem may be obtained by solving Eq. (48) in a one-dimensional infinite domain (a one-dimensional description is typically appropriate because the secretion geometry renders the problem effectively one-dimensional) subject to a radiation-type boundary condition (constant morphogen influx $j_0$ at the origin $x = 0$) [7]. Proceeding as in Sec. 4.3, we arrive at a particular case of Eq. (52), namely,

$$[u + k(x)]^\gamma \tilde{v}(x,u) - \delta(x) = D_\gamma \nabla^2 \tilde{v}(x,u). \tag{55}$$

4.4.1. *Constant reactivity*

Let us first address the case of homogeneous reactivity, i.e., $k(x) = k$. The solution in Laplace space is

$$\tilde{c}(x,u) = \frac{j_0}{2} \frac{(u+k)^{\gamma/2-1}}{u\sqrt{D_\gamma}} \exp\left[-|x|(u+k)^{\gamma/2}/\sqrt{D_\gamma}\right]. \tag{56}$$

The stationary solution is obtained from the final value theorem for the Laplace transform:

$$c_s(x) = \lim_{u \to 0} u\tilde{c}(x,u) = \frac{j_0}{2} \frac{k^{\gamma/2-1}}{\sqrt{D_\gamma}} \exp\left[-|x| k^{\gamma/2}/\sqrt{D_\gamma}\right]. \tag{57}$$

4.4.2. *Piecewise constant reactivity*

Here we assume that the reactivity is given by a superposition of Heaviside functions, i.e., $k(x) = k_0 H(R - |x|) + k_1 H(|x| - R)$. This profile provides a simple way to model a change in the degradation mechanism across an interface or beyond a certain distance from the source. In region 0 ($0 \le |x| < R$) one has $k(x) = k_0 > 0$, whereas in region 1 ($|x| \ge R$) one has $k(x) = k_1 \ge 0$. Let us respectively denote by $\tilde{v}_0(x,u)$ and $\tilde{v}_1(x,u)$ the solutions of Eq. (55) in the regions 0 and 1. These functions and their spatial derivatives must fulfill the continuity conditions

$$\tilde{v}_0(R,u) = \tilde{v}_1(R,u) \tag{58}$$

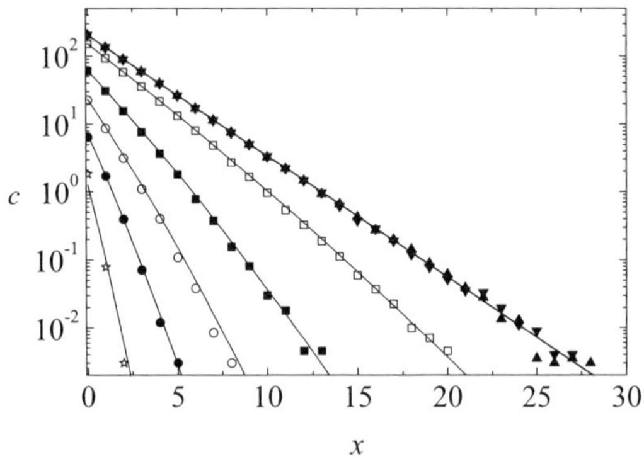


Fig. 1. Simulation results over 1000 realizations (symbols) for $c(x,t)$ with $k = 1/1000$, $\gamma = 1/2$, and $j_0 = 1$ for (from bottom to top) $t = 1, 10, 50, 200, 10^3, 10^4, 2 \times 10^4$. The particles are simulated by means of the CTRW model where the waiting time distribution is a Pareto law, $\psi(t) = \gamma/(1+t)^{1+\gamma}$, and the jumps $\{-1, 0, 1\}$ are equiprobable. These parameters lead to the $D_\gamma$-value $1/\sqrt{9\pi}$. The thin lines are the profiles obtained from the inverse numerical transformation of Eq. (56). The thick line is the theoretical prediction for the steady-state profile, Eq. (56). There are no adjustable parameters.

and

$$\left.\frac{\partial \tilde{v}_0(x,u)}{\partial x}\right|_{|x|=R} = \left.\frac{\partial \tilde{v}_1(x,u)}{\partial x}\right|_{|x|=R}. \tag{59}$$

In contrast, an integration of Eq. (55) across the origin shows that the solution must be discontinuous there,

$$\left.\frac{\partial \tilde{v}_0(x,u)}{\partial x}\right|_{x=0+} - \left.\frac{\partial \tilde{v}_0(x,u)}{\partial x}\right|_{x=0-} = -\frac{1}{D_\gamma}. \tag{60}$$

Using Eqs. (58)–(60) one gets for $|x| < R$ the stationary biexponential solution

$$c_s(x) = j_0\, k_0^{\gamma-1} \tilde{v}_0(x, u=0) \tag{61}$$

with

$$\tilde{v}_0(x,u) = A_0(u)e^{-\alpha_0 x} + B_0(u)e^{\alpha_0 x}, \quad \alpha_0^2 = \frac{(u+k_0)^\gamma}{D_\gamma}. \tag{62}$$

The characteristic coefficients $A_0(u)$ and $B_0(u)$ are determined from the conditions imposed on $\tilde{v}(x,u)$ at $|x| = 0$ and $|x| = R$. For $|x| > R$, we shall distinguish two subcases with different physical behavior. For $k_1 > 0$, one asymptotically gets the exponential decay law

$$c_s(x) = c(x, t \to \infty) \propto e^{-k_1^{\gamma/2}(x-R)/\sqrt{D_\gamma}}. \tag{63}$$

The behavior is radically different for $k_1 = 0$. For normal diffusion the profile becomes constant for large $|x|$, i.e.,

$$c_s(x) \propto j_0, \quad k_1 = 0, \gamma = 1, |x| \geq R. \tag{64}$$

However, when the diffusion is anomalous, degradation is globally too slow and one has

$$c(x, t \to \infty) \propto j_0\, t^{1-\gamma}, \quad k_1 = 0, \gamma < 1, |x| \geq R, \tag{65}$$

i.e., the profile at $x = |R|$ becomes discontinuous and there is no steady state. This behavior is confirmed by numerical CTRW simulations (see Fig. 2), indicating that secreted morphogens which escape region 0 are not as efficiently dispersed back into the reactive region as they are in the case of normal diffusion, and they thus accumulate in region 1.

### 4.4.3. *Exponentially decaying reactivity*

Here, we assume the decay law $k(x) = k_0\, e^{-\beta|x|}$ ($\beta > 0$). While in this case Eq. (55) does not appear exactly solvable for finite $u$, in the limit $u \to 0$ it is

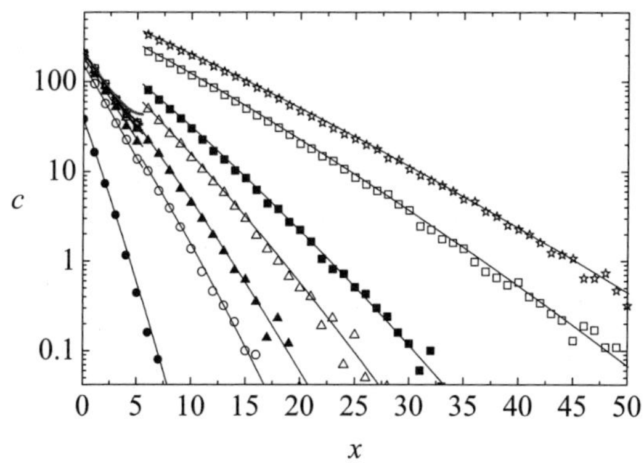


Fig. 2. Simulation results (symbols) of $c(x,t)$ for a step reactivity $[k(x) = k_0 H(R-x)]$ with $k_0 = 1/1000$ and $R = 5.5$ for $\gamma = 1/2$ and (from bottom to top) $t = 100, 10^3, 2\times 10^3, 5\times 10^3, 10^4, 5\times 10^4, 10^5$ (only values for $x > 0$ are shown). The particles are simulated as in Fig. 1. The solid line corresponds to the theoretical prediction for the steady-state profile when $x < R$. For $x > R$ no stationary profile develops. The convergence of the simulation results to the stationary profile for $x < R$ is very slow for values of $x$ close to the discontinuity at $x = R$. No adjustable parameters were used.

possible to find an exact expression for the steady state profile by techniques similar to those used above. The final result is

$$c_s(x) = j_0 \frac{k_0^{\gamma/2-1}}{2D_\gamma^{1/2}} \frac{I_0\left(\alpha k_0^{\gamma/2} e^{-\beta\gamma|x|/2}\right)}{I_1\left(\alpha k_0^{\gamma/2}\right)} e^{-(\gamma-1)\beta|x|}, \tag{66}$$

where the $I_n$'s are modified Bessel functions and $\alpha = 2/(\beta\gamma\sqrt{D_\gamma})$. As in the previous case of piecewise reactivity with $k_1 = 0$, the right-hand side of Eq. (66) displays different qualitative behaviors for normal and anomalous diffusion. In the normal diffusion case ($\gamma = 1$) one gets a monotonically decreasing profile from the concentration value at the origin to the limiting value $c_s(x \to \pm\infty) = j_0/\left[\sqrt{4k_0 D_\gamma} I_1\left(\alpha k_0^{1/2}\right)\right]$. Instead, for $\gamma < 1$ the concentration first decreases as one moves away from the source until it reaches a minimum and then it increases as $\exp[(1-\gamma)\beta\,|x|$, a vestige of the divergence found for piecewise constant reactivity with $k_1 = 0$. This result is again confirmed by CTRW simulations (cf. Fig. 3). The nonmonotonicity of the profile may have interesting biological consequences: the combined action of subdiffusive transport and a rapid spatial decrease of the reactivity may induce complex tissue patterning where spatially distant cells follow a similar developmental pathway.

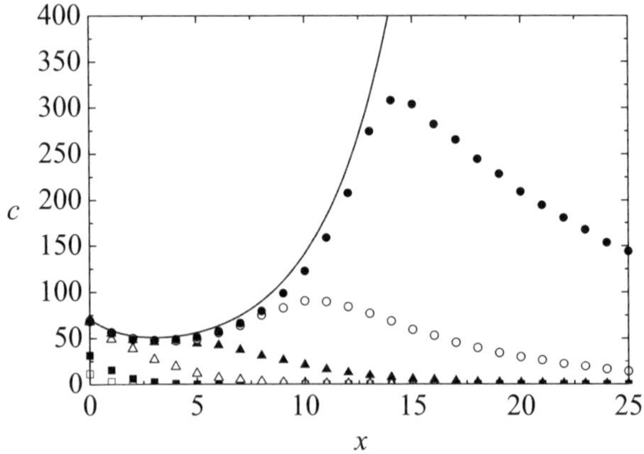


Fig. 3. Convergence of CTRW simulation results for (from bottom to top) $t = 20, 100, 10^3, 10^4, 10^5, 10^6$ (symbols) to the stationary profile predicted by the formula (66) for $j_0 = 1$, $\gamma = 0.5$, the corresponding value $D_\gamma = 1/\sqrt{9\pi}$ and exponentially decaying reactivity $k(x) = k_0 \exp(-\beta|x|)$ with $k_0 = 1/200$ and $\beta = 0.6$ (solid line). The particles are simulated as in Fig. 1. The simulation results clearly tend to the stationary solution as time increases, although the convergence is slow for large $x$.

## 5. Final Remarks

Anomalous diffusion such as the subdiffusive processes that we have discussed in this chapter is found everywhere in nature. It is perhaps not surprising that anomalous diffusion in the form of subdiffusion is actually a normal state of affairs in biological environments [4]. Although a great deal of effort has been invested in understanding how reactant and product concentrations evolve in space and time in subdiffusive media [70, 71, 74], the subject is still in its infancy. One of the great difficulties is that a broad macroscopic phenomenology such as the one we are used to in the case of normal diffusion (in the form of reaction-diffusion equations) is simply not applicable in the presence of subdiffusion. This seems to signify that every mesoscopic model needs to be dealt with separately, and that the associated macroscopic reaction-subdiffusion models are neither particularly intuitive nor broadly applicable to many mesoscopic models.

In this chapter we have chosen a particular mesoscopic starting point, namely, a continuous time random walk, and have derived and solved the associated fractional reaction-subdiffusion equations. These equations have been obtained for a few very specific models and cannot be translated to even other CTRW models without going back to the proverbial drawing board. Parenthetically, we might note that it is not yet known how to include reactions in other mesoscopic subdiffusion models, e.g., the fractional Brownian motion model.

More specifically, in this chapter we have constructed and solved a number of reaction-subdiffusion equations using the tools of fractional calculus. We have separated our discussion into two parts. In one, we have studied reactions that take place at spatially fixed locations such as the surface of an immobile particle (trap) or the hypersurface of a volume that encloses the particle whose fate we are following. In these problems the equation to be dealt with is a pure fractional subdiffusion equation, and the reactions are entirely built into boundary conditions. In the second part we have dealt with reactions that take place at random locations. Now in general the reaction has to be built into the fractional equation. We have applied the resultant equation to two examples for which we have obtained explicit solutions. In this section we have also studied a third problem, the classic $A+B \to B$ problem where the particle $A$ and the traps $B$ are all mobile and thus reactions occur at random locations. For this problem we have not constructed a reaction-subdiffusion equation. Instead, we have discussed a way to obtain asymptotic results using bounds that involve only pure fractional subdiffusion equations plus boundary conditions. These three examples, in addition to their inherent interest, serve to make the point that in the world of subdiffusion every single scenario seems to require an analysis that is suitable only for that particular scenario. Every reaction in a subdiffusive medium requires its own full analysis starting from a mesoscopic description. Whether it will always be this way, or whether some future insight will help pull together many mesoscopic cases into a macroscopic description as broad and useful as a reaction-diffusion equation in the case of normal diffusion, remains to be seen.

## Acknowledgments

K. L. gratefully acknowledges the NSF under Grant No. PHY-0855471. S. B. Y. and E. A. gratefully acknowledge the Ministerio de Educación y Ciencia under Grant No. FIS2010-16587 (partially financed with FEDER funds) and the Junta de Extremadura under Grant No. GRU10158.

## References

1. S. A. Rice, *Diffusion-limited Reactions* (Elsevier, 1985).
2. E. Kotomin and V. Kuzovkov, *Modern Aspects of Diffusion-Controlled Reactions* (Elsevier, 1996).
3. Eds. R. Klages, G. Radons and I. M. Sokolov, *Anomalous Transport* (Wiley-VCH, 2008).
4. J. A. Dix and A. S. Verkman, *Ann. Rev. Biophys.* **37**, 247 (2008).
5. E. Barkai, *Phys. Rev. E* **63**, 046118 (2001).
6. R. Metzler and J. Klafter, *Phys. Rep.* **339**, 1 (2000).
7. S. B. Yuste, E. Abad and K. Lindenberg, *Phys. Rev. E* **82**, 061123 (2010).
8. I. Podlubny, *Fractional Differential Equations* (Academic Press, 1999).
9. A. A. Kilbas H. M. Srivastava and J. J. Trujillo, *Theory and Applications of Fractional Differential Equations* (Elsevier, 2006).
10. K. Seki, M. Wojcik and M. Tachiya, *J. Chem. Phys.* **119**, 2165 (2003).
11. J. D. Eaves and D. R. Reichman, *J. Phys. Chem. B* **112**, 4283 (2008).
12. M. A. Lomholt, I. M. Zaid and R. Metzler, *Phys. Rev. Lett.* **98**, 200603 (2007).
13. R. Metzler and J. Klafter, *J. Phys. A: Math. Gen.* **37**, R161 (2004).
14. O. Bénichou, C. Loverdo, M. Moreau and R. Voituriez, *Phys. Rev. E* **74**, 020102 (2006).
15. J. Rojo, J. Revelli, C. E. Budde, H. S. Wio, G. Oshanin and K. Lindenberg, *J. Phys. A: Math. Theor.* **43**, 345001 (2010).
16. K. Seki, M. Wojcik and M. Tachiya, *J. Chem. Phys.* **119**, 7525 (2003).
17. H. Sano and M. Tachiya, *J. Chem. Phys.* **71**, 1276 (1979).
18. A. V. Barzykin and M. Tachiya, *Phys. Rev. Lett.* **73**, 3479 (2004).

19. S. Condamin, O. Benichou, V. Tejedor, R. Voituriez and J. Klafter, *Nature* **450**, 77 (2007).
20. S. Condamin, O. Benichou and J. Klafter, *Phys. Rev. Lett.* **98**, 250602 (2007).
21. S. B. Yuste and K. Lindenberg, *Phys. Rev. E* **76**, 051114 (2007).
22. A. V. Barzykin and M. Tachiya, *J. Chem. Phys.* **99**, 9591 (1993).
23. D. S. Grebenkov, *J. Chem. Phys.* **132**, 034104 (2010).
24. M. Abramowitz and I. A. Stegun, *Handbook of Mathematical Functions* (Dover, 1965).
25. A. M. Mathai and R. K. Saxena, *The H-function with Applications in Statistics and Other Disciplines* (Wiley, 1978).
26. I. M. Sokolov, *Phys. Rev. E* **63**, 056111 (2001).
27. A. I. Saichev and G. M. Zalavsky, *Chaos* **7**, 753 (1997).
28. E. Barkai and R. Silbey, *J. Phys. Chem. B* **104**, 3866 (2000).
29. G. V. Buxton, F. C. R. Cattell and F. S. Dainton, *J. Chem. Soc., Faraday Trans. 1* **71**, 115 (1975).
30. P. Clifford and N. J. B. Green, in *Contemporary Problems in Statistical Physics*, ed. G. Weiss (SIAM, 1994).
31. M. Tachiya, *Chem. Phys. Lett.* **69**, 605 (1980).
32. R. Kopelman, J. Hoshen, J. S. Newhouse and P. Argyrakis, *J. Stat. Phys.* **30**, 335 (1983).
33. P. Argyrakis, D. Hooper and R. Kopelman, *J. Phys. Chem.* **87**, 1467 (1983).
34. S. T. Gentry and R. Kopelman, *J. Chem. Phys.* **81**, 3014 (1984).
35. P. H. von Hippel and O. G. Berg, *J. Biol. Chem.* **264**, 675 (1989).
36. F. Mainardi, *Chaos Soliton. Fract.* **7**, 1461 (1996).
37. F. Mainardi and R. Gorenflo, *J. Comput. Appl. Math.* **118**, 283 (2000).
38. S. Redner, *A Guide to First-Passage Processes.* (Cambridge Univ. Press, 2001).
39. D. S. Grebenkov, *Phys. Rev. E* **81**, 021128 (2010).
40. S. B. Yuste and L. Acedo, *Phys. Rev. E* **68**, 036134 (2003).
41. L. Acedo and S. B. Yuste, *Phys. Rev. E* **66**, 011110 (2002).
42. S. B. Yuste, L. Acedo and K. Lindenberg, *Phys. Rev. E* **64**, 052102 (2001).
43. S. B. Yuste and L. Acedo, *Phys. Rev. E* **64**, 061107 (2001).
44. S. B. Yuste and L. Acedo, *Physica A* **297**, 321 (2001).
45. N. G. van Kampen, *Stochastic Processes in Physics and Chemistry*, 3rd edn. (Elsevier, 2007).
46. R. Metzler, E. Barkai and J. Klafter, *Phys. Rev. Lett.* **82**, 3563 (1999).
47. J. Crank, *The Mathematics of Diffusion*, 2nd edn. reprinted (Oxford Univ. Press, 2004).
48. H. S. Carslaw and J. C. Jaeger, *Conduction of Heat in Solids*, 2nd Edn. (Oxford Univ. Press, 1959).
49. G. Arfken and H. J. Weber, *Mathematical Methods for Physicists* (Academic Press, 2005).
50. S. B. Yuste, R. Borrego and E. Abad, *Phys. Rev. E* **81**, 021105 (2010).
51. S. B. Yuste and E. Abad, *J. Phys. A: Math. Theor.* **44**, 075203 (2011).
52. R. Borrego, E. Abad and S. B. Yuste, *Phys. Rev. E* **80**, 61121 (2009).

53. J. J. Ruiz-Lorenzo, S. B. Yuste and K. Lindenberg, *J. Phys.: Condens. Matter.* **19**, 065120 (2007).
54. G. Oshanin, O. Bénichou, M. Coppey and M. Moreau, *Phys. Rev. E* **66**, 060101R (2002).
55. S. B. Yuste and K. Lindenberg, *Phys. Rev. E* **72**, 061103 (2005).
56. S. B. Yuste, G. Oshanin, K. Lindenberg, O. Bénichou and J. Klafter, *Phys. Rev. E* **78**, 021105 (2008).
57. D. V. Donsker and S. R. S. Varadhan, *Commun. Pure Appl. Math.* **28**, 525 (1975).
58. A. M. Berezhkovskii, Yu. A. Makhnovskii and R. A. Suris, *Chem. Phys.* **137** 061103 (1989); A. M. Berezhkovskii, V. Zaloj and Yu. A. Makhnovskii, *Philos. Mag. B* **77**, 1277 (1998).
59. M. Bramson and J. L. Lebowitz, *Phys. Rev. Lett.* **61**, 2397, (1988); *J. Stat. Phys.* **62**, 297 (1991).
60. A. J. Bray and R. A. Blythe, *Phys. Rev. Lett.* **89**, 150601 (2002).
61. R. A. Blythe and A. J. Bray, *Phys. Rev. E* **67**, 041101 (2003).
62. M. Moreau, G. Oshanin. O. Bénichou and M. Coppey, *Phys. Rev. E* **67**, 045104R (2003).
63. K. Seki, A. I. Shushin, M. Wojcik and M. Tachiya, *J. Phys.: Condens. Matter.* **19**, 065117 (2007).
64. D. Froemberg and I. M. Sokolov, *Phys. Rev. Lett.* **100**, 108304 (2008).
65. B. I. Henry, T. A. M. Langlands and S. L. Wearne, *Phys. Rev. E* **74**, 031116 (2006).
66. I. M. Sokolov, M. G. W. Schmidt and F. Sagués, *Phys. Rev. E* **73**, 031102 (2006).
67. M. O. Vlad and J. Ross, *Phys. Rev. E* **66**, 061908 (2002).
68. A. Yadav and W. Horsthemke, *Phys. Rev. E* **74**, 066118 (2006).
69. T. A. M. Langlands, B. I. Henry and S. L. Wearne, *Phys. Rev. E* **77**, 021111 (2008).
70. V. Méndez, S. Fedotov and W. Horsthemke, *Reaction-Transport Systems: Mesoscopic Foundations, Fronts, and Spatial Instabilities* (Springer-Verlag, 2010).
71. A. Blumen, J. Klafter and G. Zumofen, in *Optical Spectroscopy of Glasses*, ed. I. Zschokke (Reidel, 1986).
72. M. Ibañes and J. C. Izpisúa Belmonte, *Molec. Syst. Biol.* **4**, 176 (2008).
73. G. Hornung, B. Berkowitz and N. Barkai, *Phys. Rev. E* **72**, 041916 (2005).
74. D. ben-Avraham and S. Havlin, *Diffusion and Reactions in Fractals and Disordered Systems* (Cambridge, Univ. Press 2000).

# Chapter 5

## Natural and Modified Forms of Distributed-Order Fractional Diffusion Equations

Aleksei Chechkin[*,‡], Igor M. Sokolov[†] and Joseph Klafter[‡]
[*] *Akhiezer Institute for Theoretical Physics, Kharkov 61108, Ukraine*
[†] *Institute of Physics, Humboldt University, Berlin, Germany*
[‡] *School of Chemistry, Tel Aviv University, Tel Aviv, Israel*

We consider diffusion-like equations with time and space fractional derivatives of distributed-order for the kinetic description of anomalous diffusion and relaxation phenomena, whose mean squared displacement does not change as a power law in time. Correspondingly, the underlying processes cannot be viewed as self-affine random processes processing a unique Hurst exponent. We show that different forms of distributed-order equations, which we call "natural" and "modified" ones, serve as a useful tool to describe the processes which become more anomalous with time (retarding subdiffusion and accelerated superdiffusion) or less anomalous demonstrating the transition from anomalous to normal diffusion (accelerated subdiffusion and truncated Lévy flights). Fractional diffusion equation with the distributed-order time derivative also accounts for the logarithmic diffusion (strong anomaly).

1. Introduction . . . . . . . . . . . . . . . . . . . . . . . . . 108
2. Two Forms of Time Fractional Diffusion Equations . . . . 110
   2.1. Riemann–Liouville form . . . . . . . . . . . . . . 110
   2.2. Caputo form . . . . . . . . . . . . . . . . . . . . 111
3. Two Forms of Space Fractional Diffusion Equations . . . 111
   3.1. Natural form . . . . . . . . . . . . . . . . . . . . 111
   3.2. Modified form . . . . . . . . . . . . . . . . . . . 112
4. Natural Form of Distributed-Order Time Fractional Diffusion Equation . . . . . . . . . . . . . . . . . . . . . . . . 112
   4.1. Properties of the solution . . . . . . . . . . . . . 113
   4.2. Generic case of double order equation: Decelerating subdiffusion . . . . . . . . . . . . . . . . . . . . . . . 113
   4.3. Relation to CTRW . . . . . . . . . . . . . . . . . . 114
   4.4. Superslow diffusion . . . . . . . . . . . . . . . . . 115

4.4.1. Probability density function and mean square displacement . . . . . . . . . . . . 115
4.4.2. Distributed-order fractional Fokker–Planck equation for superslow processes . 117
4.4.3. Relation to CTRW . . . . . . . . . . . . . 118
5. Modified Form of Distributed-Order Time Fractional Diffusion Equation . . . . . . . . . . . . . . . . . . . . . . 119
5.1. Thermodynamical interpretation . . . . . . . . . . 120
5.2. Generic case of double order equation: Accelerating subdiffusion . . . . . . . . . . . . . . . . . . . . . . 120
6. Natural Form of Distributed-Order Space Fractional Diffusion Equation . . . . . . . . . . . . . . . . . . . . . . 122
6.1. Generic case of double-order equation: Accelerating superdiffusion . . . . . . . . . . . . . . . . . . . . . 123
7. Modified Form of Distributed-Order Space Fractional Diffusion Equation . . . . . . . . . . . . . . . . . . . . . . 123
7.1. Generic case of double-order equation: Retarding superdiffusion . . . . . . . . . . . . . . . . . . . . . 124
7.2. Fractional diffusion equation for a power law truncated Lévy process . . . . . . . . . . . . . . . . 124
8. Conclusions and Outlook . . . . . . . . . . . . . . . . . . . 125

## 1. Introduction

"Normal" diffusion processes are described by the Fick's diffusion equation, a parabolic partial differential equation for the probability density function (PDF) $f(x,t)$ of finding a particle at site $x$ at time $t$,

$$\frac{\partial f}{\partial t} = K\frac{\partial^2 f}{\partial x^2}. \tag{1}$$

This equation was first derived by Fick in 1855, and the underlying microscopic picture was understood by Einstein 50 years later. The Fick's equation is invariant under the scale transformation $x \to \lambda x$, $t \to \lambda^2 t$, so that the characteristic scale of its solution (diffusion length) grows like $L \propto t^{1/2}$. In many cases, however, the characteristic displacement scales as $L \propto t^{1/\mu}$ with $\mu \neq 1/2$ or does not scale at all. In these cases one often speaks about anomalous kinetics. The prominent examples are the Lévy flights and the continuous-time random walks with power-law waiting-time distributions [1, 2]. These are described by fractional kinetics, i.e. through diffusion with fractional, instead of the first- or second-order derivative in its temporal or spatial variable, respectively.

On the other hand, many systems demonstrate anomalous *non-scaling* behavior, which corresponds either to crossover between different power laws, or to a non-power-law behavior as exemplified by the logarithmic growth of the distribution width. As illuminating examples of such nonscaling situations we can refer to truncated Lévy flights in the superdiffusive case [3] and to Sinai-like superslow diffusion [4] in the subdiffusive case. In this chapter we show that such a non-scaling behavior can be described by diffusion equations with *distributed-order derivatives.* The distributed-order derivative is a linear operator defined as a weighted sum of different fractional derivatives or as an integral of such over their order, $\int_a^b d\beta p(\beta)(d^\beta/dz^\beta)$, acting on the function of corresponding variable $z$ which in case of distributed-order diffusion equations means time or space variable. For *ordinary differential equations* distributed-order derivatives were introduced by Caputo in [5] for generalizing stress-strain relation of inelastic media, and in [6, 7] to find the eigenfunctions of the torsional models of inelastic or dielectric spherical shells and infinite plates. Independently, distributed-order derivative with a constant weight, based on the Riemann–Liouville fractional derivative, was introduced and studied by Nakhushev [8, 9]. Fractional differential equations containing sums of fractional order derivatives have been extensively studied by Podlubny [10]. The distributed order integration and differentiation within the functional calculus technique was described in [11]. Bagley and Torvik [12, 13] proposed the method of the solution of the ordinary distributed-order equation, which is based on generalized Taylor series representation. Diethelm and Ford [14] introduced a basic framework for the numerical solution based on approximation of the distributed-order equation by a multi-term equation, and further developed their method in detail [15]. Hartley and Lorenzo [16] used distributed-order differential equations to model thermorheological behavior and proposed two different approaches to treat distributed-order fractional operators.

For partial differential equations, the generalization of the Fick's law by the use of a distributed-order time derivative was proposed by Caputo in [17]. Fractional *diffusion equations* with time and space derivatives of distributed order were introduced in [18]. In this chapter we give a brief review of the different forms of distributed-order diffusion equations with time and space fractional derivatives respectively. We show that these equations can be classified depending on the side of the equation where the fractional derivative stays. If it stays at the same side as in case of the "normal" diffusion equation (1) (that is on the left in time-fractional or in the right in space-fractional equation) we call such distributed-order equations the "*natural*" ones. If the distributed-order derivative stays on the "wrong" side (that is on the right side in time fractional equation or in the left side in space-fractional equation) we call such equations "*modified*". We analyze the properties of natural and modified equations and present generic examples elucidating the behavior of the corresponding random diffusion processes. In this review we omit mathematical details for which we address a reader to the original papers; instead we try to show the richness of nonscaling anomalous behaviors described by distributed-order diffusion equations which appeared to be versatile instrument for the description of anomalous diffusion processes.

The plan of the paper is the following. In Secs. 2 and 3, we analyze natural and modified forms of (mono)fractional diffusion equations (time-fractional and space-fractional, respectively) and show their equivalence. Sections 4–7 are devoted to the corresponding distributed-order generalizations. The results are summarized in Sec. 8.

## 2. Two Forms of Time Fractional Diffusion Equations

### 2.1. *Riemann–Liouville form*

The time fractional diffusion equation in the Riemann–Liouville form (RL-form), which in our terminology corresponds to a "modified" form of the fractional diffusion equation, reads [1]:

$$\frac{\partial}{\partial t} f(x,t) = K_\beta D_t^{1-\beta} \frac{\partial^2}{\partial x^2} f(x,t), \tag{2}$$

$f(x, t = 0) = \delta(x)$, $0 < \beta < 1$. Here $K_\beta$ is a positive constant, $[K_\beta] = \mathrm{cm}^2/\mathrm{sec}^\beta$, and $D_t^\mu$ is the Riemann–Liouville derivative on the right semi-axis which, for a "sufficiently well-behaved" function $\phi(t)$ is defined as follows [19]:

$$D_t^\mu \phi = \frac{1}{\Gamma(1-\mu)} \frac{d}{dt} \int_0^t d\tau \frac{\phi(\tau)}{(t-\tau)^\mu}, \quad 0 \le \mu < 1. \tag{3}$$

Applying the Laplace and the Fourier transforms in succession,

$$\mathbf{F}\{\mathbf{L}\{f(x,t\}\} = f(k,s) = \int_{-\infty}^{\infty} dx e^{ikx} \int_0^\infty dt e^{-st} f(x,t),$$

and using the Laplace transform of the Riemann–Liouville derivative, $\mathbf{L}\{D_t^\mu \phi\} = s^\mu \phi(s)$, we get from (2) the form of the Laplace-transformed characteristic function $f(k,s)$ of the PDF as

$$f(k,s) = \frac{s^{\beta-1}}{s^\beta + K_\beta k^2}. \tag{4}$$

### 2.2. *Caputo form*

The time fractional diffusion equation in the Caputo form (C-form, corresponding to a "natural" form of the fractional diffusion equation) is written as follows:

$$\frac{\partial^\beta}{\partial t^\beta} f(x,t) = K_\beta \frac{\partial^2}{\partial x^2} f(x,t), \tag{5}$$

$f(x, t = 0) = \delta(x)$, $0 < \beta < 1$, where $K_\beta$ is the same constant as in Eq. (2), and the time fractional derivative is understood in the Caputo sense,

$$\frac{\partial^\beta}{\partial t^\beta}\phi = \frac{1}{\Gamma(1-\beta)} \int_0^t d\tau (t-\tau)^{-\beta} \frac{d}{d\tau}\phi(\tau). \tag{6}$$

Here, the sequence of temporal integration and differentiation is reversed with respect to a Riemann–Liouville operator. Recalling the Laplace transform of the Caputo derivative, $L\{d^\mu \phi/dt^\mu\} = s^\mu \phi(s) - s^{\mu-1}\phi(t=0)$, $0 < \mu < 1$, and making the Fourier–Laplace transform of Eq. (5), we again arrive at Eq. (4). Thus, both forms of time fractional diffusion equations are equivalent. In what follows we are interested in the mean square displacement (MSD) whose Laplace transform is given by

$$\langle x^2(s) \rangle = \left( -\frac{\partial^2 f(k,s)}{\partial k^2} \right)_{k=0}. \tag{7}$$

From Eq. (4), we get

$$\langle x^2(t) \rangle = \frac{2K_\beta t^\beta}{\Gamma(1+\beta)}. \tag{8}$$

## 3. Two Forms of Space Fractional Diffusion Equations

### 3.1. *Natural form*

The "natural" form of space fractional diffusion equation reads as

$$\frac{\partial f}{\partial t} = K_\alpha \frac{\partial^\alpha f}{\partial |x|^\alpha}, \tag{9}$$

$f(x, t = 0) = \delta(x)$, $0 < \alpha \leq 2$, where $K_\alpha$ is a positive constant, $[K_\alpha] = \text{cm}^\alpha/\text{sec}$, and the Riesz fractional derivative $\partial^\alpha/\partial|x|^\alpha$ (also denoted as fractional Laplacian $-(-\Delta)^{\alpha/2}$) is defined through its Fourier transform $\mathbf{F}(\ldots)$ as [19, 20]

$$\mathbf{F}\left( \frac{d^\alpha \phi(x)}{d|x|^\alpha} \right) = -|k|^\alpha \phi(k). \tag{10}$$

Applying the Fourier transform to Eq. (9) we get the characteristic function for the PDF of Lévy flights, [21]

$$f(k,t) = \exp(-K_\alpha |k|^\alpha t). \tag{11}$$

### 3.2. *Modified form*

Let us turn to the fractional equation for superdiffusion with the additional spatial derivative on its left-hand side,

$$\frac{\partial^{2-\alpha}}{\partial |x|^{2-\alpha}} \frac{\partial f}{\partial t} = -K_\alpha \frac{\partial^2}{\partial x^2} f, \tag{12}$$

where $K_\alpha$ is the same as in Eq. (9). Note the minus sign on the right-hand side of Eq. (12). This sign gets clear when turning to a Fourier representation: applying Fourier transform and using Eq. (10) we arrive at Eq. (11). Thus, both forms of space fractional diffusion equations, Eqs. (9) and (12), are equivalent.

Since the mean square displacement diverges for Lévy flights [21], their anomalous nature can be characterized by a typical displacement of the diffusing particle, $\delta x \propto \langle |x|^q \rangle^{1/q}$, $q < \alpha$. Thus, for both forms we get

$$\langle |x|^q \rangle \propto (K_\alpha t)^{q/\alpha}, \quad q < \alpha. \tag{13}$$

## 4. Natural Form of Distributed-Order Time Fractional Diffusion Equation

The natural form is the generalization of the Caputo form, see Eq. (5) [18, 22]:

$$\int_0^1 d\beta \tau^{\beta-1} p(\beta) \frac{\partial^\beta f}{\partial t^\beta} = K \frac{\partial^2 f}{\partial x^2}, \tag{14}$$

$f(x,0) = \delta(x)$, where $\tau$ is a positive constant representing some characteristic time of the problem, $K$ is the diffusion coefficient, $[K] = \text{cm}^2/\text{sec}$, $p(\beta)$ is dimensionless non-negative integrable function which can be normalized, $\int_0^1 d\beta p(\beta) = 1$, by the proper choice of the constant $\tau$, see discussion in [23]. Applying the Laplace and Fourier transforms in succession to Eq. (14), we get the solution in Fourier–Laplace space,

$$f(k,s) = \frac{1}{s} \frac{I_C(s\tau)}{I_C(s\tau) + k^2 K \tau}, \tag{15}$$

where

$$I_C(s\tau) = \int_0^1 d\beta (s\tau)^\beta p(\beta). \tag{16}$$

Of course, putting $p(\beta) = \delta(\beta - \beta_0)$ in Eq. (14), we arrive at the (mono) fractional diffusion Eq. (2) and its solution (4) in Fourier-Laplace space. The Laplace transform of the MSD, Eq. (7), takes the form

$$\langle x^2(s) \rangle = \frac{2K\tau}{sI_C(s\tau)}. \tag{17}$$

Then, from (17) we can derive the asymptotic behaviors of the MSD for $t \to 0$ and $t \to \infty$ from the asymptotic behaviors of $I_C(s\tau)$ for $s \to \infty$ and $s \to 0$, respectively, by virtue of the Tauberian theorems [24].

### 4.1. *Properties of the solution*

To justify the generalization (14), it is important to prove that the solution of Eq. (14) is indeed a PDF, i.e. it is a non-negative function. The proof is given in [18], where it is shown that the random process whose PDF obeys Eq. (14) is subordinated to the Wiener process. The fundamental solution of Eq. (14) is obtained in [25, 26] on the basis of its Mellin–Barnes integral representation. Such solution is proved to be related via a Laplace-type integral to the Fox-Wright functions. In [27] the theory of pseudo-differential operators was applied to Eq. (14), and solvability and uniqueness issues of the initial and boundary value problems in different function spaces were studied. The theory of Eq. (14) similar to the classical theory of the Cauchy problem for heat equation was developed in [28]. Solutions for different boundary conditions were obtained and studied in [29].

### 4.2. *Generic case of double order equation: Decelerating subdiffusion*

Here we illustrate the interesting property of the natural form (14) by taking a simple example of two fractional exponents, namely

$$p(\beta) = B_1\delta(\beta - \beta_1) + B_2\delta(\beta - \beta_2), \tag{18}$$

where $0 < \beta < 1,\ B_1 > 0,\ B_2 > 0,\ B_1 + B_2 = 1$. The corresponding diffusion equation is called *double order time fractional diffusion equation.* This choice (which we make below for the other distributed-order diffusion equations as well) allows us to show in a simple way the property of the diffusive behavior governed by this class of fractional order diffusion equations. After substituting Eq. (18) into Eq. (17) and using the Tauberian theorems, we calculate the MSD and get for the asymptotics at small and large arguments [18],

$$\langle x^2(t)\rangle \propto \begin{cases} t^{\beta_2}, & t \to 0, \\ t^{\beta_1}, & t \to \infty. \end{cases} \tag{19}$$

Since $\beta_1 < \beta_2$, we have the *diffusion with retardation* or *decelerating subdiffusion*. The crossover from the faster intermediate to the slower terminal behavior is clearly seen in numerical simulations of a Grunwald–Letnikov random walk, see the details in [22]. In [18] the authors also provided analytical expression for MSD in terms of a two-parameter Mittag-Leffler function.

### 4.3. *Relation to CTRW*

The physical relevance of the fixed-$\beta$ time-fractional diffusion equations is to a large extent due to the fact that they can be viewed as describing a "long-time" limit of continuous-time random walks (CTRW), a model which was successfully applied for the description of anomalous diffusion phenomena in many areas, e.g., turbulence [30], disordered media [31], intermittent chaotic systems [32], contaminant dispersion in catchments [33], to name a few. The CTRW models described by the fixed-$\beta$ equations correspond to the power law distributions of waiting times between the subsequent steps [1]. Here the power law behavior $\langle x^2(t)\rangle \propto t^{\beta}$ of the mean square displacement typically corresponds to the waiting-time distributions $\psi(t) \propto t^{-1-\beta}$ with $0 < \beta < 1$. Let us recall the basic formula of the CTRW in the Fourier–Laplace space [1]:

$$f(k,s) = \frac{1-\psi(s)}{s}\frac{1}{1-\psi(k,s)}, \tag{20}$$

where $\psi(s)$ is the Laplace transform of the waiting-time PDF $\psi(t)$, and $\psi(k,s)$ is the Fourier–Laplace transform of the joint PDF of jumps and waiting times $\psi(\xi,t)$. In the simplest case one assumes the decoupled form: $\psi(\xi,t) = \lambda(\xi)\psi(t)$. Moreover, the jump length variance is considered to be finite, so that $\lambda(k)$, the Fourier transform of $\lambda(\xi)$, is $\lambda(k) \simeq 1 - K\tau k^2$ to the lowest order in $k$.

First, let us consider the cases for which at large $t$ the waiting time PDF $\psi(t)$ possesses long power law tails, such that the mean waiting time diverges, that is $\psi(t) = \psi(t|\beta) \propto \tau^{\beta}/t^{1+\beta}$, $0 < \beta < 1$, and consequently, $\psi(s) \simeq 1 - (s\tau)^{\beta}$ at small $s$. With these assumptions, taking the inverse

Fourier–Laplace transform of Eq. (20) we arrive at Eq. (2). Now let us consider the case when $\beta$ *fluctuates.* Indeed, for example, in the model called the Arrhenius cascade, which is inspired by the studies of disordered systems, the unique $\beta$ appears only under the assumption that the random trapping time is related to the random depth of the potential well by the Arrhenius law [34]. In a more realistic model this law gives only the average value of the trapping time. Thus, we may speculate that in order to take into account the fluctuations of the trapping time, we can introduce *conditional* PDF $\psi(t|\beta) \propto \tau^\beta/t^{1+\beta}$, and the PDF $p(\beta)$ as well. Now, we have a relation

$$\psi(t) = \int_0^1 d\beta p(\beta)\psi(t|\beta), \tag{21}$$

where $[0;1]$ is the whole interval for variations of $\beta$. We note that all waiting-time distributions with $\beta > 1$ correspond to a similar behavior described by the first-order derivative. Then, for the $\psi(s)$, we have

$$\psi(s) \simeq 1 - \int_0^1 d\beta (s\tau)^\beta p(\beta). \tag{22}$$

Inserting Eq. (22) into Eq. (20) we arrive at Eq. (15).

The model with fluctuating $\beta$ is, of course, not the only possible interpretation of Eq. (14): the nonexact power-law behavior of the waiting-time PDF can physically have very different reasons. In particular, the representation (14) allows us to consider regularly varying waiting-time PDF which behave as $\psi(t) \propto t^{-1-\beta}g(t)$ as $t \to \infty$, where $g(t)$ is a slowly varying function, e.g., any power of $\ln t$ [24]. We are also able to consider waiting-time PDFs $\psi(t)$ that show an approximately scaling behavior with the exponents *changing with time.* For such distributions the effective PDFs $p(\beta)$ can be determined, and thus such nonperfectly scaling CTRWs can be described through distributed-order diffusion equations. A more detailed discussion of these points is contained in [18]. The CTRW scheme corresponding to a double order time fractional diffusion equation is studied in [35].

### 4.4. *Superslow diffusion*

#### 4.4.1. *Probability density function and mean square displacement*

The natural form (14) is a versatile tool for describing superslow kinetic processes, when the MSD grows as some power of logarithm of time, $\langle x^2(t)\rangle \propto \ln^\nu t$. The most known example of such a logarithmic, ultra-slow diffusion is given by Sinai diffusion [36], for which $\nu = 4$, a case when the particle moves in a quenched random force field. An even slower behavior with $\nu = 4/3$ is encountered in polymer physics (a polyampholyte hooked around an obstacle [37]). Similar behaviors were observed for motion in aperiodic environments [38] and in a family of iterated maps [39]. Strong anomaly in diffusion was also found numerically in an area-preserving parabolic map on a cylinder [40]. Some of the ultraslow systems share the interesting property that the PDF $f(x,t)$ of the displacements scales as a function of $\xi = x/(\ln t)^{\nu/2}$, and displays exponential decay in its wings, $f(\xi) \propto \exp[-A|\xi|]$ for large $|\xi|$. As it is shown in [41], the ultraslow case corresponds to a class of $p(\beta)$ in Eq. (14) having a simple power law form,

$$p(\beta) = \nu \beta^{\nu - 1}. \tag{23}$$

The normalization condition for $p(\beta)$ on $[0,1]$ assumes $\nu > 0$. The choice (23) captures effects found in different models exhibiting ultra-slow kinetics. We hasten to note that only the behavior of $p(\beta)$ in the vicinity of $\beta = 0$ is of significance for the long-time behavior of the overall process. The PDF in the long-time limit takes the form

$$f(x,t) \propto \exp\left[-\left(\frac{\Gamma(\nu+1)}{K\tau}\right)^{1/2} \frac{|x|}{\ln^{\nu/2}(t/\tau)}\right], \tag{24}$$

thus, the long-time asymptotic behavior shows exactly the double-sided exponential behavior found in models exhibiting ultraslow kinetics. The MSD is obtained with the use of (16), (17),

$$\langle x^2(t)\rangle \approx \begin{cases} 2K\nu^{-1} t \ln(t/\tau), & t/\tau \ll 1, \\ 2K\tau(\Gamma(\nu+1)) \ln^{\nu}(t/\tau), & t/\tau \gg 1. \end{cases} \tag{25}$$

Thus, strong diffusion anomalies are described within the proposed formalism. We also note the universality of the time dependence for small $t$, apart from a numerical factor. The behavior of the MSD at large $t$ coincides with the result obtained in [39] by using the CTRW approach for the diffusion generated by iterated maps. Mathematical aspects of the superslow diffusion described by the normal form (14) with generalization to a multidimensional space are considered in [42]. We also note that the distributed order diffusion equation with the constant weight was investigated in [43].

Our results here can be contrasted with the known behavior of the solution of the fixed-$\beta$ diffusion Eq. (5). The PDF is expressed via the M-function (of the Wright type) which exhibits stretched exponential behavior differing considerably from the double-sided exponential behavior discussed above.

### 4.4.2. *Distributed-order fractional Fokker–Planck equation for superslow processes*

Parallel to Eq. (14) the distributed-order fractional Fokker–Planck equation was introduced in [41],

$$\int_0^1 d\beta \tau^{\beta-1} p(\beta) \partial^\beta f/\partial t^\beta = L_{FP} f(x,t) \quad f(x,0) = \delta(x), \tag{26}$$

where $L_{\mathrm{FP}}$ is a regular Fokker–Planck operator under an external potential $U$,

$$L_{\mathrm{FP}} = \frac{\partial}{\partial x}\frac{U'(x)}{m\gamma} + K\frac{\partial^2 f}{\partial x^2}. \tag{27}$$

Many physical situations can be adequately described in the language of decaying modes, corresponding to the eigenfunctions of the Fokker–Planck operator. Introducing a separation ansatz:

$$f(x,t) = T(t)\varphi(x), \tag{28}$$

we see that Eq. (26) reduces to

$$\begin{gathered} L_{\mathrm{FP}}\varphi_n + \lambda_n \varphi_n = 0, \\ \int_0^1 d\beta \tau^{\beta-1} p(\beta) \frac{d^\beta T_n}{dt^\beta} + \lambda_n T_n = 0. \end{gathered} \tag{29}$$

Taking the Laplace transform of Eq. (29) and using Tauberian theorems we get

$$T_n(t) \approx \frac{T_n(0)}{\lambda_n \tau \ln^\nu(t/\tau)}, \quad t/\tau \gg 1. \tag{30}$$

Thus, the modes show an ultraslow, logarithmic, decay pattern. This can again be contrasted with the behavior found in fixed-$\beta$ equations for subdiffusion, where the relaxation of single modes is governed by a Mittag-Leffler pattern, with the power-law decay, $T_n(t) \propto t^{-\beta}$ for $0 < \beta < 1$, and exponential decay for $\beta = 1$ [1].

Let us now consider the response of our system to an external field (bias). While the behavior of a Sinai model in the absence of external bias is adequately described by a distributed-order fractional equation, the behavior of the system under such bias is quite different. Thus, the genuine Sinai process under the action of a homogeneous external force $F$ shows the mean displacement that grows as a power of time, $\langle x\rangle \propto t^{\mu}$ [44]. In the case of systems described by the distributed-order fractional equation, this displacement grows considerably slower, namely as a power of logarithm of time. After employing Eq. (26) with $F = -U' = \text{const}$, applying the Laplace transform and using Tauberian theorem, we get

$$\langle x(t)\rangle_F \approx \begin{cases} F(m\gamma\nu)^{-1} t\ln(\tau/t), & t/\tau \ll 1, \\ F\tau(m\gamma\Gamma(\nu+1)^{-1}\ln^{\nu}(t/\tau), & t/\tau \gg 1. \end{cases} \tag{31}$$

By comparing Eqs. (25) and (31) we get the second Einstein relation:

$$\langle x(t)\rangle_F = \frac{F\langle x^2(t)\rangle_0}{2k_BT}, \tag{32}$$

where $k_BT = m\gamma K$.

The difference between the behavior of the genuine Sinai model and the above result for distributed-order FFPE is connected with the fact that in the Sinai model (being a random force model) the external bias strongly perturbs the overall potential in which the particle moves. The model described by the distributed-order FFPE is essentially a (high-dimensional) trap model with a broad distribution of trapping times due to very deep traps. Thus, it should not be surprising that in our case the perturbation introduced by an external force can be considered as weak, so that the second fluctuation-dissipation relation holds.

### 4.4.3. *Relation to CTRW*

We now turn to the relation of the ultraslow kinetics and the CTRW processes. As in Sec. 4.3, we consider the case when there is no simple power-law waiting-time PDF but instead a weighted mixture of power-law functions, see Eqs. (21) and (22). Then, with $p(\beta)$ given by Eq. (23) we have for the waiting time PDF,

$$\psi(t) \propto \frac{1}{t[\log(t/\tau)]^{\nu+1}}. \tag{33}$$

Thus, the distributed-order time fractional diffusion equation with $p(\beta)$ being a power law corresponds to the limit of the CTRW with an extremely broad waiting-time PDF, such that there are no finite moments of a whatever (even fractional) order [45]. The stochastic foundation of ultraslow diffusion was developed in [46, 47]. It is based on random walks with a random waiting time between jumps whose probability tail falls off at a logarithmic rate. In particular, it was shown that the time process (sums of the independent waiting times) converges to an extremal process. In turn, in the case of finite variance of particle jumps, the continuous time random walk limit process has a double-sided exponential Laplace distribution. Further developments in the theory of distributed-order time fractional diffusion equation in the normal form include the concept of generalized distributed order derivative which lead to a class of anomalous diffusion equations with scalings including slowly varying functions at infinity [48].

## 5. Modified Form of Distributed-Order Time Fractional Diffusion Equation

The modified form is the generalization of the Riemann–Liouville form (2),

$$\frac{\partial f}{\partial t} = \int_0^1 d\beta p(\beta)\tau^{1-\beta} K D_t^{1-\beta} \frac{\partial^2 f}{\partial x^2}, \tag{34}$$

where $K$ is the diffusion coefficient, $[K] = \mathrm{cm}^2/\mathrm{sec}$ and similar to the normal form, $p(\beta)$ is normalized and has a meaning of the probability density. Applying to Eq. (34) the Laplace and Fourier transforms in succession, we get

$$f(k, s) = \frac{I_{RL}}{s(I_{RL} + k^2 K\tau)}, \tag{35}$$

where

$$I_{RL}(s\tau) = \left[\int_0^1 d\beta (s\tau)^{-\beta} p(\beta)\right]^{-1}. \tag{36}$$

Note that Eq. (35) is exactly the same as the equation for the Fourier–Laplace transform of the solution for the natural form, if we replace $I_{RL}(s\tau)$ by $I_C(s\tau)$, see Eqs. (15) and (16). This gives us a solution in terms of the integral formula for subordination [35, 49]. In both cases, the one described by the modified as well as by the natural distributed-order form, the parent process is the Wiener process; the properties of the operational time (directing process), however, are different in these two cases.

### 5.1. *Thermodynamical interpretation*

We note that Eq. (34), just like the time fractional equation (2) in the RL form admits a thermodynamical interpretation as a combination of a continuity equation $\partial f/\partial t = -\partial j/\partial x$ and a retarded linear response equation $j(x,t) = \Phi_t\{\partial f(x,t)/\partial x\}$ for the flux exhibiting memory of the process at previous times $t' < t$ [50]. This flux dependent on the past history is analogous to the flux for a space-fractional diffusion process [51]. We also note that the natural form of a distributed-order time-fractional diffusion equation does not immediately allow for such a thermodynamical interpretation.

### 5.2. *Generic case of double order equation: Accelerating subdiffusion*

Here, we again consider the case of two fractional exponents, Eq. (18). Inserting Eq. (18) into Eq. (36) gives

$$I_{RL}(s\tau) = (b_1 s^{-\beta_1} + b_2 s^{-\beta_2})^{-1}, \tag{37}$$

where $b_{1,2} = B_{1,2}/\tau^{\beta_{1,2}}$. The positivity of the solution of the double-order time-fractional diffusion equation in the modified form was proved in [52]. The solution itself was obtained analytically in [53] in terms of infinite series of the Fox functions. The corresponding random process can be considered as an approximation for a continuous-time random walk process with the waiting-time cumulative distribution function $\Psi(t) = 1-(b_1\tau^{\beta_1} + b_2\tau^{\beta_2})^{-1}$, and simulated accordingly. The derivation of the corresponding modified diffusion equation from the CTRW theory is considered in [35].

The MSD is easily calculated by the use of the Laplace transform,

$$\langle x^2(t)\rangle = \frac{2D_1}{\Gamma(1+\beta_1)}\left(\frac{t}{\tau}\right)^{\beta_1} + \frac{2D_2}{\Gamma(1+\beta_2)}\left(\frac{t}{\tau}\right)^{\beta_2}, \tag{38}$$

where $D_1 = B_1 K\tau$ and $D_2 = B_2 K\tau$. Since $0 < \beta_1 < \beta_2 \leq 1$, at small times the first term on the right-hand side of Eq. (40) prevails, whereas at large times the second one dominates. Thus our Eq. (40) describes *accelerating subdiffusion*. This behavior is opposite to the case of retarded (decelerated) subdiffusion described by the natural form which is considered above in Sec. 4.2. Moreover, it was shown in [23] that *the system governed by the*

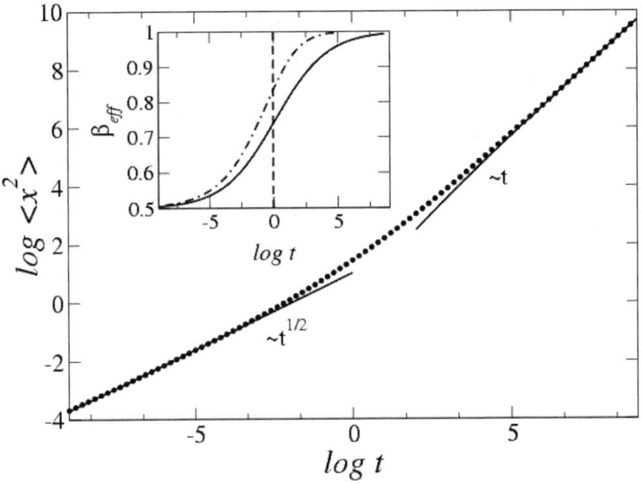


Fig. 1. Main figure: MSD $\langle x^2 \rangle$ vs. $t$ is shown by dots in the log–log scale. Parameters: $\beta_1 = 0.5$, $\beta_2 = 1$, $b_1 = b_2 = 0.5$, $K = \tau = 1$. The solid lines show asymptotics $\propto t^{-1/4}$ and $\propto t^{-1/2}$, for small and large times, respectively. The transition from subdiffusion to normal diffusion behavior is clearly seen. Inset: the effective diffusion exponent $\beta_{\text{eff}}$ obtained from the local slope of the MSD curve is shown by the solid line. For $t = 1$, $\beta_{\text{eff}} = 0.74$. Dash-dotted line demonstrates $\beta_{\text{eff}}$ calculated from the probability to stay at the origin [23].

*distributed-order time-fractional diffusion equation with two diffusion exponents has the properties very similar to the system whose exponent varies in time.* $\langle x^2(t) \rangle \propto t^{\beta_{\text{eff}}(t)}$. Two different recipes were proposed to get the function $\beta_{\text{eff}}(t)$. For example, it can be defined by the local slope of the whole MSD curve, showing the dependence of MSD on time in a log–log scale; this slope governs evolution exactly at time $t$; see Fig. 1. Another possibility is to use the analogous behavior of probability to stay at the origin [23]. Interestingly, it is also possible to approximate not only MSD, but the solution of double-order diffusion equation by the solution of a single-order with time-dependent exponent. To construct the solution, it is also possible to use $\beta_{\text{eff}}(t)$ estimated from the MSD curve or probability to stay at the origin. Even better fit can be achieved if the diffusion coefficient is varied simultaneously; see Fig. 2. Such type of accelerating subdiffusion is observed in experiments with intermittent turbulence, single-file diffusion and in some biological systems; for the examples see [23].

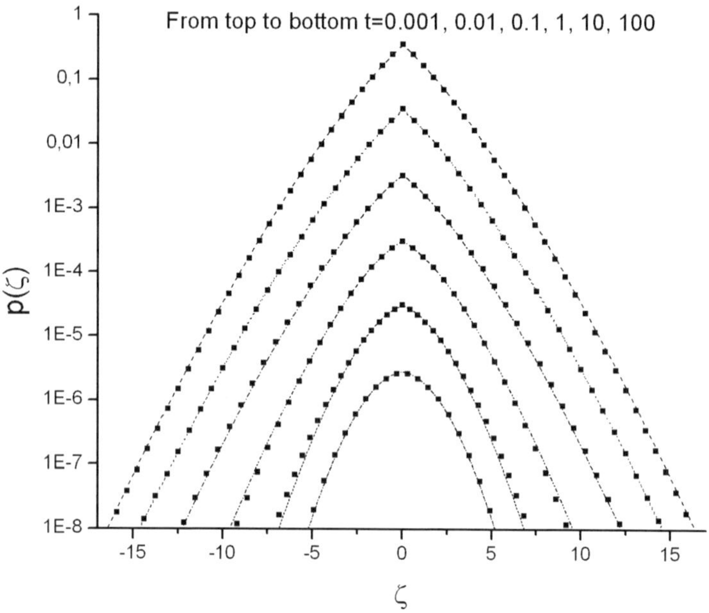


Fig. 2. The rescaled PDF $p(\zeta) = W(t)f(x,t)$, where $f(x,t)$ is obtained by inverse Laplace transform, and $W(t)$ is defined by $\int_0^{W(t)} f(x,t)dx = 1/4$, is shown for different values of $\zeta$ as a function of rescaled displacement $\zeta = x/W(t)$, for different time instants. The corresponding dashed lines demonstrate the single-order solution with the parameters, from top to bottom, $\beta_{\text{eff}} = 0.5, 0.5, 0.6, 0.73, 0.8, 0.95$; $K_{\text{eff}} = 1, 1.05, 1.7, 1.95, 2.0, 1.4$; $\tau = 1$.

## 6. Natural Form of Distributed-Order Space Fractional Diffusion Equation

The natural form is the generalization of Eq. (9) [18]:

$$\frac{\partial f}{\partial t} = \int_0^2 d\alpha p(\alpha) l^{\alpha-2} K \frac{\partial^\alpha f}{\partial |x|^\alpha}, \quad p(x,0) = \delta(x), \tag{39}$$

where $l$ (in cm) and $K$ (in $\text{cm}^2/\text{sec}$) are dimensional positive constants, $p$ is a normalized dimensionless non-negative function of $\alpha$. The equation that follows for the characteristic function from Eq. (39) has the solution

$$f(k,t) = \exp\left\{-\frac{Kt}{l^2}\int_0^2 d\alpha p(\alpha)(|k|l)^\alpha\right\}. \tag{40}$$

Note that the normalization condition $\int_{-\infty}^{\infty} dx f(x,t) = f(k=0,t) = 1$ is fulfilled.

### 6.1. *Generic case of double-order equation: Accelerating superdiffusion*

We again consider an important case with two fractional exponents,

$$p(\alpha) = A_1\delta(\alpha - \alpha_1) + A_2\delta(\alpha - \alpha_2), \tag{41}$$

where $0 < \alpha_1 < \alpha_2 \leq 2,\ A_1 > 0,\ A_2 > 0$. After substituting Eq. (41) into Eq. (40) we get

$$f(k,t) = \exp\{-a_1|k|^{\alpha_1}t - a_2|k|^{\alpha_2}t\}, \tag{42}$$

that is the characteristic function which is the product of two characteristic functions of the Lévy stable PDFs. Therefore, the inverse Fourier transformation of Eq. (42) gives the PDF which is the convolution of the two stable PDFs. This PDF is, obviously, positive. The PDF will also be positive, if the function $p(\alpha)$ is represented as a sum of $N\delta$-functions multiplied by positive constants, $N$ is a positive integer. Moreover, if $p(\alpha)$ is a continuous positive function, then discretizing the integral in Eq. (40) by a Riemann sum and passing to the limit we can also conclude on the positivity of the PDF whose evolution obeys the natural form (39).

To characterize anomalous superdiffusion governed by the double order equation, we introduce a typical displacement, see Sec. 3.2. As a result, we have [18]

$$\delta x \propto \langle |x|^q \rangle^{1/q} \propto \begin{cases} t^{1/\alpha_2}, & t \to 0, \\ t^{1/\alpha_1}, & t \to \infty. \end{cases} \tag{43}$$

Since $\alpha_1 < \alpha_2$, the proposed equation describes *accelerating* superdiffusion. We also note that the constitutive equation with space fractional distributed order derivative in the Caputo form was used to generalize the Darcy's law [54].

## 7. Modified Form of Distributed-Order Space Fractional Diffusion Equation

Let us now consider the space fractional diffusion equation with a distributed order operator on the left-hand side:

$$\int_0^2 d\alpha p(\alpha) l^{2-\alpha} \frac{\partial^{2-\alpha}}{\partial |x|^{2-\alpha}} \frac{\partial f}{\partial t} = -K \frac{\partial^2 f}{\partial x^2}, \quad f(x,0) = \delta(x), \tag{44}$$

where $l$, $K$ and $p$ have the same meaning as in Eq. (39). This form of equation was analyzed in [52] in detail.

### 7.1. *Generic case of double-order equation: Retarding superdiffusion*

Restricting ourselves to the case with two fractional exponents, see Eq. (41), we get for a typical displacement,

$$\delta x \propto \langle |x|^q \rangle^{1/q} \propto \begin{cases} t^{1/\alpha_1}, & t \to 0, \\ t^{1/\alpha_2}, & t \to \infty \end{cases} \tag{45}$$

thus, in contrast to the natural form, the modified form of the distributed order space fractional diffusion equation leads to *retarding superdiffusion.*

### 7.2. *Fractional diffusion equation for a power law truncated Lévy process*

An interesting particular example of the modified form was proposed in [55] to describe a power-law truncated Lévy process:

$$\left(1 - C_\alpha \frac{\partial^{2-\alpha}}{\partial |x|^{2-\alpha}}\right) \frac{\partial f(x,t)}{\partial t} = D \frac{\partial^2 f(x,t)}{\partial x^2}, \tag{46}$$

where $D$ is the diffusion coefficient governing the long-time asymptotic behavior, and the scale factor $C_\alpha$ is a coefficient governing the intermediate time Lévy-like behavior. It is proved that the solution of Eq. (46) is a PDF [55]. The characteristic function has the form

$$f(k,t) = \exp\left(-\frac{Dk^2 t}{1 + C_\alpha |k|^{2-\alpha}}\right). \tag{47}$$

The inverse Fourier transform of Eq. (47) cannot be performed analytically. However, we observe that at $|k|$ large enough (for small $|x|$) the characteristic function corresponds to the characteristic function of the Lévy stable distribution with the index $\alpha$. However, the asymptotics of the PDF at large $x$ is determined by the first non-analytical term of the expansion of Eq. (47) as $k \to 0$ and reads

$$f(x,t) \simeq \frac{\Gamma(5-\alpha)\sin(\pi\alpha/2)}{\pi} \frac{DC_\alpha t}{x^{5-\alpha}}, \quad x \to \infty. \tag{48}$$

Thus, the Lévy stable PDF with the Lévy index $\alpha$ is truncated by a power law of exponent 5-$\alpha$. In course of time the PDF slowly converges to a Gaussian in its central part showing however a power-law far asymptotics. The proposed Eq. (46) was shown in [23] to capture the features of the time evolution of density fluctuations measured in the tokamak plasmas [56], the behavior of the PDF of commodity prices [57], and nicely fits the empirical findings obtained from a high-resolution analysis of a database of the S&P 500 index [58].

## 8. Conclusions and Outlook

We proposed different forms of distributed-order time and space fractional diffusion equations for the description of the systems demonstrating non-scaling behavior. We showed that equations with the distributed-order fractional operator replacing the corresponding integer-number derivative (natural form of distributed-order diffusion equations) describe processes getting more anomalous in course of time, i.e. the accelerating superdiffusion and retarding subdiffusion. On the opposite, equations with distributed-order fractional operators on the "wrong" side of diffusion equation (modified form) describe processes getting less anomalous, i.e. retarded superdiffusion and accelerating subdiffusion.

## Acknowledgments

The authors are grateful to R. Gorenflo, V. Gonchar and N. Korabel, with whom many of the original results presented in this Chapter have been obtained. The authors acknowledge financial support from EC within the MC IIF grant 219966 LeFrac.

## References

1. R. Metzler and J. Klafter, *Phys. Rep.* **339**, 1 (2000).
2. I. M. Sokolov, J. Klafter and A. Blumen, *Phys. Today* **55**, 48 (2002).
3. R. N. Mantegna and H. E. Stanley, *Phys. Rev. Lett.* **73**, 2946 (1994).
4. Ya. G. Sinai, *Theor. Prob. Appl.* **27**, 256 (1983).
5. M. Caputo, *Geophys. J. Roy. Astronom. Soc.* **13**, 529 (1967).
6. M. Caputo, *Elasticitá e Dissipazione* (Zanichelli Printer, 1969).
7. M. Caputo, *Ann. Univ. Ferrara* — Sez.VII — Sc. Mat. **XLI**, 73 (1995).
8. A. M. Nakhushev, *Differential Equations* **34**, 101 (1998) (in Russian).
9. A. M. Nakhushev, *Fractional Calculus and Its Applications* (Fizmatlit, 2003) (in Russian).
10. I. Podlubny, *Fractional Differential Equations* (Academic Press, 1999).
11. A. N. Kochubei, *Ukrainian Math. J.* **60**, 550 (2008).
12. R. L. Bagley and P. J. Torvik, *Int. J. Appl. Math.* **2**, 865 (2000).
13. R. L. Bagley and P. J. Torvik, *Int. J. Appl. Math.* **2**, 965 (2000).
14. K. Diethelm and N. J. Ford, *Fract. Calc. Appl. Anal.* **4**, 531 (2001).
15. K. Diethelm and N. J. Ford, *J. Comput. Appl. Math.* **225**, 96 (2009).
16. C. F. Lorenzo and T. T. Hartley, *Nonlinear Dynamics* **29**, 57 (2002).

17. M. Caputo, *Frac. Calc. Appl. Anal.* **4**, 421 (2001).
18. A. V. Chechkin, R. Gorenflo and I. M. Sokolov, *Phys. Rev. E* **66**, 046129 (2002).
19. S. G. Samko, A. A. Kilbas and O. I. Marichev, *Fractional Integrals and Derivatives: Theory and Applications* (Gordon and Breach, 1993). (Russian Edition: 1987).
20. A. Saichev and M. Zaslavsky, *Chaos* **7**, 753 (1997).
21. G. Samorodnitsky and M. S. Taqqu, *Stable Non-Gaussian Random Processes: Stochastic Models with Infinite Variance* (Chapman and Hall, 1994).
22. A. V. Chechkin, R. Gorenflo, I. M. Sokolov and V. Yu. Gonchar, *Frac. Cal. Appl. Anal.* **6**, 259 (2003).
23. A. V. Chechkin, V. Yu. Gonchar, R. Gorenflo, N. Korabel and I. M. Sokolov, *Phys. Rev. E* **78**, 021111 (2008).
24. W. Feller, *Introduction to Probability Theory and Its Applications*, 2nd edn. (Wiley, 1971), Vol. II.
25. F. Mainardi and G. Pagnini, *J. Comput. Appl. Math.* **207**, 245 (2007).
26. F. Mainardi, G. Pagnini and R. Gorenflo, *Appl. Math. Comput.* **187**, 295 (2007).
27. S. Umarov and R. Gorenflo, *J. Anal. Appl.* **24**, 449 (2005).
28. A. N. Kochubei, *J. Math. Anal. Appl.* **340**, 252 (2008).
29. M. Naber, *Fractals* **12**, 23 (2004).
30. J. Klafter, A. Blumen and M. F. Schlesinger, *Phys. Rev. A* **35**, 3081 (1987).
31. J.-P. Bouchaud and A. Georges, *Phys. Rep.* **195**, 127 (1990).
32. G. Zumofen and J. Klafter, *Phys. Rev. E* **47**, 851 (1993).
33. R. Metzler, G. Margolin, H. Scher, B. Berkowitz and J. Klafter, *Geophys. Res. Lett.* **29**, 101029 (2002).
34. F. Bardou, in *Proc of the Conf. on Lévy Processes: Theory and Applications.* eds. by O. E. Barndorff–Nielsen, S. E. Graversen and T. Mikosch (Univ. Aarhus, 1999), pp. 22–38 (downloadable from http:/www.maphysto.dk/publications/MPS-misc/1999/11.pdf).
35. I. M. Sokolov and J. Klafter, *Chaos* **15**, 026103 (2005).
36. Ya. G. Sinai, *Theory Probab. Appl.* **27**, 256 (1983).
37. H. Schiessel, I. M. Sokolov and A. Blumen, *Phys. Rev. E* **56**, 2390 (1997).
38. F. Igloi, L. Turban and H. Rieger, *Phys. Rev. E* **59**, 1465 (1999).
39. J. Dräger and J. Klafter, *Phys. Rev. Lett.* **84**, 5998 (2000).
40. T. Prosen and M. Žnidaric, *Phys. Rev. Lett.* **87**, 114101 (2001).
41. A. V. Chechkin, J. Klafter and I. M. Sokolov, *Europhys. Lett.* **63**, 326 (2003).
42. A. N. Kochubei, *J. Math. Anal. Appl.* **340**, 252 (2008).
43. A. V. Pskhu, *Equations with Partial Derivatives of Fractional Order* (Nauka, 2005) (in Russian).
44. P. Doussal, C. Monthus and D. S. Fisher, *Phys. Rev. E* **59**, 4795 (1999).
45. S. Havlin and G. Weiss, *J. Stat. Phys.* **58**, 1267 (1990).
46. M. M. Meerschaert and H.-P. Scheffler, *Statistics Prob. Lett.* **71**, 15 (2005).
47. M. M. Meerschaert and H.-P. Scheffler, *Stoch. Proc. Appl.* **116**, 1215 (2006).
48. A. Hanyga, *J. Phys. A: Math. Theor.* **40**, 5551 (2007).
49. I. M. Sokolov and A. V. Chechkin, *Fluct. Noise Lett.* **5**, L275 (2005).

50. I. M. Sokolov, *Phys. Rev. E* **63**, 056111 (2001).
51. P. Paradisi, R. Cesari, F. Mainardi and F. Tampieri, *Physica A* **293**, 130 (2001).
52. I. M. Sokolov, A. V. Chechkin and J. Klafter, *Acta Phys. Pol. B* **35**, 1323 (2004).
53. T. A. M. Langlands, *Physica A* **367**, 136 (2006).
54. M. Caputo, *Ann. Geophys.* **46**, 223 (2003).
55. I. M. Sokolov, A. V. Chechkin and J. Klafter, *Physica A* **336**, 245 (2004).
56. R. Jha, P. K. Kaw, D. R. Kulkarni, J. C. Parikh and ADITYA Team, *Phys. Plasmas* **10**, 699 (2003).
57. H. E. Stanley, *Physica A* **318**, 279 (2003).
58. M. H. Cohen and P. Venkatesh, in Practical Fruits of Econophysics, *Proc. of the Third Nikkei Econophysics Symposium*, ed. H. Takayasu (Springer–Verlag, 2006), pp. 147–151.

# Chapter 6

## Anomalous Transport in the Presence of Truncated Lévy Flights

D. del-Castillo-Negrete
*Oak Ridge National Laboratory, Oak Ridge, TN, USA*

The role of truncated Lévy flights on anomalous diffusion and front propagation is studied. Starting from the Continuous Time Random Walk for general Lévy stochastic processes, an integro-differential equation describing the dynamics in the long-wavelength limit is obtained. In the case of exponentially tempered Lévy flights, the transport equation involves exponentially truncated fractional derivatives describing the interplay between memory, long jumps, and truncation effects in the intermediate asymptotic regime. The dynamics exhibits a transition from super-diffusion to sub-diffusion. The decay of the tail of the Green's function changes from algebraic at short times, to stretched exponential at long times. The role of truncation in the super-diffusive propagation of fronts in the tempered fractional Fisher–Kolmogorov equation is also studied. In the absence of truncation, the fronts have exponential acceleration and algebraic decaying tails. In the presence of truncation, this phenomenology prevails in an intermediate asymptotic regime. Outside this regime, the front's tail exhibits tempered decay, the acceleration is transient, and the front velocity exhibits an algebraically slow convergence to a terminal velocity. In the over-truncated regime, fronts have exponential tails and move at a constant velocity.

| | | |
|---|---|---|
| 1. | Introduction | 130 |
| 2. | Continuous Time Random Walk for General Lévy Processes | 132 |
| 3. | Green's Function of Tempered Fractional Diffusion Equation | 137 |
| | 3.1. Symmetric solution | 139 |
| | 3.2. Asymmetric solution | 141 |
| 4. | Transition from Super-diffusive to Sub-diffusive Transport | 142 |
| 5. | Truncation Effects in Super-diffusive Front Propagation | 145 |
| | 5.1. Diffusive case | 147 |
| | 5.2. Fractional case | 148 |
| | 5.3. Truncated case | 150 |

5.4. $\nu > \lambda$ . . . . . . . . . . . . . . . . . . . . . . . . 151
5.5. $\nu < \lambda$ . . . . . . . . . . . . . . . . . . . . . . . . 154
6. Conclusions . . . . . . . . . . . . . . . . . . . . . . . . . . 155

## 1. Introduction

Anomalous transport is a topic of current significant research interest in many areas of science and engineering [1]. Of particular interest is the construction of models describing transport processes that depart from the standard diffusion paradigm. This paradigm is based on the Fourier–Fick's prescription which assumes that the fluxes are proportional to the local gradients. In the case of a single scalar field, $\phi$, in a one-dimensional domain, the Fourier–Fick's prescription reads

$$q = -\chi \partial_x \phi, \tag{1}$$

where $q$ is the flux, and $\chi$ is the diffusivity that in principle can depend on $x$, $t$ and in the case of nonlinear models, in $\phi$ and $\partial_x \phi$. Equation (1), together with the continuity equation

$$\partial_t \phi = -\partial_x q, \tag{2}$$

leads to the well-known diffusion equation

$$\partial_t \phi = \partial_x (\chi \partial_x \phi). \tag{3}$$

Although the diffusion model in Eq. (3), and its two- and three-dimensional extensions involving multiple fields, has been successful used to describe a wide range of transport problems, there are cases in which it does not provide the correct description. The shortcomings of the diffusion model can be traced back to its foundations. From the statistical mechanics point of view, diffusion assumes an underlying Gaussian, Markovian stochastic process. Under this perspective, it is natural to expect that the diffusion model might breakdown when transport involves stochastic processes that deviate from these restrictive assumptions. An example of particular interest here, are transport processes involving Lévy statistics. In this case, it has been shown that, in the context of the Continuous Time Random Walk (CTRW) [2–4], the Fourier–Fick's prescription is replaced by a nonlocal relation of the form

$$q = -\chi \int \mathcal{K}(x - x')\, \partial_{x'} \phi \, dx' , \tag{4}$$

which, when substituted into Eq. (2) assuming a constant $\chi$, leads to the fractional-diffusion equation

$$\partial_t \phi = \chi \left[ l \, {}_{-\infty}D_x^\alpha + r \, {}_x D_\infty^\alpha \right] \phi , \tag{5}$$

where ${}_{-\infty}D_x^\alpha$ and ${}_x D_\infty^\alpha$ are integro-differential operators known as fractional derivatives of order $1 < \alpha \leq 2$, because in Fourier space they satisfies

$$\mathcal{F}\left[ {}_{-\infty}D_x^\alpha \phi \right] = (-ik)^\alpha \, \hat{\phi}, \qquad \mathcal{F}\left[ {}_x D_\infty^\alpha \, \phi \right] = (ik)^\alpha \, \hat{\phi}, \tag{6}$$

which is the fractional extension of the familiar transform properties of order-$n$ derivatives, where $\mathcal{F}[\phi] = \hat{\phi}$ denotes the Fourier transform. The fractional diffusion model has been used to model a wide variety of anomalous transport problems see for example [1, 5, 6] and references therein.

In the context of the CTRW the fractional diffusion model assumes the existence of Lévy flights. That is, it assumes that the size of the individual particle jumps are distributed according to a probability density function (pdf) of the form $\eta \sim |x|^{-(1+\alpha)}$. The slow algebraic decay of $\eta$ allows the possibility of assigning a significant probability to very long jumps, or rare events, which play a key role on anomalous transport. In fact, for $0 < \alpha < 2$, which is the regime of interest to transport, the second moment of the jumps pdf diverges, $\langle x^2 \rangle = \int x^2 \eta \, dx \rightarrow \infty$, which implies a lack of a characteristic transport scale. Although the existence of large displacements has been documented in experimental and theoretical studies of super-diffusive transport, from the physics point of view, it is clear that particle displacements cannot be arbitrarily large. For example, in the study of transport in fluids [7, 8] and plasmas [9], it is expected that decorrelation of particle displacements due to "non-ideal" effects or boundary effects in finite size systems will eventually lead to the truncation of otherwise arbitrarily large Lévy flights. Accordingly, it is of theoretical and practical interest to construct models of anomalous transport that incorporate large events while keeping the moments finite. In this paper, we study this problem in the context of truncated Lévy processes. Following [10] we discuss the construction of transport operators corresponding to general Lévy jump pdfs and exponentially truncated Lévy jump pdfs in particular. Starting from the CTRW, we derive a tempered fractional diffusion equation and apply it to study the interplay between memory effects and truncation. In particular, we show that truncation leads to a super-diffusive to sub-diffusive transition. As a second application, following [11], we study the role of truncation in the super-diffusive propagation of fronts in reaction-diffusion systems.

The rest of the chapter is organized as follows. Section 2 discusses the construction of macroscopic transport models corresponding to the continuum, fluid limit of CTRWs with memory and jump pdfs characterized by general Lévy densities. As a specific example, we consider in detail the case of exponentially tempered Lévy densities. Section 3 discusses the scaling properties and moments of the Green's function of tempered fractional diffusion equation. The Green functions for the symmetric and asymmetric cases are explicitly computed. The transition from super-diffusive transport to sub-diffusive transport resulting from the interplay of memory effects and tempering is discussed in Sec. 4. The role of truncation in the super-diffusive propagation of fronts in the tempered fractional Fisher–Kolmogorov equation is discussed in Sec. 5. Section 6 presents the conclusions.

## 2. Continuous Time Random Walk for General Lévy Processes

The CTRW model consists of an ensemble of particles that at times, $t_1, t_2, \ldots, t_i, \ldots$, experience displacements, or jumps, $x_1, x_2, \ldots, x_i, \ldots$. Both, the waiting times between jumps, $\tau_i = t_i - t_{i-1}$, and the size of the jumps, $x_i$, are independent, identically distributed random variables drawn from a waiting-time probability density function (pdf), $\psi(\tau)$, and a jump pdf, $\eta(x)$, respectively. Given $\eta$ and $\psi$, the probability of finding a particle at position $x$ and time $t$ is determined by the Montroll–Weiss (MW) equation [2–4, 12] $\psi$,

$$\phi(x,t) = \delta(x)\Psi(t) + \int_0^t \psi(t-t') \left[\int_{-\infty}^{\infty} \eta(x-x')\,\phi(x',t')\,dx'\right] dt', \tag{7}$$

where

$$\Psi(t) = \int_t^{\infty} \psi(t')\,dt' \tag{8}$$

is the survival probability distribution function giving the probability of waiting more than a time $t$ before jumping. Assuming that the initial condition is a Dirac delta function, $\phi(x, t = 0) = \delta(x)$, the first term on the right-hand side of Eq. (7) represents the persistence of the initial condition.

The second term on the right-hand side of Eq. (7) represents the contributions of particles moving from $x'$ to $x$. This term includes, in addition to the transition probability $\eta(x - x')$, the time integral of $\psi(t - t')$ to guarantee the arrival of the particles at location $x$ at time $t$.

The integral operators in the Montroll–Weiss Eq. (7) take a particularly simple form in Fourier–Laplace variables. Introducing

$$\mathcal{F}[\eta] = \hat{\eta}(k) = \int_{-\infty}^{\infty} e^{ikx}\eta(x)dx, \qquad \mathcal{L}[\psi] = \tilde{\psi}(s) = \int_{0}^{\infty} e^{-st}\psi(t)dt, \quad (9)$$

and using the convolution theorem of Fourier–Laplace transforms, Eq. (7) reduces to an algebraic equation that can be trivially solved to yield

$$\hat{\tilde{\phi}}(k,s) = \frac{1-\tilde{\psi}}{s}\frac{1}{1-\tilde{\psi}\,\hat{\eta}}. \quad (10)$$

Of particular interest in modeling of anomalous transport is the incorporation of memory effects due to long waiting times. A natural way to accomplish this is by using pdfs with slowly decaying algebraic tails to allow the possibility of arbitrary long waiting periods at all scales. Here we consider the Mittag-Leffler function, which can be viewed as fractional generalization of the exponential,

$$E_\beta(z) = \sum_{n=0}^{\infty} \frac{z^n}{\Gamma(\beta n + 1)}, \quad (11)$$

where $0 < \beta < 1$. A key property of $E_\beta$, which makes it a good candidate for modeling process with anomalously long waiting limes, is its asymptotic behavior,

$$E_\beta(z) \sim -\sum_{j=1}^{\infty} \frac{z^{-j}}{\Gamma(1-\beta j)}, \qquad |z| \gg 1, \qquad z < 0, \quad (12)$$

which for $z = -t^\beta$ gives $E_\beta(-t^\beta) \sim t^{-\beta}$. Motivated by this, we consider [13–15]

$$\Psi(t) = E_\beta(-t^\beta)\,. \quad (13)$$

For $\beta = 1$, $E_\beta$ reduces to the exponential, $E_1(z) = e^z$, and $\Psi$ reduces to the exponential decaying distribution characteristics of standard diffusive processes. Using the Laplace transform property,

$$\mathcal{L}\left\{t^{n\beta} E_{\beta}^{(n)}\left[\pm a t^{\beta}\right]\right\} = \frac{n! s^{\beta-1}}{\left(s^{\beta} \mp a\right)^{n+1}}, \qquad \mathrm{Re}(s) > |a|, \tag{14}$$

where $E_{\beta}^{(n)}(z) = \frac{d^n}{dz^n} E_{\beta}(z)$, Eq. (10) becomes

$$\tilde{\hat{\phi}} = \frac{s^{\beta-1}}{s^{\beta} + 1 - \hat{\eta}}. \tag{15}$$

The inversion of the Laplace transform in Eq. (15) can be formally accomplished by introducing the integro-differential operator

$${}_{0}^{c}D_t^{\beta}\phi = \frac{1}{\Gamma(1-\beta)} \int_0^t \frac{\partial_\tau \phi}{(t-\tau)^{\beta}}\, d\tau, \tag{16}$$

known as the Caputo fractional derivative of order $\beta$ which satisfies

$$\mathcal{L}\left[{}_{0}^{c}D_t^{\beta}\phi\right] = s^{\beta}\tilde{\phi} - s^{\beta-1}\phi(0). \tag{17}$$

Using Eq. (17), the Montroll–Weiss Eq. (15) gives

$${}_{0}^{c}D_t^{\beta}\hat{\phi}(k,t) = [\hat{\eta} - 1]\,\hat{\phi}. \tag{18}$$

An advantage of using the Mittag-Leffler function to model $\Psi$ is the possibility of inverting exactly the Laplace transform in the master equation in terms of fractional time derivatives. However, it is important to point out that any other choice of $\Psi$ will yield Eq. (18) in the time asymptotic limit provided $\Psi$ exhibits the algebraic decay $\Psi \sim t^{-\beta}$. That is, in the time asymptotic limit of interest to transport modeling, what matters are not the details of $\Psi$ but its asymptotic properties. In a similar way, concerning the spatial scales, what is important to macroscopic transport modeling is the large scale dynamics, i.e. the long wavelength limit. In Fourier space this corresponds to the limit $k \to 0$, that allows the approximation of the characteristic function of the jump pdf, $\eta$, as

$$\hat{\eta}(k) = e^{\Lambda(k)} \approx 1 + \Lambda(k) + \cdots, \tag{19}$$

where $\Lambda(k)$ is the characteristic exponent. Substituting Eq. (19) into Eq. (18) leads to the equation

$${}_{0}^{c}D_t^{\beta}\hat{\phi}(k,t) = \Lambda\hat{\phi}, \tag{20}$$

that models macroscopic transport in the small-$k$ (fluid) limit of the CTRW with a *general* probability distribution function of jumps with characteristic function $\hat{\eta} = e^{\Lambda}$.

Following [10] we assume that the characteristic exponent belongs to the family defined by the Lévy–Khintchine representation [16]

$$\Lambda = \ln \hat{\eta} = aik - \frac{1}{2}\sigma^2 k^2 + \int_{-\infty}^{\infty} \left[e^{ikx} - 1 - iku(x)\right] w(x) dx \,, \tag{21}$$

where $a$ is a constant and $\sigma \geq 0$. The function $u(x)$ is a truncation function used to remove the singularity of the integrand at the origin and to guarantee the convergence of the integral. It can be shown that the specific form of this function is irrelevant and that different choices only manifest as different rescalings of the constant $a$. The function $w(x)$ is called the Lévy density and it satisfies $\int \min\{1, x^2\} w(x) dx < \infty$. Substituting Eqs. (21) into (20) and taking the inverse Fourier transform yields [10]

$$\begin{aligned} {}_0^c D_t^\beta \phi = & -a\partial_x \phi + \frac{1}{2}\sigma^2 \partial_x^2 \phi \\ & + \int_{-\infty}^{\infty} \left[\phi(x-y,t) - \phi(x,t) + u(y)\partial_x \phi\right] w(y) dy \,. \end{aligned} \tag{22}$$

Equation (22) is the macroscopic transport equation describing the continuum, fluid limit of a CTRW with a general jump distribution function $\eta$ characterized by a general Lévy density $w(y)$. For an alternative derivation of Eq. (22) see [17–19].

In the case of $\alpha$-stable Lévy processes, for which $w(x) \sim |x|^{-(1+\alpha)}$, Eq. (22) gives the well-known fractional diffusion equation,

$${}_0^c D_t^\beta \phi(x,t) = -a\partial_x \phi + \frac{1}{2}\sigma^2 \partial_x^2 \phi + c\left[l \, {}_{-\infty}D_x^\alpha + r \, {}_x D_\infty^\alpha\right] \phi \,, \tag{23}$$

where

$${}_a D_x^\alpha \phi = \frac{1}{\Gamma(m-\alpha)} \frac{\partial^m}{\partial x^m} \int_a^x \frac{\phi(y)}{(x-y)^{\alpha+1-m}} dy \,, \tag{24}$$

$${}_x D_b^\alpha \phi = \frac{(-1)^m}{\Gamma(m-\alpha)} \frac{\partial^m}{\partial x^m} \int_x^b \frac{\phi(y)}{(y-x)^{\alpha+1-m}} dy \,, \tag{25}$$

with $m - 1 \leq \alpha < m$, are the Riemann–Liouville fractional derivatives [15, 20], and $l$ and $r$ are the weighting factors defined as

$$l = -\frac{(1-\theta)}{2\cos(\alpha\pi/2)} \,, \qquad r = -\frac{(1+\theta)}{2\cos(\alpha\pi/2)} \,. \tag{26}$$

Truncated Lévy distributions were originally introduced in [21, 22] as a natural prescription to guarantee the finiteness of the second moments. The role of general Lévy processes, and in particular truncated Lévy processes, in the CTRW model was studied in [10]. At a more formal level, [24] studied a general class of multivariate tempered stable Lévy processes and established their parametrization and probabilistic representations. Truncated Lévy processes have shown applicability in many areas. Some examples include: plasma physics [25], fluid mechanics [26], space plasmas [27], and finance [28].

For tempered Lévy processes [22] the Lévy density is given by

$$w_{ET}(x) = \begin{cases} c\dfrac{(1+\theta)}{2}\,|x|^{-(1+\alpha)}\,e^{-\lambda|x|} & \text{for } x<0, \\ c\dfrac{(1-\theta)}{2}x^{-(1+\alpha)}e^{-\lambda x} & \text{for } x>0, \end{cases} \tag{27}$$

$0<\alpha\le 2$, $c>0$, $-1\le\theta\le 1$ and $\lambda\ge 0$. In this case, the integral in Eq. (21) leads to the characteristic exponent

$$\begin{aligned} \Lambda_{ET} = iak - \frac{\sigma^2}{2}k^2 - \frac{c}{2\cos(\alpha\pi/2)} \\ \times \begin{cases} (1+\theta)(\lambda+ik)^\alpha + (1-\theta)(\lambda-ik)^\alpha - 2\lambda^\alpha, \\ (1+\theta)(\lambda+ik)^\alpha + (1-\theta)(\lambda-ik)^\alpha - 2\lambda^\alpha - 2ik\alpha\theta\lambda^{\alpha-1}, \end{cases} \end{aligned} \tag{28}$$

for $0<\alpha<1$ and $1<\alpha\le 2$ respectively. Substituting Eq. (28) into Eq. (20) and using

$$\mathcal{F}\left[e^{-\lambda x}\,{}_{-\infty}D_x^\alpha\,e^{\lambda x}\,\phi\right] = (\lambda-ik)^\alpha\,\hat{\phi}\,, \tag{29}$$

$$\mathcal{F}\left[e^{\lambda x}\,{}_xD_\infty^\alpha\,e^{-\lambda x}\,\phi\right] = (\lambda+ik)^\alpha\,\hat{\phi}\,, \tag{30}$$

we obtain, after inverting the Fourier transform, the tempered fractional diffusion equation [10]

$${}_0^cD_t^\beta\phi = -V\partial_x\phi + \frac{\sigma^2}{2}\partial_x^2\phi + c\mathcal{D}_x^{\alpha,\lambda}\phi - \mu\phi\,, \tag{31}$$

where the drift velocity in Eq. (31) is given by

$$V = \begin{cases} a & \text{for } 0<\alpha<1, \\ a - \dfrac{c\alpha\theta\lambda^{\alpha-1}}{|\cos(\alpha\pi/2)|} & \text{for } 1<\alpha<2, \end{cases} \tag{32}$$

$$\mu = -\frac{c\lambda^\alpha}{\cos(\alpha\pi/2)} \tag{33}$$

and the $\lambda$-truncated fractional derivative operator of order $\alpha$, $\mathcal{D}_x^{\alpha,\lambda}$, is defined as

$$\mathcal{D}_x^{\alpha,\lambda} = l e^{-\lambda x} \, {}_{-\infty}D_x^\alpha \, e^{\lambda x} + r e^{\lambda x} \, {}_xD_\infty^\alpha \, e^{-\lambda x} . \tag{34}$$

A different approach to construct macroscopic transport models incorporating truncated Lévy processes was proposed in [29] where a special case of a distributed order fractional diffusion equation was used to describe a power-law truncated Lévy process. In this case, a Lévy distribution with a power law decay of order $1+\alpha$ is truncated by a steeper power law distribution with decay $5-\alpha$. This guarantees the convergence of the moments of order lower than $4-\alpha$.

## 3. Green's Function of Tempered Fractional Diffusion Equation

The solution of the initial value problem of Eq. (31) is

$$\phi(x,t) = \frac{1}{2\pi}\int_{-\infty}^{\infty} G(x-y,t)\phi_0(y)dy\,, \tag{35}$$

where $\phi_0$ denotes the initial condition, $G$ is the Green's function or propagator given by

$$G(x,t) = \frac{1}{2\pi}\int_{-\infty}^{\infty} e^{-ikx} E_\beta\left[t^\beta \Lambda(k)\right] dk\,, \tag{36}$$

and $\Lambda$ is the characteristic function of the general Lévy process in Eq. (21). For the case of tempered Lévy processes,

$$G_{\alpha,\beta,\theta,\lambda}(x,t) = \frac{1}{2\pi}\int_{-\infty}^{\infty} e^{-ikx} E_\beta\left[t^\beta \Lambda_{ET}(k)\right] dk\,, \tag{37}$$

where $\Lambda_{ET}$ is given in Eq. (28), and the sub-indices of $G$ denote the explicit dependence on: the order of the fractional derivative in space, $\alpha$, the order of the fractional time derivative, $\beta$, the asymmetry of the fractional operator, $\theta$, and the truncation $\lambda$.

For $a=0$ and $\sigma=0$, the characteristic exponent in Eq. (28) satisfies the scaling relation $\Lambda_{ET}(\mu k;\mu\lambda) = \mu^\alpha \Lambda_{ET}(k;\lambda)$. Using Eq. (37), this scaling implies the following scaling for the propagator $G_{\alpha,\beta,\theta,\lambda}$

$$G_{\alpha,\beta,\theta,\lambda}(x,\mu t;\lambda) = \mu^{-\beta/\alpha} G_{\alpha,\beta,\theta,\lambda}(\mu^{-\beta/\alpha} x, t; \mu^{\beta/\alpha}\lambda)\,. \tag{38}$$

The dependence on $\lambda$ indicates that there is no spacetime self-similarity; the truncation introduces the preferred length scale $1/\lambda$ that breaks the scale invariance. That is, the Green's function at a time $\mu t$ cannot be obtained from a simple rescaling in $x$ of the Green's function at $t$, unless $\lambda$ is also rescaled.

One of the key issues of the tempered Lévy distributions is the finiteness of all the moments. The moment of order $n$ is given

$$\langle x^n \rangle = \int_{-\infty}^{\infty} x^n G_{\alpha,\beta,\theta,\lambda}(x,t)dx = (-i)^n \left. \frac{\partial^n}{\partial k^n} \hat{G}_{\alpha,\beta,\theta,\lambda}(k,t) \right|_{k=0}, \tag{39}$$

where according to Eq. (37), $\hat{G}_{\alpha,\beta,\theta,\lambda} = E_\beta[t^\beta \Lambda_{ET}]$. For $\lambda = 0$, only the moments of order $n < \alpha$ exist, which as discussed before, is one of the drawbacks of the Lévy $\alpha$-stable distributions from the applications point of view. On the other hand, for $\lambda \neq 0$, the function $\Lambda_{ET}(k)$ is $C^\infty$ at $k = 0$, and *all* the moments of the distribution exist. The first order moment exhibits the usual scaling

$$\langle x \rangle = \frac{V}{\Gamma(\beta+1)}\, t^\beta. \tag{40}$$

However, it is interesting to note that, according to Eq. (32), for $1 < \alpha < 2$, an asymmetry, $\theta \neq 0$, in the fractional operators can decrease of increase the effective transport velocity $V$. For the time evolution of the variance the second moment gives [10]

$$\sigma^2 = \left\langle [x - \langle x \rangle]^2 \right\rangle = \left[ \frac{2}{\Gamma(2\beta+1)} - \frac{1}{\Gamma^2(\beta+1)} \right] V^2 t^{2\beta} + \frac{2\,\chi}{\Gamma(\beta+1)}\, t^\beta, \tag{41}$$

where we have introduced the effective diffusivity

$$\chi = \frac{\sigma^2}{2} + \frac{c\alpha|\alpha-1|}{2|\cos(\alpha\pi/2)|\lambda^{2-\alpha}}. \tag{42}$$

The truncation increases the Gaussian diffusivity $\sigma^2/2$ by a term inversely proportional to $\lambda$ and directly proportional to $c$, the strength of the truncated Lévy density in Eq. (27). Furthermore, in the limit $\lambda \to 0$, $\chi \to \infty$, a result consistent with the divergence of the second moment in the absence of truncation.

### 3.1. *Symmetric solution*

Here we compute the solution in the case $\theta = a = \sigma = 0$ and $1 < \alpha < 2$. For symmetric processes, $\theta = 0$, Eq. (37) gives

$$G_{\alpha,\beta,0,\lambda}(x,t) = \frac{1}{\pi}\int_0^\infty \cos(kx) E_\beta\left[t^\beta \Lambda_{ET}(k)\right] dk. \tag{43}$$

For $\theta = a = \sigma = 0$, Eq. (28) can be rewritten as

$$\Lambda_{ET} = \frac{-c\lambda^\alpha}{\cos(\alpha\pi/2)}\left\{\left[1+\left(\frac{k}{\lambda}\right)^2\right]^{\alpha/2}\cos\left[\alpha\tan^{-1}\left(\frac{k}{\lambda}\right)\right]-1\right\}. \tag{44}$$

Substituting Eq. (44) into Eq. (43), assuming $1 < \alpha < 2$, and making the change of variables

$$\eta = \lambda x, \qquad \tau = t/t_c, \tag{45}$$

where

$$t_c = c^{-1/\beta}\lambda^{-\alpha/\beta}, \tag{46}$$

we get

$$\frac{\pi}{\lambda} G_{\alpha,\beta,0,\lambda}(\eta,\tau) = \int_0^\infty \cos(\eta u)\, E_\beta\left[\tau^\beta \Phi(u)\right] du, \tag{47}$$

$$\Phi(u) = \frac{1}{|\cos(\alpha\pi/2)|}\left\{\left(1+u^2\right)^{\alpha/2}\cos\left[\alpha\tan^{-1}(u)\right]-1\right\}. \tag{48}$$

Since Eq. (47) is not analytically tractable, we have to resort to a numerical evaluation of the integral. One might expect that this could be achieved by a direct quadrature on a finite domain. However, as Fig. 1 shows, the slow algebraic decay of the Mittag-Leffler function can introduce significant numerical errors. To eliminate these errors, the asymptotic contribution of the integral should be taken into consideration. To do this, we introduce a cutoff, $u_m \gg 1$ and consider the integral

$$I_\infty = \int_{u_m}^\infty \cos(\eta u)\, E_\beta[\tau^\beta \Phi(u)]\, du\,. \tag{49}$$

For $u \gg 1$, $\Phi(u)$ is a large negative number and we can use the asymptotic expansion of the Mittag-Leffler function in Eq. (12), and write to leading

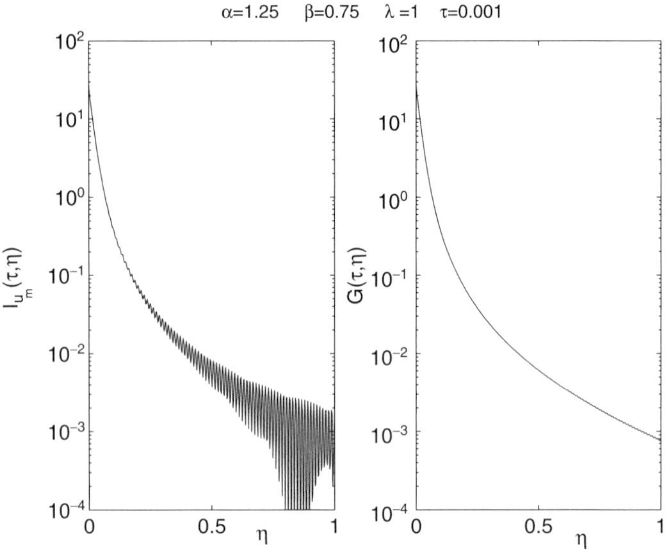


Fig. 1. Numerical computation of the symmetric Green's function, Eqs. (47) and (48), at $\tau = 0.001$, of the tempered fractional diffusion Eq. (31) for $a = \sigma = 0$, $\alpha = 1.25$, $\beta = 0.5$ and $\lambda = 1$. The left panel shows the significant numerical errors introduced in the direct numerical quadrature of the Green's function integral in Eq. (52). The right panel shows the Green's function evaluated using Eq. (51). It is observed that the analytic asymptotic term corrects the errors introduced by the slow algebraic decay of the Mittag-Leffler function.

order

$$I_\infty \sim \frac{-\tau^{-\beta}}{\Gamma(1-\beta)} \int_{u_m}^{\infty} \frac{\cos(\eta u)}{\Phi(u)} du, \tag{50}$$

for $\tau^\beta |\Phi(u_m)| \gg 1$. Approximating, $\Phi(u) \sim -u^\alpha$, and integrating by parts we get the leading contribution of $I_\infty$ for $\eta \neq 0$ which allow to approximate [10]

$$G_{\alpha,\beta,0,\lambda}(\eta,\tau) = I_{u_m} - \frac{\lambda \tau^{-\beta}}{\pi \Gamma(1-\beta)} \frac{\sin(u_m \eta)}{u_m^\alpha\, \eta}, \tag{51}$$

$$I_{u_m} = \frac{\lambda}{\pi} \int_0^{u_m} \cos(\eta u) E_\beta[\tau^\beta \Phi(u)]\, du \tag{52}$$

for

$$u_m \gg \left(\frac{t_c}{t}\right)^{\beta/\alpha}. \tag{53}$$

The right panel on Fig. 1 shows how the incorporation of the asymptotic term, $I_\infty$, corrects the rapidly oscillating errors introduced in the numerical evaluation of the Green's function integral in a finite domain.

### 3.2. *Asymmetric solution*

As a particular example that will play an important role in the study of fronts, we consider the solutions of Eq. (31) with $1 < \alpha < 2$, $\beta = 1$, and $\theta = -1$. We will assume an advection velocity $a = v$, which results in the following asymmetric, truncated fractional equation

$$\partial_t \phi = \chi \left[ e^{-\lambda x} \, {}_{-\infty}D_x^\alpha \left( e^{\lambda x} \phi \right) - \lambda^\alpha \phi \right] , \tag{54}$$

with $\chi = c/|\cos(\alpha\pi/2)|$. In this case, the Green's function in Eq. (37) reduces to

$$G_{\alpha,1,-1,\lambda} = \frac{1}{2\pi} \int_{-\infty}^{\infty} e^{-ikx + \chi t[(\lambda - ik)^\alpha - \lambda^\alpha]} \, dk \, , \tag{55}$$

which can be equivalently written as

$$G_{\alpha,1,-1,\lambda} = e^{-\lambda x - \chi \lambda^\alpha t} \, (\chi t)^{-1/\alpha} \, G_{\alpha,1,-1,0}(\eta) \, , \tag{56}$$

where

$$G_{\alpha,1,-1,0}(\eta) = \frac{1}{2\pi} \int_{-\infty}^{\infty} e^{i^\alpha k^\alpha + ik\eta} \, dk \, , \tag{57}$$

is the Green's function of the asymmetric, $\alpha$-stable ($\lambda = 0$) fractional diffusion equation, in terms of the similarity variable

$$\eta = x \, (\chi t)^{-1/\alpha} \, . \tag{58}$$

Using the asymptotic expression of the $\alpha$-stable Lévy distributions [16, 30]

$$G_{\alpha,1,-1,0}(\eta) \sim \eta^{-1-\alpha} \, , \qquad \eta > 0 \, , \tag{59}$$

it follows from Eq. (55) that

$$G_{\alpha,1,-1,\lambda} \sim \chi t e^{-\chi \lambda^\alpha t} \, \frac{e^{-\lambda x}}{x^{1+\alpha}} \, , \qquad x \gg (\chi t)^{1/\alpha} \, . \tag{60}$$

For the decay of the left tail, we use the asymptotic expansion

$$G_{\alpha,1,-1,0}(\eta) \sim |\eta|^{a_2} \, e^{-b_2 |\eta|^{c_2}} \, , \qquad \eta < 0 \, , \tag{61}$$

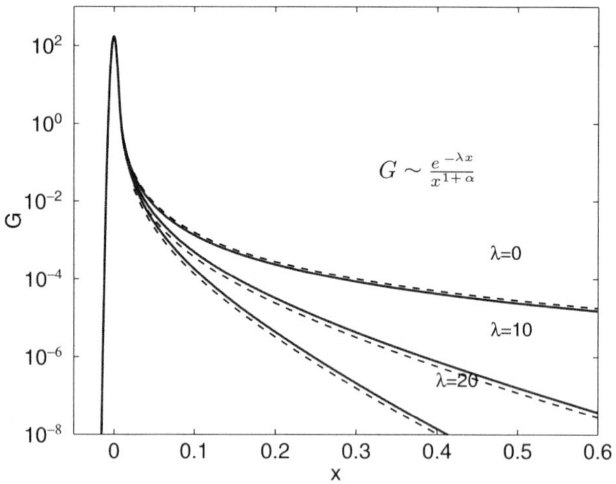


Fig. 2. Green's functions according to Eq. (55), of the asymmetric truncated fractional diffusion equation as function of $x$ at fixed $t$, for $\alpha = 1.5$, $\beta = 1$, $\theta = -1$, and $\lambda = 0$, 10, and 20. For $x > 0.1$, there is very good agreement with the asymptotic expansion in Eq. (60) shown in dashed lines and slightly shifted downwards for visualization purposes.

where $a_2 = (2-\alpha)/(2(\alpha-1))$, $b_2 = (\alpha-1)\alpha^{\alpha/(\alpha-1)}$ and $c_2 = \alpha/(\alpha-1)$ [30], to conclude

$$G_{\alpha,1,-1,\lambda} \sim (\chi t)^{-(a_2+1)/\alpha} e^{-\chi\lambda^\alpha t} |x|^{a_2} e^{-b_2(\chi t)^{-c_2/\alpha} |x|^{c_2} + \lambda|x|}, \tag{62}$$

for $x < 0$ and $|x| \gg (\chi t)^{1/\alpha}$. Since we are assuming that $1 < \alpha < 2$, it follows that $c_2 > 1$ and the $-|x|^{c_2}$ term in the exponent dominates the $\lambda|x|$ term, leading to a faster than exponential decay of the left tail for any value of $\lambda$. Figure 2 shows plots of the Green's function in Eq. (55) for $\alpha = 1.5$ and different values of $\lambda$, along with the asymptotic approximation in Eq. (60).

## 4. Transition from Super-diffusive to Sub-diffusive Transport

As discussed before, the spatio-temporal evolution of the Green's function is not self-similar in the presence of truncation. That is, according to Eq. (38), for $\lambda \neq 0$, the Green's function at successive times cannot be obtained from a simple spatial re-scaling of the Green's function at earlier times. This breakdown of self-similarity is illustrated in Fig. 3 that shows $G_{\alpha,\beta,0,\lambda}$, in Eq. (43) with $\alpha = 1.25$, $\beta = 0.75$, $\lambda = 1$, for $\tau = 10^{-4}$, $\tau = 1$ and $\tau = 10^3$. At short times, compared with the crossover time, $t_c$ in Eq. (46), $G_{\alpha,\beta,0,\lambda}$

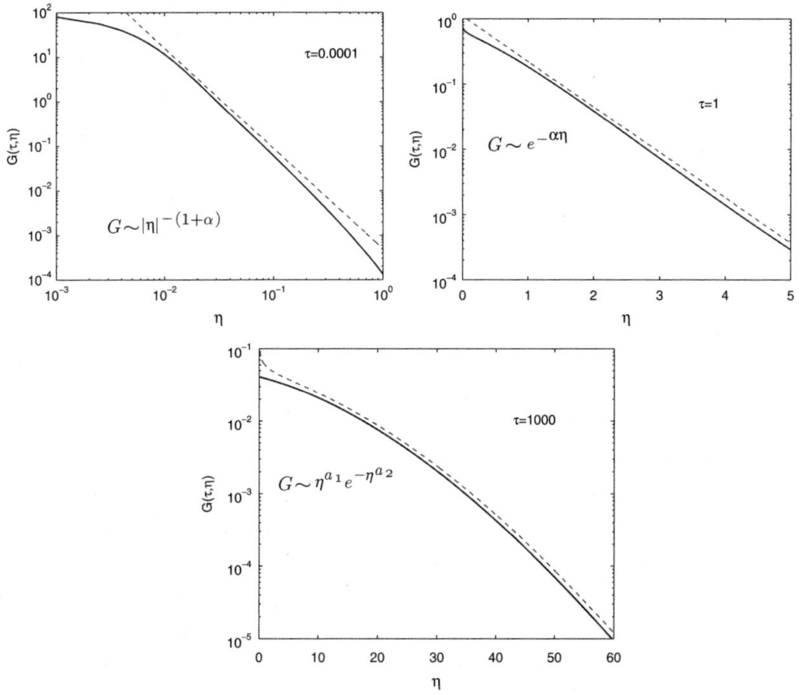


Fig. 3. (Color online) Super-diffusive to sub-diffusive transition in the Green's function of the tempered fractional diffusion Eq. (31) for $a = \sigma = \theta = 0$, $\alpha = 1.25$, $\beta = 0.75$ and $\lambda = 1$. Each panel shows the Green's function at different times. The tempering introduces a preferred length scale and the Green's function does not exhibits a spatio-temporal evolution for $\lambda \neq 0$. At short times, $\tau = 10^{-4}$, (top-left panel) the Green's function follows the algebraic decay in Eq. (63), shown with a dashed red line, characteristic of super-diffusive transport. At the cross-over time, $\tau = 1$, (top-right panel) the Green's function follows the exponential decay in Eq. (65), shown with a dashed red line. At long times, $\tau = 10^3$, (bottom panel) the Green's function exhibits a transition to the stretched exponential decay in Eq. (66), shown with a dashed red line, characteristic of sub-diffusive transport.

exhibits the asymptotic algebraic decay characteristic of the Green's function of the fractional diffusion equation with $\beta < 1 < \alpha < 2$, i.e.

$$G_{\alpha,\beta,0,\lambda}(\eta, t \ll t_c) \sim |\eta|^{-(1+\alpha)}\,, \tag{63}$$

for $|\eta \gg 1$. This result is expected because Eq. (38) implies

$$G_{\alpha,\beta,0,\lambda}(x, t; \lambda) = \lambda G_{\alpha,\beta,0,\lambda}(\lambda x, \lambda^{\alpha/\beta} t; 1)\,, \tag{64}$$

which indicates that the Green's function for finite $\lambda$ in the limit $t \to 0$ corresponds (up to a re-scaling) to the Green's function for finite time in the limit $\lambda \to 0$. The top-right panel in Fig. 3 shows the Green's function for $\tau = 1$, i.e. $t \approx t_c$. In this case, the distribution (shown in a log-normal scale) exhibits the exponential dependence

$$G_{\alpha,\beta,0,\lambda}(x, t \approx t_c) \sim e^{-a\eta}\,, \tag{65}$$

As Fig. 3 shows, for times much longer than the cross-over time, the Green's function transitions into a stretched exponential of the form

$$G_{\alpha,\beta,0,\lambda}(x, t \gg t_c) \sim \eta^{a_1} \exp\left(-\eta^{a_2}\right)]\,, \tag{66}$$

with $a_1 = (\beta - 1)/(2 - \beta)$ and $a_2 = 2/(2 - \beta)$, which corresponds to the Green's function of the subdiffusive ($\alpha = 2$, $0 < \beta < 1$) fractional equation [4].

The transition from super-diffusive transport to sub-diffusive transport also manifests in the time evolution of the decay of the Green's function at the origin $G_{\alpha,\beta,0,\lambda}(0,t)$ [10]. From Eq. (38), it follows that, $G_{\alpha,\beta,0,\lambda}(0,t;\lambda) = t^{-\beta/\alpha} G_{\alpha,\beta,0,\lambda}(0,1;t^{\beta/\alpha}\lambda)$. This implies that, at short times, the Green's function of the tempered fractional diffusion equation exhibits the same decay as the Green's function of the $\alpha$-stable, fractional diffusion equation, i.e.

$$G_{\alpha,\beta,0,\lambda}(0, t \ll 1; \lambda) \sim t^{-\beta/\alpha}. \tag{67}$$

This scaling is verified in Fig. 4 for the case $\alpha = 1.5$, $\beta = 0.75$ and $\theta = 0$. To study the decay at long times, we use stationary phase arguments and get the leading order asymptotic behavior of $G_{\alpha,\beta,0,\lambda}(0, t \gg 1; \lambda)$ by Taylor expanding the argument of the Mittag-Leffler function around $k = 0$ in Eq. (37). Note, however, that care must be taken since, contrary to the usual situation involving exponentials, $E_\beta$ decays algebraically for $\beta \neq 1$. Restricting attention to the symmetric case, $\theta = 0$, with no drift $a = 0$, and substituting the small $k$ expansion $\Lambda_{ET} = -\chi k^2 + \cdots$ into Eq. (37) we get

$$\int_0^\infty E_\beta[t^\beta \Lambda_{ET}]dk \sim \int_0^\infty E_\beta[-\chi t^\beta k^2]dk = \frac{1}{\sqrt{\chi t^\beta}} \int_0^\infty E_\beta(-u^2)du\,, \tag{68}$$

and therefore,

$$G_{\alpha,\beta,0,\lambda}(0, t \gg 1; \lambda) \sim t^{-\beta/2}\,. \tag{69}$$

That is, as shown in Fig. 4, the truncation gives rise to a transition in the scaling from $\sim t^{-\beta/\alpha}$, for $t \ll 1$, to $\sim t^{-\beta/2}$, for $t \gg 1$. In the case $2\beta/\alpha > 1$ this corresponds to a transition from super-diffusive scaling to sub-diffusive

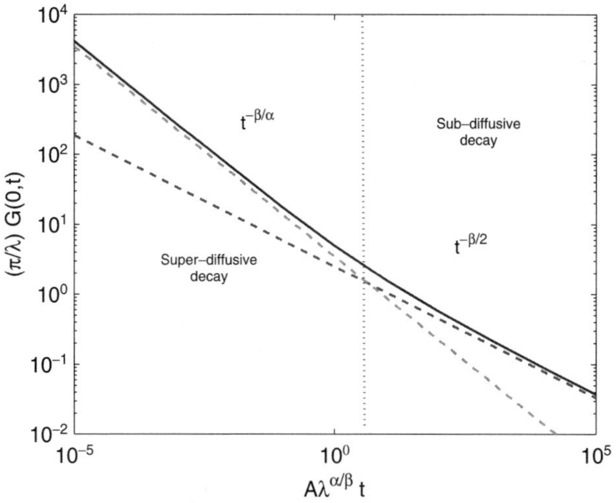


Fig. 4. Super-diffusive to sub-diffusive transition in the decay of the Green's function at the origin, $G(0,t)$, in the truncated fractional diffusion equation Eq. (31) with $\theta = 0$, $\lambda = 1$, $\alpha = 1.25$ and $\beta = 0.75$. The solid line shows $G(0,t)$, the steeper dashed line corresponds to the super-diffusive scaling in Eq. (67), and the less steep dashed line shows the sub-diffusive scaling in Eq. (69). The scaling constant in the horizontal axis is $A = [c/|\cos(\alpha\pi/2)|]^{1/\beta}$.

scaling. Consistent with Fig. 3, the transition occurs at the crossover time, $t \sim t_c$. The case, $\beta = 1$, corresponds to the ultraslow Gaussian convergence of truncated Lévy processes reported in [21, 23].

## 5. Truncation Effects in Super-diffusive Front Propagation

In the Fisher–Kolmogorov model,

$$\partial_t \phi = \chi \partial_x^2 \phi + \gamma \phi (1 - \phi) , \tag{70}$$

where $\chi$ denotes the diffusivity and $\gamma$ is a constant, the competition between the diffusivity and the nonlinearity leads to the propagation of fronts in which the stable, $\phi = 1$, state advances at a constant speed through the destabilization of the $\phi = 0$ unstable state [31].

Early work on reaction-anomalous-diffusion systems include [32–35]. Reference [32] studied bistable reaction processes and anomalous diffusion caused by Lévy flights. The interplay of sub-diffusion and Turing instabilities was discussed in [35]. The acceleration and algebraic decay of fronts due to super-diffusive transport was originally studied in the context of the fractional Fisher–Kolmogorov equation in [34] and in [33] in the context of an equivalent probabilistic model. Similar phenomenology was also observed in chaotic couple map lattices with long-range couplings in [36]. Other works include: the study of analytic solutions of fractional reaction-diffusion [37]; the study of a reaction-diffusion system with a bistable reaction term and directional anomalous diffusion [38]; the study of the construction of reaction-sub-diffusion equations [39]; the study of Turing instabilities [40]; the study of the effect of super-diffusion on pattern formation selection in the Brusselator model [41]; and the study of the fractional Ginzburg–Landau and Kuramoto–Sivashinsky equations [42] among others.

In this section, following [11], we explore the role of truncation in the super-diffusive propagation of fronts by modeling transport using the truncated fractional derivative operator discussed in Sec. 2. In the most general setting, the equation is [11]

$$ {}_0^cD_t^\beta \phi = -V\partial_x \phi + c\mathcal{D}_x^{\alpha,\lambda}\phi - \mu\phi + \gamma\phi(1-\phi) \, . \tag{71} $$

We will restrict attention to asymmetric, truncated fractional diffusion operators of the form in Eq. (54) and consider

$$ \partial_t \phi = \chi \left[ e^{-\lambda x} \, {}_{-\infty}D_x^\alpha \left( e^{\lambda x}\phi \right) - \lambda^\alpha \phi \right] + \gamma\phi(1-\phi), \tag{72} $$

with initial conditions of the form

$$ \phi(x, t=0) = \begin{cases} A & \text{for } x < 0, \\ B\, e^{-\nu x} & \text{for } x > 0, \end{cases} \tag{73} $$

where $A$, $B$ and $\nu$ are constants.

To study the asymptotic behavior of fronts in the truncated fractional Fisher–Kolmogorov equation, we use the leading edge approximation. This approximation is based on the idea that, at the leading edge of the front, $\phi \ll 1$, and therefore in this region the nonlinear reaction term can be linearized around $\phi = 0$, resulting in the linear equation

$$ \partial_t \phi = \chi e^{-\lambda x} \, {}_{-\infty}D_x^\alpha \left( e^{\lambda x}\phi \right) + (\gamma - \chi\lambda^\alpha)\phi \, , \tag{74} $$

where it is observed that the truncation has a direct effect on the growth rate through the term $-\chi\lambda^\alpha$. There are two characteristic time scales in this problem. One is the crossover time scale, $t_c$ in Eq. (46), and the other is the reaction time scale,

$$t_r = 1/\gamma. \tag{75}$$

In what follows we will assume that $t_c > t_r$ to guarantee that the effective reaction constant, $\gamma_{\text{eff}} = \gamma - \chi\lambda^\alpha$, is positive as needed for the excitation and propagation of "pull" type fronts.

Substituting

$$\phi = e^{-\lambda x + (\gamma - \chi\lambda^\alpha)t}\psi(x,t) \tag{76}$$

into Eq. (74) gives the asymmetric, $\alpha$-stable fractional diffusion equation

$$\partial_t \psi = \chi\,_{-\infty}D_x^\alpha \psi\,, \tag{77}$$

whose general solution for an initial condition $\psi(x, t = 0) = \psi_0(x)$ is

$$\psi(x,t) = \int_{-\infty}^{\infty} G_{\alpha,1,-1,0}(\eta)\,\psi_0[x - (\chi t)^{1/\alpha}\eta]d\eta\,, \tag{78}$$

where $G_{\alpha,1,-1,0}$ given in Eq. (57). For the initial condition in Eq. (73), Eqs. (78) and (76) give the following expression for the spatio-temporal evolution of the leading edge of the front [11]

$$\phi = e^{-\nu x + (\gamma - \chi\lambda^\alpha)t}\,\mathcal{I}_1 + Ae^{(\gamma - \chi\lambda^\alpha)t}\mathcal{I}_2\,, \tag{79}$$

where

$$\mathcal{I}_1(x,t) = \int_{-\infty}^{x/(\chi t)^{1/\alpha}} e^{(\nu-\lambda)(\chi t)^{1/\alpha}\eta} G_{\alpha,1,-1,0}(\eta)d\eta\,, \tag{80}$$

$$\mathcal{I}_2(x,t) = \int_{x/(\chi t)^{1/\alpha}}^{\infty} G_{\alpha,1,-1,0}(\eta)e^{-\lambda(\chi t)^{1/\alpha}\eta}d\eta\,. \tag{81}$$

### 5.1. *Diffusive case*

In the special case of diffusive transport $(\alpha, \theta, \lambda) = (2, 0, 0)$, $G_{2,1,0,0}$ is given by the Gaussian propagator

$$G_{2,1,0,0} = \frac{1}{2\sqrt{\pi}}e^{-\eta^2/4}\,, \tag{82}$$

and the leading edge solution in Eq. (79) reduces to

$$\phi = e^{-\nu x + (\gamma + \nu^2\chi)t}P_n\left(\frac{x - 2\nu\chi t}{\sqrt{2\chi t}}\right) + Ae^{\gamma t}\left[1 - P_n\left(\frac{x}{\sqrt{2\chi t}}\right)\right], \tag{83}$$

where $P_n(z) = (1/\sqrt{2\pi}) \int_{-\infty}^{z} e^{-u^2/2} du$ is the normal probability distribution. Using the asymptotic expansion $P_n(z) \sim 1 - (1/\sqrt{2\pi}) e^{-z^2/2}/z$, we conclude that, for $x > 2\nu\chi t$, in the limit

$$x/(\chi t)^{1/2} \gg 1 \,, \tag{84}$$

$$\phi \sim e^{-\nu(x-ct)} + \sqrt{\frac{\chi t}{\pi}} \left[ \frac{A}{x} - \frac{1}{x - 2\nu\chi t} \right] e^{-\frac{\gamma}{c_m^2 t}(x - c_m t)(x + c_m t)} \,, \tag{85}$$

where

$$c = \frac{\gamma}{\nu} + \nu\chi \,. \tag{86}$$

As expected, in this case, the leading edge exhibits the well-known asymptotic exponential dependence, $\phi \sim e^{-\nu(x-ct)}$, with front speed $c$ depending on the decay rate, $\nu$, of the initial condition. The minimum speed,

$$c_m = 2(\gamma\chi)^{1/2} \,, \tag{87}$$

is achieved for $\nu = \sqrt{\gamma/\chi}$. Note that, according to Eq. (85), in this case the convergence to constant speed is exponentially fast, i.e. the second term in the asymptotic expansion scales as $\sim e^{-a(x/\tau)^2}$. This result will be contrasted below with the much slower convergence in the case of truncated Lévy flights.

### 5.2. *Fractional case*

As discussed in [34], in the fractional ($\alpha$-stable) case, the leading edge solution is given by Eqs. (79) and (80) with $\lambda = 0$ and, in the limit,

$$x/(\chi t)^{1/\alpha} \gg 1 \,, \tag{88}$$

the leading asymptotic behavior of $\phi$ is

$$\phi \sim \chi t e^{\gamma t} \left[ \frac{A}{\alpha} x^{-\alpha} + \frac{1}{\nu} x^{-1-\alpha} + \cdots \right] . \tag{89}$$

The key difference with the Gaussian case, is the algebraic decay of the leading edge accompanied by the exponential acceleration of the front. Note that when $A \neq 0$, the front tail exhibits the decay $\phi \sim 1/x^{\alpha}$. However, when $A = 0$, the front decays faster, $\phi \sim 1/x^{\alpha+1}$. The top panel of Fig. 5 shows snapshots of the front profile $\phi$ as function of $x$ at different times in the $\alpha$-stable case obtained from the numerical solution of Eq. (72) with $\lambda = 0$.

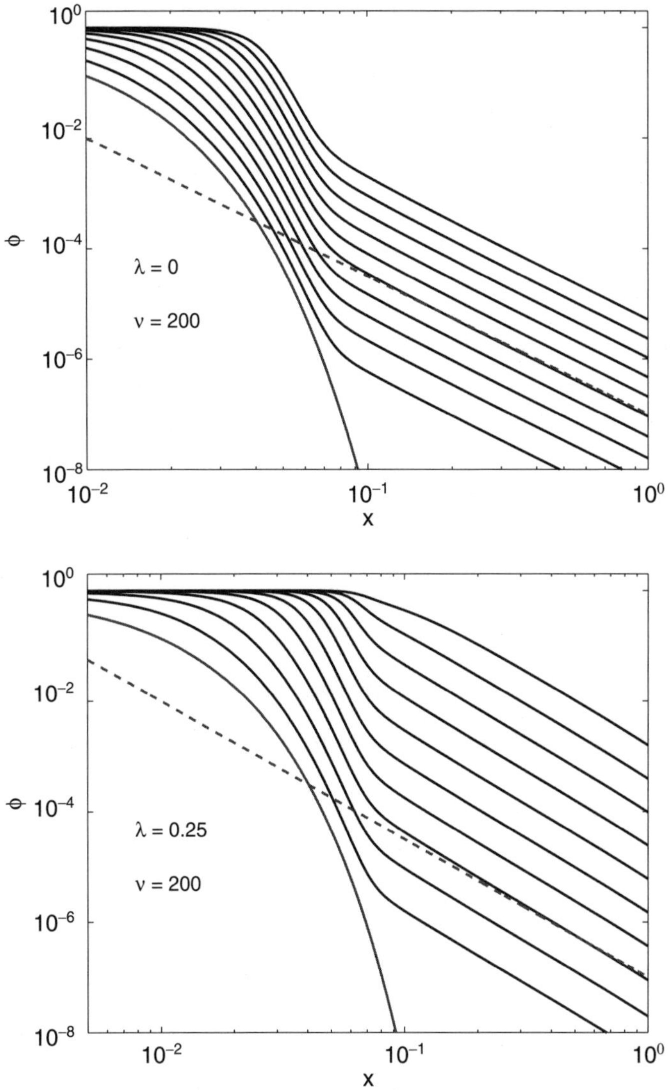


Fig. 5. Effect of truncation in the super-diffusive propagation of fronts with $\alpha = 1.5$, $\theta = -1$, $\gamma = 1$ and $\chi = 5 \times 10^{-7}$. The solid lines show the front profiles at different times, with the left-most curve corresponding to the exponential decaying initial condition. In the absence of truncation, $\lambda = 0$, (top panel) the front exhibits algebraic decaying tails. For $\lambda = 0.25$ and $\nu = 200$, (bottom panel) the front continues to exhibit algebraic decaying tails in the intermediate asymptotic regime $(\chi t)^{1/\alpha} \ll x \ll 1/\lambda$. In both plots, the dashed lines correspond to the scaling in Eq. (89).

In this case the front exhibits an algebraically decaying tail which in log–log scale manifests as a straight line [34]. The superimposed dashed line in the figure verifies the scaling in Eq. (89). In this, and all the subsequent simulations, we discretized the fractional derivative in the $x \in (0,1)$ domain using the Grunwald–Letnikov representation, and considered $\alpha = 1.5$, $\theta = -1$, $\gamma = 1$, $\chi = 5 \times 10^{-7}$.

In the Gaussian case, the exponential tail of the front propagates "rigidly", i.e. maintaining the exponential decaying shape, at a constant speed $c$. However, in the presence of fractional diffusion, each point of the leading edge moves at a different speed and the tail does not translate rigidly. In this case, it is convenient to define the Lagrangian trajectory of a point in the leading edge of the front as $x_L = x_L(t;\phi_0)$ where $\phi(x_L(t),t) = \phi_0$ with $\phi_0 \ll 1$. According to Eq. (89), in the $\alpha$-stable case, for $A = 0$,

$$x_L(t) = C \exp\left[\frac{1}{1+\alpha}(\gamma t + \ln t)\right], \tag{90}$$

where $C$ is a constant that depends on $\phi_0$, $\chi$ and $\nu$, and

$$v_L(t) = \frac{\gamma}{1+\alpha}\left(\frac{\chi t}{\phi_0 \nu}\right)^{\frac{1}{1+\alpha}}\left[\frac{1}{\gamma t}+1\right] e^{\left(\frac{\gamma}{1+\alpha}\right)t}, \tag{91}$$

which implies an unbounded, exponential growth of the front speed. For large $t$, the corresponding leading-order behavior of the front acceleration is

$$a_L(t) = \left(\frac{\gamma}{1+\alpha}\right)^2\left(\frac{\chi t}{\phi_0 \nu}\right)^{\frac{1}{1+\alpha}} e^{\left(\frac{\gamma}{1+\alpha}\right)t}. \tag{92}$$

The exponential unbounded growth of the Lagrangian velocity in Eq. (91) is shown in the $\lambda = 0$ curve in Fig. 7.

### 5.3. *Truncated case*

As discussed in [11], it can be shown that in the limit in Eq. (88),

$$\mathcal{I}_1 \sim \mathcal{C} + \frac{\tau^\alpha}{\nu - \lambda}\frac{e^{(\nu-\lambda)x}}{x^{1+\alpha}} + \cdots, \tag{93}$$

where $\mathcal{C}$ is a constant, and

$$\mathcal{I}_2 \sim \frac{\tau^\alpha}{\lambda}\frac{e^{-\lambda x}}{x^{1+\alpha}}\left[1 - \frac{(\alpha+1)}{\lambda x} + \cdots\right]. \tag{94}$$

Substituting Eqs. (93) and (94) into Eq. (79) we get

$$\phi \sim Ce^{-\nu x+(\gamma-\chi\lambda^\alpha)t} + \left(\frac{1}{\nu-\lambda} + \frac{A}{\lambda}\right)\frac{\chi t}{x^{\alpha+1}} e^{-\lambda x+(\gamma-\chi\lambda^\alpha)t} \qquad (95)$$

for $\nu \neq \lambda$. The leading term on the right-hand side of Eq. (95) depends on the relative values of $\nu$ and $\lambda$.

### 5.4. $\nu > \lambda$

In this case the initial condition decays faster than the truncation, and for large $x$, the leading order term in Eq. (95) is

$$\phi \sim \left[\frac{1}{\nu-\lambda} + \frac{A}{\lambda}\right]\frac{t\chi}{x^{1+\alpha}}\, e^{-\lambda x+(\gamma-\chi\lambda^\alpha)t}. \qquad (96)$$

The role of the truncation is clearly seen in the exponential factor $e^{-\lambda x}$ that dominates the decay for $x \gg 1/\lambda$. Figures 5 and 6 show a very good agreement between the numerical result and the scaling in Eq. (96).

When $x \ll 1/\lambda$, we can expand the exponential in Eq. (96) and write

$$\phi \sim \left[\frac{1}{\nu-\lambda} + \frac{A}{\lambda}\right] t\chi\, e^{(\gamma-\chi\lambda^\alpha)t}\, \frac{1}{x^{1+\alpha}} \left[1 - \lambda x + \frac{\lambda^2}{2}x^2 + \cdots\right]. \qquad (97)$$

According to Eq. (97) in the intermediate asymptotic regime,

$$(\chi t)^{1/\alpha} \ll x \ll 1/\lambda\,, \qquad (98)$$

the front exhibits to leading order the ideal (untruncated) Lévy flight algebraic scaling, $\phi \sim 1/x^{1+\alpha}$. This scaling is numerically verified in the bottom panel of Fig. 5 that shows that for small values of $\lambda$ the algebraic decay of the tail remains in the intermediate asymptotic regime where the role of truncation is negligible. Outside the intermediate asymptotic regime, the effect of the truncation depends critically on the ratio of the length scale of the truncation, $1/\lambda$, and length scale, $1/\nu$, of the initial condition. When, $\lambda < \nu$, i.e. when the initial condition decays faster than the truncation, the tail of the front follows the scaling in Eq. (96) as observed in the top panel of Fig. 5.

Moreover, as Fig. 7 shows, in the intermediate asymptotic regime (which in the numerical simulations roughly corresponds to $0 < \lambda \leq 1$) the front's velocity and acceleration exhibit unbounded monotonic growth and follow

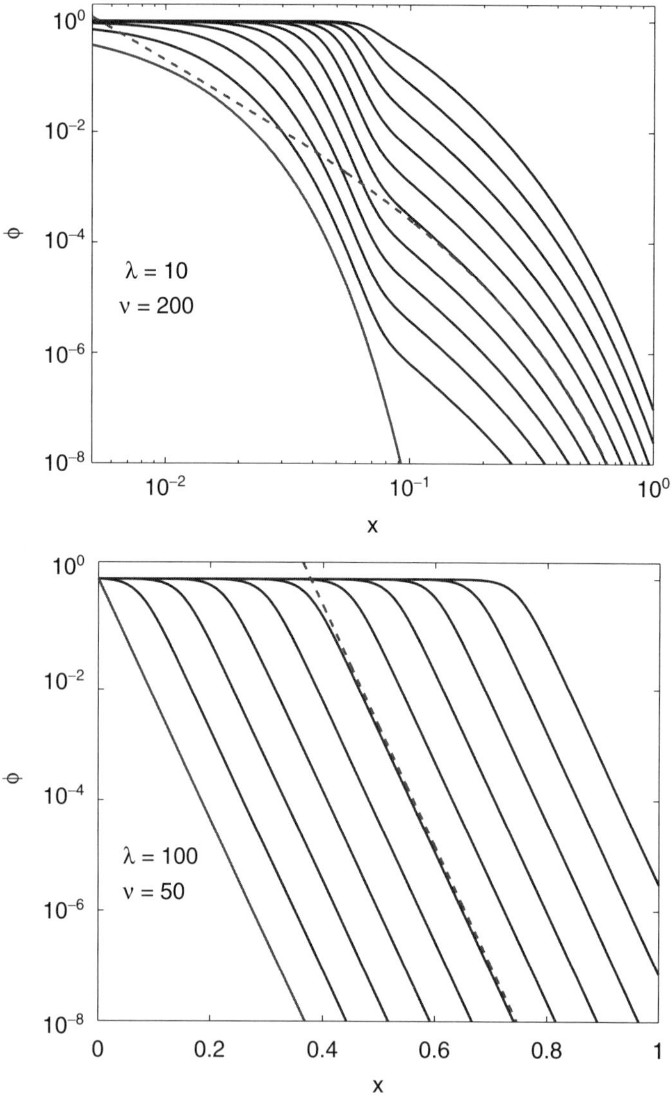


Fig. 6. Effect of truncation in the super-diffusive propagation of fronts with $\alpha = 1.5$, $\theta = -1$, $\gamma = 1$, and $\chi = 5\times 10^{-7}$. The solid lines show the front profiles at different times, with the left-most curve corresponding to the exponential decaying initial condition. As shown in the top panel, for large enough truncation ($\lambda = 10$ in this case), the algebraic decay is exponentially tempered, and the tails follow the scaling in Eq. (96) as indicated by the dashed line fitting. In the over-truncated, $\lambda > \nu$, case shown in the bottom panel, the tails of the front follow the standard diffusive (Fisher–Kolmogorov) exponential decay in Eq. (105).

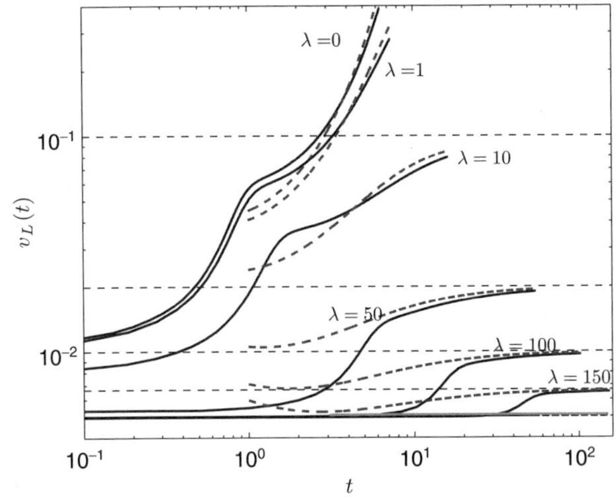


Fig. 7. (Color online) Time dependence of Lagrangian front velocities, $v_L(t) = dx_L(t)/dt$, according to the asymmetric truncated fractional Fisher–Kolmogorov Eq. (72) with $\alpha = 1.5$, $\theta = -1$, $\gamma = 1$, $\chi = 5 \times 10^{-7}$ and different values of $\lambda$. The solid (blue) curves denote the numerical results and the dashed (red) curves the asymptotic result according to Eq. (100). The horizontal (black) dash lines denote the corresponding terminal velocities according to Eq. (102). The solid (green) line at the bottom denotes the corresponding front speed in the Gaussian diffusive case according to Eq. (86).

to a good approximation the ideal Lévy flight scaling in Eqs. (90)–(92). Outside the intermediate asymptotic regime, i.e. for $x > 1/\lambda$, the Lagrangian trajectory of the front is given by

$$-\lambda x_L(t) + (\gamma - \chi\lambda^\alpha)\, t + \ln t - (\alpha+1) \ln x_L(t) = M\,, \tag{99}$$

where $M$ is a constant that depends on $\phi_0$. From Eq. (99) we obtain the following expression for the Lagrangian velocity of the front, $v_L = dx_L(t)/dt$,

$$v_L(t) = \frac{\gamma - \chi\lambda^\alpha + \frac{1}{t}}{\lambda + \frac{\alpha+1}{x_L(t)}}\,. \tag{100}$$

In the limit $t \to \infty$

$$v_L \sim v_* - \frac{\alpha}{\lambda t} + \cdots\,, \tag{101}$$

where the terminal velocity is given by

$$v_* = \frac{\gamma - \lambda^\alpha \chi}{\lambda}\,, \tag{102}$$

which is positive since it has been assumed that $\gamma > \lambda^\alpha \chi$ [11]. The asymptotic approach to the terminal velocity is clearly observed in Fig. 7 where the horizontal dashed lines show the terminal velocity in Eq. (102) for the values of $\lambda$ considered in the numerical simulations. According to Eq. (101) the time required for the front velocity to approach the terminal velocity within a given margin $v_* - v_L$ scales as $t \sim 1/\lambda$. The corresponding Lagrangian acceleration of the front, $a_L = dv_L(t)/dt$, is given by

$$a_L(t) = \frac{v_L(t)}{t\,(\lambda t v_* + 1)} \left\{ (\alpha+1) \left[ \frac{v_L(t)}{x_L(t)/t} \right]^2 - 1 \right\} . \tag{103}$$

From Eqs. (100) and (103) it follows that for large times,

$$a_L(t) \sim \frac{\alpha}{\lambda t^2} . \tag{104}$$

The Lagrangian velocity of the leading edge of the front ($\sim \phi_0 = 10^{-6}$) is shown in Fig. 7 for different values of $\lambda$. The solid lines denote the numerical results, and the dashed lines denote the asymptotic result according to Eq. (100). The horizontal dashed lines denote the terminal velocity according to Eq. (102). Note that as $\lambda$ increases, the terminal velocity is smaller and the convergence is faster.

### 5.5. $\nu < \lambda$

When $\lambda > \nu$, i.e. when the initial condition decays slower than the truncation, the leading order term in Eq. (95) is

$$\phi \sim \mathcal{C} e^{-\nu x + (\gamma - \chi \lambda^\alpha) t} , \tag{105}$$

which indicates that the front moves with the constant velocity

$$v = \frac{\gamma - \lambda^\alpha \chi}{\nu} . \tag{106}$$

An example of this over-truncated regime is presented in the bottom panel of Fig. 6 that shows a very good agreement with the exponential decay (note the log-normal scale) in Eq. (105) for $\lambda = 100$ and $\nu = 50$. The remaining case to consider is $\lambda = \nu$. In this case, the asymptotic approximation in Eq. (94) still holds. However, the expression in Eq. (93) cannot be used, and we have to go back to the integral $\mathcal{I}_1$ Eq. (80)

$$\mathcal{I}_1 = \mathcal{C} + \int_\Omega^{x/\tau} G_{\alpha,1,-1,0}(\eta) d\eta , \tag{107}$$

where this time the constant on the right-hand side is defined as $\mathcal{C} = \int_{-\infty}^{\Omega} G_{\alpha,1,-1,0}(\eta)d\eta$. Using the asymptotic expression for $G_{\alpha,1,-1,0} \sim \eta^{-1-\alpha}$ and integrating, it is concluded that the leading order term in Eq. (79) is $\phi \sim \mathcal{C}e^{-\lambda x+(\gamma-\chi\lambda^{\alpha})t}$ which implies that the constant velocity of the front is given by Eq. (106) with $\nu = \lambda$ [11].

## 6. Conclusions

Among the main drawbacks of Lévy statistics in applied problems, is the divergence of the moments due to the slow algebraic decay of the tails of the pdfs. This problem can be circumvented by using truncated pdfs. Of particular interest is the exponentially tempered case in which the algebraic decay is damped by an exponential factor leading to finite moments of all orders. This type of pdfs has found applicability in many areas including plasma physics, fluid mechanics, and finance.

In this paper we studied the role of exponentially tempered Lévy flights in the long-wavelength limit of the CTRW, and constructed a macroscopic transport equation in terms of truncated fractional derivatives. This equation describes the interplay between truncation and memory in the intermediate asymptotic regime not captured by the fractional diffusion equation based on Riemann–Liouville derivatives.

The Green's function of the tempered fractional diffusion equation was explicitly computed in the symmetric and asymmetric cases. At short times, the decay of the Green's function at the origin shows the characteristic decay of the fractional diffusion equation, $G(0,t) \sim t^{-\beta/\alpha}$. However, at long times, the decay exhibits a transition to the scaling $G(0,t) \sim t^{\beta/2}$. For $2\beta/\alpha > 1$ this corresponds to a transition from super-diffusive to sub-diffusive decay. The crossover time to sub-diffusive decay scales as $\tau \sim \lambda^{-\alpha/\beta}$ where $1/\lambda$ is the truncation length scale. This transition also manifests in the tails of the Green's function that change from algebraic (characteristic of super-diffusive processes) for $\tau \ll \lambda^{-\alpha/\beta}$, to stretched exponential (characteristic of sub-diffusive processes) for $\tau \gg \lambda^{-\alpha/\beta}$.

We have also presented a numerical and analytical study of the role of truncated Lévy flights in the propagation of fronts in reaction super-diffusion systems. The study was based on the tempered fractional Fisher–Kolmogorov model. Depending on the level of truncation, $\lambda$, four front propagation regimes were identified:

- For $\lambda = 0$ the fronts have algebraic decaying tails and show exponential acceleration.

- For $\lambda \neq 0$ the algebraic decay of the tail, and the acceleration of the front, prevail in the intermediate asymptotic regime, $(\chi t)^{1/\alpha} \ll x \ll 1/\lambda$.
- Outside the intermediate asymptotic regime, i.e. for $x > 1/\lambda$ and $(\chi t)^{1/\alpha} \ll x$, the tail of the front exhibits the tempered decay $\phi \sim e^{-\lambda x}/x^{1+\alpha}$, the acceleration is transient, and the front asymptotically approaches the terminal speed, $v_*$, at the rate $v_L \sim v_* - \alpha/(\lambda t)$.
- In the over-truncated regime, $\lambda > \nu$, (where $\nu$ is the decay rate of the exponential tail of the initial condition) Lévy flights have no qualitative effect on the asymptotic dynamics of the front that exhibits a diffusive-type exponential tail, and constant propagation velocity.

## Acknowledgments

This work has been supported by the Oak Ridge National Laboratory, managed by UT-Battelle, LLC, for the U.S. Department of Energy under contract DE-AC05-00OR22725.

## References

1. Eds. G. Radons, R. Klages and I. M. Sokolov, *Anomalous Transport: Foundations and Applications* (Wiley-VCH, 2008).
2. E. W. Montroll and G. H. Weiss, *J. Math. Phys.* **6**, 167 (1965).
3. E. W. Montroll and M. F. Shlesinger, in *Nonequilibrium Phenomena II. From Stochastics to Hydrodynamics*, eds. J. L. Lebowitz and E. W. Montroll (Elsevier, 1984).
4. R. Metzler and J. Klafter, *Phys. Rep.* **339**, 1 (2000).
5. D. del-Castillo-Negrete, B. A. Carreras and V. E. Lynch, *Phys. Plasmas* **11**, 3854 (2004).
6. D. del-Castillo-Negrete, *Phys. Plasmas* **13**, 082308 (2006).
7. D. del-Castillo-Negrete, *Phys. Fluids* **10**, 576 (1998).
8. D. del-Castillo-Negrete, *Nonlin. Process. Geophys.* **17**, 795 (2010).
9. D. del-Castillo-Negrete, B. A. Carreras and V. E. Lynch, *Phys. Rev. Lett.* **94**, 065003 (2005).
10. A. Cartea and D. del-Castillo-Negrete, *Phys. Rev. E* **76**, 041105 (2007).
11. D. del-Castillo-Negrete, *Phys. Rev. E* **79**, 031120 (2009).
12. V. M. Kenkre, E. W. Montroll and M. F. Shlesinger, *J. Stat. Phys.* **9**, 45 (1973).
13. R. Hilfer and L. Anton, *Phys. Rev. E* **51**, R848 (1995).
14. E. Scalas. R. Gorenflo and F. Mainardi, *Phys. Rev. E* **69**, 011107 (2004).
15. I. Podlubny, *Fractional Differential Equations* (Academic Press, 1999).
16. G. Samorodnitsky and M. S. Taqqu, *Stable Non-Gaussian Random Processes* (Chapman & Hall, 1994).

17. M. M. Meerschaert, D. A. Benson, H.-P. Scheffler and B. Baeumer, *Phys. Rev. E* **65**, 041103 (2002).
18. H.-P. Scheffler and M. M. Meerschaert, *J. Appl. Probab.* **41**, 623 (2004).
19. A. Piryatinska, A. I. Saichev and W. A. Woyczynski, *Physica A* **349**, 375 (2005).
20. S. G. Samko, A. A. Kilbas and O. I. Marichev, *Fractional Integrals and Derivatives* (Gordon and Breach, 1993).
21. R. N. Mantegna and H. E. Stanley, *Phys. Rev. Lett.* **73**, 2946 (1994).
22. I. Koponen, *Phys. Rev. E* **52**, 1197 (1995).
23. M. F. Shlesinger, *Phys. Rev. Lett.* **74**, 4959 (1995).
24. J. Rosinski, *Stoch. Process. Appl.* **117**, 677 (2007).
25. R. Jha, P. K. Kaw, D. R. Kulkami, J. C. Parikh and the ADITYA Team, *Phys. Plasmas* **10**, 699 (2003).
26. B. Dubrulle and J.-Ph. Laval, *Eur. Phys. J. B* **4**, 143 (1998).
27. R. Bruno, L. Sorriso-Valvo, V. Carbone and B. Bavassano, *Europhys. Lett.* **66**, 146 (2004).
28. R. N. Mantegna and H. E. Stanley, *Nature* **376**, 46 (1995).
29. I. M. Sokolov, A. V. Chechkin and J. Klafter, *Physica A* **336**, 245 (2004).
30. F. Mainardi, Y. Luchko and G. Pagnini, *Fract. Calculus Appl. Anal.* **4**, 153 (2001).
31. J. D. Murray, *Mathematical Biology* (Springer-Verlag, 1989).
32. D. H. Zanette, *Phys. Rev. E* **55**, 1181 (1997).
33. R. Mancinelli, D. Vergni and A. Vulpiani, *Europhys. Lett.* **60**, 532 (2002).
34. D. del-Castillo-Negrete, B. A. Carreras and V. E. Lynch, *Phys. Rev. Lett.* **91**, 018302 (2003).
35. B. I. Henry and S. L. Wearne, *Phys. A* **276**, 448 (2000).
36. A. Torcini and S. Lepri, *Phys. Rev. E* **55**, R3805 (1997).
37. R. K., Saxena, A. M. Mathai and H. J. Haubold, *Astrophys. Space Sci.* **305**, 289 (2006).
38. D. Hernandez, R. Barrio and C. Varea, *Phys. Rev. E* **74**, 046116 (2006).
39. I. M. Sokolov, M. G. W. Schmidt and F. Sagués, *Phys. Rev. E* **73**, 031102 (2006).
40. T. A. M. Langlands, B. I. Henry and S. L. Wearne, *J. Phys. Cond. Matter* **19**, 065115 (2007).
41. A. A. Golovin, B. J. Matkowsky and V. A. Volpert, *SIAM J. Appl. Math.* **69**, 251 (2008).
42. Y. Nec, A. A. Nepomnyashchy and A. A. Golovin, *Euro. Phys. Lett.* **82**, 58003 (2008).

# Chapter 7

## Anomalous Diffusion: From Fractional Master Equations to Path Integrals

R. Friedrich

*Institute of Theoretical Physics,*
*Westfälische Wilhelms-Universität,*
*Wilhelm Klemm Str. 9*
*48149 Münster, Germany*

We formulate the fractional master equation for a class of anomalous diffusion processes with drift and discuss applications to the anomalous diffusion of particles in external fields, the anomalous diffusion of particles with inertia, scattering processes in landscapes of randomly distributed scatterers, and the treatment of reaction-subdiffusive systems. Furthermore, we present a path integral representation for these genuinely non-Markovian processes, which allows one to assess multiple time properties.

1. Introduction . . . . . . . . . . . . . . . . . . . . . . . . . . 160
2. The Master Equation . . . . . . . . . . . . . . . . . . . . . . 161
3. Fractional Master Equations . . . . . . . . . . . . . . . . . . 163
4. Anomalous Diffusion via Subordination . . . . . . . . . . . 166
5. Single-Time Distributions: Examples . . . . . . . . . . . . . 167
   5.1. Generalized Ornstein–Uhlenbeck process . . . . . . 167
   5.2. Advection of a particle . . . . . . . . . . . . . . . . . 168
   5.3. Particles with inertia . . . . . . . . . . . . . . . . . . 169
   5.4. Ballistic motion in a landscape with random scatterers . . . . . . . . . . . . . . . . . . . . . . . . 169
6. Approximate Solutions of the Master Equation . . . . . . 171
7. Subdiffusive Reaction-Diffusion Equations . . . . . . . . . 172
8. Path-Integral Representation . . . . . . . . . . . . . . . . . 173
   8.1. Generalized Ornstein–Uhlenbeck processes . . . . . 177
9. Master Equation from Path Integrals . . . . . . . . . . . . 178
10. Summary and Outlook . . . . . . . . . . . . . . . . . . . . . 179

## 1. Introduction

A major challenge of modern science is the exploration of stochastic processes with nonnormal statistics arising, e.g., in the context of turbulent flows in fluids and plasmas, option pricing in economics or the occurrence of extreme events like earth quakes or epileptic seizures. Non-Gaussian and non-Markovian statistics are signatures of complex systems as a direct consequence of nonlinear cooperative evolution under far from equilibrium conditions. Though the theory of linear, Markovian and Gaussian systems is well-established, our understanding of complex behavior in the inanimate as well as the animate world is still rather incomplete [1].

A successful tool to investigate complexity on the basis of stochastic processes is the analysis of diffusive behavior of a state variable $q$. A first characterization is based on the notions of normal, sub- and superdiffusive behavior, which is related to the assessment of power-law behavior of the second moment, $\langle (q - \langle q \rangle)^2 \rangle \approx t^\gamma$, $\gamma = 1$, $\gamma < 1$, $\gamma > 1$. However, it has become evident that a more refined analysis is necessary and interest has been devoted to the characterization of the probability distribution $f(q,t)$. A complete characterization, clearly, is the specification of a suitable stochastic process for the state variable $q$ under consideration, which takes into account both deterministic as well as stochastic aspects of time evolution.

Subdiffusive motion has been investigated on the basis of so-called continuous time random walks (for reviews see [2–4]). These processes are characterized by a random walk with a certain jump length and waiting time distribution. Processes with a diverging mean waiting time are related with fractional evolution equations for the single time probability distribution. If, in addition, deterministic drifts have to be included, various different fractional diffusion equations can be formulated, each representing a different type of stochastic process.

In the present contribution, we shall investigate a class of stochastic processes for the state vector $\mathbf{q}(t)$ described by

$$\mathbf{q}_{i+1} = \mathbf{q}_i + \mathbf{D}^1(\mathbf{q}_i)\tau + \sqrt{D^2(\mathbf{q}_i)\tau}\mathbf{r}_i + \mathbf{R}(\mathbf{q}_i, \tau)\alpha_i. \tag{1}$$

Here, $\mathbf{r}_i$ and $\mathbf{R}$ as well as $\alpha_i$ are random processes. For all $\alpha_i = 0$, we obtain a discrete Langevin model of diffusion processes with drift $\mathbf{D}^1(\mathbf{q})$. Multiplicative noise is included (using the Itô intepretation) through the symmetric matrix $D^2(\mathbf{q})$. Continuous time random walks are modeled by the process $\alpha_i$ formed by strings of zeros and ones, where the random vector $\mathbf{R}(\mathbf{q}, \tau)$ determines the jump length distributions. In the simplest case, the sequence of zero's and one's can be assumed to undergo a renewal process. (For the definition of renewal processes we refer to the monograph [5].)

We derive the generalized master equation determining the single time probability distribution $f(\mathbf{q}, t)$ of the class of stochastic processes defined in Eq. (1) and discuss various applications to problems like generalized Ornstein–Uhlenbeck processes, diffusion and advection of particles, anomalous diffusion of particles with inertia and the formulation of reaction-subdiffusion equations. In a second part we extend our attempts to determine multiple time probability distribution for these genuinely non-Markovian processes by a formulation of a path integral. This chapter focuses on our own research contributions, and, therefore, the selection of topics and their presentation is subjective. Results concerning the path integral formulation are new.

## 2. The Master Equation

In this secion we derive a master equation determining the probability distribution for the process of Eq. (1), which is composed of two parts. The first part is a Markovian stochastic evolution described by the operator $L$. The second part is a fast (jump) process changing the state vector instantaneously to a new value. In between the jumps, the probability density $f(\mathbf{q}, t)$ evolves according to the Markov process with generator $L$, i.e.

$$f(\mathbf{q}, t) = e^{L(t-t')} f(\mathbf{q}, t') = \int d\mathbf{q}' S(\mathbf{q}, \mathbf{q}', t-t') f(\mathbf{q}', t'). \tag{2}$$

The generator L is a stochastic operator, e.g., of the form of a Fokker–Planck operator [8, 9], or, in case of a purely deterministic evolution, a Liouville operator.

The jump process is a renewal process [5] which is characterized by a waiting time distribution

$$W(\mathbf{q}|\mathbf{q}', t) dt\, d\mathbf{q}\, d\mathbf{q}'. \tag{3}$$

This quantity defines the probability that a time interval $t$ elapses between two consecutive jumps, where a jump at the end of the interval instantaneously drives the state from $\mathbf{q}'$ to the interval $d\mathbf{q}$ located at $\mathbf{q}$.

The following master equation defining the temporal evolution of the single time probability distribution $f(\mathbf{q}, t)$ can be derived for this class of stochastic processes

$$\begin{aligned}\frac{\partial}{\partial t} f(\mathbf{q}, t) &= f(\mathbf{q}, 0)\delta(t) + Lf(\mathbf{q}, t) \\ &\quad + \int_0^t dt' \int d\mathbf{q}' d\mathbf{q}'' Q(\mathbf{q}|\mathbf{q}'', t - t') S(\mathbf{q}'', \mathbf{q}', t - t') f(\mathbf{q}', t') \\ &\quad - \int_0^t dt' \int d\mathbf{q}' d\mathbf{q}'' Q(\mathbf{q}''|\mathbf{q}, t - t') S(\mathbf{q}, \mathbf{q}', t - t') f(\mathbf{q}', t'). \end{aligned} \tag{4}$$

The kernel $Q(\mathbf{q}|\mathbf{q}', t - t')$, thereby, is related to the distribution $W(\mathbf{q}|\mathbf{q}', t)$.

The derivation of the master equation is operational: We first construct the probability distribution $f(\mathbf{q}, t)$ and then verify that it obeys the above master equation.

We define the probability distribution $\eta(\mathbf{q}, t)$, which is the probability distribution that a jump has occurred at times $t - 0$ and the state vector has changed to $\mathbf{q}$. The probability distribution $\eta(\mathbf{q}, t)$ evolves according to the equation

$$\begin{aligned}\eta(\mathbf{q}, t) &= \int_0^t dt' \int d\mathbf{q}' d\mathbf{q}'' W(\mathbf{q}|\mathbf{q}'', t - t') S(\mathbf{q}'', \mathbf{q}', t - t') \eta(\mathbf{q}', t') \\ &\quad + f(\mathbf{q}, 0)\delta(t). \end{aligned} \tag{5}$$

This relation follows from the consideration that the probability distributions $\eta(\mathbf{q}, t)$ and $\eta(\mathbf{q}', t')$ correspond to two consecutive jumps occurring at times $t'$ and $t$. The last term on the right-hand side takes into account the initial condition.

The probability distribution $f(\mathbf{q}, t)$ of the random variable at time $t$ is obtained from the distribution $\eta(\mathbf{q}, t')$ according to

$$f(\mathbf{q}, t) = \int_0^t dt'' \int d\mathbf{q}' w(t - t'') S(\mathbf{q}, \mathbf{q}', t - t'') \eta(\mathbf{q}', t''). \tag{6}$$

Thereby, the quantity $w(t - t'')$ is the probability that no jump process occurs in the time interval $t - t''$ after the jump at time $t''$. This quantity is given in terms of $W(\mathbf{q}'|\mathbf{q}, t - t')$ by

$$w(t) = 1 - \int_0^t dt' \int d\mathbf{q}' W(\mathbf{q}'|\mathbf{q}, t'). \tag{7}$$

We observe that $w(0) = 1$.

The master equation is obtained by differentiating of Eq. (6) with respect to time and elimination of the probability distribution $\eta(\mathbf{q},t)$. The calculation is outlined in the Appendix.

A comparision with the class of continuous time random walks (we refer the reader to the excellent reviews [2] and book [6]) shows that we consider an extension of these processes by allowing for a deterministic or stochastic evolution of the state vector in between the individual jumps. This extension shows up in the appearance of the propagator $S(\mathbf{q},\mathbf{q}',t-t')$ in the master equation.

The master equation (4) is an integro-differential equation which allows one to determine the single time probability distribution from the initial condition $f(\mathbf{q},0)$. We mention that, on the other hand, we are able to numerically generate realizations of the stochastic processes under consideration by a combination of the Markovian evolution, generated by $L$, and the jump processes with the waiting time distribution $W(\mathbf{q}|\mathbf{q}',t-t')$. This fact will allow us to present a path integral representation of the stochastic process.

## 3. Fractional Master Equations

Fractional master equations arise for certain limiting cases of the kernel $Q(\mathbf{q}|\mathbf{q}',t-t')$, which is directly related to the waiting time distribution $W(\mathbf{q}|\mathbf{q}',t-t')$ according to Eq. (A.10). This kernel can be simplified by performing the following assumptions.

The first simplification is concerned with the statistical dependency between the jump length and the waiting times [7]. Statistical independence is expressed by the factorization

$$W(\mathbf{q}|\mathbf{q}',t-t') = W^{\mathbf{q}}(\mathbf{q}|\mathbf{q}')\Phi(t-t') \tag{8}$$

which immediately turns over to the factorization of the kernel

$$Q(\mathbf{q}|\mathbf{q}',t-t') = W^{\mathbf{q}}(\mathbf{q}|\mathbf{q}')Q(t-t'). \tag{9}$$

The probability distribution $\Phi(t)$ is the distribution of waiting times between two consecutive jumps. From the appendix, Eq. (A.10), we conclude that the kernel $Q(t-t')$ is defined via the Laplace transform of $\Phi(t)$ according to

$$Q(\lambda) = \frac{\lambda\Phi(\lambda)}{1-\Phi(\lambda)}. \tag{10}$$

The limiting case, where the mean waiting time diverges,

$$\int_0^\infty \Phi(t) t \, dt \longrightarrow \infty \tag{11}$$

leads to the kernel $Q(\lambda)$

$$Q(\lambda) = \Gamma(\alpha)\lambda^{-\alpha} \tag{12}$$

and we arrive at the fractional integral

$$\int_0^t dt' Q(t-t')G(t') \longrightarrow \frac{1}{\Gamma(\alpha)} \frac{\partial}{\partial t} \int_0^t \frac{dt'}{(t-t')^{1-\alpha}} G(t') = D_t^{1-\alpha} G(t) \tag{13}$$

($\Gamma(\alpha)$ denotes the Gamma-function, $\alpha$ is a parameter from the range $0 \leq \alpha \leq 1$). It is instructive to perform the limiting procedure starting from a physical representation of $Q(t-t')$ as described in the monograph [6], as well as the presentation in [13]. Furthermore, we point out that the presence of the additional process generated by $L$ enforces us to introduce a new fractional integral

$$D_t^{1-\alpha}(L) = \frac{1}{\Gamma(\alpha)} \int_0^t \frac{dt'}{(t-t')^{1-\alpha}} e^{L(t-t')}. \tag{14}$$

A second simplification is concerned with an approximation of the kernel $W^{\mathbf{q}}(\mathbf{q}|\mathbf{q}')$. An approximation of the form

$$W^{\mathbf{q}}(\mathbf{q}|\mathbf{q}') = [1 + L_{\mathrm{FP}}]\delta(\mathbf{q} - \mathbf{q}') \tag{15}$$

leads to a generalized Fokker–Planck equation

$$\begin{aligned} \frac{\partial}{\partial t} f(\mathbf{q}, t) &= Lf(\mathbf{q}, t) + L_{\mathrm{FP}} \int_0^t dt' Q(t-t') e^{L(t-t')} f(\mathbf{q}, t') \\ &= Lf(\mathbf{q}, t) \\ &\quad + L_{\mathrm{FP}} \int_0^t dt' Q(t-t') \int d\mathbf{q}' S(\mathbf{q}, \mathbf{q}', t-t') f(\mathbf{q}', t'). \end{aligned} \tag{16}$$

Here, we have introduced the additional Fokker–Planck operator [8, 9] $L_{\mathrm{FP}}$

$$L_{\mathrm{FP}} = -\nabla_{\mathbf{q}} \cdot \mathbf{D}^1(\mathbf{q}) + \nabla_{\mathbf{q}} \nabla_{\mathbf{q}} \mathbf{D}^2(\mathbf{q}) \tag{17}$$

which is related to the statistics of the jump process. Furthermore, we point out that the infinitesimal generators $L$ and $L_{\mathrm{FP}}$ of our stochastic processes need not be Fokker–Planck operators. In order to deal with Lévy flight behavior one may also use generators of the form $L = (-\Delta)^\beta$. At this point we refer the reader to [10], where Lévy type diffusion has been treated in the framework of continuous Markov processes.

Let us reiterate our approach. The operator $L$ is the generator of a Markov process evolving in time, which is interrupted by events whose transitions are specified by the generator $L_{\mathrm{FP}}$ and are characterized by a waiting time distribution $\Phi(t)$.

The occurrence of two Fokker–Planck operators $L$, $L_{\mathrm{FP}}$ lead to a variety of different stochastic processes. Let us exemplify this fact by considering the operators $L = -\frac{\partial}{\partial q}K(q)$ and $L_{\mathrm{FP}} = \frac{\partial^2}{\partial q^2}$. The first operator belongs to deterministic evolution of a particle in a potential $K(q) = -\frac{\partial}{\partial q}U(q)$, the second operator belongs to random diffusional motion. Starting from the simple Fokker–Planck process

$$\frac{\partial}{\partial t}f(q,t) = [L + L_{\mathrm{FP}}]f(q,t) \tag{18}$$

we can formulate the fractional Fokker–Planck equation

$$\frac{\partial}{\partial t}f(q,t) = \int dt' Q(t-t')\{L + L_{\mathrm{FP}}\}f(q,t'). \tag{19}$$

As we shall discuss in the next section both processes (18), (19) are related via subordination. This process has been introduced by Metzler *et al.* [18, 19].

A further process is described by the stochastic equation

$$\frac{\partial}{\partial t}f(q,t) = Lf(q,t) + L_{\mathrm{FP}}\int dt' Q(t-t')e^{L(t-t')}f(q,t'), \tag{20}$$

where the propagator $e^{L(t-t')}$ is related to the deterministic overdamped motion of the particle in the potential $U(q)$, $e^{L(t-t')} = \delta\left(q - Q(q',t-t')\right)$ and $\dot{Q}(q,t) = -\frac{\partial}{\partial Q}U(Q(q,t))$ with initial condition $Q(q',0) = q'$. This equation belongs to the class of processes investigated in this article.

The generalized Fokker–Planck equation,

$$\frac{\partial}{\partial t}f(q,t) = Lf(q,t) + \int dt' Q(t-t')L_{\mathrm{FP}}f(q,t') \tag{21}$$

which is frequently encountered in the literature, has no realization in the class of stochastic processes under consideration. This has been recognized for the case of the anomalous diffusion in [12, 13].

## 4. Anomalous Diffusion via Subordination

In this section we consider only the jump processes, which amounts to taking $L = 0$. In this case we end up with the well-known Fokker–Planck equation of the continuous time random walks,

$$\frac{\partial}{\partial t} f(\mathbf{q}, t) = L_{\mathrm{FP}} \int_0^t dt' Q(t - t') f(\mathbf{q}, t'). \tag{22}$$

As has been discussed extensively [14, 15], this generalized Fokker–Planck equation can be solved on the basis of the solution of the proper Fokker–Planck equation

$$\frac{\partial}{\partial s} f_0(\mathbf{q}, s) = L_{\mathrm{FP}} f_0(\mathbf{q}, s) \tag{23}$$

via the integral transform

$$f(\mathbf{q}, t) = \int_0^\infty ds p(s, t) f_0(\mathbf{q}, s). \tag{24}$$

Here, the quantity $p(s, t)$ can be considered as the probability distribution of a random process relating an *internal time* $s$ with the *physical time* $t$. The probability distribution $p(s, t)$ is defined by the following master equation with the initial condition

$$\begin{aligned} \frac{\partial}{\partial t} p(s, t) &= -\frac{\partial}{\partial s} \int_0^t dt' Q(t - t') p(s, t'), \\ \int_0^t dt' Q(t - t') p(0, t') &= \delta(t). \end{aligned} \tag{25}$$

The Laplace transform of $p(s, t)$ can be easily evaluated

$$p(s, \lambda) = \frac{e^{-\frac{\lambda}{Q(\lambda)} s}}{Q(\lambda)}. \tag{26}$$

In the fractional case $p(s, t) = \frac{t}{\alpha s^{1+1/\alpha}} L_\alpha(\frac{t}{s^{1/\alpha}})$, where $L_\alpha(x)$ denotes the one-sided Lévy stable distribution of order $\alpha$ [14].

The introduction of the *internal time* $s$ is usually denoted as subordination: A random process $\mathbf{q}(t)$ in *physical time* $t$ is constructed from a process $\mathbf{Q}(s)$ generated by the Fokker–Planck operator $L_{\mathrm{FP}}$ in *internal time* $s$ via a random transformation $s = S(t)$:

$$\mathbf{q}(t) = \mathbf{Q}\left[S(t)\right]. \tag{27}$$

For an outline of the procedure we refer to [16, 17].

The procedure of subordination allows one to evaluate multiple time statistics $f(\mathbf{q}_N, t_N; \ldots; \mathbf{q}_1, t_1)$, as has been outlined in [20, 21]. Subordination yields

$$f(\mathbf{q}_N, t_N; \ldots; \mathbf{q}_1, t_1) = \int_0^\infty ds_N \cdots \int_0^\infty ds_1 p(s_N, t_N; \ldots; s_1, t_1) f_0(\mathbf{q}_N, s_N; \ldots; \mathbf{q}_1, s_1), \quad (28)$$

where $f_0(\mathbf{q}_N, s_N; \ldots; \mathbf{q}_1, s_1)$ is the multiple times probability distribution in internal time $s$ and $p(s_N, t_N; \ldots; s_1, t_1)$ denotes the joint probability distribution of the random process $s = S(t)$. In [22] we have presented a path-integral formulation. Below, we shall explicitly obtain this path integral from our treatment of the process (1).

## 5. Single-Time Distributions: Examples

In this section we list several examples of processes described by Eq. (1).

### 5.1. *Generalized Ornstein–Uhlenbeck process*

We start by addressing an Ornstein–Uhlenbeck process modified by the inclusion of a renewal process. The generator $L$ of the Ornstein–Uhlenbeck process takes the form

$$L = \gamma \frac{\partial}{\partial q} + \frac{1}{2} Q_0 \frac{\partial^2}{\partial q^2} \quad (29)$$

and the corresponding propagator $S(q, q', t - t')$ is given by the Gaussian

$$S(q, q', t - t') = \frac{e^{-\frac{\gamma(q - q' e^{-\gamma(t-t')})^2}{2Q_0\left(1 - e^{-2\gamma(t-t')}\right)}}}{\sqrt{2\pi Q_0/\gamma(1 - e^{-2\gamma(t-t')})}}. \quad (30)$$

The generalized Fokker–Planck equation including jumps takes the form

$$\frac{\partial}{\partial t} f(q,t) = Lf(q,t) \frac{\partial}{\partial q^2} \int dt' Q(t-t') \int dq' S(q, q', t-t') f(q', t'). \quad (31)$$

It is straightforward to derive the following chain of evolution equations for the moments $\langle q^n(t) \rangle$ of the process:

$$\frac{\partial}{\partial t} \langle q^n(t) \rangle = -\gamma n \langle q^n(t) \rangle + \frac{n(n-1)}{2} Q_0 \langle q^{n-2}(t) \rangle + n(n-1) \int_0^t dt' Q(t-t') \langle q^{n-2}(t') \rangle_{\mathrm{OU}}. \quad (32)$$

Thereby, the quantity $\langle q^{n-2}(t')\rangle_{\rm OU}$ denotes the moments of the Ornstein–Uhlenbeck process, generated by $L$. Explicitly, the second-order moment (initial condition $x(0)=0$) takes the form

$$\langle x(t)^2\rangle = e^{-2\gamma t}\left(1-\frac{Q_0}{\gamma}\right)+\frac{Q_0}{\gamma}+2\int_0^t dt' e^{-2\gamma(t-t')}\int_0^{t'} dt'' Q(t'-t''). \tag{33}$$

Similarly, higher order moments can be evaluated. Multiple time distributions have been calculated in [20, 21].

### 5.2. *Advection of a particle*

Advection and anomalous diffusion of particles is of central interest for a variety of physical problems, see e.g., [23–25].

We consider the advection of particles, i.e. the operators $L=-v\frac{\partial}{\partial q}$, $L_{\rm FP}=\frac{\partial^2}{\partial q^2}$. The generalized Fokker–Planck equation, which belongs to the case of permanent advection reads [26]

$$\frac{\partial}{\partial t}f(q,t) = -v\frac{\partial}{\partial q}f(q,t)+\int_0^t dt' Q(t-t')\frac{\partial^2}{\partial q^2}f(q-v(t-t'),t'). \tag{34}$$

This equation can be solved by transformation to a comoving frame,

$$f(q,t)=\int_0^\infty ds p(s,t)\frac{e^{-\frac{(q-vt)^2}{2s}}}{\sqrt{2\pi s}}. \tag{35}$$

Advection introduces a pure shift $f(q,t)=f(q-vt,t)$ of the probability distribution.

In contrast, the solution of the generalized Fokker–Planck equation

$$\frac{\partial}{\partial t}f(q,t)=\int_0^t dt' Q(t-t')\left\{-v\frac{\partial}{\partial q}+\frac{\partial^2}{\partial q^2}\right\}f(q,t') \tag{36}$$

reads

$$f(q,t)=\int_0^t ds p(s,t)\frac{e^{-\frac{(q-vs)^2}{2s}}}{2\pi s}. \tag{37}$$

We remind the reader that this pdf does not exhibit the form $f(q-vt,t)$ due to the fact that, interpreting the result from the point of view of subordination, advection now occurs with respect to internal time.

### 5.3. *Particles with inertia*

In this section we review an application to particles with inertia. For details of the calculations we refer the reader to [12, 13]. The state vector $\mathbf{q}$ is composed of the position and the velocity of a particle (with unit mass). We consider the case where the particles are freely streaming interrupted by instantaneously occurring jumps. The evolution in between two consecutive jumps is given by $\mathbf{x}(t) = \mathbf{x}(t') - \mathbf{v}(t-t')$, $\mathbf{v}(t) = \mathbf{v}(t')$. As a consequence, the evolution operator $S(\mathbf{x}, \mathbf{v}; \mathbf{x}', \mathbf{v}'; t-t')$ takes the form

$$S(\mathbf{x}, \mathbf{v}; \mathbf{x}', \mathbf{v}'; t-t') = \delta(\mathbf{v}-\mathbf{v}')\delta\left(\mathbf{x}-\mathbf{x}'-\mathbf{v}(t-t')\right). \tag{38}$$

Assuming that the kernel can be approximated by

$$Q(\mathbf{x}, \mathbf{v}|\mathbf{x}', \mathbf{v}'; t-t') = \tilde{Q}(t-t')\delta(\mathbf{x}-\mathbf{x}')[1+\Delta_v]\delta(\mathbf{v}-\mathbf{v}'), \tag{39}$$

we end up with the generalized diffusion equation describing anomalous diffusion of particles with inertia for our class of random jump processes.

$$\left(\frac{\partial}{\partial t} + \mathbf{v}\cdot\nabla_{\mathbf{x}}\right) = \int_0^t dt' Q(t-t')[\Gamma\nabla_{\mathbf{v}}\cdot\mathbf{v} + K\Delta_v] f\left(\mathbf{x}-\mathbf{v}(t-t'), t'\right). \tag{40}$$

This fractional Kramers equation has to be distinguished from the equation of Metzler and Klafter [27], which belongs to an anomalous process generated from a Kramers process by the procedure of subordination, and the fractional Kramers equation by Barkai and Silbey [28], for which we were not able to formulate a corresponding stochastic process.

The anomalous statistics of the corresponding stochastic processes have been analyzed in details in [12, 13]. As a main result, we mention the long time asymptotics of the second moments for the case where the kernel $Q(t-t')$ is approximated by a fractional derivative, (cf. Eq. (25))

$$\langle \mathbf{x}(t)^2\rangle = (1-\alpha)t^2, \quad \langle x(t)v_x(t)\rangle = (1-\delta)\frac{K}{\Gamma}t, \quad \langle \mathbf{v}(t)^2\rangle = \frac{K}{\Gamma}. \tag{41}$$

The motion is ballistic, however there are subleading terms $\langle \mathbf{x}(t)^2\rangle \approx t^{2-\alpha}$.

### 5.4. *Ballistic motion in a landscape with random scatterers*

The treatment of the motion of particles with inertia under the impact of randomly occurring scattering events can be used to investigate the statistics of particles moving in a landscape with a random spatial distribution of localized scatterers. For a detailed analysis, we refer the reader to the treatment in [29]. We introduce the probability distribution $P(\mathbf{v}|\mathbf{v}', R)$, which is related to the probability that two scatterers are separated by a distance $R$, whereby the first scatterer has changed the velocity of the particle to $\mathbf{v}'$ and, after a period of free flight $t$, is changed to $\mathbf{v}$ by the impact of the second scatterer.

Apparently, this problem can be mapped onto the class of processes investigated in this contribution. We note that the waiting time distribution $W(\mathbf{v}|\mathbf{v}', t-t')$ is related to $P(\mathbf{v},|\mathbf{v}', R)$ according to

$$W(\mathbf{v}|\mathbf{v}', t-t') = \int dRP(\mathbf{v}|\mathbf{v}', R), \delta\left(t - t' - \frac{R}{v'}\right) \tag{42}$$

since the particle travels the distance $R = v'(t-t')$, $v = \sqrt{\mathbf{v}^2}$ in the time interval $t - t'$. Assuming statistical independency of the transition of the velocities and the distribution of relative separation of scatterers, $P(\mathbf{v}|\mathbf{v}', R) = p(\mathbf{v}|\mathbf{v}')g(R)$, we obtain the kernel

$$Q(\mathbf{v}|\mathbf{v}', t-t') = p(\mathbf{v}|\mathbf{v}')Q(t-t') \tag{43}$$

and the corresponding master equation takes the form

$$\begin{aligned}\left[\frac{\partial}{\partial t} + \mathbf{v}\cdot\nabla_{\mathbf{x}}\right] f(\mathbf{x},\mathbf{v},t) = \int_0^t dt' \int d\mathbf{v}' \{p(\mathbf{v}|\mathbf{v}')v'Q[v'(t-t')] \\ - \delta(\mathbf{v}-\mathbf{v}')vq[v(t-t')]\} f(\mathbf{x}-, \mathbf{v}'(t-t'), \mathbf{v}', t').\end{aligned} \tag{44}$$

The time kernel $Q(t-t')$ is defined in the usual way in terms of its Laplace transform, $Q(\lambda) = \lambda g(\lambda)/(1-g(\lambda))$.

It is of great interest to consider the case of a scale-free distribution of scatterers,

$$g(R) \approx R^{d-1}\frac{1}{R^\gamma} \tag{45}$$

($d$ denotes the dimension of the space). For an extended discussion, we refer the reader to [29], where especially the probability distribution of velocity, $h(v,t)$, is addressed. Using the usual diffusion approximation for the kernel $p(\mathbf{v}, \mathbf{v}')$ we end up with the fractional diffusion equation

$$\frac{\partial}{\partial t}h(\mathbf{v},t) = D_t^{1-\alpha}\nabla_{\mathbf{v}}v^{\alpha-1}\nabla_{\mathbf{v}}h(\mathbf{v},t) \tag{46}$$

with $\alpha = \gamma - d$. The equation with $\alpha = 1$ has been used to describe diffusion (of velocity) on a fractal structure [30]. In this interpretation, the diffusion equation could be related to subdiffusive motion on a fractal structure.

(We remind the reader that diffusion here applies to the velocity.) A solution of this generalized Fokker–Planck equation can be obtained in terms of the integral representation

$$f(v,t) = \int_0^\infty dsp(s,t)N\frac{v^{1-\alpha}}{[(3-\alpha)s]^{(d+a-\alpha)/(3-\alpha)}}e^{-\frac{v^{3-\alpha}}{(3-\alpha)^2 s}}. \tag{47}$$

$N$ denotes a normalization constant.

Moreover, we point out that the generalized Fokker–Planck equation arises in the context of Lagrangian fluid turbulence [31–33].

## 6. Approximate Solutions of the Master Equation

A general method for the solution of Fokker–Planck equation consists in a series expansion of the probability distribution into the set of eigenvectors of the Fokker–Planck operator [8, 9]. It is evident that a similar treatment can be pursued for the generalized Fokker–Planck equation under consideration.

We use the eigenvectors of the operator $L$

$$\begin{aligned} L\varphi_= &- \lambda_j\varphi_j, \\ \langle\varphi_j^\dagger|\varphi_i\rangle &= \delta_{ij}, \end{aligned} \tag{48}$$

where the modes $\varphi_j^\dagger$ are the eigenvectors of the adjoint operator $L^\dagger$. $L^\dagger$ denotes the backward Fokker–Planck operator. We perform the following ansatz for the probability distribution $f(\mathbf{q},t)$

$$f(\mathbf{q},t) = \sum_j f_j(t)\varphi_j(\mathbf{q}). \tag{49}$$

The resulting evolution equations for the amplitudes $f_j(t)$ read

$$\left[\frac{d}{dt} + \lambda_j\right] f_j(t) = \int_0^t dt' W_{jk}(t-t')e^{-\lambda_k(t-t')} f_k(t'). \tag{50}$$

A further step is to turn to an interaction picture performing the transformation

$$f_j(t) = e^{-\lambda_j t} g_j(t) \tag{51}$$

which leads us to

$$\frac{d}{dt} g_j(t) = \int_0^t dt' W_{jk}(t-t')e^{(\lambda_j-\lambda_k)(t-t')} g_k(t'). \tag{52}$$

Let us consider the case, where the operators $L$ and $L_{\mathrm{FP}}$ commute. In this case, there are common eigenfunctions, and the matrix elements $W_{jk}(t-t') = \delta_{jk} w_k(t-t')$ become diagonal leading to uncoupled fractional differential equation for the mode amplitudes.

The eigenfunction expansion can be used to assess relaxational behavior towards a stationary state. The treatment is straightforward.

## 7. Subdiffusive Reaction-Diffusion Equations

It is possible to formulate reaction-diffusion equations for subdiffusive chemical reactions based on the class of stochastic processes under consideration. An extended discussion can be found in [26].

As usual we divide space into cells and introduce a random walk between the different cells. The subdiffusive transport is modeled by the introduction of random waiting times for the jumps of particles between the cells. The reaction among the chemical species, however, evolves continuously in time, changing the concentrations $\mathbf{C} = [C_1, \dots, C_d]$ also in between the jumps, according to the nonlinear reaction kinetics $\dot{\mathbf{C}} = \mathbf{R}(\mathbf{C})$. The reaction-subdiffusion equations are then derived along the lines of the derivation of reaction-diffusion equations invoking a mean field approximation.

$$\frac{\partial}{\partial t}\mathbf{C}(\mathbf{x},t) = \mathbf{R}(\mathbf{C}(\mathbf{x},t)) + D\Delta_x \int_0^t dt' Q(t-t')\mathbf{G}\left(\mathbf{C}(\mathbf{x},t'), t-t'\right). \tag{53}$$

The nonlinear function $\mathbf{G}\left(\mathbf{C}(\mathbf{x},t'), t-t'\right)$ is the solution of the deterministic reacion kinetics

$$\frac{d}{dt}\mathbf{G}\left(\mathbf{C}(\mathbf{x},t'), t-t'\right) = \mathbf{R}\left(\mathbf{G}\left(\mathbf{C}(\mathbf{x},t'), t-t'\right)\right) \tag{54}$$

with initial condition $\mathbf{G}\left(\mathbf{C}(\mathbf{x},t'), 0\right) = \mathbf{C}(\mathbf{x},t')$. As a main result we observe that the diffusion term, in general, is nonlinear.

There have been several attempts to establish reaction-diffusion equations for subdiffusive chemical species (see [34, 35] and references therein). *Ad hoc* formulations can be obtained from Eq. (53) by replacing $G(C(\mathbf{x},t'), t-t')$ with $C(\mathbf{x},t')$. However, such equations show serious inconsistencies; in some cases, even mass conservation is violated. The correct way to proceed was discussed in [34] for a special example of a linear bimolecular reaction scheme $A \rightleftharpoons B$. This case is contained in our general subdiffusive reaction-diffusion equation if one restricts the treatment to a linear system.

As a nonlinear example, we consider the subdiffusive Fisher–Kolmogorov–Petrovskii–Piskunov (FKPP) equation, which has been proposed as a model equation for the propagation of favorable gene in a population [36]. The diffusive version of the FKPP equation is one of the basic equations for the investigation of reaction fronts in nonlinear reaction kinetics. Our subdiffusive version reads

$$\frac{\partial}{\partial t}C(\mathbf{x},t) = C(\mathbf{x},t)\,(1-C(\mathbf{x},t)) + \Delta_{\mathbf{x}}\int_0^t dt'\,Q(t-t')\,\frac{C(\mathbf{x},t')}{[1-e^{-(t-t')}]C(\mathbf{x},t')+e^{-(t-t')}}. \tag{55}$$

We point out the emergence of a nonlinear diffusion term in addition to a temporal memory. A study of front propagation can be found in [37].

## 8. Path-Integral Representation

In this section we shall develop a path-integral representation of the class of stochastic processes under consideration. We mention that a Feynman–Kac formula for the calculation of functionals of anomalous diffusion has been presented in [38].

We begin by constructing a single realization of the process and subsequently perform the average with respect to the occurring random processes. We introduce a time discretization denoting $\mathbf{q}_i = \mathbf{q}(t_i)$. Starting at times $t_i$ with the state vector $\mathbf{q}_i$, we proceed to $\mathbf{q}_{i+1}$ at time $t_{i+1}$ according to

$$\mathbf{q}_{i+1} = \mathbf{q}_1 + \tau\mathbf{D}^1(\mathbf{q}_i) + \sqrt{\tau D^2(\mathbf{q}_i)}\mathbf{r}_i + \alpha_i\mathbf{R}_i(\mathbf{q}_i,\tau). \tag{56}$$

Thereby, the variable $\alpha_i$ may take the values zero and one, indicating the occurrence of a jump, which is characterized by the random vector $\mathbf{R}(\mathbf{q}_i,\tau)$. The random vector is a function of time increment $\tau$ and may depend on $\mathbf{q}_i$. The vector $\mathbf{r}_i$ in addition is a random vector, with Gaussian distributed components. For $\alpha_i = 0$, we obtain the discretized version of a proper Langevin process.

In case of continuous time random walks the process $\alpha_i$ is a renewal process, which is characterized by the strings $[\alpha_N, \alpha_{N-1}, \ldots, \alpha_0] = [0,1,0,0,\ldots,0,1,0,\ldots,0,1]$. The probability of a string can be expressed in terms of waiting times, i.e. the number of zero's embedded between two 1's. The probability $P(1,0,\ldots,0,1,\alpha_{N'-1},\ldots,\alpha_0)$ of the occurrence of a string $1, 0, \ldots, 0, 1, \alpha_{N'-1}, \ldots, \alpha_0$ is given by

$$P(1, 0, \ldots, 0, 1, \alpha_{N'-1}, \ldots, \alpha_0) = W_{N-N'} P(1, \alpha_{N'-1}, \ldots, \alpha_0). \tag{57}$$

The quantity $W_{N-N'}$ denotes the probability to find a sequence of $N-1-N'$ zero's in between two one's.

Our treatment can be easily extended to other class of processes. We generalize to random variables $\alpha_i$ whih can take continuous values and characterize their statistics by the probability density $p(\alpha_N, \ldots, \alpha_0)$. This allows us to include also processes which are discussed in the realm of so-called *superstatistics* [39].

The path integral representation is based on the *fine-grained* transition probability from $\mathbf{q}_i, t_i$ to $\mathbf{q}_{i+1}, t_{i+1}$

$$\begin{aligned} p(\mathbf{q}_{i+1}, t_{i+1}|\mathbf{q}_i, t_i) \\ = \delta\left(\mathbf{q}_{i+1} - \mathbf{q}_i - \tau \mathbf{D}^1(\mathbf{q}_i) - \sqrt{\tau D^2(\mathbf{q}_i)}\mathbf{r}_i - \alpha_i \mathbf{R}_i(\mathbf{q}_i, \tau)\right). \end{aligned} \tag{58}$$

We use the representation of the $\delta$-function

$$\delta(\mathbf{q}) = \int \frac{d\hat{\mathbf{q}}}{(2\pi)^n} e^{i\hat{\mathbf{q}}\cdot\mathbf{q}} \tag{59}$$

and obtain

$$p(\mathbf{q}_{i+1}, t_{i+1}|\mathbf{q}_i, t_i) = \int \frac{d\hat{\mathbf{q}}_i}{(2\pi)^n} e^{iS[(\mathbf{q}_{i+1}|\mathbf{q}_i, \hat{\mathbf{q}}_i, \alpha_i)} \tag{60}$$

with

$$\begin{aligned} S(\mathbf{q}_{i+1}|\mathbf{q}_i, \hat{\mathbf{q}}_i, \alpha_i) = i\tau\hat{\mathbf{q}}_i \cdot \left\{\frac{\mathbf{q}_{i+1} - \mathbf{q}_i}{\tau} - \mathbf{D}^1(\mathbf{q}_i)\right\} \\ - i\hat{\mathbf{q}}_i\sqrt{\tau D^2(\mathbf{q}_i)}\mathbf{r}_i - i\alpha_i\hat{\mathbf{q}}_i \cdot \mathbf{R}_i(\mathbf{q}_i, \tau). \end{aligned} \tag{61}$$

The fine grained probability distribution of a single realization of the process is obtained by iteration

$$\begin{aligned} \tilde{f}(\mathbf{q}_N, t_N; \ldots \mathbf{q}_1, t_1; \mathbf{q}_0, t_0) \\ = p(\mathbf{q}_N, t_N; \mathbf{q}_{N-1}, t_{N-1}) \cdots p(\mathbf{q}_1, t_1; \mathbf{q}_0, t_0) f(\mathbf{q}_0, t_0). \end{aligned} \tag{62}$$

Now, we average over the random variables $\mathbf{r}_i$ and $\mathbf{R}(\mathbf{q}_i, \tau)$ to obtain the *coarse grained probability*

$$f(\mathbf{q}_N, t_N; \ldots; \mathbf{q}_0, t_0) = \int \prod_{i=0}^{N-1} \frac{d\hat{\mathbf{q}}_i}{(2\pi)^n} e^{iS_0} \langle e^{iS_r + iS_R} \rangle \tag{63}$$

with

$$S_0 = \sum_{k=0}^{N-1} \tau \hat{\mathbf{q}}_k \cdot \left[ \frac{\mathbf{q}_{k+1} - \mathbf{q}_k}{\tau} - \mathbf{D}^1(\mathbf{q}_k) \right],$$

$$S_r = - \sum_{k=0}^{N-1} \hat{\mathbf{q}}_k \sqrt{\tau D^2(\mathbf{q}_k)} \mathbf{r}_k, \tag{64}$$

$$S_R = - \sum_{k=0}^{N-1} \alpha_k \hat{\mathbf{q}}_k \cdot \mathbf{R}_k(\mathbf{q}_k, \tau).$$

For the averaging process we invoke several assumptions leading to the simplest case. The extension is straightforward and obvious. First, we assume statistical independence of the random process $\mathbf{r}_i$ and the process $\alpha_k$, $\mathbf{R}(\mathbf{q}_k, \tau)$, as well as Gaussian statistics, characterized by the correlation matrix

$$\langle \mathbf{r}_i \mathbf{r}_k \rangle = E \delta_{ik}. \tag{65}$$

Due to statistical independence, the expectation

$$\begin{aligned} \langle e^{iS_r} e^{iS_R} \rangle &= \langle e^{iS_r} \rangle \langle e^{iS_R} \rangle \\ &= e^{-\frac{1}{2} \sum_{k=0}^{N-1} \tau \hat{\mathbf{q}}_k D^2(\mathbf{q}_k) \hat{\mathbf{q}}_k} \langle e^{iS_R} \rangle \end{aligned} \tag{66}$$

factorizes.

Again, we consider Gaussian statistics of the random process $\mathbf{R}$ and assume statistical independence of the process $\alpha_k$. Averaging with respect to the random variable $\mathbf{R}_k$ yields

$$\langle e^{iS_R} \rangle_R = e^{-\frac{1}{2} \sum_{k=0}^{N-1} \tau \alpha_k^2 \hat{\mathbf{q}}_k Q \hat{\mathbf{q}}_k}, \tag{67}$$

where

$$\langle \mathbf{R}_k \mathbf{R}_j \rangle = \tau Q \delta_{kj}. \tag{68}$$

We can now summarize: The statistical properties of the discretized random process $\mathbf{q}_i$, $i = 0, \ldots, N$ are given in terms of the path integral

$$f(\mathbf{q}_N, t_N; \ldots; \mathbf{q}_0, t_0) = \int D\hat{\mathbf{q}} D\alpha e^{iS[\mathbf{q}, \hat{\mathbf{q}}, \alpha]} p(\alpha_{N-1}; \ldots; \alpha_1; \alpha_0). \tag{69}$$

We have used the abreviations

$$\prod_{i=0}^{N-1} \frac{d\hat{\mathbf{q}}_i}{(2\pi)^n} = D\hat{\mathbf{q}}, \quad \prod_{i=0}^{N-1} d\mathbf{q}_i = D\mathbf{q}, \quad \prod_{i=0}^{N-1} d\alpha_i = D\alpha. \tag{70}$$

This path integral contains a modified Martin–Siggia–Rose action [40] of the form

$$S[\mathbf{q},\hat{\mathbf{q}},\alpha] = \sum_{k=0}^{N-1} \tau\hat{\mathbf{q}}_k \cdot \left[\frac{\mathbf{q}_{k+1}-\mathbf{q}_k}{\tau} - \mathbf{D}^1(\mathbf{q}_k)\right] + \frac{i}{2}\sum_{k=0}^{N-1} \tau\hat{\mathbf{q}}_k D^2(\mathbf{q}_k)\hat{\mathbf{q}}_k + \frac{i}{2}\sum_{k=0}^{N-1} \tau\alpha_k^2\hat{\mathbf{q}}_k Q\hat{\mathbf{q}}_k. \tag{71}$$

The first two terms belong to the well-known Martin–Siggia–Rose action of a Fokker–Planck process. The additional contribution is related to the stochastic process $\{\alpha_n\}$ and, in turn, is characterized by the joint probability density $p(\alpha_{N-1};\dots;\alpha_1;\alpha_0)$. In the realm of the theory of continuous time random walks this joint probability density can be related to the waiting time statistics of a renewal process. In the realm of the theory of *superstatistics*, a stochastic process has to be specified for $\alpha_i$.

In both cases the probability distribution $f(\mathbf{q}_N,t_N;\dots;\mathbf{q}_0,t_0)$ can be obtained from a distribution

$$f(\mathbf{q}_N,\alpha_N,t_N;\dots;\mathbf{q}_0,\alpha_0,t_0) = \int D\hat{\mathbf{q}} e^{iS[\mathbf{q},\hat{\mathbf{q}},\alpha]} \tag{72}$$

by averaging over the process $\alpha_i$:

$$f(\mathbf{q}_N,t_N;\dots;\mathbf{q}_0,t_0) = \int D\alpha f(\mathbf{q}_N,\alpha_N,t_N;\dots;\mathbf{q}_0,\alpha_0,t_0)p(\alpha_{N-1};\dots;\alpha_1;\alpha_0). \tag{73}$$

A similar relation holds for the corresponding characteristic functions, which are defined according to

$$Z(\hat{\mathbf{q}}_N,t_N;\dots;\hat{\mathbf{q}}_0,t_0) = \int D\mathbf{q} e^{i\sum_{k=0}^{N}\tau\hat{\mathbf{q}}_i\cdot\mathbf{q}_i} f(\mathbf{q}_N,t_N;\dots;\mathbf{q}_0,t_0) \tag{74}$$

and a similar definition for $Z(\hat{\mathbf{q}}_N,\alpha_N,t_N;\dots;\hat{\mathbf{q}}_0,\alpha_0,t_0)$.

This property allows one to calculate the path integrals in two steps. First, one can determine the characteristic function $Z(\hat{\mathbf{q}}_N,\alpha_N,t_N;\dots;\hat{\mathbf{q}}_0,\alpha_0,t_0)$ and then evaluate the characteristic function $Z(\hat{\mathbf{q}}_N,t_N;\dots;\hat{\mathbf{q}}_0,t_0)$ by a subsequent evaluation of the expectation value with respect to the process $\{\alpha_i\}$.

### 8.1. *Generalized Ornstein–Uhlenbeck processes*

As an example we shall now determine the characteristic function of the generalized Ornstein–Uhlenbeck process defined by

$$\mathbf{q}_{i+1} = \mathbf{q}_i + (1 - \gamma\tau)\mathbf{q}_i + \sqrt{\tau}[\sqrt{Q_0}\mathbf{r}_i + \alpha_i\sqrt{Q}\mathbf{R}_i]. \tag{75}$$

Due to linearity of the evolution, one can calculate $\mathbf{q}_N$ for a single realization and determine the first and second moments of the process. Using the initial condition $\mathbf{q}_0 = 0$ we obtain

$$\mathbf{q}_N = \sum_{N'=0}^{N} (1 - \gamma\tau)^{N-N'}\sqrt{\tau}[\sqrt{Q_0}\mathbf{r}_{N'} + \alpha_{N'}\sqrt{Q}\mathbf{R}_{N'}] \tag{76}$$

and the characteristic function reads

$$Z(\hat{\mathbf{q}}_N, \alpha_N, t_N; \ldots; \hat{\mathbf{q}}_0, \alpha_0, t_0) = e^{-\frac{1}{2}\sum_{k,l=0}^{N}\tau^2\hat{\mathbf{q}}_k C_{kl}(\alpha)\hat{\mathbf{q}}_m} \tag{77}$$

with the correlation functions

$$C_{kl} = C_{kl}^0 + C_{kl}^R = \sum_{m=0}^{\mathrm{Min}(k,l)} \tau[Q_0 + \alpha_m Q](1 - \tau\gamma)^{k+l-2m}. \tag{78}$$

Averaging with respect to the process $\alpha$ yields the characteristic function

$$Z(\hat{\mathbf{q}}_N, t_N; \ldots; \hat{\mathbf{q}}_0, t_0) = e^{-\frac{1}{2}\sum_{kl}\tau^2\hat{\mathbf{q}}_k C_{kl}^0\hat{\mathbf{q}}_l} Z_R(\hat{\mathbf{q}}_N, t_N; \ldots; \hat{\mathbf{q}}_0, t_0) \tag{79}$$

with

$$Z_R = \int D\alpha e^{-\frac{1}{2}\sum_{kl}\tau^2\hat{\mathbf{q}}_k C_{kl}^R(\alpha)\hat{\mathbf{q}}_l} p(\alpha_{N-1}, \ldots, \alpha_0). \tag{80}$$

We can perform the limit $N \to \infty$, $\tau \to 0$ and obtain the path integral representation of the characteristic functional

$$Z(\hat{\mathbf{q}}) = e^{-\frac{1}{2}\int dt_1 \int dt_2 \hat{\mathbf{q}}(t_1)C^0(t_1,t_2)\hat{\mathbf{q}}(t_2)} \times \int D\alpha(t) e^{-\frac{1}{2}\int dt_1 \int dt_2 \hat{\mathbf{q}}(t_1)C^R(t_1,t_2,\alpha(t))\hat{\mathbf{q}}(t_2)} p(\alpha(t)). \tag{81}$$

The correlation functions are given by the continuum limit of (78):

$$C(t_1,t_2) = \frac{Q_0}{\gamma}e^{-2\gamma|t_1-t_2|},$$
$$C^R(t_1,t_2) = Q\int_{-\infty}^{\mathrm{Min}(t_1,t_2)} dt' e^{-\gamma(t_1+t_2-2t')}\alpha(t')^2. \tag{82}$$

The correlation function $C^R(t_1,t_2)$ is a functional of the process $\alpha(t)$, whose statistics is given by the probability functional $p(\alpha(t))$ obtained from $p(\alpha_N,\cdots,\alpha_0)$. We mention that the two-time probability distributions $p(s_1,s_2;t_1,t_2)$ has been evaluated in [20, 21] for a renewal process.

The expression becomes especially simple if we consider the case without damping, $\gamma = 0$. Here, we obtain

$$C^R_{kl} = Q\tau\sum_{m=0}^{\mathrm{Min}(k,l)}\alpha_m = QS_{\mathrm{Min}(k,l)} \tag{83}$$

with the random, internal time $S_n = \sum_{m=0}^n \alpha_n\tau$.

We end up with the representation

$$Z = e^{-\frac{1}{2}\sum_{kl}\tau^2\hat{\mathbf{q}}_k C^0_{kl}\hat{\mathbf{q}}_l}$$
$$\int DS\tilde{p}(S_{N-1}\cdots S_0)e^{-\frac{1}{2}\sum_{l=0}^N\sum_{k=0}^N\tau^2\hat{\mathbf{q}}_k QS_{\mathrm{Min}(l,k)}\hat{\mathbf{q}}_l}. \tag{84}$$

with the definition of the probability distribution of the internal time $S_i$

$$\tilde{p}(S_{N-1}\cdots S_0) = \int D\alpha\prod_{l=0}^{N-1}\delta(S_l - \sum_{k=0}^{l}\tau\alpha_k)p(\alpha_{N-1}\cdots\alpha_0). \tag{85}$$

The relationship with subordination procedures is obvious.

## 9. Master Equation from Path Integrals

In this section we shall show how the generalized master equation can be derived from the path integral representation. For the sake of simplicity we restrict our attention to the case of generalized Ornstein–Uhlenbeck process, whose characteristic functional has been derived above. We consider the characteristic function of the probability distribution $f(\mathbf{q},N\tau)$ which we denote by $Z(\hat{\mathbf{q}},N\tau)$. This quantity is obtained from (79) using

$\hat{\mathbf{q}}_0 = \cdots = \hat{\mathbf{q}}_{N-1} = 0,\ \hat{\mathbf{q}} = \mathbf{q}_N$

$$Z(\hat{\mathbf{q}}, N\tau) = \int D\alpha p(\alpha_{N-1}, \ldots \alpha_0) e^{-\frac{1}{2}\hat{\mathbf{q}}[Q_j N\tau + Q\sum_{j=1}^{N}\alpha_j^2\tau]\hat{\mathbf{q}}}. \tag{86}$$

Again, we introduce the probability distribution

$$p(s,t) = \int D\alpha\delta\left(s - \sum_{j=0}^{N}\tau\alpha_j^2\right)p(\alpha_{N-1}, \ldots, \alpha_0) \tag{87}$$

and end up with the characteristic function

$$Z(\hat{\mathbf{q}}, t) = e^{-\frac{1}{2}\hat{\mathbf{q}}Q_0\hat{\mathbf{q}}t}\int_0^\infty dsp(s,t)e^{-\frac{1}{2}\hat{\mathbf{q}}Q\hat{\mathbf{q}}s}. \tag{88}$$

The generalized master equation is easily obtained by calculating the derivative with respect to $t$:

$$\begin{aligned}\frac{d}{dt}Z(\hat{\mathbf{q}}, t) = &-\frac{1}{2}\hat{\mathbf{q}}Q_0\hat{\mathbf{q}}Z(\hat{\mathbf{q}}, t)\\ &-\frac{1}{2}\hat{\mathbf{q}}Q\hat{\mathbf{q}}\int_0^t dt'Q(t-t')e^{-\frac{1}{2}\hat{\mathbf{q}}Q_0\hat{\mathbf{q}}(t-t')}Z(\hat{\mathbf{q}}, t').\end{aligned} \tag{89}$$

Thereby, we have used the fact that the probability distribution $p(s,t)$ obeys Eq. (25). Calculating the corresponding equation for the probability distribution we end up with the master equation (31) of the generalized Ornstein–Uhlenbeck process discussed above, for the case $\gamma = 0$. It is obvious that the derivation can be extended to the general case.

## 10. Summary and Outlook

We have considered a class of anomalous diffusion processes, which are generated by a combination of a diffusion and a renewal process. The processes can be numerically generated by iterating the generalized Langevin equation (1) [42, 43]. For this class we have been able to derive a master equation, which contains a new type of fractional integral. This new type of fractional integral arises due to the fact that the process evolves in time in between two consecutive jumps. In general, the considered processes cannot be discussed in the framework of subordination.

In the first part of this chapter we have considered various examples, ranging from Ornstein–Uhlenbeck processes, the diffusion of particles with inertia up to reaction subdiffusion systems. The example of the FKPP equation has shown that due to this generalized fractional integral the diffusion term actually becomes nonlinear.

In a second part we have established a path integral formulation for the processes under consideration. This path integral can be used to calculate functionals of anomalous diffusion processes, an issue which has recently been discussed by Turgmen *et al.* [38]. We have been able to derive the generalized master equation from our path integral. The derivation of a Feynman–Kac formula, as outlined in [38], is straightforward. It is expected that our path integral formula can be successfully used to solve advanced problems for the class of generalized Langevin equations (1). Apparently, processes belonging to the class of continuous time random walks and processes arising due to superstatistical mechanisms can be treated by a unified approach. Finally, we mention that it would be interesting to extend methods of data analysis, which are available for Fokker–Planck processes [41], to these classes of stochastic processes.

## Appendix. Derivation of Master Equation

We notice that

$$\frac{\partial}{\partial t} S(\mathbf{q}, \mathbf{q}', t-t') = LS(\mathbf{q}, \mathbf{q}', t-t') \tag{A.1}$$

and obtain by differentiating Eq. (6) with respect to time

$$\begin{aligned}\frac{\partial}{\partial t} f(\mathbf{q}, t) &= Lf(\mathbf{q}, t) + \eta(\mathbf{q}, t) \\ &\quad + \int_0^t dt' \int d\mathbf{q}' \frac{\partial}{\partial t} w(t-t') S(\mathbf{q}, \mathbf{q}', t-t') \eta(\mathbf{q}', t'). \end{aligned} \tag{A.2}$$

Now, we invoke the relation

$$\frac{dw(t-t')}{dt} = -\int d\mathbf{q}' W(\mathbf{q}'|\mathbf{q}, t-t') \tag{A.3}$$

which is a result of the definition of $w(t)$

$$w(t) = 1 - \int_0^t dt' \int d\mathbf{q}' W(\mathbf{q}'|\mathbf{q}, t') \tag{A.4}$$

and obtain under consideration of Eq. (5)

$$\frac{\partial}{\partial t} f(\mathbf{q}, t) = f(q, 0)\delta(t) + Lf(\mathbf{q}, t)$$
$$+ \int_0^t dt' \int d\mathbf{q}'' \int d\mathbf{q}' W(\mathbf{q}|\mathbf{q}'', t-t') S(\mathbf{q}'', \mathbf{q}', t-t')\eta(\mathbf{q}', t')$$
$$- \int_0^t dt' \int d\mathbf{q}' \int d\mathbf{q}'' W(\mathbf{q}''|\mathbf{q}, t-t') S(\mathbf{q}, \mathbf{q}', t-t')\eta(\mathbf{q}', t'). \tag{A.5}$$

It is convenient to transform our equations to Laplace space. This yields

$$\lambda f(\mathbf{q}, \lambda) = Lf(\mathbf{q}, \lambda) + \int d\mathbf{q}' W(\mathbf{q}|\mathbf{q}', \lambda - L)\eta(\mathbf{q}', \lambda)$$
$$- \int d\mathbf{q}' W(\mathbf{q}'|\mathbf{q}, \lambda - L)\eta(\mathbf{q}, \lambda). \tag{A.6}$$

Furthermore,

$$f(\mathbf{q}, \lambda) = w(\lambda - L)\eta(\mathbf{q}, \lambda). \tag{A.7}$$

Thereby, we have taken into account that the Laplace transform of the operator

$$L\{W(\mathbf{q}|\mathbf{q}', t)e^{Lt}\} = W(\mathbf{q}|\mathbf{q}', \lambda - L) \tag{A.8}$$

can be obtained from the Laplace transform $W(\mathbf{q}|\mathbf{q}', \lambda)$ by the shift $\lambda \to \lambda - L$. We can now invert Eq. (A.7) to obtain the master equation in Laplace space

$$\lambda f(\mathbf{q}, \lambda) = Lf(\mathbf{q}, \lambda) + \int d\mathbf{q}' Q(\mathbf{q}|\mathbf{q}', \lambda - L) f(\mathbf{q}', \lambda)$$
$$- \int d\mathbf{q}' Q(\mathbf{q}'|\mathbf{q}, \lambda - L) f(\mathbf{q}, \lambda). \tag{A.9}$$

Thereby, we have defined

$$Q(\mathbf{q}|\mathbf{q}', \lambda) = W(\mathbf{q}|\mathbf{q}', \lambda) w^{-1}(\lambda) = \frac{\lambda W(\mathbf{q}|\mathbf{q}', \lambda)}{1 - \int d\mathbf{q} W(\mathbf{q}|\mathbf{q}', \lambda)}. \tag{A.10}$$

The inverse Laplace transform of Eq. (A.9) yields the final formula for the master equation (4).

## References


1. Ed. R. A. Meyers, *Encyclopedia of Complexity and System Science* (Springer, 2009).
2. R. Metzler and J. Klafter, *Phys. Rep.* **339**, 1 (2000); *J. Phys. A, Math. Gen.* **37**, R161 (2004).
3. M. F. Shlesinger, G. M. Zaslavsky and J. Klafter, *Nature* **363**, 31 (1999).
4. G. M. Zaslavsky, *Phys. Today* **52**, 39 (1999).
5. W. Feller, *An Introduction to Probability Theory*, Vols. I–III (John Wiley and Sons, 1968).
6. R. Balescu, *Aspects of Anomalous Transport in Plasmas* (IOP, 2005).
7. A. Compte, *Phys. Rev. E* **55**, 6821 (1997).
8. H. Risken and T. Frank, *The Fokker–Planck Equation*, Springer Series in Synergetics, Vol. 18 (Springer, 1989).
9. C. W. Gardiner, *Stochastic Methods*, Springer Series in Synergetics, Vol. 13 (Springer, 2009).
10. I. Lubashevsky, R. Friedrich and A. Heuer, *Phys. Rev. E* **79**, 011110 (2009); *ibid.* **80**, 031148 (2009).
11. R. Metzler and A. Compte, *J. Phys. Chem. B* **104**, 3858 (2000).
12. R. Friedrich, F. Jenko, A. Baule and S. Eule, *Phys. Rev. Lett.* **96**, 230601 (2006).
13. R. Friedrich, F. Jenko, A. Baule and S. Eule, *Phys. Rev. E* **74**, 041103 (2006).
14. E. Barkai, *Phys. Rev. E* **63**, 046118 (2001).
15. I. M. Sokolov, *Phys. Rev. E* **66**, 041101 (2002).
16. H. C. Fogedby, *Phys. Rev. E* **50**, 1657 (1994).
17. A. Baule and R. Friedrich, *Phys Rev. E* **71**, 026101 (2005).
18. R. Metzler, E. Barkai and J. Klafter, *Phys. Rev. Lett.* **82**, 3563 (1999).
19. E. Barkai, R. Metzler and J. Klafter, *Phys. Rev. E* **61**, 132 (2000).
20. A. Baule and R. Friedrich, *Eur. Phys. Lett.* **77**, 10002 (2007).
21. A. Baule and R. Friedrich, *Eur. Phys. Lett.* **79**, 60004 (2007).
22. S. Eule and R. Friedrich, Towards a path integral formulation of continuous time random walks, in *Path Integrals, New Trends and Perspectives*, eds. W. Janke and A. Pelster (World Scientific, 2008).
23. B. Berkowitz *et al.*, *Water Resour. Res.* **38**, 1191 (2002).
24. R. Metzler and A. Comte, *J. Phys. B* **104**, 3858 (2000).
25. A. Compte and M. O. Caceres, *Phys. Rev. Lett.* **81**, 3140 (1998).
26. S. Eule, R. Friedrich, F. Jenko and I. M. Sokolov, *Phys. Rev. E* **78**, 060102 (2008).
27. R. Metzler and J. Klafter, *J. Chem. Phys. B* **104**, 3851 (2000); *Phys. Rev. E* **61**, 6308 (2000).
28. E. Barkai and R. Silbey, *J. Chem Phys. B* **104**, 3866 (2000).
29. H. Affan, R. Friedrich and S. Eule, *Phys. Rev. E* **80**, 011137 (2009).
30. B. O' Shaughnessy and I. Procaccia, *Phys. Rev. Lett.* **54**, 0455 (1985).
31. R. Friedrich, *Phys. Rev. Lett.* **90**, 084501 (2003).
32. R. Friedrich, Turbulent Fields, in *Advances in Physics and Astrophysics of the 21st Century*, ed. I. Zhelyazkov, *Suppl. Bulgarian J. Phys.* **33**, (Heron Press, 2005).

33. R. Friedrich and J. Peinke, *Fluid Dynamics: Turbulence, Springer Encyclopedia of Complexity and Systems Science*, ed. R. A. Meyers, (Springer, 2009).
34. F. Saques, V. P. Shkilev and I. M. Sokolov, *Phys. Rev. E* **77**, 032102 (2008).
35. D. Froemberg, H. Schmidt-Martens, I. M. Sokolov and F. Sagues, *Phys. Rev. E* **78**, 011128 (2008).
36. R. A. Fisher, *Ann. Eugenics* **7**, 353, (1937); A. Kolmogorov, I. Petrovskii and N. Piskounov, *Bull Moscow Univ. Math. Mech.* **1**, 1 (1937).
37. R. Hahn, Subdiffusive Reaction-Diffusion Equations, Diploma Thesis, Münster (2010).
38. L. Turgmen, S. Carmi and E. Barkai, *Phys. Rev. Lett.* **103**, 190201 (2010).
39. C. Beck and E. G. D. Cohen, *Physica A* **322**, 267 (2003).
40. P. C. Martin, E. D. Siggia and H. A. Rose, *Phys. Rev. A* **8**, 423 (1973).
41. R. Friedrich, J. Peinke, M. Sahimi and M. R. R. Tabar, *Fluctuations, Springer Encyclopedia of Complexity and Systems Science*, ed. R. A. Meyers (Springer, 2009).
42. S. Eule, R. Friedrich and F. Jenko, *J. Phys. Chem. B* **111**, 13041 (2007).
43. S. Eule, R. Friedrich, F. Jenko and D. Kleinhans, *J. Phys. Chem. B* **111**, 11474 (2007).

# Chapter 8

## Fractional Feynman–Kac Equation for Anomalous Diffusion Functionals

Shai Carmi and Eli Barkai

*Department of Physics and Advanced Materials and Nanotechnology Institute, Bar-Ilan University, Ramat Gan 52900, Israel*

Functionals of the path of a Brownian particle are known to obey the celebrated Feynman–Kac equation. We develop here a fractional Feynman–Kac equation for functionals of the sub-diffusive continuous-time random walk process. Several simple examples of functionals are explicitly treated such as the occupation time, the first passage time, and the maximum of the walk. In the presence of a binding field, the fractional Feynman–Kac equation describes the route to weak ergodicity breaking.

1. Introduction . . . . . . . . . . . . . . . . . . . . . . . . 185
2. Derivation of the Equations . . . . . . . . . . . . . . . . . . . 188
   2.1. Fractional Feynman–Kac equation . . . . . . . . . 188
   2.2. A backward equation . . . . . . . . . . . . . . . . . 194
3. Functionals of Free Particles . . . . . . . . . . . . . . . . 195
   3.1. Occupation time in half-space . . . . . . . . . . . 195
   3.2. First passage time . . . . . . . . . . . . . . . . . . 197
   3.3. The maximal displacement . . . . . . . . . . . . . 198
4. Functionals of Bounded Particles: Weak Ergodicity Breaking . . . . . . . . . . . . . . . . . . . . . . 199
   4.1. The fluctuations of the time-averaged position . . . 199
   4.2. Simulation of trajectories . . . . . . . . . . . . . 203
5. Summary . . . . . . . . . . . . . . . . . . . . . . . . . . 204

### 1. Introduction

In nature, a noisy signal $x(t)$ represents a physical observable that is difficult to predict. However, the time-average of the signal, $\int_0^t x(\tau)d\tau/t$, is in many cases non-random. Consider for example a Brownian particle attached to a spring and embedded in a thermal heat bath. The trajectory of the particle $x(t)$ is rather noisy and erratic, but yet, we expect the time-averages

$$\overline{x^n(t)} = \int_0^t x^n(\tau)d\tau/t$$

to converge, in the long time limit, to their corresponding ensemble averages

$$\lim_{t\to\infty} \overline{x^n(t)} = \langle x^n \rangle\,,$$

where in thermal equilibrium, ensemble averages are described by Boltzmann statistics

$$\langle x^n \rangle = \frac{1}{Z}\int_{-\infty}^{\infty} x^n e^{-V(x)/(k_B T)} dx. \tag{1}$$

Here, $Z$ is the normalizing partition function and $V(x)$ is a binding potential, for example, $V(x) = \frac{1}{2}m\omega^2 x^2$ for a particle attached to a spring. Boltzmann statistics is beautiful since it is not sensitive to the dynamics of the particle. For example, the underlying dynamics can be either Markovian or non-Markovian [1] (e.g., a Langevin equation with either white or colored noise); in both cases, the ensemble average will be distributed according to Boltzmann statistics.

In recent years there is growing interest in anomalous diffusion processes modeled with fractional kinetic equations, such as the fractional Fokker–Planck equation for subdiffusion in a binding potential [2–9]. Fractional kinetics exhibits ergodicity breaking: time averages remain random variables, even in the long time limit [10–16]. This ergodicity breaking is weak; there are no inaccessible regions in the phase space, as in strong ergodicity breaking.

To investigate weak ergodicity breaking, we study in this chapter the more general problem of functionals of the process $x(t)$, defined as

$$A = \int_0^t U[x(\tau)]d\tau, \tag{2}$$

where $U(x)$ is a prescribed function. For example, with $U(x) = x^n$, the time-averages are $\overline{x^n(t)} = A/t$. We consider paths of sub-diffusive continuous-time random walk (CTRW), an important and widely investigated process that is frequently used to describe the motion of particles in disordered systems [17–20]. We then obtain an equation for the probability density function (PDF) of functionals of these paths. Our new equation turns to be a powerful tool when studying functionals of subdiffusive CTRW processes. In particular, using our equation we obtain the fluctuations of the time-average of $x(t)$ and thereby learn about the kinetics of the weak ergodicity breaking.

When $x(t)$ describes a trajectory of a Brownian particle, $A$ is called a Brownian functional [21]. Functionals of Brownian motion arise in numerous problems across a variety of scientific fields from condensed matter physics [22–25], to hydrodynamics [26], meteorology [27], and finance [28, 29]. The PDF of $A$ satisfies a Schrödinger-like equation, derived in 1949 by Kac inspired by Feynman's path integrals [30].

Denote by $G(x, A, t)$ the joint PDF of finding, at time $t$, the particle at $x$ and the functional at $A$. The Feynman–Kac theory shows that (for a free Brownian particle and for $U(x) \geq 0$) [21, 30],

$$\frac{\partial}{\partial t} G(x, p, t) = K \frac{\partial^2}{\partial x^2} G(x, p, t) - pU(x) G(x, p, t), \tag{3}$$

where the equation is in Laplace space, $A \to p$, and $K$ is the diffusion coefficient.

We generalize the Feynman–Kac equation to non-Brownian functionals. We consider CTRW as the underlying process leading to anomalous diffusion. In the continuum limit of this process, we derive the following *fractional* Feynman–Kac equation [31],

$$\frac{\partial}{\partial t} G(x, p, t) = K_\alpha \mathcal{L}_{\mathrm{FP}} \mathcal{D}_t^{1-\alpha} G(x, p, t) - pU(x) G(x, p, t), \tag{4}$$

where $K_\alpha$ is a generalized diffusion coefficient and the symbol $\mathcal{D}_t^{1-\alpha}$ is Friedrich's *substantial fractional derivative* and is equal in Laplace space $t \to s$ to $[s + pU(x)]^{1-\alpha}$ [32]. $\mathcal{L}_{\mathrm{FP}} = \frac{\partial^2}{\partial x^2} - \frac{\partial}{\partial x} \frac{F(x)}{k_{\mathrm{B}} T}$ is the Fokker–Planck operator, where $F(x) = -V'(x)$ is the force at $x$ and $T$ is the temperature. When $\alpha = 1$ and $F(x) = 0$, Eq. (4) reduces to the celebrated Feynman–Kac equation (3).

With our fractional Feynman–Kac equation (4), we investigate functionals of subdiffusive CTRW processes both in the absence and in the presence of a binding force field. In the force-free case, questions of ergodicity are irrelevant since $Z \to \infty$. There are however several functionals of interest. For example, the time spent by a particle in a given domain, or the occupation time, is given by the functional $A = \int_0^t U[x(\tau)] d\tau$, where $U(x) = 1$ in the domain and is zero otherwise [33, 34]. This functional is useful in kinetic studies of chemical reactions that take place exclusively in a domain [35, 36]. In the presence of a binding force, we use our equation to study the time average of $x(t)$, a random variable that exhibits weak ergodicity breaking.

The rest of the chapter is organized as follows. In Sec. 2, we derive Eq. (4) and a *backward* equation that depends on the initial position rather than the final one. In Sec. 3, we demonstrate solutions of our equations for a few functionals of interest of free particles. In Sec. 4, we investigate functionals of bounded particles and illustrate how our equation leads to weak ergodicity breaking. We summarize in Sec. 5. A short description of our work appeared in [31] more details can be found in [37].

## 2. Derivation of the Equations

### 2.1. *Fractional Feynman–Kac equation*

The continuous-time random walk model [2,17–20] considers an infinite one-dimensional lattice with spacing $a$ on which a particle is allowed to jump to its nearest neighbors only. The probabilities of jumping left $L(x)$ and right $R(x)$ depend on the force at the point. Denote by $V(x)$ the potential at point $x$. We assume thermal equilibrium at temperature $T$ and detailed balance. Namely, that

$$L(x)e^{-V(x)/(k_B T)} = R(x-a)e^{-V(x-a)/(k_B T)} \tag{5}$$

holds true for every lattice point $x$ [14].

Waiting times between jump events are independent identically distributed random variables with PDF $\psi(\tau)$, and the process starts with a particle at $x = x_0$. The particle waits at $x_0$ for time $\tau$ drawn from $\psi(\tau)$ and then jumps to either $x_0 + a$ (with probability $R(x)$) or $x_0 - a$ (with probability $L(x)$), after which the process is renewed. If $F(x) = 0$, then $R(x) = L(x) = 1/2$. We assume that for long waiting times,

$$\psi(\tau) \sim \frac{B_\alpha}{|\Gamma(-\alpha)|}\tau^{-(1+\alpha)}. \tag{6}$$

For $0 < \alpha < 1$, the average waiting time is infinite and the process is sub-diffusive with (for $F(x) = 0$) $\langle x^2 \rangle = 2K_\alpha t^\alpha/\Gamma(1+\alpha)$ ($K_\alpha = a^2/(2B_\alpha)$, units $\mathrm{m}^2/\mathrm{sec}^\alpha$) [8]. We look for the differential equation that describes the distribution of functionals in the continuum limit of this model, focusing on $0 < \alpha < 1$.

Recall that the functional is defined as $A = \int_0^t U[x(\tau)]d\tau$ with the arbitrary function $U(x)$ and that $G(x, A, t)$ is the joint PDF of $x$ and $A$ at time $t$. For the particle to be at $(x, A)$ at time $t$, it must have been at $[x, A - \tau U(x)]$ at the time $t - \tau$ when the last jump was made. Let $\chi(x, A, t)dt$ be the probability of the particle to jump into $(x, A)$ in the time interval $[t, t + dt]$. We have,

$$G(x, A, t) = \int_0^t W(\tau)\chi[x, A - \tau U(x), t - \tau]d\tau, \tag{7}$$

where $W(\tau) = 1 - \int_0^\tau \psi(\tau')d\tau'$ is the probability for *not* moving in a time interval of length $\tau$.

To calculate $\chi$, we note that to *arrive to* $(x, A)$ at time $t$, the particle must have arrived to either $[x - a, A - \tau U(x - a)]$ or $[x + a, A - \tau U(x + a)]$ at the time $t - \tau$ when the previous jump was made. Given the probabilities of jumping left and right,

$$\begin{aligned}\chi(x, A, t) = &\int_0^t \psi(\tau)\{L(x + a)\chi[x + a, A - \tau U(x + a), t - \tau] \\ &+ R(x - a)\chi[x - a, A - \tau U(x - a), t - \tau]\}d\tau \\ &+ \delta(x - x_0)\delta(A)\delta(t),\end{aligned} \tag{8}$$

where $\psi(\tau)$ is the PDF of $\tau$, the time between jumps. The term $\delta(x - x_0)\delta(A)\delta(t)$ corresponds to no jumps.

Assume that $U(x) \geq 0$ for all $x$ and thus $A \geq 0$ (an assumption we will relax later). Let $\chi(x, p, t) = \int_0^\infty e^{-pA}\chi(x, A, t)dA$ be the Laplace transform $A \to p$ of $\chi(x, A, t)$, we use along this work the convention that the variables in parentheses define the space we are working in. We note that

$$\begin{aligned}\int_0^\infty e^{-pA}\chi[x, A - \tau U(x), t]dA &= e^{-p\tau U(x)} \int_0^\infty e^{-pA'}\chi(x, A', t)dA' \\ &= e^{-p\tau U(x)}\chi(x, p, t),\end{aligned}$$

where we used the fact that $\chi(x, A, t) = 0$ for $A < 0$. Thus, Laplace transforming $(A \to p)$ Eq. (8), we find

$$\begin{aligned}\chi(x, p, t) = L(x + a)&\int_0^t \psi(\tau)e^{-p\tau U(x+a)}\chi(x + a, p, t - \tau)d\tau \\ &+ R(x - a)\int_0^t \psi(\tau)e^{-p\tau U(x-a)}\chi(x - a, p, t - \tau)d\tau \\ &+ \delta(x - x_0)\delta(t).\end{aligned} \tag{9}$$

Laplace transforming $t \to s$ Eq. (9) using the convolution theorem,

$$\chi(x,p,s) = L(x+a)\hat{\psi}[s+pU(x+a)]\chi(x+a,p,s) + R(x-a)\hat{\psi}[s+pU(x-a)]\chi(x-a,p,s) + \delta(x-x_0), \quad (10)$$

where $\hat{\psi}(s)$ is the Laplace transform of the waiting time PDF. Let $\chi(k,p,s) = \int_{-\infty}^{\infty} e^{ikx}\chi(x,p,s)dx$ be the Fourier transform $x \to k$ of $\chi(x,p,s)$. Fourier transforming Eq. (10) and changing variables $x \to x \pm a$,

$$\chi(k,p,s) = e^{-ika}\int_{-\infty}^{\infty} e^{ikx}L(x)\hat{\psi}[s+pU(x)]\chi(x,p,s)dx + e^{ika}\int_{-\infty}^{\infty} e^{ikx}R(x)\hat{\psi}[s+pU(x)]\chi(x,p,s)dx + e^{ikx_0}.$$

If the lattice spacing is small enough such that $aF(x) \ll k_BT$ and $R(x-a) \approx R(x)$, we can use $L(x) = 1 - R(x)$ and the detailed balance equation (5) to show that [8]

$$R(x) \approx \frac{1}{2}\left[1 + \frac{aF(x)}{2k_BT}\right] \quad L(x) \approx \frac{1}{2}\left[1 - \frac{aF(x)}{2k_BT}\right]. \quad (11)$$

Using (11) and applying the Fourier transform identity $\mathcal{F}\{xf(x)\} = -i\frac{\partial}{\partial k}f(k)$,

$$\chi(k,p,s) = \left[\cos(ka) + i\sin(ka)\frac{aF\left(-i\frac{\partial}{\partial k}\right)}{2k_BT}\right] \times\hat{\psi}\left[s + pU\left(-i\frac{\partial}{\partial k}\right)\right]\chi(k,p,s) + e^{ikx_0}. \quad (12)$$

Note that the order of the terms is important: for example, $\cos(ka)$ does not commute with $\hat{\psi}\left[s+pU\left(-i\frac{\partial}{\partial k}\right)\right]$. The formal solution of Eq. (12) is

$$\chi(k,p,s) = \left\{1 - \left[\cos(ka) + i\sin(ka)\frac{aF\left(-i\frac{\partial}{\partial k}\right)}{2k_BT}\right] \times\hat{\psi}\left[s + pU\left(-i\frac{\partial}{\partial k}\right)\right]\right\}^{-1} e^{ikx_0}. \quad (13)$$

We next use our expression for $\chi$ to calculate $G(x,A,t)$. Transforming Eq. (7) $(x,A,t) \to (k,p,s)$,

$$G(k,p,s) = \frac{1 - \hat{\psi}\left[s + pU\left(-i\frac{\partial}{\partial k}\right)\right]}{s + pU\left(-i\frac{\partial}{\partial k}\right)}\chi(k,p,s), \quad (14)$$

where we used $\hat{W}(s) = \int_0^\infty e^{-st}W(t)dt = [1-\hat{\psi}(s)]/s$. Substituting Eq. (13) into (14), we find

$$G(k,p,s) = \frac{1-\hat{\psi}\left[s+pU\left(-i\frac{\partial}{\partial k}\right)\right]}{s+pU\left(-i\frac{\partial}{\partial k}\right)} \times \left\{1-\left[\cos(ka)+i\sin(ka)\frac{aF\left(-i\frac{\partial}{\partial k}\right)}{2k_{\mathrm{B}}T}\right] \times \hat{\psi}\left[s+pU\left(-i\frac{\partial}{\partial k}\right)\right]\right\}^{-1} e^{ikx_0}. \tag{15}$$

To derive a differential equation for $G(x,p,t)$, we recall the waiting time distribution is $\psi(t) \sim B_\alpha t^{-(1+\alpha)}/|\Gamma(-\alpha)|$ (Eq. (6)), and write its Laplace transform $\hat{\psi}(s)$ for $s \to 0$ as [2]

$$\hat{\psi}(s) \sim 1 - B_\alpha s^\alpha, \qquad 0<\alpha<1, \quad s\to 0. \tag{16}$$

Substituting Eq. (16) into Eq. (15), applying the small $k$ expansions $\cos(ka) \sim 1 - k^2a^2/2$ and $\sin(ka) \sim ka$, and neglecting the high order terms,

$$G(k,p,s) = \left[s+pU\left(-i\frac{\partial}{\partial k}\right)\right]^{\alpha-1} \times \left\{K_\alpha\left[k^2 - ik\frac{F\left(-i\frac{\partial}{\partial k}\right)}{k_{\mathrm{B}}T}\right] + \left[s+pU\left(-i\frac{\partial}{\partial k}\right)\right]^{\alpha}\right\}^{-1} e^{ikx_0},$$

where we used the generalized diffusion coefficient [8]

$$K_\alpha \equiv \lim_{a^2, B_\alpha \to 0} \frac{a^2}{2B_\alpha}. \tag{17}$$

By neglecting the high order terms in $s$ and $k$, we effectively reached the continuum limit of the lattice walk [38]. Rearranging the expression in the last equation we find

$$sG(k,p,s) - e^{ikx_0} = -K_\alpha\left[k^2 - ik\frac{F\left(-i\frac{\partial}{\partial k}\right)}{k_B T}\right]\left[s+pU\left(-i\frac{\partial}{\partial k}\right)\right]^{1-\alpha} \times G(k,p,s) - pU\left(-i\frac{\partial}{\partial k}\right)G(k,p,s).$$

Inverting $k \to x, s \to t$, we finally obtain our fractional Feynman–Kac equation

$$\frac{\partial}{\partial t} G(x,p,t) = K_\alpha \mathcal{L}_{\text{FP}} \mathcal{D}_t^{1-\alpha} G(x,p,t) - pU(x)G(x,p,t), \tag{18}$$

where

$$\mathcal{L}_{\text{FP}} = \frac{\partial^2}{\partial x^2} - \frac{\partial}{\partial x}\frac{F(x)}{k_{\text{B}}T} \tag{19}$$

is the Fokker–Planck operator. The initial condition is $G(x,A,t=0) = \delta(x-x_0)\delta(A)$, or $G(x,p,t=0) = \delta(x-x_0)$.

The symbol $\mathcal{D}_t^{1-\alpha}$ represents the fractional substantial derivative operator introduced in [32]:

$$\mathcal{L}\{\mathcal{D}_t^{1-\alpha} G(x,p,t)\} = [s + pU(x)]^{1-\alpha} G(x,p,s), \tag{20}$$

where $\mathcal{L}\{f(t)\} = \int_0^t e^{-st} f(t)dt$ is the Laplace transform $t \to s$ (see also [39]). In $t$ space,

$$\mathcal{D}_t^{1-\alpha} G(x,p,t) = \frac{1}{\Gamma(\alpha)}\left[\frac{\partial}{\partial t} + pU(x)\right] \int_0^t \frac{e^{-(t-\tau)pU(x)}}{(t-\tau)^{1-\alpha}} G(x,p,\tau)d\tau. \tag{21}$$

Thus, due to the long waiting times, the evolution of $G(x,p,t)$ is non-Markovian and depends on the entire history.

In $s$ space, the fractional Feynman–Kac equation reads

$$\begin{aligned} sG(x,p,s) - \delta(x-x_0) = K_\alpha \mathcal{L}_{\text{FP}}[s + pU(x)]^{1-\alpha} G(x,p,s) \\ - pU(x)G(x,p,s). \end{aligned} \tag{22}$$

A few remarks should be made.

(i) *The integer Feynman–Kac equation.* Consider the force-free case $F(x) = 0$. As expected, for $\alpha = 1$ our fractional equation (18) reduces to the (integer) Feynman–Kac equation (3).

(ii) *The fractional Fokker–Planck equation.* For $p = 0$, $G(x, p = 0, t) = \int_0^\infty G(x,A,t)dA$ reduces to $G(x,t)$, the marginal PDF of finding the particle at $x$ at time $t$ regardless of the value of $A$. Correspondingly, Eq. (18) reduces to the fractional Fokker–Planck equation:

$$\frac{\partial}{\partial t}G(x,t) = K_\alpha \mathcal{L}_{\rm FP}\mathcal{D}_{{\rm RL},t}^{1-\alpha}G(x,t), \tag{23}$$

where $\mathcal{D}_{{\rm RL},t}^{1-\alpha}$ is the Riemann–Liouville fractional derivative operator ($\mathcal{D}_{{\rm RL},t}^{1-\alpha}G(x,s) \to s^{1-\alpha}G(x,s)$ in Laplace $t \to s$ space) [2, 7, 8].

(iii) *A general functional.* When the functional is not necessarily positive, the Laplace transform $A \to p$ must be replaced by a Fourier transform $G(x,p,t) = \int_{-\infty}^{\infty} e^{ipA}G(x,A,t)dA$. It can be shown [37] that the fractional Feynman–Kac equation looks like (18), but with $p$ replaced by $-ip$,

$$\frac{\partial}{\partial t}G(x,p,t) = K_\alpha \mathcal{L}_{\rm FP}\mathcal{D}_t^{1-\alpha}G(x,p,t) + ipU(x)G(x,p,t), \tag{24}$$

where $\mathcal{D}_t^{1-\alpha} \to [s - ipU(x)]^{1-\alpha}$ in Laplace $s$ space.

(iv) *Lévy flights.* Consider CTRW with $F(x) = 0$ and displacements $\Delta_x$ distributed according to a symmetric PDF $f(\Delta_x) \sim |\Delta_x|^{-(1+\mu)}$, with $0 < \mu < 2$. The Fourier transform of this distribution is $f(k) \sim 1 - C_\mu |k|^\mu$ [2]. This process is known as a Lévy flight, and it can be shown to obey the following fractional Feynman–Kac equation (for $U(x) \geq 0$) [37]

$$\frac{\partial}{\partial t}G(x,p,t) = K_{\alpha,\mu}\nabla_x^\mu \mathcal{D}_t^{1-\alpha}G(x,p,t) - pU(x)G(x,p,t), \tag{25}$$

where $K_{\alpha,\mu} = C_\mu/B_\alpha$ (units m$^\mu$/sec$^\alpha$), and $\mathcal{D}_t^{1-\alpha}$ is the substantial fractional derivative operator defined above (Eqs. (20) and (21)). $\nabla_x^\mu$ is the Riesz spatial fractional derivative operator defined in Fourier $x \to k$ space as $\nabla_x^\mu \to -|k|^\mu$ [2].

(v) *Fractional Schrödinger equations.* Our fractional Feynman–Kac equation (18) has the form of a fractional Schrödinger equation in imaginary time. Real time, fractional Schrödinger equations for the wave function have also been recently proposed [40–44]. However, these are different from our fractional Feynman–Kac equation because they are either based on a fractional *spatial* derivative, or on a *non-substantial* temporal fractional derivative that leads to non-Hermitian evolution and hence non-normalizable quantum mechanics. Whether these fractional Schrödinger equations actually describe any physical reality is still unclear [45].

(vi) *How to solve the fractional Feynman–Kac equation.* To obtain the PDF of a functional $A$, the following recipe could be followed [21]:

(1) Solve Eq. (22), the fractional Feynman–Kac equation in $(x,p,s)$ space. Equation (22) is a second order, ordinary differential equation in $x$.

(2) Integrate the solution over all $x$ to eliminate the dependence on the final position of the particle.
(3) Invert the solution $(p, s) \to (A, t)$, to obtain $G(A, t)$, the PDF of $A$ at time $t$.

In the next section we will see that the second step can be circumvented by using a backward equation.

### 2.2. *A backward equation*

In many cases we are only interested in the distribution of the functional, $A$, regardless of the final position of the particle, $x$. Therefore, it turns out quite convenient (see Sec. 3) to obtain an equation for $G_{x_0}(A, t)$ — the PDF of $A$ at time $t$, given that the process has started at $x_0$.

According to the CTRW model, the particle, after its first jump at time $\tau$, is at either $x_0 - a$ or $x_0 + a$. Alternatively, the particle does not move at all during the measurement time $[0, t]$. Hence,

$$\begin{aligned} G_{x_0}(A, t) = \int_0^t & \psi(\tau)\{R(x_0)G_{x_0+a}[A - \tau U(x_0), t - \tau] \\ & + L(x_0)G_{x_0-a}[A - \tau U(x_0), t - \tau]\}d\tau \\ & + W(t)\delta[A - tU(x_0)]. \end{aligned} \tag{26}$$

Here, $\tau U(x_0)$ is the contribution to $A$ from the pausing time at $x_0$ in the time interval $[0, \tau]$. The last term on the right-hand side of Eq. (26) describes a motionless particle, for which $A(t) = tU(x_0)$. We now transform Eq. (26) $(x_0, A, t) \to (k_0, p, s)$, using techniques similar to those used in the previous subsection. This leads to, for $U(x) \geq 0$,

$$\begin{aligned} G_{k_0}(p, s) = \hat{\psi}\left[s + pU\left(-i\frac{\partial}{\partial k_0}\right)\right] & \left[\cos(k_0 a) - \frac{aF\left(-i\frac{\partial}{\partial k_0}\right)}{2k_B T} i\sin(k_0 a)\right] \\ \times G_{k_0}(p, s) & + \frac{1 - \hat{\psi}\left[s + pU\left(-i\frac{\partial}{\partial k_0}\right)\right]}{s + pU\left(-i\frac{\partial}{\partial k_0}\right)}\delta(k_0). \end{aligned}$$

To derive the last equation, we used the approximations (11). To reach the continuum limit we expand $\hat{\psi}(s) \sim 1 - B_\alpha s^\alpha$, $\cos(k_0 a) \sim 1 - k_0^2 a^2/2$, and

$\sin(k_0 a) \sim k_0 a$. After some rearrangements,

$$sG_{k_0}(p,s) - \delta(k_0) = -K_\alpha \left[s + pU\left(-i\frac{\partial}{\partial k_0}\right)\right]^{1-\alpha} \left[{k_0}^2 + \frac{F\left(-i\frac{\partial}{\partial k_0}\right)}{k_B T} ik_0\right]$$
$$\times G_{k_0}(p,s) - pU\left(-i\frac{\partial}{\partial k_0}\right) G_{k_0}(p,s).$$

Inverting $k_0 \to x_0$ and $s \to t$, we obtain the backward fractional Feynman–Kac equation:

$$\frac{\partial}{\partial t} G_{x_0}(p,t) = K_\alpha \mathcal{D}_t^{1-\alpha} \mathcal{L}_{\mathrm{FP}}^{(\mathrm{B})} G_{x_0}(p,t) - pU(x_0) G_{x_0}(p,t), \tag{27}$$

where

$$\mathcal{L}_{\mathrm{FP}}^{(\mathrm{B})} = \frac{\partial^2}{\partial x_0^2} + \frac{F(x_0)}{k_B T}\frac{\partial}{\partial x_0} \tag{28}$$

is the *backward* Fokker–Planck operator. Here, $\mathcal{D}_t^{1-\alpha}$ equals in Laplace $t \to s$ space $[s + pU(x_0)]^{1-\alpha}$. The initial condition is $G_{x_0}(A, t = 0) = \delta(A)$, or $G_{x_0}(p, t = 0) = 1$. In Eq. (18) the operators depend on $x$ while in Eq. (27) they depend on $x_0$. Therefore, Eq. (18) is called the forward equation while Eq. (27) is called the backward equation. Notice that here, the fractional derivative operator appears to the left of the Fokker–Planck operator $\mathcal{L}_{\mathrm{FP}}^{(\mathrm{B})}$, in contrast to the forward equation (18). Also, note the different sign and order of terms in $\mathcal{L}_{\mathrm{FP}}^{(\mathrm{B})}$ (Eq. (28)) compared to $\mathcal{L}_{\mathrm{FP}}$ (Eq. (19)).

## 3. Functionals of Free Particles

In this Section, we illustrate a number of applications of our equation in the force-free case, $F(x) = 0$. For more applications, see [37].

### 3.1. *Occupation time in half-space*

The occupation time of a particle in the positive half-space has wide applications in both physics [16,46,47] and mathematics [48]. Formally, the occupation time is defined as $T_+ = \int_0^t \Theta[x(\tau)]d\tau$, where $\Theta(x) = 1$ for $x \geq 0$ and is zero otherwise. To find the PDF of the occupation time, consider the backward equation ((27), transformed $t \to s$):

$$sG_{x_0}(p,s) - 1 = \begin{cases} K_\alpha s^{1-\alpha} \dfrac{\partial^2}{\partial {x_0}^2} G_{x_0}(p,s) & x_0 < 0, \\ K_\alpha (s+p)^{1-\alpha} \dfrac{\partial^2}{\partial {x_0}^2} G_{x_0}(p,s) - pG_{x_0}(p,s) & x_0 > 0. \end{cases} \tag{29}$$

These are second order, ordinary differential equations in $x_0$. Solving the equations in each half-space separately, demanding that $G_{x_0}(p,s)$ is finite for $|x_0| \to \infty$,

$$G_{x_0}(p,s) = \begin{cases} C_0 \exp(x_0 s^{\alpha/2}/\sqrt{K_\alpha}) + \dfrac{1}{s} & x_0 < 0, \\ C_1 \exp[-x_0 (s+p)^{\alpha/2}/\sqrt{K_\alpha}] + \dfrac{1}{s+p} & x_0 > 0. \end{cases} \tag{30}$$

For $x_0 \to -\infty$, the particle is never at $x > 0$ and thus $G_{x_0}(T_+, t) = \delta(T_+)$ and $G_{x_0}(p,s) = s^{-1}$, in accordance with Eq. (30). Similarly, for $x_0 \to +\infty$, $G_{x_0}(T_+, t) = \delta(T_+ - t)$ and $G_{x_0}(p,s) = (s+p)^{-1}$. Demanding that $G_{x_0}(p,s)$ and its first derivative are continuous at $x_0 = 0$, we obtain a pair of equations for $C_0, C_1$:

$$C_0 + s^{-1} = C_1 + (s+p)^{-1}, \quad C_0 s^{\alpha/2} = -C_1 (s+p)^{\alpha/2},$$

whose solution is

$$C_0 = -\frac{p(s+p)^{\alpha/2-1}}{s[s^{\alpha/2} + (s+p)^{\alpha/2}]}, \quad C_1 = \frac{p s^{\alpha/2-1}}{(s+p)[s^{\alpha/2} + (s+p)^{\alpha/2}]}. \tag{31}$$

We assume the process starts at $x_0 = 0$. $G_0(p,s) = C_1 + (s+p)^{-1}$, or, after some simplifications,

$$G_0(p,s) = \frac{s^{\alpha/2-1} + (s+p)^{\alpha/2-1}}{s^{\alpha/2} + (s+p)^{\alpha/2}}. \tag{32}$$

Using [49], the PDF of $\lambda \equiv T_+/t$, for long times, is the symmetric *Lamperti* PDF [50, 51]:

$$f(\lambda) = \frac{\sin(\pi\alpha/2)}{\pi} \frac{\lambda^{\alpha/2-1}(1-\lambda)^{\alpha/2-1}}{\lambda^\alpha + (1-\lambda)^\alpha + 2\lambda^{\alpha/2}(1-\lambda)^{\alpha/2}\cos(\pi\alpha/2)}. \tag{33}$$

This equation has also been derived using other methods [47]. Naively, one expects the particle to spend about half the time at $x > 0$. In contrast, we learn from Eq. (33) that the particle tends to spend most of the time at either $x > 0$ or $x < 0$: $f(\lambda)$ has two peaks at $\lambda = 1$ and $\lambda = 0$. This is exacerbated in the limit $\alpha \to 0$, where the distribution converges to two delta functions at $T^+/t = 1$ and at $T^+/t = 0$. For $\alpha = 1$ (Brownian motion) we recover the well-known arcsine law of Lévy [21, 46, 47].

### 3.2. *First passage time*

The time $t_f$ when a particle starting at $x_0 = 0$ first hits $x = b$ is called the first passage time and is a quantity subject to many studies in physics and other fields [52]. The distribution of first passage times for anomalous paths can be obtained from the fractional Feynman–Kac equation using an identity due to Kac [53]:

$$\Pr\{t_f > t\} = \Pr\left\{\max_{0\le\tau\le t} x(\tau) < b\right\} = \lim_{p\to\infty} G_{x_0}(p,t), \tag{34}$$

where $G_{x_0}(p,t)$ describes the functional $A_f = \int_0^t U[x(\tau)]d\tau$ with

$$U(x) = \begin{cases} 0 & x < b, \\ 1 & x > b. \end{cases} \tag{35}$$

This is true since $G_{x_0}(p,t) = \int_0^\infty e^{-pA_f} G_{x_0}(A_f,t)dA_f$, and thus, if the particle has never crossed $x = b$, we have $A_f = 0$ and $e^{-pA_f} = 1$, while otherwise, $A_f > 0$ and for $p \to \infty$, $e^{-pA_f} = 0$. The functional (35) is similar to the functional studied in the previous section. We obtain for $x_0 = 0$,

$$G_0(p,s) = \frac{1}{s}\left[1 - e^{-\frac{b}{\sqrt{K_\alpha}}s^{\alpha/2}} \frac{p(s+p)^{\alpha/2-1}}{s^{\alpha/2} + (s+p)^{\alpha/2}}\right].$$

To find the first passage time distribution we take the limit of infinite $p$,

$$\lim_{p\to\infty} G_0(p,s) = \frac{1}{s}\left(1 - e^{-\frac{b}{\sqrt{K_\alpha}}s^{\alpha/2}}\right). \tag{36}$$

Defining $\tau_f = (b^2/K_\alpha)^{1/\alpha}$, we invert $s \to t$:

$$\lim_{p\to\infty} G_0(p,t) = \Pr\{t_f > t\} = 1 - \int_0^t \frac{1}{\tau_f} l_{\alpha/2}\left(\frac{\tau}{\tau_f}\right) d\tau,$$

where $l_{\alpha/2}(t)$ is the one-sided Lévy distribution of order $\alpha/2$, whose Laplace transform is $l_{\alpha/2}(s) = e^{-s^{\alpha/2}}$. The PDF of the first passage times, $f(t)$, satisfies $f(t) = \frac{\partial}{\partial t}(\Pr\{t_f < t\}) = \frac{\partial}{\partial t}(1 - \Pr\{t_f > t\})$. Thus,

$$f(t) = \frac{1}{\tau_f} l_{\alpha/2}\left(\frac{t}{\tau_f}\right). \tag{37}$$

A different derivation of this equation appears in [54]. The long times behavior of $f(t)$ is obtained from the $s \to 0$ limit: $f(s) \sim 1 - \frac{b}{\sqrt{K_\alpha}} s^{\alpha/2}$. Therefore, for long times

$$f(t) \sim \frac{b}{|\Gamma(-\frac{\alpha}{2})|\sqrt{K_\alpha}} t^{-(1+\alpha/2)}. \tag{38}$$

For $\alpha = 1$, we reproduce the famous $t^{-3/2}$ decay law of a one-dimensional random walk [52].

### 3.3. *The maximal displacement*

The maximal displacement of a diffusing particle is a random variable whose study has been of recent interest (see, e.g., [55–58] and references therein). To obtain the distribution of this variable, we use the functional defined in the previous section (Eq. (35)). Let $x_m \equiv \max_{0 \le \tau \le t} x(\tau)$, and recall from Eq. (34) that $\Pr\{x_m < b\} = \lim_{p \to \infty} G_{x_0}(p, t)$. From the previous section we have, for $x_0 = 0$ (Eq. (36)),

$$\Pr\{x_m < b\} = \frac{1}{s}\left(1 - e^{-\frac{b}{\sqrt{K_\alpha}} s^{\alpha/2}}\right).$$

Hence, the PDF of $x_m$ is

$$P(x_m, s) = \frac{s^{\alpha/2-1}}{\sqrt{K_\alpha}} e^{-\frac{x_m}{\sqrt{K_\alpha}} s^{\alpha/2}}.$$

Inverting $s \to t$, we obtain

$$P(x_m, t) = \frac{2}{\alpha\sqrt{K_\alpha}} \frac{t}{(x_m/\sqrt{K_\alpha})^{1+2/\alpha}} l_{\alpha/2}\left[\frac{t}{\left(x_m/\sqrt{K_\alpha}\right)^{2/\alpha}}\right], \quad x_m > 0. \tag{39}$$

This PDF has the same shape as the PDF of $x$ up to a scale factor of 2 [8], and it is in agreement with the very recent result of [57], derived using a renormalization group method.

## 4. Functionals of Bounded Particles: Weak Ergodicity Breaking

### 4.1. *The fluctuations of the time-averaged position*

In this section we use our equation for non-Brownian functionals in a force field to characterize weak ergodicity breaking. Consider the fractional Fokker–Planck equation (23) in the harmonic potential $V(x) = m\omega^2 x^2/2$. In thermal equilibrium, the position of an ensemble of independent particles follows the Boltzmann distribution (see Eq. (1)),

$$G(x) = \frac{1}{Z}\exp[-V(x)/(k_{\mathrm{B}}T)] = \frac{1}{Z}\exp[-m\omega^2 x^2/(2k_{\mathrm{B}}T)]. \tag{40}$$

We are interested in the *time-averaged* position:

$$\overline{x} = \frac{\int_0^t x(\tau)d\tau}{t}.$$

Clearly, the numerator is a functional of the anomalous path, $A_x = \int_0^t x(\tau)d\tau$, or in other words, $U(x) = x$. When the motion is Brownian, the process is ergodic since the field is binding. Due to the symmetry of the harmonic field, in the long time limit $\overline{x}$ is statistically equal to zero, namely,

$$\begin{aligned}\lim_{t\to\infty}\overline{x} = \langle x\rangle &= \int_{-\infty}^{\infty} xG(x)dx \\ &= \frac{1}{Z}\int_{-\infty}^{\infty} x\exp\left[-\frac{m\omega^2 x^2}{2k_{\mathrm{B}}T}\right]dx = 0.\end{aligned} \tag{41}$$

Put differently, in an ergodic system the fluctuations of the time-averaged position vanish:

$$\lim_{t\to\infty}\left\langle(\Delta\overline{x})^2\right\rangle = \lim_{t\to\infty}\left[\left\langle\overline{x}^2\right\rangle - \langle\overline{x}\rangle^2\right] = 0. \tag{42}$$

For anomalous subdiffusion ergodicity is broken [11, 13, 14] and the fluctuations of the time-averaged position do not vanish even for long times. This is illustrated in Fig. 1.

With our equation for anomalous functionals, we now treat the problem of the fluctuations of the time-average of $x(t)$,

$$\left\langle(\Delta\overline{x})^2\right\rangle = \left\langle A_x^2\right\rangle/t^2,$$

where we assumed $x_0 = 0$ such that $\langle\overline{x}\rangle = 0$ at all times due to symmetry. The equation that describes the distribution of the functional $A_x$ is the

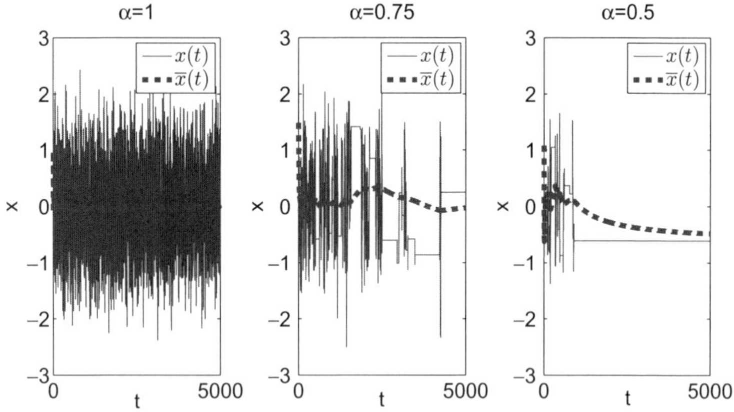


Fig. 1. Trajectories of particles in a binding harmonic field and the time-averaged position. Paths were generated for $\alpha = 1, 0.75, 0.5$ using the method described in Sec. 4.2 [59] with $a = 0.01$, $K_\alpha = 1/2$, $k_B T/(m\omega^2) = 1/2$, and $x_0 = 0$. The solid lines are the trajectories and the dashed line correspond to the time-averaged position $\overline{x} = \int_0^t x(\tau)d\tau/t$. For $\alpha = 1$, the time-average of $x(t)$ quickly converges to zero because ergodicity holds. For $\alpha < 1$, the process is non-ergodic since the average waiting time is infinite. For small enough $\alpha$, the time-average approaches the position where the particle is stuck for the most time.

forward Eq. (24):

$$\frac{\partial}{\partial t}G(x,p,t) = K_\alpha \left[\frac{\partial^2}{\partial x^2} + \frac{\partial}{\partial x}\frac{m\omega^2 x}{k_B T}\right] \mathcal{D}_t^{1-\alpha} G(x,p,t) + ipxG(x,p,t),$$

where we used $U(x) = x$, and $p$ is the Fourier pair of $A_x$. Since $G(x,p,t = 0) = \delta(x)$, we have in $s$ space,

$$sG(x,p,s) - \delta(x) = K_\alpha \left[\frac{\partial^2}{\partial x^2} + \frac{\partial}{\partial x}\frac{m\omega^2 x}{k_B T}\right] [s - ipx]^{1-\alpha} \times G(x,p,s) + ipxG(x,p,s). \tag{43}$$

To find $\langle A_x^2 \rangle$, we use the relation

$$\langle A_x^2 \rangle_s = \int_{-\infty}^{\infty}\int_{-\infty}^{\infty} A_x^2 G(x, A_x, s) dA_x dx = -\int_{-\infty}^{\infty} \frac{\partial^2}{\partial p^2} G(x,p,s)\bigg|_{p=0} dx. \tag{44}$$

Operating on both sides of (43) with $-\frac{\partial^2}{\partial p^2}$, substituting $p = 0$, and integrating over all $x$, we obtain, in $s$ space,

$$s\langle A_x^2 \rangle_s = 2\langle A_x x \rangle_s , \tag{45}$$

where we used the fact that the integral over the $\mathcal{L}_{\rm FP}$ term vanishes because $x^n G$ and $x^n \frac{\partial G}{\partial x}$ are zero for $|x| \to \infty$ for all $n$. Thus, the problem of finding $\langle A_x^2 \rangle_s$ reduces to that of finding $\langle A_x x \rangle_s$. We next use

$$\begin{aligned} \langle A_x x \rangle_s &= \int_{-\infty}^{\infty} \int_{-\infty}^{\infty} A_x x G(x, A_x, s) dA_x dx \\ &= -i \int_{-\infty}^{\infty} x \left. \frac{\partial}{\partial p} G(x,p,s) \right|_{p=0} dx. \end{aligned} \tag{46}$$

Operating on both sides of (43) with $-i\frac{\partial}{\partial p}$, substituting $p = 0$, and integrating over $\int_{\infty}^{\infty} \cdots x\, dx$, we obtain,

$$s \langle A_x x \rangle_s = [1 + (1-\alpha)(s\tau)^{-\alpha}] \langle x^2 \rangle_s - s(s\tau)^{-\alpha} \langle A_x x \rangle_s ,$$

where we defined the relaxation time $\tau^\alpha = k_{\rm B}T/(K_\alpha m\omega^2)$. Thus,

$$s \langle A_x x \rangle_s = \frac{(1-\alpha) + (s\tau)^\alpha}{1 + (s\tau)^\alpha} \langle x^2 \rangle_s . \tag{47}$$

Finally, to find $\langle x^2 \rangle_s$, we use

$$\langle x^2 \rangle_s = \int_{-\infty}^{\infty} \int_{-\infty}^{\infty} x^2 G(x, A_x, s) dA_x\, dx = \int_{-\infty}^{\infty} x^2 G(x, p=0, s) dx. \tag{48}$$

Substituting $p = 0$ in (43) and integrating over $\int_{\infty}^{\infty} \cdots x^2 dx$,

$$s \langle x^2 \rangle_s = 2K_\alpha s^{-\alpha} - 2s(s\tau)^{-\alpha} \langle x^2 \rangle_s ,$$

where we used the normalization condition $\int\int G(x, A_x, s) dA_x dx = 1/s$. Thus,

$$s \langle x^2 \rangle_s = \frac{2 \langle x^2 \rangle_{\rm th}}{2 + (s\tau)^\alpha} . \tag{49}$$

where $\langle x^2 \rangle_{\rm th} = K_\alpha \tau^\alpha = k_{\rm B}T/(m\omega^2)$ is the second moment of $x$ in thermal equilibrium (see Eq. (40)). Combining Eqs. (45), (47), and (49), we find in

Laplace $s$ space,

$$\left\langle A_x^2 \right\rangle_s = \frac{4}{s^3} \frac{(1-\alpha) + (s\tau)^\alpha}{1 + (s\tau)^\alpha} \frac{\left\langle x^2 \right\rangle_{\text{th}}}{2 + (s\tau)^\alpha}.$$

To invert to the time domain, we first write $\left\langle A_x^2 \right\rangle_s$ as partial fractions:

$$\left\langle A_x^2 \right\rangle_s = \frac{2\left\langle x^2 \right\rangle_{\text{th}}}{s^3} \left[ (1-\alpha) + 2\alpha \frac{(s\tau)^\alpha}{1 + (s\tau)^\alpha} - (1+\alpha) \frac{(s\tau)^\alpha}{2 + (s\tau)^\alpha} \right]. \tag{50}$$

The inverse of the last equation is

$$\left\langle A_x^2 \right\rangle = \left\langle x^2 \right\rangle_{\text{th}} t^2 \{(1-\alpha) + 4\alpha E_{\alpha,3}[-(t/\tau)^\alpha] - 2(1+\alpha) E_{\alpha,3}[-2(t/\tau)^\alpha]\},$$

where we used the Laplace transform relation [60]

$$\int_0^\infty e^{-st} t^2 E_{\alpha,3}[-c(t/\tau)^\alpha] dt = \frac{1}{s^3} \frac{(s\tau)^\alpha}{c + (s\tau)^\alpha}$$

and $E_{\alpha,3}(z)$ is the Mittag-Leffler function, defined as [60]

$$E_{\alpha,3}(z) = \sum_{n=0}^{\infty} \frac{z^n}{\Gamma(3 + \alpha n)}. \tag{51}$$

The Mittag-Leffler function arises frequently in the context of fractional calculus as the fractional extension of the exponential function [7]. Using $\left\langle \overline{x}^2 \right\rangle = \left\langle A_x^2 \right\rangle / t^2$ we finally have

$$\left\langle \overline{x}^2 \right\rangle = \left\langle x^2 \right\rangle_{\text{th}} \{(1-\alpha) + 4\alpha E_{\alpha,3}[-(t/\tau)^\alpha] - 2(1+\alpha) E_{\alpha,3}[-2(t/\tau)^\alpha]\}. \tag{52}$$

Equation (52) is illustrated in Fig. 2.

To find the long times behavior, we expand Eq. (50) for small $s$, invert, and divide by $t^2$ (Fig. 2),

$$\left\langle \overline{x}^2 \right\rangle \sim (1-\alpha) \left\langle x^2 \right\rangle_{\text{th}} + \frac{(3\alpha - 1)\left\langle x^2 \right\rangle_{\text{th}}}{\Gamma(3-\alpha)} \left( \frac{t}{\tau} \right)^{-\alpha} + \mathcal{O}\left[ \left( \frac{t}{\tau} \right)^{-2\alpha} \right]. \tag{53}$$

Thus, for $\alpha < 1$

$$\lim_{t \to \infty} \left\langle (\Delta \overline{x})^2 \right\rangle = (1-\alpha) \left\langle x^2 \right\rangle_{\text{th}} > 0 \tag{54}$$

and ergodicity is broken [13, 14]. Only when $\alpha = 1$, we have ergodic behavior $\lim_{t\to\infty} \left\langle (\Delta \overline{x})^2 \right\rangle = 0$. Equations (52) and (53) thus exhibit $t^{-\alpha}$ convergence of the fluctuations of the time-averaged position to its asymptotic

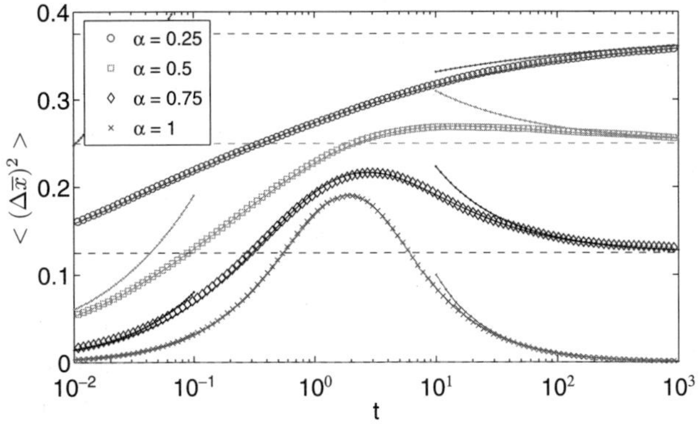


Fig. 2. Fluctuations of the time-average of $x(t)$, $\langle(\Delta\overline{x})^2\rangle$, for a particle in a binding harmonic field. Except for $\alpha = 1$, the fluctuations do not vanish when $t \to \infty$ and thus ergodicity is broken. Symbols represent simulation results for (top to bottom) $\alpha = 0.25, 0.5, 0.75, 1$ (see also legend). CTRW trajectories were generated using the method described in the main text (Sec. 4.2) with parameter values $a = 0.02$, $K_\alpha = 1/2$, $\langle x^2\rangle_{\text{th}} = 1/2$ (or $\tau^\alpha = 1$), and $x_0 = 0$. Simulations ended at $t = 10^3$ and included at least $10^4$ trajectories for each $\alpha$. Theory, Eq. (52), is plotted in solid lines, and agrees well with simulations without any fit. The straight dashed lines are the infinite time limit $\langle(\Delta\overline{x})^2\rangle = (1-\alpha)\langle x^2\rangle_{\text{th}}$ (Eq. (54)). The convergence to this limit is slowest for $\alpha = 0.25$. The dotted lines represent the short-times and long-times approximations, Eqs. (53) and (55), respectively.

value. The convergence to the long time limit is extremely slow for $\alpha \to 0$ as expected. For short times [7]

$$E_{\alpha,3}[-(t/\tau)^\alpha] \sim \frac{1}{2} - \frac{(t/\tau)^\alpha}{\Gamma(3+\alpha)}.$$

Therefore, for short times,

$$\langle\overline{x}^2\rangle \approx \frac{4\langle x^2\rangle_{\text{th}}}{\Gamma(3+\alpha)}\left(\frac{t}{\tau}\right)^\alpha. \tag{55}$$

This is demonstrated in Fig. 2. For $t = 0$, $\langle\overline{x}^2\rangle = 0$ because $x_0 = 0$.

### 4.2. *Simulation of trajectories*

The fractional Feynman–Kac equation describes the joint PDF of $x$ and $A$ in the continuum limit of the CTRW. In this limit, $a \to 0$ and $B_\alpha \to 0$ but the generalized diffusion coefficient $K_\alpha = a^2/(2B_\alpha)$ (Eq. (17)) is kept finite [8]. We simulate trajectories of this process as follows [59].

We place a particle on a one-dimensional lattice in initial position $x_0$. We set the lattice spacing $a$ and the generalized diffusion coefficient $K_\alpha$ and determine $B_\alpha = a^2/(2K_\alpha)$. Waiting times are then drawn for $\alpha = 1$ from an exponential distribution $\psi(\tau) = e^{-\tau/\tau_0}/\tau_0$ with mean $\tau_0 = B_1$. This is implemented by setting $\tau = -\tau_0 \ln(u)$, where $u$ is a number uniformly distributed in $[0, 1]$. For $\alpha < 1$, we set $\tau_0 = [B_\alpha/\Gamma(1-\alpha)]^{1/\alpha}$ and $\tau = \tau_0 u^{-1/\alpha}$. This corresponds to $\psi(\tau) = \frac{B_\alpha}{|\Gamma(-\alpha)|}\tau^{-(1+\alpha)}$ $(\tau \geq \tau_0)$ as in Sec. 2. After waiting time $\tau$, we move the particle right or left (to $x_0 + a$ or $x_0 - a$, respectively) with probabilities $R(x)$ and $L(x)$, respectively, as given in Eq. (11). Specifically, for the harmonic potential we conveniently set $\langle x^2 \rangle_{\text{th}} = k_{\text{B}}T/(m\omega^2) = 1/2$ and thus $R(x) = (1 - ax)/2$ and $L(x) = (1 + ax)/2$. We then repeatedly wait and move the particle until the end of the simulation.

We implemented this algorithm to study the time-average of $x(t)$ for a particle in a harmonic potential. For each trajectory, we calculated $\overline{x}(t)$ for various time points. For each time point, we then computed the fluctuations $\langle (\Delta \overline{x})^2 \rangle$ over many trajectories. Convergence to the theoretical result was observed when we used smaller and smaller lattice spacing $a$, while keeping fixed the generalized diffusion coefficient $K_\alpha$. The final results are presented in Fig. 2 and indeed agree with our theory, Eq. (52) without any fit. We also implemented the alternative simulation method of [61,62], to reach the same results (not shown).

## 5. Summary

Functionals of the path of a Brownian particle have been investigated in numerous studies since the development of the Feynman–Kac equation in 1949 [30]. We developed forward and backward equations (Eqs. (18) and (27), respectively) for functionals of anomalous diffusion based on fractional calculus [31]. In the force-free case, we studied a few functionals of interest, including the occupation time in half-space and the maximal displacement; further examples can be found in [37]. For a particle in a harmonic field, we calculated the fluctuations of the time-averaged position and demonstrated weak ergodicity breaking.

The fractional Feynman–Kac equation (18) can be obtained from the integer equation (3) by insertion of a substantial fractional derivative operator [32]. In that sense, it is a natural generalization of that of Kac's. The distributions we obtained for specific functionals are also the expected extensions of their Brownian counterparts: the arcsine law for the occupation time in half-space [52] was replaced by Lamperti's PDF (Eq. (33)) [50], and the famous $t^{-3/2}$ decay of the one-dimensional first passage time PDF [52] became $t^{-(1+\alpha/2)}$ (Eq. (38)). Instead of the usual ergodic behavior of the Brownian motion, we observed weak ergodicity breaking. Thus, our analysis supports the notion that CTRW and the emerging fractional paths are elegant generalizations of ordinary Brownian motion. Nevertheless, other non-Brownian processes are also important. For example, it would be interesting to find an equation for the PDF of anomalous functionals when the underlying process is fractional Brownian motion [63].

## Acknowledgments

S.C. is supported by the Adams Fellowship Program of the Israel Academy of Sciences and Humanities. E.B. thanks the Israel Science Foundation for financial support.

## References

1. P. Siegle, I. Goychuk and P. Hänggi, *Phys. Rev. Lett.* **105**, 100602 (2010).
2. R. Metzler and J. Klafter, *Phys. Rep.* **339**, 1 (2000).
3. I. M. Sokolov and J. Klafter, *Chaos* **15**, 026103 (2005).
4. I. Sokolov, J. Klafter and A. Blumen, *Phys. Today* **55**, 48 (2002).
5. R. Gorenflo and F. Mainardi, *Anomalous Transport: Foundations and Applications*, (Wiley-VCH, 2008), pp. 93–128.
6. R. Hilfer, *Anomalous Transport: Foundations and Applications*, (Wiley-VCH, 2008), pp. 17–74.
7. R. Metzler, E. Barkai and J. Klafter, *Phys. Rev. Lett.* **82**, 3563 (1999).
8. E. Barkai, R. Metzler and J. Klafter, From continuous time random walks to the fractional Fokker–Planck equation, *Phys. Rev. E* **61**, 132 (2000).
9. Eds. R. Klages, G. Radons and I. M. Sokolov, *Anomalous Transport: Foundations and Applications* (Wiley-VCH, 2008).
10. J. P. Bouchaud, *J. Phys. I* **2**, 1705 (1992).
11. G. Bel and E. Barkai, *Phys. Rev. Lett.* **94**, 240602 (2005).
12. G. Margolin and E. Barkai, *Phys. Rev. Lett.* **94**, 080601 (2005).
13. A. Rebenshtok and E. Barkai, *Phys. Rev. Lett.* **99**, 210601 (2007).
14. A. Rebenshtok and E. Barkai, *J. Stat. Phys.* **133**, 565 (2008).
15. Y. He, S. Burov, R. Metzler and E. Barkai, *Phys. Rev. Lett.* **101**, 058101 (2008).
16. F. D. Stefani, J. P. Hoogenboom and E. Barkai, *Phys. Today* **62**, 34 (2009).
17. S. Havlin and D. ben-Avraham, *Adv. Phys.* **36**, 695 (1987).
18. J. P. Bouchaud and A. Georges, *Phys. Rep.* **195**, 127 (1990).
19. E. W. Montroll and G. H. Weiss, *J. Math. Phys.* **6**, 167 (1965).
20. H. Scher and E. Montroll, *Phys. Rev. B* **12**, 2455 (1975).
21. S. N. Majumdar, *Curr. Sci.* **89**, 2076 (2005).

22. A. Comtet, J. Desbois and C. Texier, *J. Phys. A: Math. Gen.* **38**, R341 (2005).
23. G. Foltin, K. Oerding, Z. Racz, R. L. Workman and R. P. K. Zia, *Phys. Rev. E* **50**, R639 (1994).
24. D. S. Grebenkov, *Rev. Mod. Phys.* **79**, 1077 (2007).
25. G. Hummer and A. Szabo, *Proc. Natl. Acad. Sci. USA* **98**, 3658 (2001).
26. A. Baule and R. Friedrich, *Phys. Lett. A* **350**, 167 (2006).
27. S. N. Majumdar and A. J. Bray, *Phys. Rev. E* **65**, 051112 (2002).
28. A. Comtet, C. Monthus and M. Yor, *J. Appl. Probab.* **35**, 255 (1998).
29. M. Yor, *On Exponential Functionals of Brownian Motion and Related Processes.* (Springer, 2001).
30. M. Kac, On distributions of certain Wiener functionals, *Trans. Am. Math. Soc.* **65**, 1 (1949).
31. L. Turgeman, S. Carmi and E. Barkai, *Phys. Rev. Lett.* **103**, 190201 (2009).
32. R. Friedrich, F. Jenko, A. Baule and S. Eule, *Phys. Rev. Lett.* **96**, 230601 (2006).
33. A. H. Gandjbakhche and G. H. Weiss, *Phys. Rev. E* **61**, 6958 (2000).
34. N. Agmon, *Chem. Phys. Lett.* **497**, 184 (2010).
35. G. Zumofen, J. Hohlbein and C. G. Hübner, *Phys. Rev. Lett.* **93**, 260601 (2004).
36. D. S. Grebenkov, *Phys. Rev. E* **76**, 041139 (2007).
37. S. Carmi, L. Turgeman and E. Barkai, *J. Stat. Phys.* **141**, 1 (2010).
38. M. M. Meerschaert and H. P. Scheffler, *J. Appl. Probab.* **41**, 623 (2004).
39. I. M. Sokolov and R. Metzler, *Phys. Rev. E* **67**, 010101(R) (2003).
40. Y. Hu and G. Kallianpur, *Appl. Math. Optim.* **42**, 281 (2000).
41. N. Laskin, *Phys. Rev. E* **62**, 3135 (2000).
42. N. Laskin, *Phys. Rev. E* **66**, 056108 (2002).
43. M. Naber, *J. Math. Phys.* **45**, 3339 (2004).
44. S. Wang and M. Xu, *J. Math. Phys.* **48**, 043502 (2007).
45. A. Iomin, *Phys. Rev. E* **80**, 022103 (2009).
46. S. N. Majumdar and A. Comtet, *Phys. Rev. Lett.* **89**, 060601 (2002).
47. E. Barkai, *J. Stat. Phys.* **123**, 883 (2006).
48. S. Watanabe, *Proc. Symp. Pure Math.* **57**, 157 (1995).
49. C. Godrèche and J. M. Luck, *J. Stat. Phys.* **104**, 489 (2001).
50. J. Lamperti, *Trans. Am. Math. Soc.* **88**, 380 (1958).
51. A. Baldassarri, J. P. Bouchaud, I. Dornic and C. Godrèche, *Phys. Rev. E* **59**, 20 (1999).
52. S. Redner, *A Guide to First-Passage Processes* (Cambridge Univ. Press, 2001).
53. M. Kac, On some connections between probability theory and differential and integral equations, in *Second Berkeley Symposium on Mathematical Statistics and Probability* (University of California Press, 1951), p. 189.
54. E. Barkai, *Phys. Rev. E* **63**, 046118 (2001).
55. A. Comtet and S. N. Majumdar, *J. Stat. Mech.*, P06013 (2005).
56. S. N. Majumdar, J. Randon-Furling, M. J. Kearney and M. Yor, *J. Phys. A: Math. Theor.* **41**, 365005 (2008).
57. G. Schehr and P. Le-Doussal, *J. Stat. Mech.*, P01009 (2010).

58. V. Tejedor, O. Bénichou, R. Voituriez, R. Jungmann, F. Simmel, C. Selhuber-Unkel, L. B. Oddershede and R. Metzler, *Biophys. J.* **98**, 1364 (2010).
59. E. Heinsalu, M. Patriarca, I. Goychuk, G. Schmid and P. Hänggi, *Phys. Rev. E* **73**, 046133 (2006).
60. I. Podlubny, *Fractional Differential Equations* (Academic Press, 1999).
61. M. Magdziarz, A. Weron and K. Weron, *Phys. Rev. E* **75**, 016708 (2007).
62. D. Kleinhans and R. Friedrich, *Phys. Rev. E* **76**, 061102 (2007).
63. B. B. Mandelbrot and J. W. Van Ness, *SIAM Rev.* **10**, 422 (1968).

# Chapter 9

## Foundations of Fractional Dynamics: A Short Account

R. Hilfer*
*Institut für Computerphysik, Universität Stuttgart, Pfaffenwaldring 27, 70569 Stuttgart, Germany*

Applications of fractional dynamics have received a steadily increasing amount of attention during the past decade. Its foundations have found less interest. This chapter briefly reviews the physical foundations of fractional dynamics.

| | | |
|---|---|---|
| 1. | Introduction | 209 |
| 2. | The Aristotelian Concept of Time | 211 |
| 3. | Time Evolution of Observables | 211 |
| 4. | Time Evolution of States | 213 |
| 5. | Conservative Systems | 214 |
| 6. | Statement of the Problem | 214 |
| 7. | Induced Measure Preserving Transformations | 215 |
| 8. | Fractional Time Evolution | 216 |
| 9. | Irreversibility | 218 |
| 10. | Infinitesimal Generators | 220 |
| 11. | Experimental Evidence | 220 |
| 12. | Dissipative Systems | 223 |

### 1. Introduction

A *fractional dynamical system* has been defined [1] as a dynamical system involving fractional (i.e. noninteger order) time derivatives instead of integer order time derivatives. Despite the long history of fractional calculus in mathematics (see [2–6] for reviews), despite numerous publications on

*Also at Institut für Physik, Universität Mainz, 55099 Mainz, Germany.

fractional powers of infinitesimal generators [7–16], and despite a rapidly growing literature on possible applications of fractional dynamical systems to physical phenomena (see [17–20] and the present volume for reviews), there seem to exist only few publications discussing the physical foundations of fractional dynamics.[a]

My objective in this chapter is to call attention to the foundations of fractional dynamics and fractional time evolution by reformulating the problem stated originally in [1,21] and briefly summarizing some known results. As everyone knows, fractional time derivatives do not appear in any established fundamental theory of physics such as classical mechanics, electrodynamics, or quantum mechanics. Instead, integer (first and second) order time derivatives occur in all fundamental theories of physics. Obviously, time is a primordial and fundamental concept from the foundations of physics. Replacing integer order with fractional order time derivatives therefore changes the fundamental concept of time and with it the concept of evolution in the foundations of physics. Evolution equations in physics do not contain fractional time derivatives, because this would contradict the deep and fundamental principle, that time evolution is time translation. Most publications on fractional dynamics proceed directly to applied problems, but do not justify, discuss or even mention, that they remove the fundamental concept of time evolution (= time translation) from the foundations of physics.

Difficulties with fractional dynamics arise also, because fractional derivative operators can be defined in numerous ways [4–6]. Embedding a conventional dynamical system into a family of fractional dynamical systems is not unique. In fact, an infinite number of choices are possible and many publications fail to justify or discuss their particular choice.

Given the need for a fundamental justification of fractional dynamics, this chapter is structured as follows. Let me first recall some basic ideas about time. Observables, states and their time evolution are discussed next. Restricting attention to conservative dynamical systems raises a fundamental problem for the time evolution of macroscopic states. Induced measure preserving transformations are then introduced to solve this problem. Averaging them shows that macroscopic states evolve in time by convolution rather than translation. My short account

[a]The term "fractional dynamics" is used synonymously with "fractional dynamical system."

of the foundations of fractional dynamics concludes with remarks about irreversibility, experimental evidence, and dissipative systems.

## 2. The Aristotelian Concept of Time

The concepts of time and time evolution are fundamental for physics (and other sciences). Aristotle [22, Δ 11] defined time as ἀριθμὸς κινήσεως (i.e. as the integer or rational number of motion),[b] and formulates the idea, that past and future are separated by a mathematical point, that he calls τὸ νῦν (the Now). Newton [24, p. 5] formulates and postulates "*Tempus absolutum verum et Mathematicum, in se et natura sua absque relatione ad externum quodvis, aequabiliter fluit, alioque nomine dicitur Duratio*".[c] The concept of time in modern physics is based on the ideas of Aristotle in their Newtonian formulation. Time is viewed as a *flux aequabilis* (uniform flow) or succession of Aristotelian time instants.

The theoretical and mathematical abstraction of this concept of time from general mathematical theories of physical phenomena has led to the fundamental principle of time translation invariance and energy conservation in modern physics. All fundamental theories of contemporary physics postulate time translation invariance as a basic symmetry of nature.

## 3. Time Evolution of Observables

Time is commonly considered as the set of Aristotelian time instants $\dot{t}$. The set of all time instants is represented mathematically by the set of real numbers $\mathbb{R}$. Time is "measured" by observing clocks. Clocks are physical systems. Let $a$ be an observable quantity (e.g., the position of the sun, the moon or some hand on a watch), and let $\mathcal{A}$ be the set of observables of such a physical system. A *dynamical system* is a triple $(\mathcal{A}, \mathbb{R}, T)$ where $\mathcal{A}$ is the set of observables of a physical system, $\mathbb{R}$ represents time, and the mapping

$$\begin{aligned} T : \mathcal{A} \times \mathbb{R} &\to \mathcal{A} \\ (a, \dot{t}) &\mapsto T(a, \dot{t}) \end{aligned} \tag{1}$$

[b]While Aristotle was perhaps counting heart beats, days, months, years, or time intervals determined with a κλεψύδρα, the idea to count periods has persisted. Today the unit of time corresponds to counting 9 192 631 770 periods of oscillation of a certain form of radiation emitted from $^{133}$Cs-atoms [23].

[c]Transl.: "Absolute, true and mathematical time flows uniformly, in itself, according to its own nature, and without relation to anything outside itself; it is also called by the name duration."

is its dynamical rule [25]. It describes the change of observable quantities with time. For the dynamical rule $T$ the following properties are postulated:

(1) For all time instants $\dot{s}, \dot{t} \in \mathbb{R}$ the dynamical rule obeys

$$T(T(a,\dot{s}),\dot{t}) = T(a,\dot{s}+\dot{t}) \tag{2}$$

for all $a \in \mathcal{A}$.

(2) There exists a time instant $\dot{t}_* \in \mathbb{R}$, called *beginning*, such that

$$T(a,\dot{t}_*) = a \tag{3}$$

holds for all $a \in \mathcal{A}$.

(3) The map $T$ is continuous in time in a suitable topology.

The set of observables reflects the kinematical structure of the physical system. The dynamical rule prescribes the *time evolution* of the system. Setting $\dot{s} = \dot{t}_*$ in Eq. (2) and using Eq. (3) shows that either $\dot{t}_* = 0$ must hold, or else the observable must be time independent. The time evolution of observables is the one-parameter family $\{{}_{\mathcal{A}}T^{\dot{t}}\}_{\dot{t}\in\mathbb{R}}$ of maps ${}_{\mathcal{A}}T^{\dot{t}} : \mathcal{A} \to \mathcal{A}$ defined by

$${}_{\mathcal{A}}T^{\dot{t}}a := T(a,\dot{t}) \tag{4}$$

for $\dot{t} \in \mathbb{R}$. The time evolution obeys the group law

$${}_{\mathcal{A}}T^{\dot{t}}\,{}_{\mathcal{A}}T^{\dot{s}} = {}_{\mathcal{A}}T^{\dot{s}+\dot{t}} \tag{5}$$

for all $\dot{t}, \dot{s} \in \mathbb{R}$, and the identity law

$${}_{\mathcal{A}}T(0) = \mathbf{1}, \tag{6}$$

where $\mathbf{1}$ is the identity on $\mathcal{A}$. The continuity law requires a topology. It is usually assumed, that $\mathcal{A}$ is a Banach space with norm $\|\cdot\|$, and that

$$\lim_{\dot{t}\to 0+} \|{}_{\mathcal{A}}T^{\dot{t}}a - a\| = 0 \tag{7}$$

holds for all $a \in \mathcal{A}$. Equations (5), (6) and (7) define a strongly continuous one parameter group of operators $\{{}_{\mathcal{A}}T^{\dot{t}}\}_{\dot{t}\in\mathbb{R}}$ on $\mathcal{A}$, called a *flow* [26,27]. For bounded linear operators strong and weak continuity are equivalent [28].

Identifying $a = a(0)$ and writing $T(a, \dot{t}) = a(\dot{t})$, the time evolution becomes time translation to the left, i.e.

$$_{\mathcal{A}}T^{\dot{t}} a(\dot{s}) = a(\dot{s} + \dot{t}) \tag{8}$$

for all $\dot{t}, \dot{s} \in \mathbb{R}$. If the arrow of time is taken into account, then the flow of time is directed, and only the time instants $\dot{t} \geq 0$ after the beginning can occur. In that case, inverse elements do not exist, and the family $\{_{\mathcal{A}}T^{\dot{t}}\}_{\dot{t} \geq 0}$ of operators forms only a semigroup [28, 29] instead of a group.

## 4. Time Evolution of States

In general, the set of observables $\mathcal{A}$ of a physical system is not only a Banach space, but forms an algebra, more specifically, a C*-algebra [30]. In classical physics this algebra is commutative. The states $\mu$ of a physical system are normalized, positive linear functionals on its algebra of observables [30]. As such they are elements of the dual space $\mathcal{A}^*$. The notation $\langle \mu, a \rangle$ is used for the value $\mu(a)$ of the observable $a$ in the state $\mu$. Convex combinations of states are again states. If a state cannot be written as a convex combination of other states, it is called pure. Because the observable algebra $\mathcal{A}$ is a subset of its bidual, $\mathcal{A} \subset \mathcal{A}^{**}$, its elements can be considered as functions on the set $X(\mathcal{A})$ of its characters,[d] i.e. $a(\chi) = \langle \chi, a \rangle$ for $a \in \mathcal{A}, \chi \in X(\mathcal{A})$. By virtue of this correspondence, known as the Gelfand isomorphism [30, 31], a commutative $C^*$-algebra is isomorphic to the algebra $C_0(X(\mathcal{A}))$ of continuous functions on the set $X(\mathcal{A})$ of its characters equipped with the weak* topology. Characters are pure states.

The time evolution of states is obtained from the time evolution of observables by passing to adjoints [28,32]. The adjoint time evolution $_{\mathcal{A}^*}T^{\dot{t}}$ : $\mathcal{A}^* \to \mathcal{A}^*$ with $\dot{t} \in \mathbb{R}$ consists of all adjoint operators $(_{\mathcal{A}}T^{\dot{t}})^*$ on the dual space $\mathcal{A}^*$ [28, 33]. If $\mu(\dot{t}_1)$ denotes the state at time $\dot{t}_1 \in \mathbb{R}$, then Eq. (8) implies

$$\begin{aligned} \langle \mu(\dot{t}_1), {}_{\mathcal{A}}T^{\dot{t}} a(\dot{t}_1) \rangle &= \langle \mu(\dot{t}_1), a(\dot{t}_1 + \dot{t}) \rangle \\ &= \langle \mu(\dot{t}_2 - \dot{t}), a(\dot{t}_2) \rangle = \langle {}_{\mathcal{A}^*}T^{\dot{t}} \mu(\dot{t}_2), a(\dot{t}_2) \rangle, \end{aligned} \tag{9}$$

where $\dot{t}_2 = \dot{t}_1 + \dot{t} \in \mathbb{R}$ is arbitrary. For left translations the adjoint group

$$_{\mathcal{A}^*}T^{\dot{t}} \mu(\dot{s}) = \mu(\dot{s} - \dot{t}) \tag{10}$$

[d] A character is an algebraic *-homomorphism from a commutative C*-algebra to $\mathbb{C}$ .

is the group of right translations with $\dot{s}, \dot{t} \in \mathbb{R}$. The adjoint semigroup is weak* continuous, but in general not strongly continuous, unless the Banach space $\mathcal{A}$ is reflexive [32].

## 5. Conservative Systems

In classical mechanics, the commutative algebra of observables $\mathcal{A} = C_0(\Gamma)$ is the algebra of continous functions on phase space $\Gamma$, that vanish at infinity.[e] The characters (pure states) are point measures on phase space $\Gamma$, and one has the isomorphism $\Gamma \equiv X(\mathcal{A})$. By the Riesz representation theorem the states $\mu \in \mathcal{A}^* \equiv C_0(X(\mathcal{A}))^* \equiv C_0(\Gamma)^*$ in classical mechanics are probability measures on phase space $\Gamma \equiv X(\mathcal{A})$. Every state $\mu \in C_0(\Gamma)^*$ gives rise to a probability measure space $(\Gamma, \mathfrak{G}, \mu)$ where $\mathfrak{G}$ is the $\sigma$-algebra of measurable subsets of phase space $\Gamma$.

Let $S : \Gamma \to \Gamma$ be an invertible map such that $S$ and $S^{-1}$ are both measurable, i.e. such that $S^{-1}\mathfrak{G} = S\mathfrak{G} = \mathfrak{G}$ where $SG := \{Sx : x \in G\}$ for $G \in \mathfrak{G}$. The map $S$ is called a *measure preserving transformation* and the measure $\mu$ on $\Gamma$ is called *invariant under* $S$, if $\mu(G) = \mu(SG) = \mu(S^{-1}G)$ for all $G \in \mathfrak{G}$. An invariant measure is called *ergodic* with respect to $S$, if it cannot be decomposed into a convex combination of $S$-invariant measures.

Here and in the following the measure preserving transformation is the adjoint time evolution ${}_{\mathcal{A}^*}T^{\dot{t}}$ which is denoted more briefly as ${}_{\Gamma}T^{\dot{t}} = {}_{\mathcal{A}^*}T^{\dot{t}}$. Pure states (characters) are not invariant under ${}_{\Gamma}T^{\dot{t}}$. Examples of invariant probability measures are furnished by the set of equilibrium states of a conservative system with Hamiltonian dynamics. If $\mu$ is an equilibrium state of a conservative system, then $(\Gamma, \mathfrak{G}, \mu, {}_{\Gamma}T^{\dot{t}})$ is a measure preserving system.

## 6. Statement of the Problem

Let $(\Gamma, \mathfrak{G}, \mu, {}_{\Gamma}T^{\dot{t}})$ be a measure preserving many-body system. The detailed microscopic time evolution ${}_{\Gamma}T^{\dot{t}} : \Gamma \to \Gamma$ is frequently not of interest in applications, because it is much too detailed to be computable. Instead one is interested in a reduced, coarse grained or averaged time evolution of macroscopic states where the system is locally or globally in equilibrium. Examples are isolated systems at phase coexistence or in metastable

[e]This means that for each $a \in C_0(\Gamma)$ and $\varepsilon > 0$ there is a compact subset $K \subseteq \Gamma$ such that $|a(x)| < \varepsilon$ for all $x \in \Gamma \setminus K$.

states.[f] This gives rise to the problem of finding the time evolution ${}_G T^{\dot{t}}$ : $G \to G$ on subsets $G \subset \Gamma$ of phase space.

It is not possible to define ${}_G T^{\dot{t}} = {}_\Gamma T^{\dot{t}}|_G$ as the restriction of ${}_\Gamma T^{\dot{t}}$ to $G$, because for fixed initial state $x(0) \in G \subset \Gamma$ the time evolution ${}_\Gamma T^{\dot{t}}$ produces states ${}_\Gamma T^{\dot{t}} x(0) = x(\dot{t}) \notin G$. Equivalently, for fixed time $\dot{t}$ the map ${}_\Gamma T^{\dot{t}}$ maps states $x \in G$ to states not in $G$. The restriction ${}_\Gamma T^{\dot{t}}|_G$ is not defined for all $\dot{t} \in \mathbb{R}$. This seems to preclude a sensible definition of ${}_G T^{\dot{t}}$.

This problem of defining an induced continuous time evolution for mixed states on subsets of small measure was introduced and solved in [1,21]. It originated from the general classification theory for phase transitions [35–39]. The solution involves discretization of ${}_\Gamma T^{\dot{t}}$, averaging Kakutani's induced measure preserving transformation [26,40] and Kac's theorem for recurrence times [1,21].

## 7. Induced Measure Preserving Transformations

Consider a subset $G \subset \Gamma$ with small but positive measure $\mu(G) > 0$ of a measure preserving many body system $(\Gamma, \mathfrak{G}, \mu, {}_\Gamma T^{\dot{t}})$. Because $\mu(G) > 0$, the subset $G$ becomes a probability measure space $(G, \mathfrak{S}, \nu)$. The induced probability measure is $\nu = \mu/\mu(G)$ and $\mathfrak{S} = \mathfrak{G} \cap G$ is the trace of $\mathfrak{G}$ in $G$ [41].

The measure preserving continuous time evolution ${}_\Gamma T^{\dot{t}}$ is discretized by setting

$$\dot{t} = k\tau \tag{11}$$

with $k \in \mathbb{Z}$ and $\tau > 0$ the discretization time step. A character $x \in G$ is called *recurrent*, if there exists an integer $k \geq 1$ such that ${}_\Gamma T^{k\tau} x \in G$. If $G \in \mathfrak{G}$ and $\mu$ is invariant under ${}_\Gamma T$, then almost every character in $G$ is recurrent by virtue of the Poincaré recurrence theorem. A subset $G$ is called recurrent, if $\mu$-almost every point $x \in G$ is recurrent. By Poincaré's recurrence theorem the recurrence time $t_G(x)$ of the character $x \in G$, defined as

$$t_G(x) = \tau \min\{k \geq 1 : {}_\Gamma T^{k\tau}(x) \in G\}, \tag{12}$$

is positive and finite for almost every $x \in G$. For every $k \geq 1$ let

$$G_k = \{x \in G : t_G(x) = k\tau\} \tag{13}$$

---

[f] This differs from relaxation to equilibrium discussed in [34].

denote the set of characters with recurrence time $k\tau$. Then the number

$$p(k) = \nu(G_k) \tag{14}$$

is the probability to find a recurrence time $k\tau$. The numbers $p(k)$ define a discrete probability density $p(k)\delta(\dot{s} - k\tau)$ on the arithmetic progression $\dot{s} - k\tau, k \in \mathbb{N}, \dot{s} \in \mathbb{R}$. Every probability measure $\varrho(\dot{s})$ on $(G, \mathfrak{G})$ at time instant $\dot{s}$ is then defined on the same arithmetic progression through

$$\varrho(B, \dot{s} - k\tau) = \varrho(B \cap G_k, \dot{s}) \tag{15}$$

for all $B \in \mathfrak{G}$ and $\dot{s} \in \mathbb{R}$. The induced time evolution ${}_G\mathcal{T}$ on the subset $G$ is defined for every $B \in \mathfrak{G}$ as the average [1, 21]

$${}_G\mathcal{T}\varrho(B, \dot{s}) = \sum_{k=1}^{\infty} p(k)\varrho(B, \dot{s} - k\tau), \tag{16}$$

where $\dot{s} \in \mathbb{R}$. For characters $\varrho = x \in G$, one recovers the first step in the discretized microscopic time evolution ${}_G\mathcal{T}x(\dot{s}) = x(\dot{s} - t_G(x))$ as expected. For mixed states $\varrho$ this formula allows the transition from the microscopic to the macroscopic time evolution. It assigns an averaged translation to the first step in the induced time evolution of mixed states.

## 8. Fractional Time Evolution

The induced time evolution is obtained from ${}_G\mathcal{T}$ by iteration. According to its definition in Eq. (16), the induced measure preserving transformation ${}_G\mathcal{T}$ acts as a convolution in time,

$${}_G\mathcal{T}\varrho = \varrho * p, \tag{17}$$

where $\varrho$ is a mixed state on $(G, \mathfrak{G})$. Iterating $N$ times gives

$${}_G\mathcal{T}^N\varrho = ({}_G\mathcal{T}^{N-1}\varrho) * p = \varrho * \underbrace{p \cdots * p}_{N \text{ factors}} = \varrho * p_N, \tag{18}$$

where $p_N(k) = p(k) * \cdots * p(k)$ is the probability density of the sum

$$\mathcal{T}_N = \tau_1 + \cdots + \tau_N \tag{19}$$

of $N$ independent and identically with $p(k)$ distributed random recurrence times $\tau_i$. Then the long time limit $N \to \infty$ for induced measure preserving transformations on subsets of small measure is generally governed by well-known local limit theorems for convolutions [42–45]. Application to the case at hand yields the following fundamental theorem of fractional dynamics [1,21].

**Theorem 1.** *Assume that $\tau > 0$ is maximal in the sense that there is no larger $\tau$ for which all recurrence times lie in $\tau\mathbb{N}$. Then the following conditions are equivalent:*

(1) *Either $\sum_{k=1}^{\infty} kp(k) < \infty$ or there exists a number $0 < \gamma < 1$ such that*

$$\gamma = \sup \left\{ 0 < \beta < 1 : \sum_{k=1}^{\infty} k^{\beta} p(k) < \infty \right\}. \tag{20}$$

(2) *There exist constants $D_N \geq 0, D \geq 0$ and $0 < \alpha \leq 1$ such that*

$$\lim_{N\to\infty} \sup_k \left| \frac{D_N}{\tau} p(k) - \frac{1}{D^{1/\alpha}} h_\alpha \left( \frac{k\tau}{D_N D^{1/\alpha}} \right) \right| = 0, \tag{21}$$

*where $\alpha = 1$, if $\sum_{k=1}^{\infty} kp(k) < \infty$, and $\alpha = \gamma$ otherwise. The function $h_\alpha(x)$ vanishes for $x \leq 0$, and is*

$$h_\alpha(x) = \frac{1}{x} \sum_{j=0}^{\infty} \frac{(-1)^j x^{-\alpha j}}{j!\,\Gamma(-\alpha j)} \tag{22}$$

*for $x > 0$.*

If the limit exists, and is nondegenerate, i.e. $D \neq 0$, then the rescaling constants $D_N$ have the form

$$D_N = (N\Lambda(N))^{1/\alpha}, \tag{23}$$

where $\Lambda(N)$ is a slowly varying function [46], i.e.

$$\lim_{x\to\infty} \frac{\Lambda(bx)}{\Lambda(x)} = 1 \tag{24}$$

for all $b > 0$.

The theorem shows that

$$p_N(k) \approx \frac{\tau}{D_N D^{1/\alpha}} h_\alpha \left( \frac{k\tau}{D_N D^{1/\alpha}} \right) \tag{25}$$

holds for sufficiently large $N$. The asymptotic behavior of the iterated induced measure preserving transformation ${}_G T^N$ for $N \to \infty$ allows to remove the discretization, and to find the induced continuous time evolution on subsets $G \subset \Gamma$. First, the definition (15) is extended from the arithmetic progression $\dot{s} - \tau\mathbb{N}$ to $\dot{t} \leq \dot{s}$ by linear interpolation. Let $\widetilde{\varrho}(\dot{t})$ denote the extended measure defined for $\dot{t} \leq \dot{s}$. Using Eq. (11) and setting

$$t = D_N D^{1/\alpha} \tag{26}$$

the summation in Eq. (16) can be approximated for sufficently large $N \to \infty$ by an integral. Then ${}_G T^N \widetilde{\varrho}(\dot{s}) \approx {}_G T^t_\alpha \widetilde{\varrho}(\dot{s})$, where

$${}_G T^t_\alpha \widetilde{\varrho}(\dot{s}) = \int_0^\infty \widetilde{\varrho}(\dot{s} - \dot{t}) h_\alpha \left(\frac{\dot{t}}{t}\right) \frac{\mathrm{d}\dot{t}}{t} \tag{27}$$

is the induced continuous time evolution. ${}_G T^t_\alpha$ is also called *fractional time evolution*. Laplace tranformation shows that ${}_G T^t_\alpha$ fulfills Eq. (5). It is an example of subordination of semigroups [7, 33, 47, 48]. Indeed

$${}_G T^t_\alpha = \frac{1}{t} \int_0^\infty T^{\dot{t}} h_\alpha \left(\frac{\dot{t}}{t}\right) \mathrm{d}\dot{t}, \tag{28}$$

where $T^{\dot{t}}$ denotes right translations on the interpolated measure. As $D_N \geq 0$ and $D \geq 0$, Eq. (26) implies $t \geq 0$. As remarked in the introduction, the induced time evolution is in general not a translation (group or semigroup), but a convolution semigroup. The fundamental classification parameter

$$\alpha = \alpha(T, G, \tau) \tag{29}$$

depends not only on the dynamical rule $T(\cdot, \dot{t})$ and the subset $G$, but also on the discretization time step $\tau$, i.e. on the time scale of interest.

## 9. Irreversibility

The result in Eq. (27) has provided new insight into the irreversibility paradox [21, p. 554]. For $\alpha \to 0^-$, one finds

$$h_1(x) = \lim_{\alpha \to 0^-} h_\alpha(x) = \delta(x - 1) \tag{30}$$

and therefore

$${}_G T^t_1 \widetilde{\varrho}(\dot{s}) = \int_0^\infty \widetilde{\varrho}(\dot{s} - \dot{t}) \delta \left(\frac{\dot{t}}{t} - 1\right) \frac{\mathrm{d}\dot{t}}{t} = \widetilde{\varrho}(\dot{s} - t) \tag{31}$$

is a right translation. Here $\dot{t} \in \mathbb{R}$ denotes a time instant, while $t \geq 0$ is a time duration. This shows, that induced right translations do not form a group, but only a semigroup.

These observations suggest a reformulation of the controversial irreversibility problem [6, 49]. The problem of irreversibility is normally formulated as:

**Definition 1 (The normal irreversibility problem).** Assume that time is reversible. Explain how and why time irreversible equations arise in physics.

The assumption that time is reversible, i.e. $\dot{t} \in \mathbb{R}$, is made in all fundamental theories of modern physics. The explanation of macroscopically irreversible behavior for macroscopic nonequilibrium states of subsystems is due to Boltzmann. It is based on the applicability of statistical mechanics and thermodynamics, the large separation of scales, the importance of low entropy initial conditions, and probabilistic reasoning [34].

The problem with assuming $\dot{t} \in \mathbb{R}$ is not the second law of thermodynamics, because the foundations of thermodynamics and statistical mechanics do not cover all dynamical systems in nature. The problem with the arrow of time is that an experiment (i.e. the preparation of certain intial conditions for a dynamical system) cannot be repeated yesterday, but only tomorrow [49]. While it is possible to translate the spatial position of a physical system, it is not possible to translate the temporal position of a physical system backwards in time.[g] This was emphasized in [6, 49]. These simple observations combined with Eqs. (30) and (31) suggest to reformulate the standard irreversibility problem:

**Definition 2 (The reversed irreversibility problem).** Assume that time is irreversible. Explain how and why time reversible equations arise in physics.

The reversed irreversibility problem was introduced in [49]. Its solution is given by Theorem 1 combined with (30) and (31). The impossibility of performing experiments in the past is fundamental and evident. Therefore, as emphasized in [49], it must be assumed that time is irreversible. The normal irreversibility problem starts from an assumption, that contradicts experiment, while the reversed problem starts from the correct assumption. Theorem 1 combined with (30) and (31) explains why time translations, i.e. the case $\alpha = 1$, arise in physics, and why it arises more frequently than the case $\alpha < 1$.

[g]Note that this is not the same as reversing the momenta of all particles in a physical system.

## 10. Infinitesimal Generators

The operators ${}_G T^t_\alpha$ form a family of strongly continous semigroups on $C_0(G)^*$ provided that the coarsened translations $T^t$ in Eq. (28) are strongly continuous [33,48] and $\int_0^\infty \|T^s\| h_\alpha(s/t)/t\,\mathrm{d}s < \infty$. In this case the infinitesimal generators for $0 < \alpha \leq 1$ are defined by

$$\mathrm{A}_\alpha \widetilde{\varrho} = \operatorname*{s\text{-}lim}_{t\to 0} \frac{{}_G T^t_\alpha \widetilde{\varrho} - \widetilde{\varrho}}{t} \tag{32}$$

for all $\widetilde{\varrho} \in C_0(G)^*$ for which the strong limit s-lim exists. In general, the infinitesimal generators are unbounded operators. If $\mathrm{A} = -\mathrm{d}/\mathrm{d}t$ denotes the infinitesimal generator of the coarsened translation, $T^t$, then

$$\mathrm{A}_\alpha = -(-\mathrm{A})^\alpha = -\left(\frac{\mathrm{d}}{\mathrm{d}t}\right)^\alpha \tag{33}$$

are fractional time derivatives [16, 50]. The action of $\mathrm{A}_\alpha$ on mixed states can be represented in different ways. Frequently an integral representation

$$\mathrm{A}_\alpha \widetilde{\varrho} = \lim_{\epsilon\to o} C \int_\epsilon^\infty t^{-\alpha-1}(\mathbf{1} - T^t)\widetilde{\varrho}\mathrm{d}t \tag{34}$$

of Marchaud type [8, 51] is used. The integral representation

$$\mathrm{A}_\alpha \widetilde{\varrho} = \lim_{\epsilon\to o} C \int_\epsilon^\infty t^{-\alpha}\mathrm{A}(\mathbf{1} - t\mathrm{A})^{-1}\widetilde{\varrho}\mathrm{d}t \tag{35}$$

in terms of the resolvent of A [12] defines the same fractional derivative operator [52]. Representations of Grünwald–Letnikov type are also well known [16].

In summary, fractional dynamical systems must be expected to appear generally in mathematical models of macroscopic phenomena. They arise as coarse grained macroscopic time evolutions from inducing a microscopic time evolution on the subsets $G \subset \Gamma$ of small measure in phase space, that are typically incurred in statistical mechanics [1, 21, 50].

## 11. Experimental Evidence

If fractional time evolutions from Eq. (27) with $\alpha < 1$ must be expected on general grounds, then they should be observable in experiment. Numerous experimental examples of anomalous dynamics or strange kinetics have been identified (see [17–20] and the present volume for reviews). Here the example of dielectric $\alpha$-relaxation in glasses is briefly discussed [53, 54], because it concerns experimental data over up to 19 decades in time [55], and because the explanation of its excess wing has been a matter of debate.

For every induced time evolution on $G$ with time scale $\tau > 0$ and fractional order $\beta(\tau)$

$$ {}_G T^{\tau t}_{\beta(\tau)} = {}_G T^{\tau_1 t}_{\beta(\tau)} {}_G T^{(\tau-\tau_1)t}_{\beta(\tau)} = {}_G T^{\tau_1 t}_{\beta(\tau)} {}_G T^{\tau_2 t}_{\beta(\tau_1+\tau_2)} \tag{36} $$

holds generally with $\tau = \tau_1 + \tau_2$. A physical system typically shows different physical phenomena on different time scales $\tau_i$. In [53, 54] it was assumed that the second factor in Eq. (36) becomes approximately fractional in the sense that

$$ {}_G T^{\tau_2 t}_{\beta(\tau_1+\tau_2)} \approx {}_G T^{\tau_2 t}_{\alpha(\tau_2)} \tag{37} $$

holds in the weak* or strong topology with

$$ \lim_{\tau_2 \to 0} \alpha(\tau_2) = \beta(\tau). \tag{38} $$

The resulting composite time evolution ${}_G T^{\tau_1 t}_{\beta(\tau)} {}_G T^{\tau_2 t}_{\alpha(\tau_2)}$ was studied in [53,54] for the case $\beta(\tau) = 1$. Rescaling this composite operator as in the case of Debye relaxation and computing the infinitesimal generator yields the fractional differential equation [53, 54]

$$ \tau_1 \frac{\mathrm{d}f}{\mathrm{d}t} + \tau_2^\alpha \mathrm{A}_\alpha f = -f \tag{39} $$

with $\mathrm{A}_\alpha$ from Eq. (33) and initial value $f(0) = 1$. Its solution is

$$ f(t) = \mathrm{E}_{(1,1-\alpha),1}\left(-\frac{t}{\tau_1}, -\frac{\tau_2^\alpha}{\tau_1} t^{1-\alpha}\right), \tag{40} $$

where

$$ \mathrm{E}_{(a_1,a_2),b}(z_1, z_2) = \sum_{k=0}^{\infty} \sum_{\substack{\ell_1 \geq 0 \\ \ell_1+\ell_2=k}} \sum_{\ell_2 \geq 0} \frac{k!}{\ell_1! \ell_2!} \frac{z_1^{\ell_1} z_2^{\ell_2}}{\Gamma(b + a_1\ell_1 + a_2\ell_2)} \tag{41} $$

with $a_1, a_2 > 0$ and $b, z_1, z_2 \in \mathbb{C}$ is the binomial Mittag-Leffler function [61].

The complex frequency dependent susceptibility is obtained from the the normalized relaxation function as $\chi(u) = 1 - u\hat{f}(u)$ where $\hat{f}(u)$ is the Laplace transform of $f(t)$ and $u = \mathrm{i}\omega$ is the imaginary circular frequency [54, p. 402, Eq. (18)]. The real part of the complex dielectric susceptibility for propylene carbonate at temperature $T = 193\,\mathrm{K}$ is plotted in Fig. 1, its imaginary part in Fig. 2. These figures are taken from [53]. Crosses

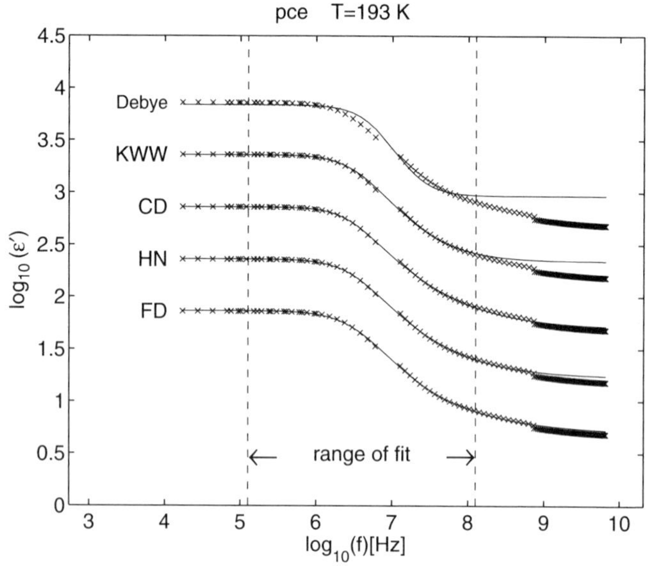


Fig. 1. Five different fits to the real part $\varepsilon'(f)$ of the complex dielectric function of propylene carbonate at $T = 193\,\mathrm{K}$ as a function of frequency $f$. Experimental data represented by crosses are from [56]. The fitting functions correspond to an exponential (Debye), stretched exponential (KWW), Cole–Davidson [57, 58] Havriliak–Negami [59, 60] and the fractional dynamics (FD) relaxation from (40). The range over which the data were fitted is indicated by dashed vertical lines in the figure. For clarity the data were displaced vertically by half a decade each. The original location of the data corresponds to the curve labelled FD.

represent experimental data. Different fit functions are shifted by half a decade for better visibility. The range over which the data were fitted is indicated by two dashed vertical lines. The curve labeled FD (fractional dynamics) is the susceptibility corresponding to the relaxation function in Eq. (40). It reproduces the high frequency wing even outside the range of its fit. This is not the case for the other four curves, labeled Debye, KWW, CD and HN. They correspond to four popular fit functions for dielectric relaxation [55, 62]. The curve Debye corresponds to a simple exponential function, KWW (Kohlrausch–Williams–Watts) is a stretched exponential relaxation function. The relaxation functions for the two remaining cases, CD (Cole–Davidson) and HN (Havriliak–Negami) were given for the first time in [58, 60].

Figure 3 from [54] shows the real and imaginary parts of the dielectric susceptibility for glycerol as its temperature varies over the glass transition

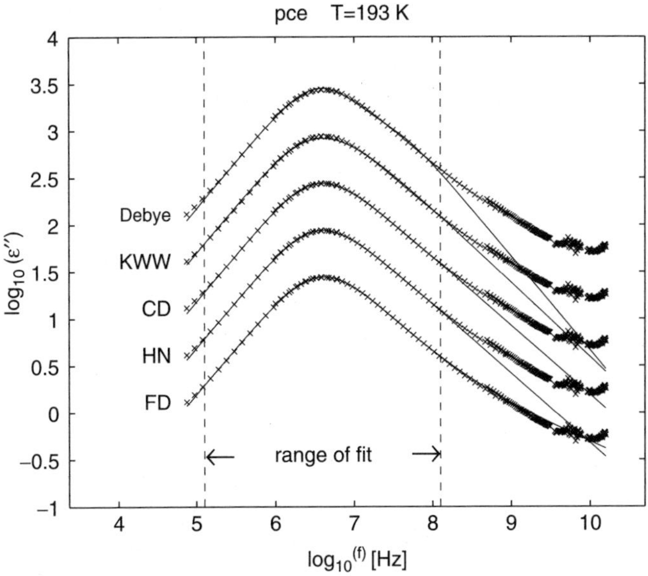


Fig. 2. Five different fits as in Fig. 1 for the imaginary part $\varepsilon''(f)$ of the complex dielectric function of propylene carbonate at $T = 193\,\mathrm{K}$ as a function of frequency $f$.

range from $T = 323\,\mathrm{K}$ to $T = 184\,\mathrm{K}$. The fits are based on a trinomial fractional relaxation function as detailed in [54, 61].

## 12. Dissipative Systems

The concept of time is the same for conservative and dissipative systems. For conservative dynamical systems a mathematically rigorous derivation of fractional dynamics from an underlying nonfractional dynamical system has remained elusive, although some authors have tried to relate $\alpha$ to invariant tori, strange attractors or other phase space structures [63, 64]. For dissipative systems the rigorous derivation has been possible for Bochner–Levy diffusion [7, 44, 47, 65] and Montroll–Weiss diffusion [66–70]. Due to restrictions on page number and preparation time only the latter case will be considered very briefly.

For diffusive dynamical systems a mathematically rigorous relation of fractional dynamics with microscopic Montroll–Weiss continuous time random walks was discovered in [71, 72]. It was shown that a diffusion (or master) equation with fractional time derivatives (i.e. a dissipative fractional dynamical system) can be related rigorously to the microscopic model

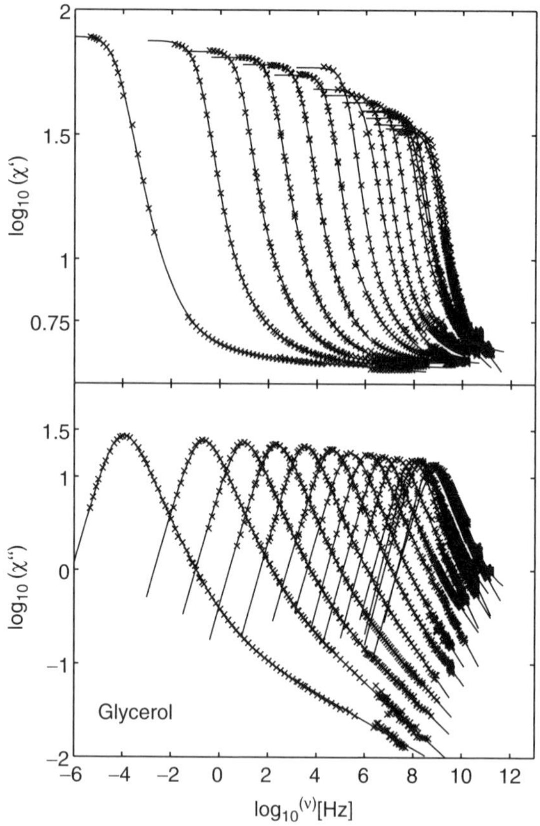


Fig. 3. Separate fits for real part (upper figure) and imaginary part (lower figure) of the complex dielectric susceptibility $\chi(\nu) = \chi'(\nu) + i\chi''(\nu)$ of glycerol for temperatures $T = 323$, 303, 295, 289, 273, 263, 253, 243, 234, 223, 213, 204, 195, 184 K (from right to left) as function of frequency $\nu$ (from [54]). The experimental data are taken from [56], the fit uses a generalized composite fractional relaxation model. For details see [54].

of Montroll–Weiss continuous time random walks (CTRW's) [66, 70] in the same way as ordinary diffusion is related to random walks [44]. This discovery became decoupled from its source in the widely cited review [19], and was later incorrectly attributed to [74] in [73].[h]

The fractional order $\alpha$ can be identified and has a physical meaning related to the statistics of waiting times in the Montroll–Weiss theory. The relation was established in two steps. First, it was shown in [71] that

[h]Contrary to [73, p. 51] fractional derivatives are never mentioned in [74].

Montroll–Weiss continuous time random walks with a Mittag-Leffler waiting time density are rigorously equivalent to a fractional master equation.

Then, in [72] this underlying random walk model was connected to the fractional time diffusion equation in the usual asymptotic sense [75] of long times and large distances.[i] For additional results see also [50, 76–78].

The relation between fractional diffusion and continuous time random walks, established in [71,72] and elaborated in [50,76–78], has initiated many subsequent investigations of fractional dissipative systems, particularly into fractional Fokker–Planck equations with drift [17–19, 73, 79–86].

## References

1. R. Hilfer, *Chaos, Solitons & Fractals* **5**, 1475 (1995).
2. K. Oldham and J. Spanier, *The Fractional Calculus* (Academic Press, 1974).
3. B. Ross, *Historia Mathematica* **4**, 75 (1977).
4. S. Samko, A. Kilbas and O. Marichev, *Fractional Integrals and Derivatives* (Gordon and Breach, 1993).
5. P. Butzer and U. Westphal, Introduction to fractional calculus, in *Applications of Fractional Calculus in Physics*, ed. R. Hilfer (World Scientific, 2000), p. 1.
6. R. Hilfer, Threefold introduction to fractional derivatives, in *Anomalous Transport: Foundations and Applications*, eds. R. Klages, G. Radons and I. Sokolov (Wiley-VCH, 2008), pp. 17–74.
7. S. Bochner, *Harmonic Analysis and the Theory of Probability* (Univ. of California Press, 1955).
8. A. Balakrishnan, *Pacific J. Math.* **10**, 419 (1960).
9. K. Yosida, *Proc. Jpn. Acad.* **36**, 86 (1960).
10. T. Kato, *Proc. Jpn. Acad.* **36**, 94 (1960).
11. H. Komatsu, *Pacific J. Math.* **19**, 285 (1966).
12. H. Komatsu, *Pacific J. Math.* **21**, 89 (1967).
13. U. Westphal, *Compositio Math.* **22**, 67 (1970).
14. U. Westphal, *Proc. London Math. Soc.* **29**, 557 (1974).
15. O. Lanford and D. Robinson, *J. Austral. Math. Soc. (A)* **46**, 473 (1989).
16. U. Westphal, Fractional powers of infinitesimal generators of semigroups, in *Applications of Fractional Calculus in Physics*, ed. R. Hilfer (World Scientific, 2000), p. 131.
17. R. Hilfer, *Applications of Fractional Calculus in Physics* (World Scientific, 2000).
18. Eds. R. Hilfer, R. Metzler, A. Blumen and J. Klafter, *Chemical Physics* **284**, (2002).
19. R. Metzler and J. Klafter, *Phys. Rep.* **339**, 1 (2000).
20. R. Klages *et al.*, *Anomalous Transport* (Wiley-VCH, 2008).

[i]This is emphasized in Eqs. (1.8) and (2.1) in [72] that are, of course, asymptotic.

21. R. Hilfer, *Fractals* **3**, 549 (1995).
22. W. Ross, *Aristotelis Physica. Recognovit Brevique Adnotatione Critica Instruxit W. D. Ross* (Oxford Univ. Press, 1950).
23. P. Mohr, B. Taylor and D. Newell, *J. Phys. Chem. Ref. Data.* **37**, 1187 (2008).
24. I. Newton, *Philosophiae Naturalis Principia Mathematica* (Societas Regia ac Joseph Streater, 1687).
25. N. Bhatia and G. Szegö, *Stability Theory of Dynamical Systems* (Springer, 1970).
26. I. Cornfeld, S. Fomin and Y. Sinai, *Ergodic Theory* (Springer, 1982).
27. B. Mandelbrot, *The Fractal Geometry of Nature* (Freeman, 1982).
28. A. Pazy, *Semigroups of Linear Operators and Applications to Partial Differential Equations* (Springer, 1983).
29. E. Hille and R. Phillips, *Functional Analysis and Semi-Groups* (Amer. Math. Soc., 1957).
30. O. Bratteli and D. Robinson, *Operator Algebras and Quantum Statistical Mechanics I* (Springer, 1979).
31. W. Thirring, *Lehrbuch der Mathematischen Physik 3: Quantenmechanik von Atomen und Molekülen* (Springer, 1979).
32. J. Neerven, *The Adjoint of a Semigroup of Linear Operators* (Springer, 1992).
33. R. Phillips, *Pacific J. Math.* **2**, 343 (1952).
34. J. Lebowitz, *Rev. Mod. Phys.* **71**, S346 (1999).
35. R. Hilfer, *Physica Scripta.* **44**, 321 (1991).
36. R. Hilfer, *Phys. Rev. Lett.* **68**, 190 (1992).
37. R. Hilfer, *Mod. Phys. Lett. B* **6**, 773 (1992).
38. R. Hilfer, *Phys. Rev. E* **48**, 2466 (1993).
39. R. Hilfer. On a new class of phase transitions, in *Random Magnetism and High-Temperature Superconductivity*, eds. W. Beyermann, N. Huang-Liu and D. MacLaughlin (World Scientific, 1994), p. 85.
40. S. Kakutani, *Proc. Jpn. Acad. Ser. A* **19**, 635 (1943).
41. H. Bauer, *Maß- und Integrationstheorie* (Walter de Gruyter, 1992).
42. B. Gnedenko and A. Kolmogorov, *Limit Distributions for Sums of Independent Random Variables* (Addison-Wesley, 1954).
43. H. Bergström, *Limit Theorems for Convolutions* (Wiley, 1963).
44. W. Feller, *An Introduction to Probability Theory and Its Applications*, Vol. II (Wiley, 1971).
45. I. Ibragimov and Y. Linnik, *Independent and Stationary Sequences of Random Variables* (Wolters-Nordhoff Publishing, 1971).
46. E. Seneta, *Regularly Varying Functions* (Springer-Verlag, 1976).
47. S. Bochner, *Proc. Natl. Acad. Sci. USA* **35**, 368 (1949).
48. E. Nelson, *Trans. Amer. Math. Soc.* **88**, 400 (1958).
49. R. Hilfer, Remarks on fractional time, in *Time, Quantum and Information*, eds. L. Castell and O. Ischebeck (Springer, 2003), p. 235.
50. R. Hilfer, Fractional time evolution, in *Applications of Fractional Calculus in Physics*, ed. R. Hilfer (World Scientific, 2000), p. 87.
51. A. Marchaud, *J. Math. Pures Appl.* **6**, 337 (1927).
52. J. Stafney, *Illinois J. Math.* **20**, 124 (1976).
53. R. Hilfer, *J. Phys.: Condens. Matter.* **14**, 2297 (2002).

54. R. Hilfer, *Chem. Phys.* **284**, 399 (2002).
55. P. Lunkenheimer, U. Schneider, R. Brand and A. Loidl, *Contemp. Phys.* **41**, 15 (2000).
56. U. Schneider, P. Lunkenheimer, R. Brand and A. Loidl, *Phys. Rev. E* **59**, 6924 (1999).
57. D. Davidson and R. Cole, *J. Chem. Phys.* **19**, 1484 (1951).
58. R. Hilfer, *J. Noncryst. Solids.* **305**, 122 (2002).
59. S. Havriliak and S. Negami, *J. Polymer Sci.: Part C* **14**, 99 (1966).
60. R. Hilfer, *Phys. Rev. E* **65**, 061510 (2002).
61. R. Hilfer, Y. Luchko and Z. Tomovski, *Fractional Calculus Appl. Anal.* **12**, 299 (2009).
62. Eds. F. Kremer and A. Schönhals, *Broad Band Dielectric Spectroscopy* (Springer-Verlag, 2003).
63. G. Zaslavsky, Fractional kinetics of Hamiltonian chaotic systems, in *Applications of Fractional Calculus in Physics*, ed. R. Hilfer (World Scientific, 2000), p. 202.
64. P. Inizan, Dynamique Fractionnaire Pour Le Chaos Hamiltonien. Ph.D. thesis, L'Observatoire de Paris, (2011).
65. M. Riesz, *Acta Math.* **81**, 1 (1949).
66. E. Montroll and G. Weiss, *J. Math. Phys.* **6**, 167 (1965).
67. M. Barber and B. Ninham, *Random and Restricted Walks* (Gordon and Breach, 1970).
68. E. Montroll and H. Scher, *J. Stat. Phys.* **9**, 101 (1973).
69. R. Hilfer and R. Orbach, *Chem. Phys.* **128**, 275 (1988).
70. B. Hughes, *Random Walks and Random Environments*, Vol. 1 (Clarendon Press, 1995).
71. R. Hilfer and L. Anton, *Phys. Rev. E* **51**, R848 (1995).
72. R. Hilfer, *Fractals* **3**, 211 (1995).
73. I. Sokolov, J. Klafter and A. Blumen, *Physics Today*, **2002**, 48 (2002).
74. A. V. Balakrishnan, *Physica* **132A**, 569 (1985).
75. M. Shlesinger, *J. Stat. Phys.* **10**, 421 (1974).
76. R. Hilfer. On fractional diffusion and its relation with continuous time random walks, in *Anomalous Diffusion: From Basis to Applications*, eds. A. P. R. Kutner and K. Sznajd-Weron (Springer, 1999), p. 77.
77. R. Hilfer, *J. Phys. Chem. B* **104**, 3914 (2000).
78. R. Hilfer, *Physica A* **329**, 35 (2003).
79. A. Compte, *Phys. Rev. E* **55**, 4191 (1996).
80. R. Metzler, J. Klafter and I. Sokolov, *Phys. Rev. E* **58**, 1621 (1998).
81. R. Metzler, E. Barkai and J. Klafter, *Phys. Rev. Lett.* **82**, 3563 (1999).
82. I. Sokolov, *Phys. Rev. E* **63**, 056111 (2001).
83. M. Meerschaert, D. Benson, H. Scheffler and P. Becker-Kern, *Phys. Rev. E* **66**, 060102 (2002).
84. F. E. Scalas and R. Gorenflo, *Phys. Rev. E* **69**, 011107 (2004).
85. R. Gorenflo, F. Mainardi, D. Moretti, G. Pagnini and P. Paradisi, *Chem. Phys.* **284**, 521 (2002).
86. G. Zaslavsky, *Phys. Rep.* **371**, 461 (2002).

# Chapter 10

## Parametric Subordination in Fractional Diffusion Processes

Rudolf Gorenflo* and Francesco Mainardi†

**Department of Mathematics and Informatics, Free University of Berlin, Arnimallee 3, D-14195 Berlin, Germany*

*†Department of Physics, University of Bologna, Via Irnerio 46, I-40126 Bologna, Italy*

1. Introduction . . . . . . . . . . . . . . . . . . . . . . . . 230
2. Notions and Notations . . . . . . . . . . . . . . . . . . . . . 231
   2.1. The Fourier transform . . . . . . . . . . . . . . . . . 231
   2.2. The Laplace transform . . . . . . . . . . . . . . . . . 232
   2.3. The auxiliary functions of Mittag-Leffler type . . . 232
   2.4. The auxiliary functions of Wright type . . . . . . . 234
   2.5. The Lévy stable distributions . . . . . . . . . . . . 237
3. The Space-Time Fractional Diffusion . . . . . . . . . . . . 241
   3.1. The Riesz–Feller space-fractional derivative . . . . 242
   3.2. The Caputo fractional derivative . . . . . . . . . . 243
   3.3. The fundamental solution of the space-time fractional diffusion equation . . . . . . . . . . . . . 243
   3.4. Alternative forms of the space-time fractional diffusion equation . . . . . . . . . . . . . . . . . . 245
4. Analytic and Stochastic Pathways to Subordination in Space-Time Fractional Diffusion . . . . . . . . . . . . 246
   4.1. The analytical interpretation via operational time . 246
   4.2. Stochastic interpretation . . . . . . . . . . . . . . . 248
   4.3. Evolution equations for the densities of physical and operational time in mutual dependence . . . . 250
   4.4. The random walks . . . . . . . . . . . . . . . . . . . 251
5. Graphical Representations and Conclusions . . . . . . . . 253

## 1. Introduction

The purpose of this chapter is to describe our method of parametric subordination to produce particle trajectories for the so-called fractional diffusion processes.

By replacing in the common diffusion equation the first-order time derivative and the second-order space derivative by appropriate fractional derivatives, we obtain a fractional diffusion equation whose solution describes the temporal evolution of the density of an extensive quantity, e.g., of the sojourn probability of a diffusing particle.

After giving a survey on analytic methods for determination of the solution (this is the macroscopic aspect), we turn our attention to the problem of simulation of particle trajectories (the microscopic aspect). By some authors such simulation is called "particle tracking", see e.g., [63].

As an approximate method among physicists, the so-called Continuous Time Random Walk (CTRW) is very popular. On the other hand, it is possible to produce a sequence of precise snapshots of a true trajectory. This is achieved by a change from the "physical time" to an "operational time" in which the simulation is carried out. By two Markov processes happening in operational time, the running of physical time and the motion in space are produced. Then, elimination of the operational time yields a picture of the desired trajectory. It is remarkable that by such combination of two Markov processes, a non-Markovian process is generated.

The two Markov processes can be obtained and analyzed in two ways:

(a) from the CTRW model by a well-scaled passage to the "diffusion limit",
(b) directly from an integral representation of the fundamental solution of the fractional diffusion equation.

We have developed way (a) in our 2007 paper [18] via passage to the diffusion limit in the Cox–Weiss solution formula for CTRW and by the technique of splitting the CTRW into three separate walks and passing in each of these to the diffusion limit in our recent paper [17].

The plan of our chapter is as follows. In Sec. 2, we provide for the reader's convenience some preliminary notions and notations as a mathematical background for our further analysis. In Sec. 3, we introduce the *space-time fractional diffusion equation*, based on the *Riesz–Feller* and *Caputo* fractional derivatives, and we present the fundamental solution. In Sec. 4, we provide the stochastic interpretation of the space-time fractional diffusion equation discussing the concepts of subordination, the main goal of this chapter. Finally, in Sec. 5, we show some graphical representations along with conclusions.

## 2. Notions and Notations

In this section we survey some preliminary notions including Fourier and Laplace transforms, special functions of Mittag-Leffler and Wright type and Lévy stable probability distributions.

Since in what follows we shall meet only real or complex-valued functions of a real variable that are defined and continuous in a given open interval $I = (a, b)$, $-\infty \le a < b \le +\infty$, except, possibly, at isolated points where these functions can be infinite, we restrict our presentation of the integral transforms to the class of functions for which the Riemann improper integral on $I$ absolutely converges. In so doing we follow Marichev [31] and we denote this class by $L^c(I)$ or $L^c(a, b)$.

### 2.1. *The Fourier transform*

Let

$$\hat{f}(\kappa) = \mathcal{F}\{f(x);\kappa\} = \int_{-\infty}^{+\infty} \mathrm{e}^{+i\kappa x} f(x)dx, \quad \kappa \in \mathbb{R}, \tag{1a}$$

be the Fourier transform of a function $f(x) \in L^c(\mathbb{R})$, and let

$$f(x) = \mathcal{F}^{-1}\{\hat{f}(\kappa);x\} = \frac{1}{2\pi}\int_{-\infty}^{+\infty} \mathrm{e}^{-i\kappa x} \hat{f}(\kappa)d\kappa, \quad x \in \mathbb{R}, \tag{1b}$$

be the inverse Fourier transform.[a]

Related to the Fourier transform is the notion of pseudo-differential operator. Let us recall that a generic pseudo-differential operator $A$, acting with respect to the variable $x \in \mathbb{R}$, is defined through its Fourier representation, namely

$$\int_{-\infty}^{+\infty} \mathrm{e}^{i\kappa x} A[f(x)]dx = \hat{A}(\kappa)\hat{f}(\kappa), \tag{2}$$

[a] If $f(x)$ is piecewise differentiable, then the formula (1b) holds true at all points where $f(x)$ is continuous and the integral in it must be understood in the sense of the Cauchy principal value.

where $\hat{A}(\kappa)$ is referred to as symbol of $A$, formally given as $\hat{A}(\kappa) = (A\mathrm{e}^{-i\kappa x})\mathrm{e}^{+i\kappa x}$.

### 2.2. *The Laplace transform*

Let

$$\tilde{f}(s) = \mathcal{L}\{f(t); s\} = \int_0^\infty \mathrm{e}^{-st} f(t)dt, \quad \mathrm{Re}(s) > a_f \tag{3a}$$

be the Laplace transform of a function $f(t) \in \mathcal{L}^c(0,T)$, $\forall T > 0$ and let

$$f(t) = \mathcal{L}^{-1}\{\tilde{f}(s); t\} = \frac{1}{2\pi i}\int_{\gamma-i\infty}^{\gamma+i\infty} \mathrm{e}^{st}\tilde{f}(s)ds, \quad \mathrm{Re}(s) = \gamma > a_f, \tag{3b}$$

with $t > 0$, be the inverse Laplace transform[b]

### 2.3. *The auxiliary functions of Mittag-Leffler type*

The Mittag-Leffler functions that we denote by $E_\alpha(z)$, $E_{\alpha,\beta}(z)$ are so named in honor of Gösta Mittag-Leffler, the eminent Swedish mathematician, who introduced and investigated these functions in a series of notes starting from 1903 in the framework of the theory of entire functions [43–45]. The functions are defined by the series representations, convergent in the whole complex plane $\mathbb{C}$

$$E_\alpha(z) := \sum_{n=0}^{\infty} \frac{z^n}{\Gamma(\alpha n + 1)}, \quad \mathrm{Re}(\alpha) > 0; \tag{4}$$

$$E_{\alpha,\beta}(z) := \sum_{n=0}^{\infty} \frac{z^n}{\Gamma(\alpha n + \beta)}, \quad \mathrm{Re}(\alpha) > 0,\ \beta \in \mathbb{C}. \tag{5}$$

Originally Mittag-Leffler assumed only the parameter $\alpha$ and assumed it to be positive, but soon later the generalization with two complex parameters was considered by Wiman [58]. In both cases the Mittag-Leffler functions are entire of order $1/\mathrm{Re}(\alpha)$. Generally $E_{\alpha,1}(z) = E_\alpha(z)$.

---

[b]A sufficient condition for the existence of the Laplace transform is that the original function is of exponential order as $t \to \infty$. This means that some constant $a_f$ exists such that the product $\mathrm{e}^{-a_f t}\,|f(t)|$ is bounded for all $t$ greater than some $T$. Then $\tilde{f}(s)$ exists and is analytic in the half plane $\mathrm{Re}(s) > a_f$. If $f(t)$ is piecewise differentiable, then the formula (3b) holds true at all points where $f(t)$ is continuous and the (complex) integral in it must be understood in the sense of the Cauchy principal value.

Using their series representations, it is easy to recognize

$$\begin{cases} E_{1,1}(z) = E_1(z) = \mathrm{e}^z, & E_{1,2}(z) = \dfrac{\mathrm{e}^z - 1}{z}, \\ E_{2,1}(z^2) = \cosh(z), & E_{2,1}(-z^2) = \cos(z), \\ E_{2,2}(z^2) = \dfrac{\sinh(z)}{z}, & E_{2,2}(-z^2) = \dfrac{\sin(z)}{z}, \end{cases} \tag{6}$$

and more generally

$$\begin{cases} E_{\alpha,\beta}(z) + E_{\alpha,\beta}(-z) = 2E_{2\alpha,\beta}(z^2), \\ E_{\alpha,\beta}(z) - E_{\alpha,\beta}(-z) = 2zE_{2\alpha,\alpha+\beta}(z^2). \end{cases} \tag{7}$$

We note that in Chap. 18 of Vol. 3 of [5] devoted to Miscellaneous Functions, we find a valuable survey of these functions, which were later recognized as belonging to the more general class of Fox $H$-functions introduced after 1960.

For our purposes, relevant roles are played by the following auxiliary functions of the Mittag-Leffler type on support $\mathbb{R}^+$ defined as follows, where $\lambda > 0$, along with their Laplace transforms

$$e_\alpha(t;\lambda) := E_\alpha(-\lambda t^\alpha) \div \frac{s^{\alpha-1}}{s^\alpha + \lambda}, \tag{8}$$

$$e_{\alpha,\beta}(t;\lambda) := t^{\beta-1}E_{\alpha,\beta}(-\lambda t^\alpha) \div \frac{s^{\alpha-\beta}}{s^\alpha + \lambda}, \tag{9}$$

$$e_{\alpha,\alpha}(t;\lambda) := t^{\alpha-1}E_{\alpha,\alpha}(-\lambda t^\alpha) = \frac{d}{dt}e_\alpha(-\lambda t^\alpha) = \div\frac{1}{s^\alpha + \lambda}. \tag{10}$$

Here we have used the sign $\div$ for the juxtaposition of a function depending on $t$ with its Laplace transform depending on $s$. Later we use this sign also for juxtaposition of a function depending on $x$ with its Fourier transform depending on $\kappa$.

**Remark.** We outline that the above auxiliary functions (for restricted values of the parameters) turn out to be *completely monotone* (CM) functions so that they enter in some types of relaxation phenomena of physical relevance.

We recall that a function $f(t)$ is CM in $\mathbb{R}^+$ if $(-1)^n f^n(t) \geq 0$. The function $\mathrm{e}^{-t}$ is the prototype of a CM function. For a Bernstein theorem, more generally they are expressed in terms of a (generalized) real Laplace transform of a positive measure

$$f(t) = \int_0^\infty \mathrm{e}^{-rt} K(r)\mathrm{d}r, \quad K(r) \geq 0. \tag{11}$$

Restricting attention to the auxiliary function in two parameters, we can prove for $\lambda > 0$ that

$$e_{\alpha,\beta}(t;\lambda) := t^{\beta-1} E_{\alpha,\beta}(-\lambda t^\alpha) \quad \text{CM} \quad \text{iff} \quad 0 < \alpha \leq \beta \leq 1. \tag{12}$$

Using the Laplace transform we can prove, following Gorenflo and Mainardi [13], that for $0 < \alpha < 1$ and $\lambda = 1$

$$E_\alpha(-t^\alpha) \simeq \begin{cases} 1 - \dfrac{t^\alpha}{\Gamma(\alpha+1)} + \dfrac{t^{2\alpha}}{\Gamma(2\alpha+1)} \cdots & t \to 0^+, \\ \dfrac{t^{-\alpha}}{\Gamma(1-\alpha)} - \dfrac{t^{-2\alpha}}{\Gamma(1-2\alpha)} \cdots & t \to +\infty, \end{cases} \tag{13}$$

and

$$E_\alpha(-t^\alpha) = \int_0^\infty \mathrm{e}^{-rt} K_\alpha(r) dr \tag{14}$$

with

$$K_\alpha(r) = \frac{1}{\pi} \frac{r^{\alpha-1} \sin(\alpha\pi)}{r^{2\alpha} + 2r^\alpha \cos(\alpha\pi) + 1} = \frac{1}{\pi r} \frac{\sin(\alpha\pi)}{r^\alpha + 2\cos(\alpha\pi) + r^{-\alpha}} > 0. \tag{15}$$

### 2.4. *The auxiliary functions of Wright type*

The Wright function, that we denote by $W_{\lambda,\mu,}(z)$, is so named in honor of E. Maitland Wright, the eminent British mathematician, who introduced and investigated this function in a series of notes starting from 1933 in the framework of the asymptotic theory of partitions, see [59–61]. The function is defined by the series representation, convergent in the whole $z$-complex plane $\mathbb{C}$,

$$W_{\lambda,\mu}(z) := \sum_{n=0}^{\infty} \frac{z^n}{n!\Gamma(\lambda n + \mu)}, \quad \lambda > -1, \ \mu \in \mathbb{C}. \tag{16}$$

Originally, Wright assumed $\lambda \geq 0$, and, only in 1940 [62], he considered $-1 < \lambda < 0$. We note that in Chap. 18 of Vol. 3 of [5] devoted to Miscellaneous Functions, we find an earlier analysis of these functions, which, similarly with the Mittag-Leffler functions, were later recognized as belonging to the more general class of Fox $H$-functions introduced after 1960. However, in that chapter, presumably for a misprint, the parameter $\lambda$ of the Wright function is restricted to be non-negative. It is possible to prove that the Wright function is entire of order $1/(1+\lambda)$, hence it is of exponential type only if $\lambda \geq 0$. For this reason we propose to distinguish the Wright functions in two kinds according to $\lambda \geq 0$ (*first kind*) and $-1 < \lambda < 0$ (*second kind*). Both kinds of functions are related to the Mittag-Leffler function via Laplace transform pairs: in fact we have, see for details appendix F of the recent book by Mainardi [26], for the case $\lambda > 0$ (Wright functions of the first kind)

$$W_{\lambda,\mu}(\pm r) \div \frac{1}{s} E_{\lambda,\mu}\left(\pm\frac{1}{s}\right), \quad \lambda > 0, \tag{17}$$

and for the case $\lambda = -\nu \in (-1, 0)$ (Wright functions of the second kind),

$$W_{-\nu,\mu}(-r) \div E_{\nu,\mu+\nu}(-s), \quad 0 < \nu < 1. \tag{18}$$

For our purposes relevant roles are played by the following auxiliary functions of the Wright type (of the second kind)

$$F_\nu(z) := W_{-\nu,0}(-z) = \sum_{n=1}^{\infty} \frac{(-z)^n}{n!\Gamma(-\nu n)}, \quad 0 < \nu < 1 \tag{19}$$

and

$$M_\nu(z) := W_{-\nu,1-\nu}(-z) = \sum_{n=0}^{\infty} \frac{(-z)^n}{n!\Gamma[-\nu n + (1-\nu)]}, \quad 0 < \nu < 1, \tag{20}$$

interrelated through $F_\nu(z) = \nu z M_\nu(z)$. The relevance of these functions was pointed out by Mainardi in his former analysis of the time fractional diffusion equation via Laplace transform. Restricting our attention to the $M$-Wright functions on support $\mathbb{R}^+$ we point out the Laplace transforms pairs

$$M_\nu(r) \div E_\nu(-s), \quad 0 < \nu < 1. \tag{21}$$

$$\frac{\nu}{r^{\nu+1}} M_\nu(1/r^\nu) \div \mathrm{e}^{-s^\nu}, \quad 0 < \nu < 1. \tag{22}$$

$$\frac{1}{r^\nu} M_\nu(1/r^\nu) \div \frac{\mathrm{e}^{-s^\nu}}{s^{1-\nu}}, \quad 0 < \nu < 1. \tag{23}$$

It was also proved in [26] that the $M$-Wright function on support $\mathbb{R}^+$ is a probability density function (pdf) that in the literature is sometimes known as the density related to Mittag-Leffler probability distribution. Its absolute moments of order $\delta > -1$ in $\mathbb{R}^+$ are finite and turn out to be

$$\int_0^\infty r^\delta M_\nu(r)dx = \frac{\Gamma(\delta+1)}{\Gamma(\nu\delta+1)}, \quad \delta > -1, \quad 0 \le \nu < 1. \tag{24}$$

We point out that in the limit $\nu \to 1^-$ the function $M_\nu(r)$, for $r \in \mathbb{R}^+$, tends to the Dirac generalized function $\delta(r-1)$.

For our next purposes, it is worthwhile to introduce the *function in two variables*

$$M_\nu(x,t) := t^{-\nu} M_\nu(xt^{-\nu}), \quad 0 < \nu < 1, \quad x,t \in \mathbb{R}^+, \tag{25}$$

which defines a spatial probability density in $x$ evolving in time $t$ with self-similarity exponent $H = \nu$. Of course for $x \in \mathbb{R}$ we can consider the symmetric version obtained from (15) multiplying by $1/2$ and replacing $x$ by $|x|$. Hereafter we provide a list of the main properties of this density, which can be derived from the Laplace and Fourier transforms for the corresponding Wright $M$-function in one variable.

From Eq. (23) we derive the Laplace transform of $M_\nu(x,t)$ with respect to $t \in \mathbb{R}^+$,

$$\mathcal{L}\{M_\nu(x,t); t \to s\} = s^{\nu-1} \mathrm{e}^{-xs^\nu}. \tag{26}$$

From Eq. (21) we derive the Laplace transform of $M_\nu(x,t)$ with respect to $x \in \mathbb{R}^+$,

$$\mathcal{L}\{M_\nu(x,t); x \to s\} = E_\nu(-st^\nu). \tag{27}$$

From the recent book by Mainardi [26] we recall the Fourier transform of $M_\nu(|x|,t)$ with respect to $x \in \mathbb{R}$,

$$\mathcal{F}\{M_\nu(|x|,t); x \to \kappa\} = 2E_{2\nu}(-\kappa^2 t^{2\nu}), \tag{28}$$

and, in particular,

$$\begin{cases} \displaystyle\int_0^\infty \cos(\kappa x) M_\nu(x,t)dx = E_{2\nu,1}(-\kappa^2 t^{2\nu}), \\ \displaystyle\int_0^\infty \sin(\kappa x) M_\nu(x,t)dx = \kappa t^\nu E_{2\nu,1+\nu}(-\kappa^2 t^{2\nu}). \end{cases} \tag{29}$$

It is worthwhile to note that for $\nu = 1/2$ we recover the Gaussian density evolving with time with variance $\sigma^2 = 2t$

$$\frac{1}{2} M_{1/2}(x,t) = \frac{1}{2\sqrt{\pi} t^{1/2}} \mathrm{e}^{-x^2/(4t)}. \tag{30}$$

## 2.5. *The Lévy stable distributions*

The term stable has been assigned by the French mathematician Paul Lévy, who, in the twenties of the last century, started a systematic research in order to generalize the celebrated *Central Limit Theorem* to probability distributions with infinite variance. For stable distributions we can assume the following

**Definition.** If two independent real random variables with the same shape or type of distribution are combined linearly with positive coefficients and the distribution of the resulting random variable has the same shape, the common distribution (or its type, more precisely) is said to be stable.

The restrictive condition of stability enabled Lévy (and then other authors) to derive the *canonic form* for the characteristic function of the densities of these distributions. Here we follow the parametrization by Feller [6] revisited in [14] and in [27]. Denoting by $L_\alpha^\theta(x)$ a generic stable density in $\mathbb{R}$, where $\alpha$ is the *index of stability* and and $\theta$ the asymmetry parameter, improperly called *skewness*, its characteristic function reads:

$$\begin{cases} L_\alpha^\theta(x) \div \hat{L}_\alpha^\theta(\kappa) = \exp\left[-\psi_\alpha^\theta(\kappa)\right], \psi_\alpha^\theta(\kappa) = |\kappa|^\alpha e^{i(\operatorname{sign}\kappa)\theta\pi/2}, \\ \qquad\qquad\qquad\qquad 0 < \alpha \le 2, |\theta| \le \min\{\alpha, 2-\alpha\}. \end{cases} \tag{31}$$

We note that the allowed region for the real parameters $\alpha$ and $\theta$ turns out to be a diamond in the plane $\{\alpha, \theta\}$ with vertices in the points $(0,0)$, $(1,1)$, $(1,-1)$, $(2,0)$, that we call the *Feller–Takayasu diamond*, see Fig. 1. For values of $\theta$ on the border of the diamond (that is $\theta = \pm\alpha$ if $0 < \alpha < 1$, and $\theta = \pm(2-\alpha)$ if $1 < \alpha < 2$) we obtain the so-called *extremal stable densities*.

We note the *symmetry relation* $L_\alpha^\theta(-x) = L_\alpha^{-\theta}(x)$, so that a stable density with $\theta = 0$ is symmetric.

Stable distributions have noteworthy properties on which the interested reader can be informed from the relevant existing literature. Hereafter we recall some peculiar

**Properties.**

- Each stable density $L_\alpha^\theta$ possesses a domain of attraction, see e.g., [6].
- Any stable density is unimodal and indeed bell-shaped, i.e. its $n$th derivative has exactly $n$ zeros in $\mathbb{R}$, see [9].
- The stable distributions are self-similar and infinitely divisible.

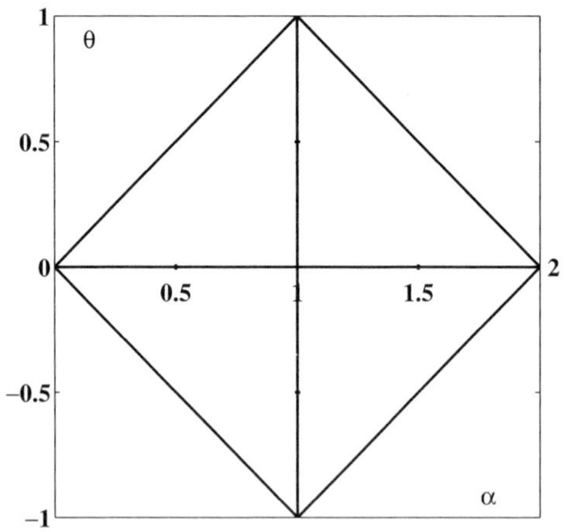


Fig. 1. The Feller–Takayasu diamond.

These properties derive from the canonic form (31) through the scaling property of the Fourier transform.

*Self-similarity* means

$$L_\alpha^\theta(x,t) \div \exp\left[-t\psi_\alpha^\theta(\kappa)\right] \Leftrightarrow L_\alpha^\theta(x,t) = t^{-1/\alpha} L_\alpha^\theta(x/t^{1/\alpha}), \tag{32}$$

where $t$ is a positive parameter. If $t$ is time, then $L_\alpha^\theta(x,t)$ is a spatial density evolving in time with self-similarity.

*Infinite divisibility* means that for every positive integer $n$, the characteristic function can be expressed as the $n$th power of some characteristic function, so that any stable distribution can be expressed as the $n$-fold convolution of a stable distribution of the same type. Indeed, taking in (31) $\theta = 0$, without loss of generality, we have

$$\mathrm{e}^{-t|\kappa|^\alpha} = \left[\mathrm{e}^{-(t/n)|\kappa|^\alpha}\right]^n \Leftrightarrow L_\alpha^0(x,t) = \left[L_\alpha^0(x,t/n)\right]^{*n}, \tag{33}$$

where

$$\left[L_\alpha^0(x,t/n)\right]^{*n} := L_\alpha^0(x,t/n) * L_\alpha^0(x,t/n) * \cdots * L_\alpha^0(x,t/n) \tag{34}$$

is the multiple Fourier convolution in $\mathbb{R}$ with $n$ identical terms.

Only in special cases the inversion of the Fourier transform in (31) can be carried out using standard tables, and provides well-known probability distributions.

For $\alpha = 2$ (so $\theta = 0$), we recover the *Gaussian pdf*, that turns out to be the only stable density with finite variance, and more generally with finite moments of any order $\delta \geq 0$. In fact

$$L_2^0(x) = \frac{1}{2\sqrt{\pi}} \mathrm{e}^{-x^2/4}. \tag{35}$$

All the other stable densities have finite absolute moments of order $\delta \in [-1, \alpha)$ as we will later show.

For $\alpha = 1$ and $|\theta| < 1$, we get

$$L_1^\theta(x) = \frac{1}{\pi} \frac{\cos(\theta\pi/2)}{[x + \sin(\theta\pi/2)]^2 + [\cos(\theta\pi/2)]^2}, \tag{36}$$

which for $\theta = 0$ includes the *Cauchy–Lorentz pdf*,

$$L_1^0(x) = \frac{1}{\pi} \frac{1}{1 + x^2}. \tag{37}$$

In the limiting cases $\theta = \pm 1$ for $\alpha = 1$ we obtain the *singular Dirac pdf's*

$$L_1^{\pm 1}(x) = \delta(x \pm 1). \tag{38}$$

In general, we must recall the power series expansions provided in [6]. We restrict our attention to $x > 0$ since the evaluations for $x < 0$ can be obtained using the symmetry relation. The convergent expansions of $L_\alpha^\theta(x) (x > 0)$ turn out to be;
for $0 < \alpha < 1$, $|\theta| \leq \alpha$ :

$$L_\alpha^\theta(x) = \frac{1}{\pi x} \sum_{n=1}^{\infty} (-x^{-\alpha})^n \frac{\Gamma(1 + n\alpha)}{n!} \sin\left[\frac{n\pi}{2}(\theta - \alpha)\right]; \tag{39}$$

for $1 < \alpha \leq 2$, $|\theta| \leq 2 - \alpha$:

$$L_\alpha^\theta(x) = \frac{1}{\pi x} \sum_{n=1}^{\infty} (-x)^n \frac{\Gamma(1 + n/\alpha)}{n!} \sin\left[\frac{n\pi}{2\alpha}(\theta - \alpha)\right]. \tag{40}$$

From the series in (39) and the symmetry relation, we note that *the extremal stable densities for* $0 < \alpha < 1$ *are unilateral*, precisely vanishing for $x > 0$ if $\theta = \alpha$, vanishing for $x < 0$ if $\theta = -\alpha$. In particular, the unilateral extremal densities $L_\alpha^{-\alpha}(x)$ with $0 < \alpha < 1$ have support $\mathbb{R}^+$ and Laplace transform $\exp(-s^\alpha)$. For $\alpha = 1/2$, we obtain the so-called *Lévy–Smirnov pdf*:

$$L_{1/2}^{-1/2}(x) = \frac{x^{-3/2}}{2\sqrt{\pi}} \mathrm{e}^{-1/(4x)}, \quad x \geq 0. \tag{41}$$

It is worth to note that the Gaussian *pdf* (35) and the Lévy–Smirnov *pdf* (41) are well known in the treatment of the Brownian motion: the former as the spatial density on an infinite real line, the latter as the first passage time density on a semi-infinite line, see e.g., [6].

As a consequence of the convergence of the series in (39) and (40) and of the symmetry relation, we recognize that the stable *pdf*'s with $1 < \alpha \leq 2$ are entire functions, whereas with $0 < \alpha < 1$ have the form

$$L_\alpha^\theta(x) = \begin{cases} 1/x\Phi_1(x^{-\alpha}) & \text{for } x > 0, \\ 1/|x|\Phi_2(|x|^{-\alpha}) & \text{for } x < 0, \end{cases} \tag{42}$$

where $\Phi_1(z)$ and $\Phi_2(z)$ are distinct entire functions. The case $\alpha = 1$ with $|\theta| < 1$ must be considered in the limit for $\alpha \to 1$ of (39) and (40), because the corresponding series reduce to power series akin with geometric series in $1/x$ and $x$, respectively, with a finite radius of convergence. The corresponding stable densities are no longer represented by entire functions, as can be noted directly from their explicit expressions (36) and (37).

From a comparison between the series expansions in (39) and (40) and in (19) and (20), we recognize that for $x > 0$ our *auxiliary functions of the Wright type are related to the extremal stable densities as follows*, see [30],

$$L_\alpha^{-\alpha}(x) = \frac{1}{x} F_\alpha(x^{-\alpha}) = \frac{\alpha}{x^{\alpha+1}} M_\alpha(x^{-\alpha}), \quad 0 < \alpha < 1, \tag{43}$$

$$L_\alpha^{\alpha-2}(x) = \frac{1}{x} F_{1/\alpha}(x) = \frac{1}{\alpha} M_{1/\alpha}(x), \quad 1 < \alpha \leq 2. \tag{44}$$

In the above equations, for $\alpha = 1$, the skewness parameter turns out to be $\theta = -1$, so we get the singular limit

$$L_1^{-1}(x) = M_1(x) = \delta(x - 1). \tag{45}$$

We do not provide here the asymptotic representations of the stable densities referring the interested reader to [27]. However, based on asymptotic representations, we can state the following: For $0 < \alpha < 2$ the stable densities exhibit *fat tails* in such a way that their absolute moment of order $\delta$ is finite only if $-1 < \delta < \alpha$. More precisely, one can show that for non-Gaussian, not extremal, stable densities the asymptotic decay of the tails is

$$L_\alpha^\theta(x) = O(|x|^{-(\alpha+1)}), \quad x \to \pm\infty. \tag{46}$$

For the extremal densities with $\alpha \neq 1$ this is valid only for one tail (as $|x| \to \infty$), the other (as $|x| \to \infty$) being of exponential order. For $1 < \alpha < 2$, the extremal *pdf*'s are two-sided and exhibit an exponential left tail (as $x \to -\infty$) if $\theta = +(2-\alpha)$, or an exponential right tail (as $x \to +\infty$) if $\theta = -(2-\alpha)$. Consequently, the Gaussian *pdf* is the unique stable density with finite variance. Furthermore, when $0 < \alpha \leq 1$, the first absolute moment is infinite so we should use the median instead of the non-existent expected value in order to characterize the corresponding *pdf*.

Let us also recall a relevant identity between stable densities with index $\alpha$ and $1/\alpha$ (a sort of reciprocity relation) pointed out in [6], that is, assuming $x > 0$,

$$\frac{1}{x^{\alpha+1}} L^{\theta}_{1/\alpha}(x^{-\alpha}) = L^{\theta^*}_{\alpha}(x), \quad 1/2 \leq \alpha \leq 1, \quad \theta^* = \alpha(\theta+1) - 1. \tag{47}$$

The condition $1/2 \leq \alpha \leq 1$ implies $1 \leq 1/\alpha \leq 2$. A check shows that $\theta^*$ falls within the prescribed range $|\theta^*| \leq \alpha$ if $|\theta| \leq 2 - 1/\alpha$. We leave as an exercise for the interested reader the verification of this reciprocity relation in the limiting cases $\alpha = 1/2$ and $\alpha = 1$. For more details on Lévy stable densities we refer the reader to specialized treatises, as [6, 23, 50, 51, 56, 64], where different notations are adopted. We would also like to refer to the 1986 paper by Schneider [53], where he first provided the Fox $H$-function representation of the stable distributions (with $\alpha \neq 1$) and to the 1990 book by Takayasu [54], where he first gave the diamond representation in the plane $\{\alpha, \theta\}$.

## 3. The Space-Time Fractional Diffusion

We now consider the Cauchy problem for the (spatially one-dimensional) *space-time fractional diffusion* (STFD) equation.

$${}_tD^{\beta}_{*} u(x,t) = {}_xD^{\alpha}_{\theta} u(x,t), \quad u(x,0) = \delta(x), \quad x \in \mathbb{R}, \quad t \geq 0, \tag{48}$$

where $\{\alpha,\, \theta,\, \beta\}$ are real parameters restricted to the ranges

$$0 < \alpha \leq 2, \quad |\theta| \leq \min\{\alpha, 2-\alpha\}, \quad 0 < \beta \leq 1. \tag{49}$$

Here ${}_xD^{\alpha}_{\theta}$ denotes the *Riesz–Feller fractional derivative* of order $\alpha$ and skewness $\theta$, acting on the space variable $x$, and ${}_tD^{\beta}_{*}$ denotes the *Caputo fractional derivative* of order $\beta$, acting on the time variable $t$. We recall the definitions of these fractional derivatives based on their representation in the Fourier and Laplace transform domain, respectively. So doing we avoid the subtleties lying in the inversion of the corresponding fractional integrals, see e.g., the 2001 survey by Mainardi *et al.* [27]. For general information on fractional integrals and derivatives we recommend the books [24, 46, 49].

### 3.1. *The Riesz–Feller space-fractional derivative*

We define the *Riesz–Feller* derivative as the pseudo-differential operator whose symbol is the logarithm of the characteristic function of a general *Lévy strictly stable* probability density with *index of stability* $\alpha$ and asymmetry parameter $\theta$ (improperly called *skewness*). As a consequence of Eq. (31), for a sufficiently well-behaved function $f(x)$, we define the *Riesz–Feller* space-fractional derivative of order $\alpha$ and skewness $\theta$ via the Fourier transform

$$\begin{cases} \mathcal{F}\{{}_xD_\theta^\alpha f(x);\kappa\} = -\psi_\alpha^\theta(\kappa)\hat{f}(\kappa), \quad \psi_\alpha^\theta(\kappa) = |\kappa|^\alpha i^{i\theta\,\mathrm{sign}\,\kappa}, \\ \qquad\qquad\qquad\qquad 0 < \alpha \leq 2, \quad |\theta| \leq \min\{\alpha, 2-\alpha\}. \end{cases} \tag{50}$$

Notice that $i^{\theta\,\mathrm{sign}\,\kappa} = \exp\left[i(\mathrm{sign}\,\kappa)\theta\pi/2\right]$. For $\theta = 0$, we have a symmetric operator with respect to $x$, which can be interpreted as

$${}_xD_0^\alpha = -\left(-\frac{d^2}{dx^2}\right)^{\alpha/2}, \tag{51}$$

as can be formally deduced by writing $-|\kappa|^\alpha = -(\kappa^2)^{\alpha/2}$. We thus recognize that the operator $D_0^\alpha$ is related to a power of the positive definitive operator $-{}_xD^2 = -\frac{d^2}{dx^2}$ and must not be confused with a power of the first-order differential operator ${}_xD = \frac{d}{dx}$ for which the symbol is $-i\kappa$. An alternative illuminating notation for the symmetric fractional derivative is due to Zaslavsky [48], and reads

$${}_xD_0^\alpha = \frac{d^\alpha}{d|x|^\alpha}. \tag{52}$$

For $0 < \alpha < 2$ and $|\theta| \leq \min\{\alpha, 2-\alpha\}$, the *Riesz–Feller* derivative reads

$$\begin{aligned} {}_xD_\theta^\alpha f(x) = \frac{\Gamma(1+\alpha)}{\pi} \bigg\{ & \sin[(\alpha+\theta)\pi/2] \int_0^\infty \frac{f(x+\xi)-f(x)}{\xi^{1+\alpha}}\,d\xi \\ & + \sin[(\alpha-\theta)\pi/2] \int_0^\infty \frac{f(x-\xi)-f(x)}{\xi^{1+\alpha}}\,d\xi \bigg\}. \end{aligned} \tag{53}$$

### 3.2. *The Caputo fractional derivative*

For a sufficiently well-behaved function $f(t)$ we define the *Caputo* time-fractional derivative of order $\beta$ with $0 < \beta \le 1$ through its Laplace transform

$$\mathcal{L}\{{}_tD_*^\beta f(t); s\} = s^\beta \tilde{f}(s) - s^{\beta-1} f(0^+), \quad 0 < \beta \le 1. \tag{54}$$

This leads us to define

$${}_tD_*^\beta f(t) := \begin{cases} \dfrac{1}{\Gamma(1-\beta)} \displaystyle\int_0^t \frac{f^{(1)}(\tau)d\tau}{(t-\tau)^\beta}, & 0 < \beta < 1, \\ \dfrac{d}{dt} f(t), & \beta = 1. \end{cases} \tag{55}$$

For the essential properties of the Caputo derivative, see [4, 13, 46].

### 3.3. *The fundamental solution of the space-time fractional diffusion equation*

Let us note that the solution $u(x,t)$ of the Cauchy problem (48) and (49), known as the *Green function* or fundamental solution of the space-time fractional diffusion equation, is a probability density in the spatial variable $x$, evolving in time $t$. In the case $\alpha = 2$ and $\beta = 1$ we recover the standard diffusion equation for which the fundamental solution is the Gaussian density with variance $\sigma^2 = 2t$. Sometimes, to point out the parameters, we may denote the fundamental solution as

$$u(x,t) = G_{\alpha,\beta}^\theta(x,t). \tag{56}$$

For our purposes let us here confine ourselves to recall the representation in the Laplace–Fourier domain of the (fundamental) solution as it results from the application of the transforms of Laplace and Fourier to Eq. (48). Using $\hat{\delta}(\kappa) \equiv 1$ we have:

$$s^\beta \hat{\tilde{u}}(\kappa, s) - s^{\beta-1} = -|\kappa|^\alpha i^{\theta \operatorname{sign} \kappa} \hat{\tilde{u}}(\kappa, s),$$

hence

$$\hat{\tilde{u}}(\kappa, s) = \widehat{\widetilde{G_{\alpha,\beta}^\theta}}(\kappa, s) = \frac{s^{\beta-1}}{s^\beta + |\kappa|^\alpha i^{\theta \operatorname{sign} \kappa}}. \tag{57}$$

For explicit expressions and plots of the fundamental solution of (48) in the space-time domain, we refer the reader to Mainardi, Luchko and Pagnini [27]. There, starting from the fact that the Fourier transform of the fundamental solution can be written as a Mittag-Leffler function with complex argument,

$$\hat{u}(\kappa,t) = \widehat{G^{\theta}_{\alpha,\beta}}(\kappa,t) = E_{\beta}(-|\kappa|^{\alpha} i^{\theta \operatorname{sign} \kappa} t^{\beta}), \tag{58}$$

these authors have derived a Mellin–Barnes integral representation of $u(x,t) = G^{\theta}_{\alpha,\beta}(x,t)$ with which they have proved the non-negativity of the solution for values of the parameters $\{\alpha, \theta, \beta\}$ in the range (49) and analyzed the evolution in time of its moments. The representation of $u(x,t)$ in terms of Fox $H$-functions can be found in Mainardi, Pagnini and Saxena [29], see also Chap. 6 in the recent book by Mathai, Saxena and Haubold [32]. We note, however, that the solution of the STFD Eq. (48) and its variants has been investigated by several authors; let us only mention some of them [1, 2, 8, 10, 19–21, 33, 34, 36, 38–42, 46, 52] where the connection with the CTRW was also pointed out.

In particular the fundamental solution for the *space fractional diffusion* $\{0 < \alpha < 2, \beta = 1\}$ is expressed in terms of a stable density of order $\alpha$ and skewness $\theta$,

$$G^{\theta}_{\alpha,1}(x,t) = t^{-1/\alpha} L^{\theta}_{\alpha}\left(\frac{x}{t^{1/\alpha}}\right), \quad -\infty < x < +\infty, \quad t \geq 0. \tag{59}$$

whereas for the *time fractional diffusion* $\{\alpha = 2, 0 < \beta < 1\}$ in terms of a (symmetric) $M$-Wright function of order $\beta/2$,

$$G^{0}_{2,\beta}(x,t) = \frac{1}{2} t^{-\beta/2} M_{\beta/2}\left(\frac{|x|}{t^{\beta/2}}\right), \quad -\infty < x < +\infty, \quad t \geq 0, \tag{60}$$

For the *standard diffusion* $\{\alpha = 2, \beta = 1\}$ we recover the Gaussian density

$$G^{0}_{2,1}(x,t) = \frac{1}{2\sqrt{\pi t}} \mathrm{e}^{-x^2/(4t)} = t^{-1/2} L^{0}_{2}\left(\frac{x}{t^{1/2}}\right) = \frac{1}{2} t^{-1/2} M_{1/2}\left(\frac{|x|}{t^{1/2}}\right).$$

Let us finally recall that the $M$-Wright function does appear also in the fundamental solution of the *rightward time fractional drift equation*,

$$_{t}D^{\beta}_{*} u(x,t) = -\frac{\partial}{\partial x} u(x,t), \quad -\infty < x < +\infty, \quad t \geq 0. \tag{61}$$

Denoting by $\mathcal{G}^{*}_{\beta}(x,t)$ this fundamental solution, we have

$$\mathcal{G}^{*}_{\beta}(x,t) = \begin{cases} t^{-\beta} M_{\beta}\left(\frac{x}{t^{\beta}}\right), & x > 0, \\ 0, & x < 0, \end{cases} \tag{62}$$

that for $\beta = 1$ reduces to the right running pulse $\delta(x - t)$ for $x > 0$. For details see [15, 28].

### 3.4. *Alternative forms of the space-time fractional diffusion equation*

We note that in the literature there exist other forms alternative and equivalent to Eq. (48) with initial condition $u(x, 0) = u_0(x)$ including the case $u_0(x) = \delta(x)$. For this purpose we must briefly recall the definitions of fractional integral and fractional derivative according to Riemann–Liouville.

The Riemann–Liouville fractional integral for a sufficiently well-behaved function $f(t)$ $(t \geq 0)$ is defined for any order $\mu > 0$ as

$$ {}_tJ^\mu f(t) := \frac{1}{\Gamma(\mu)} \int_0^t (t - \tau)^{\mu - 1} f(\tau) d\tau. \tag{63}$$

We note the convention ${}_tJ^0 = I$ (identity) and the semigroup property

$$ {}_tJ^\mu {}_tJ^\nu = {}_tJ^\nu {}_tJ^\mu = {}_tJ^{\mu+\nu}, \quad \mu \geq 0, \quad \nu \geq 0. \tag{64}$$

The fractional derivative of order $\mu > 0$ in the *Riemann–Liouville* sense is defined as the operator ${}_tD^\mu$ which is the left inverse of the Riemann–Liouville integral of order $\mu$ (in analogy with the ordinary derivative),

$$ {}_tD^\mu {}_tJ^\mu = I, \quad \mu > 0. \tag{65}$$

If $m$ denote the positive integer such that $m - 1 < \mu \leq m$, we recognize from Eqs. (63)–(65):

$$ {}_tD^\mu f(t) := {}_tD^m {}_tJ^{m-\mu} f(t). \tag{66}$$

Then, restricting our attention to an order $\beta$ with $0 < \beta \leq 1$ (namely $m = 1$) the corresponding Riemann–Liouville fractional derivative turns out

$$ {}_tD^\beta f(t) = \begin{cases} \dfrac{d}{dt}\left[\dfrac{1}{\Gamma(1-\beta)} \displaystyle\int_0^t \dfrac{f(\tau) d\tau}{(t-\tau)^\beta}\right], & 0 < \beta < 1, \\ \dfrac{d}{dt} f(t), & \beta = 1. \end{cases} \tag{67}$$

Then we get the relationship among the Caputo fractional derivative with the classical Riemann–Liouville fractional integral and derivative:

$$ {}_tD_*^\beta f(t) := {}_tJ^{1-\beta} {}_tD^1 f(t) = {}_tD^\beta [f(t) - f(0)] = {}_tD^\beta f(t) - \frac{f(0)}{\Gamma(1-\beta) t^\beta}, \tag{68}$$

and, as a consequence, the equivalence of (48) with the following problems

$$u(x,t) - u(x,0) = {}_tJ^{\beta}\, {}_xD_{\theta}^{\alpha} u(x,t), \quad u(x,0) = \delta(x), \tag{69}$$

$$\frac{\partial}{\partial t} u(x,t) = {}_tD^{1-\beta}\, {}_xD_{\theta}^{\alpha} u(x,t), \quad u(x,0) = \delta(x). \tag{70}$$

## 4. Analytic and Stochastic Pathways to Subordination in Space-Time Fractional Diffusion

Our starting keypoint to introduce the analytical and stochastic approaches to subordination in space-time fractional diffusion processes is the fundamental solution of the space-time fractional diffusion equation in the Laplace–Fourier domain given by (57).

### 4.1. *The analytical interpretation via operational time*

Separating variables in (57) and using the trick to write $1/(z+a)$ for $\mathrm{Re}(z+a) > 0$ as a Laplace integral

$$\frac{1}{z+a} = \int_0^{\infty} \mathrm{e}^{-z\rho} \mathrm{e}^{-a\rho} d\rho$$

we have, identifying $\rho := t_*$ as *operational time*, the following instructive expression for (57):

$$\hat{\tilde{u}}(\kappa, s) = \int_0^{\infty} [\exp(-t_* |\kappa|^{\alpha} i^{\theta \operatorname{sign} \kappa})][s^{\beta-1} \exp(-t_* s^{\beta})] dt_*. \tag{71}$$

We note that the first factor in (71)

$$\hat{f}_{\alpha,\theta}(\kappa, t_*) := \exp(-t_* |\kappa|^{\alpha} i^{\theta \operatorname{sign} \kappa}) \tag{72}$$

is the Fourier transform of a skewed stable density in $x$, evolving in operational time $t_*$, of a process $x = y(t_*)$ along the real axis $x$ happening in operational time $t_*$, that we write as

$$f_{\alpha,\theta}(x, t_*) = t_*^{-1/\alpha} L_{\alpha}^{\theta}\left(\frac{x}{t_*^{1/\alpha}}\right). \tag{73}$$

We can interpret the second factor

$$\tilde{q}_{\beta}(t_*, s) := s^{\beta-1} \exp(-t_* s^{\beta}) \tag{74}$$

as Laplace representation of the probability density in $t_*$ evolving in $t$ of a process $t_* = t_*(t)$, generating the operational time $t_*$ from the physical time $t$, that is expressed via a fractional integral of a skewed Lévy density as

$$q_\beta(t_*, t) = t_*^{-1/\beta} {}_tJ^{1-\beta} L_\beta^{-\beta}\left(\frac{t}{t_*^{1/\beta}}\right) = t^{-\beta} M_\beta\left(\frac{t_*}{t^\beta}\right), \qquad (75)$$

see Eq. (26). To prove that $q_\beta(t_*, t)$ (surely positive for $t_* > 0$) is indeed a probability density, we must further prove that it is normalized, $\int_{t_*=0}^{\infty} q_\beta(t_*, t)dt_* = 1$. For this purpose it is sufficient to prove that its Laplace transform with respect to $t_*$ is equal to 1 for $s_* = 0$. To get this Laplace transform $\tilde{q}_\beta(s_*, t)$, we proceed as follows. Starting from the known Laplace transform with respect to $t$,

$$\tilde{q}_\beta(t_*, s) = s^{\beta-1} \exp\left(-t_* s^\beta\right), \qquad (76)$$

we apply a second Laplace transformation with respect to $t_*$ with parameter $s_*$ to get

$$\tilde{\tilde{q}}_\beta(s_*, s) = \frac{s^{\beta-1}}{s_* + s^\beta}, \qquad (77)$$

so, by inversion with respect to $t$

$$\tilde{q}_\beta(s_*, t) = \int_{t_*=0}^{\infty} \mathrm{e}^{-s_* t} q_\beta(t_*, t)dt_* = E_\beta(-s_* t^\beta), \qquad (78)$$

and setting $s_* = 0$

$$\int_{t_*=0}^{\infty} q_\beta(t_*, t)dt_* = E_\beta(0) = 1. \qquad (79)$$

Weighting the density of $x = y(t_*)$ with the density of $t_* = t_*(t)$ over $0 \le t < \infty$ yields the density $u(x, t)$ in $x$ evolving with time $t$.

In physical variables $\{x, t\}$, using Eqs. (71)–(75), we have the *subordination integral formula*

$$u(x, t) = \int_{t_*=0}^{\infty} f_{\alpha,\theta}(x, t_*) q_\beta(t_*, t)dt_*, \qquad (80)$$

where $f_{\alpha,\theta}(x, t_*)$ (density in $x$ evolving in $t_*$) refers to the process $x = y(t_*)$ ($t_* \to x$) generating in "operational time" $t_*$ the spatial position $x$, and $q_\beta(t_*, t)$ (density in $t_*$ evolving in $t$) refers to the process $t_* = t_*(t)$ ($t \to t_*$) generating from physical time $t$ the "operational time" $t_*$.

Our aim is to construct a process $x = x(t)$ whose probability density is $u(x, t)$, density in $x$, evolving in $t$. We will soon find justification for denoting the variable of integration by $t_*$. We will exhibit it as the "operational time" for our fractional diffusion process, and for distinction we will call the variable $t$ its "physical time". In fact $f_{\alpha,\theta}(x, t_*)$ is a probability density in $x \in \mathbb{R}$, evolving in operational time $t_* > 0$ and $q_\beta(t_*, t)$ is a probability density in $t_* \geq 0$, evolving in physical time $t > 0$.

### 4.2. *Stochastic interpretation*

Clearly $f_{\alpha,\theta}(x, t_*)$ characterizes a stochastic process describing a trajectory $x = y(t_*)$ in the $(t_*, x)$-plane, that can be visualized as a particle traveling along space $x$, as operational time $t_*$ is proceeding. Is there also a process $t_* = t_*(t)$, a particle moving along the positive $t_*$ axis, happening in physical time $t$? Naturally we want $t_*(t)$ increasing, at least in the weak sense,

$$t_2 > t_1 \Rightarrow t_*(t_2) \geq t_*(t_1).$$

We answer this question in the affirmative by showing that, by inverting the stable process $t = t(t_*)$ whose probability density (in $t$, evolving in operational time $t_*$) is the extremely positively skewed stable density

$$r_\beta(t, t_*) = t_*^{-1/\beta} L_\beta^{-\beta}\left(\frac{t}{t_*^{1/\beta}}\right). \tag{81}$$

In fact, recalling from Sec. 2.5,

$$\tilde{r}_\beta(s, t_*) = \exp(-t_* s^\beta), \tag{82}$$

there exists the stable process $t = t(t_*)$, weakly increasing, with density in $t$ evolving in $t_*$ given by (81). We call this process *the leading process.*

Happily, we can invert this process. Inversion of a weakly increasing trajectory means that horizontal segments are converted to vertical segments and vice versa jumps (as vertical segments) to horizontal segments (in graphical visualization).

Consider a fixed sample trajectory $t = t(t_*)$ and also its fixed inversion $t_* = t_*(t)$. Fix an instant $T$ of physical time and an instant $T_*$ of operational time. Then, because $t = t(t_*)$ is increasing, we have the equivalence

$$t_*(T) \leq T_* \Leftrightarrow T \leq t(T_*),$$

which, with notation slightly changed by

$$t_*(T) \to t'_*, \quad T_* \to t_*, \quad T \to t, \quad t(T_*) \to t',$$

implies

$$\int_0^{t_*} q(t'_*, t)dt'_* = \int_t^\infty r_\beta(t', t_*)dt', \tag{83}$$

for the probability density $q(t_*, t)$ in $t_*$ evolving in $t$. It follows

$$q(t_*, t) = \frac{\partial}{\partial t_*}\int_t^\infty r_\beta(t', t_*)dt' = \int_t^\infty \frac{\partial}{\partial t_*} r_\beta(t', t_*)dt'.$$

We continue in the $s_*$-Laplace domain assuming $t > 0$,

$$\tilde{q}(s_*, t) = \int_t^\infty (s_* \tilde{r}_\beta(t', s_*) - \delta(t'))dt'.$$

It suffices to consider $t > 0$, so that we have $\delta(t') = 0$ in this integral. Observing from (82)

$$\tilde{\tilde{r}}_\beta(s, s_*) = \frac{1}{s_* + s^\beta}, \tag{84}$$

we find

$$\tilde{r}_\beta(t, s_*) = \beta t^{\beta-1} E'_\beta(-s_* t^\beta), \tag{85}$$

so that

$$\tilde{q}(s_*, t) = \int_t^\infty s_* \beta {t'}^{\beta-1} E'_\beta(-s_* {t'}^\beta)dt' = E_\beta(-s_* t^\beta), \tag{86}$$

finally

$$q(t_*, t) = t^{-\beta} M_\beta(t_*/t^\beta). \tag{87}$$

From (86) we also see that

$$\tilde{\tilde{q}}(s_*, s) = \frac{s^{\beta-1}}{s_* + s^\beta} = \tilde{\tilde{q}}_\beta(s_*, s), \tag{88}$$

implying (76) and, see (77),

$$q(t_*, t) \equiv q_\beta(t_*, t), \tag{89}$$

so that indeed the process $t_* = t_*(t)$ is the inverse to the stable process $t = t(t_*)$ and has density $q_\beta(t_*, t)$.

**Remark.** The process at hand, $t_* = t_*(t)$, which is referred to as the inverse stable subordinator, is honored with the name "Mittag-Leffler process" by Meerschaert *et al.* [33, 35]. Honoring this process by the name of Mittag-Leffler can be justified by the fact that by (86), the Laplace transform of its density is a Mittag-Leffler type function or by the fact that it is a properly scaled diffusion limit of the counting function $N(t)$ of the fractional generalization of the Poisson process whose residual waiting time probability is the Mittag-Leffler type function $E_\beta(-t^\beta)$, see recent papers of ours [11, 15]. In view of its probability density it may also be called the $M$-Wright process.

Stipulating that there exists a weakly increasing process $t_* = t_*(t)$ with density $q_\beta(t_*, t)$, we can analogously find the density of its inverse $t = t(t_*)$ which comes just as $r_\beta(t, t_*)$. However, in the context of our here presented considerations not being allowed to know that such process $t_* = t_*(t)$ exists, we have taken as a gift from God the process $t = t(t_*)$ and shown by its inversion that there exists a process $t_* = t_*(t)$ with the desired properties.

From the density $r_\beta(t, t_*)$ of *the leading process* $t = t(t_*)$, we have found the density of *the directing process* $t_* = t_*(t)$ as given by the Laplace transform pair (76), that is

$$q_\beta(t_*, t) \div \tilde{q}_\beta(t_*, s) = s^{\beta-1} \exp(-t_* s^\beta).$$

In physical coordinates we have (75) and (87), so also an expression through an $M$-Wright function,

$$q_\beta(t_*, t) = {}_tJ^{1-\beta} r_\beta(t, t_*) = t^{-\beta} M_\beta\left(\frac{t_*}{t^\beta}\right), \tag{90}$$

see Eq. (26).

### 4.3. *Evolution equations for the densities of physical and operational time in mutual dependence*

The Laplace–Laplace representation of the density $r_\beta(t, t_*)$ of the process $t = t(t_*)$ is, according to (84),

$$\tilde{\tilde{r}}_\beta(s, s_*) = \frac{1}{s^\beta + s_*}.$$

This implies

$$s_* \tilde{\tilde{r}}_\beta(s, s_*) - 1 = -s^\beta \tilde{\tilde{r}}_\beta(s, s_*),$$

and by inverting the transforms and observing the initial condition $r_\beta(t, t_* = 0) = \delta(t)$, we arrive at the Cauchy problem

$$\frac{\partial}{\partial t_*} r_\beta(t, t_*) = -{}_tD_*^\beta r_\beta(t, t_*), \quad r_\beta(t, t_* = 0) = \delta(t). \tag{91}$$

Because it suffices to consider only $t > 0$ where $\delta(t) = 0$, we need not introduce a singular term on the right-hand side.

The Laplace–Laplace representation of the density $q_\beta(t_*, t)$ of the process $t_* = t_*(t)$ is, according to (88),

$$\tilde{\bar{q}}_\beta(s_*, s) = \frac{s^{\beta-1}}{s_* + s^\beta}.$$

This implies

$$s^\beta \tilde{\bar{q}}_\beta(s_*, s) - s^{\beta-1} = -s_* \tilde{\bar{q}}_\beta(s_*, s),$$

and by inverting the transforms and observing the initial condition $q_\beta(t_*, t = 0) = \delta(t_*)$, we arrive at the Cauchy problem

$$ {}_tD_*^\beta q_\beta(t_*, t) = -\frac{\partial}{\partial t_*} q_\beta(t_*, t), \quad q_\beta(t_*, t = 0) = \delta(t_*). \tag{92}$$

Because it suffices to consider only $t_* > 0$ where $\delta(t_*) = 0$, we can ignore the delta function on the right-hand side.

**Remark.** The fractional differential equations in the above Cauchy problems have the same form. By replacing $t$ by $t_*$ and $r$ by $q$, one of them goes over into the other. However, in the first problem the delta initial condition refers to the fractional derivative (of order $\beta$), in the second problem to the ordinary (first-order) derivative. These equations are akin with the time-fractional drift equation treated in (61) and (62), with different coordinates and proper initial conditions, as explained above. The process $t = t(t_*)$ of the first problem is a positive-oriented (extreme) stable process, whereas the process $t_* = t_*(t)$ is a fractional drift process, see (61) and (62) with $x$ replaced by $t_*$. The reason for the two evolution equations to have the same form is that the two described processes are inverse to each other, their graphical representations coincide just by interchanging the coordinate axes. The delta initial condition for each equation is given at value zero of the evolution variable for the variable in which the solution is a density.

### 4.4. *The random walks*

We can now construct the process $x = x(t)$ for the position $x$ of the particle depending on physical time $t$ as follows in two ways. With the variable $t$ (physical time), $t_*$ (operational time), $x$ (position), we have the processes (i), (ii) and (iii), as follows:

(i) $t = t(t_*)$ with density $r_\beta(t,t_*)$ in $t$, evolving in $t_*$, *the leading process*,
(ii) $x = y(t_*)$ with density $f_{\alpha,\theta}(x,t_*)$ in $x$, evolving in $t_*$, *the parent process*,
(iii) $t_* = t_*(t)$ with density $q_\beta(t_*,t)$ in $t_*$, evolving in $t$, *the directing process.*

Observing that the processes (i) and (iii) are inverse to each other, and taking account of the subordination integral (80), we define *the space-time fractional diffusion process* as *the subordinated process*

$$x = x(t) = y(t_*(t)). \tag{93}$$

Simulation of a trajectory for the subordinated process means: generate in running physical time $t$ the operational time $t_*$, then the operational process $y(t_*)$.

Now, the Mittag-Leffler (or $M$-Wright) process $t_* = t_*(t)$ is non-Markovian and not so easy to simulate. The alternative (we call it "parametric subordination") is to produce in dependence of the operational time $t_*$ the processes (i) and (ii) and then eliminate $t_*$ from the system

$$t = t(t_*), \quad x = y(t_*), \tag{94}$$

to get $x = x(t)$ from $x = y(t_*)$ by change of time from $t_*$ to $t$.

We can produce a sequence of precise snapshots of $t = t(t_*)$ and $x = y(t_*)$ in the $(t_*,t)$ plane and the $(t_*,x)$ plane by setting, with a step-size $\tau_* > 0$,

$$t_{*,n} = n\tau_*, \quad \bar{t}_n = T_{*,1} + T_{*,2} + \cdots + T_{*,n}, \quad \bar{x}_n = X_{*,1} + X_{*,2} + \cdots + X_{*,n}, \tag{95}$$

taking for $k = 1,2,\ldots,n$ each $T_{*,k}$ as a random number with density $\tau_*^{-1/\beta} L_\beta^{-\beta}(t/\tau_*^{1/\beta})$ and each $X_{*,k}$ as a random number with density $\tau_*^{-1/\alpha} L_\alpha^\theta(x/\tau_*^{1/\alpha})$, corresponding by self-similarity to the step $\tau_*$.

We can do this by taking random numbers $T_k$ and $X_k$ with density $L_\beta^{-\beta}(t)$ and $L_\alpha^\theta(x)$, respectively, and then with $\tau = \tau_*^{1/\beta}$ and $h = \tau_*^{1/\alpha}$, setting $T_{*,k} = \tau\, T_k$, $X_{*,k} = h\, X_k$. In other words, we produce (a renewal process at equidistant times with reward) a positively oriented random walk on the half-line $t \geq 0$ and a random walk on $-\infty < x < +\infty$ with jumps at equidistant operational time instants $t_* = n\tau_*$. We recognize the scaling relation $\tau^\beta/h^\alpha \equiv 1$, analogous to that used by us in earlier papers of ours on well-scaled passage to the diffusion limit in CTRW under power law regime, see [12,15,16]. Methods for producing stable random deviates can be found in the books [22,23].

Finally, we transfer into the $(t,x)$-plane the points with coordinates $\bar{t}_n, \bar{x}_n = \bar{y}_n$ and so obtain a sequence of precise snapshots of a true process $x = x(t)$. Finer details of the process $x = x(t)$ become visible by using smaller values of the operational step-size $\tau_*$.

## 5. Graphical Representations and Conclusions

We recall that, denoting the physical time with $t$, the operational time with $t_*$, the physical space with $x$, the density of the fractional diffusion process turns out to be given by the following subordination integral, see (80),

$$u(x,t) = \int_{t_*=0}^{\infty} f_{\alpha,\theta}(x,t_*) q_\beta(t_*,t) dt_*, \tag{96}$$

where $f_{\alpha,\theta}(x,t_*)$, is the density (in $x$ evolving in $t_*$) of the parent process $x = y(t_*) = x(t(t_*))$ and $q_\beta(t_*,t)$ is the density (in $t_*$ evolving in $t$) of the directing process $t_* = t_*(t)$.

By using the Fourier–Laplace pathway we recall the two densities related to the parameters $\alpha, \theta, \beta$ from (73)–(75),

$$f_{\alpha,\theta}(x,t_*) = t_*^{-1/\alpha} L_\alpha^\theta \left(\frac{x}{t^{1/\alpha}}\right), \tag{97}$$

$$q_\beta(t_*,t) = t_*^{-1/\beta} {}_tJ^{1-\beta} L_\beta^{-\beta} \left(\frac{t}{t_*^{1/\beta}}\right) = t^{-\beta} M_\beta \left(\frac{t_*}{t^\beta}\right), \tag{98}$$

where $L$ refers to the Lévy stable density and $M$ to the Wright function, both introduced in Sec. 2. But for the *parametric subordination*, the relevant density is $r_\beta(t,t_*)$ governing the *leading process*, a density in the physical time $t$ evolving with the operational time $t_*$: it turns out to be the unilateral Lévy density of order $\beta$, namely, see (81),

$$r_\beta(t,t_*) = t_*^{-1/\beta} L_\beta^{-\beta} \left(\frac{t}{t_*^{1/\beta}}\right). \tag{99}$$

We have shown in Sec. 4.4 that in our approach (referred to as *parametric subordination*) the process of space-time fractional diffusion (non-Markovian for $\beta < 1$) can be simulated by two Markovian processes governed by stable densities, provided by $f_{\alpha,\theta}(x,t_*)$ and $r_\beta(t,t_*)$, as pointed out in our paper with Vivoli [18], where we have dealt with the CTRW model. There, before passing to the diffusion limit, we have two Markov processes happening on a discrete set of equidistant instants (for simplicity the non-negative integers) $n = 0, 1, 2, \ldots$, meaning $\tau_* = 1$, one of them moving randomly rightwards along $t \geq 0$, the other moving randomly on the real line $-\infty < x < \infty$ with jumps $T_n$ and $X_n$, respectively, for $n \geq 1$. By summing the "waiting times" $T_k$ and the jumps $X_k$ from 1 to $n$ we obtain sequences of jump instants $t = t_n$ and positions $x = x_n$, that we display in the $(t, x)$-plane. In fact, CTRW is the virgin form of parametric subordination.

We note that our approach is akin to that based on two stochastic differential equations, known in physics as Langevin equations, see [7,25,63]. We have indicated these two stochastic differential equations in [18]. Here now we content ourselves with referring to the above cited papers.

In Sec. 4.4 we have split the fractional diffusion process into three processes (i), (ii), (iii), each of them containing two of the three coordinates: space $x$, physical time $t$, operational time $t_*$. We simulate *the leading process* by a random walk ($rw_1$), *the parent process* by a random walk ($rw_2$), and *the subordinated process* (which yields the desired trajectory) by a random walk ($rw$). The inversion of ($rw_1$) gives us a random walk ($rw_3$) for simulation of *the directing process* $t_* = t_*(t)$. Essentially, we need to carry out only ($rw_1$) and ($rw_2$) according to Eq. (95). By transferring the points $(\bar{t}_n, \bar{x}_n)$ into the $(t, x)$-plane, we get the random walk ($rw$) as visualization of a random trajectory $x = x(t) = y(t_*(t))$ according to the subordinated process which is our space-time fractional diffusion process of interest.

To make transparent the situation we display as a diagram in Fig. 2 the connections between the four random walks. It is now instructive to show some numerical realizations of these random walks for two case studies of symmetric ($\theta = 0$) fractional diffusion processes: $\{\alpha = 2, \beta = 0.80\}$, $\{\alpha = 1.5, \beta = 0.90\}$. As explained in a previous subsection, for each case we need to construct the sample paths for three distinct processes, the leading process $t = t(t_*)$, the parent process $x = y(t_*)$ (both in the operational time) and, finally, the subordinated process $x = x(t)$, corresponding to the required fractional diffusion process.

We shall depict the above sample paths in Figs. 3–5 respectively, devoting the left and the right plates to the different case studies.

Plots in Fig. 3 (devoted to the leading process, the limit of ($rw_1$)) thus represent sample paths in the $(t_*, t)$-plane of unilateral Lévy motions of order $\beta$. By interchanging the coordinate axes we can consider Fig. 3 as representing sample paths of the directing process, the limit of ($rw_3$).

Plots in Fig.4 (devoted to the parent process, the limit of ($rw_2$)) represent sample paths in the $(t_*, x)$-plane, produced in the way explained

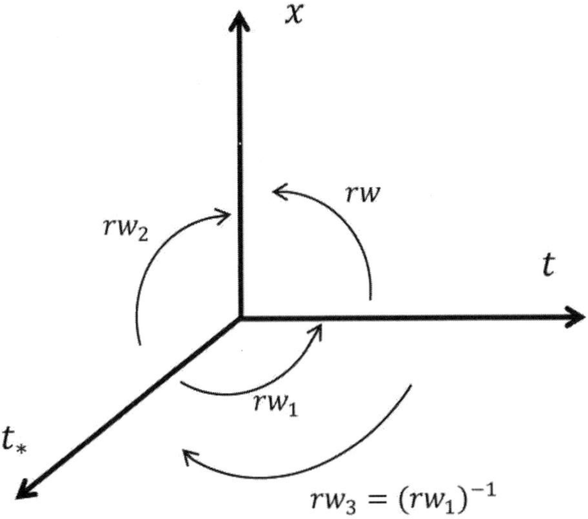


Fig. 2. Diagram for the connections between the four random walks $(rw_1)$, $(rw_2)$, $(rw_3)$ and $(rw)$, related to the leading, parent, directing and subordinated processes, respectively.

above, for Lévy motions of order $\alpha$ and skewness $\theta = 0$ (symmetric stable distributions).

By the indicated method, see (95), we have with (for simplicity) $\tau_* = 1$, $\theta = 0$ (symmetry) produced 10,000 numbers $\bar{t}_n$ and corresponding numbers $\bar{y}_n$. Plotting the points $(t_{*,n}, \bar{t}_n)$ into the $(t_*, t)$ (operational time, physical time) plane, the points $(t_{*,n}, \bar{y}_n)$ into the $(t_*, x)$ (operational time, position in space) plane we get Figs. 3 and 4 for visualization of $(rw_1)$ and $(rw_2)$, respectively. Figure 5 (as a visualization of $(rw)$ is obtained by plotting the points $(\bar{t}_n, \bar{y}_n)$ into the $(t, x)$ (physical time and space) plane.

Actually, we have invested slightly more work in producing the figures. Namely, to make visible the jumps as vertical segments, we have in Fig. 3 connected the points $(t_{*,n}, \bar{t}_n)$ and $(t_{*,n+1}, \bar{t}_n)$ by a horizontal segment, the points $(t_{*,n+1}, \bar{t}_n)$ and $(t_{*,n+1}, \bar{t}_{n+1})$ by a vertical segment. Analogously in Fig. 4 with the indexed $\bar{t}$ replaced by indexed $\bar{y}$. In Fig. 5 we have connected the points $(\bar{t}_n, \bar{y}_n)$ and $(\bar{t}_{n+1}, \bar{y}_n)$ by a horizontal segment, the points $(\bar{t}_{n+1}, \bar{y}_n)$ and $(\bar{t}_{n+1}, \bar{y}_{n+1})$ by a vertical segment.

Resuming, we can consider Fig. 3 as a representation of $(rw_1)$ for the leading process, or by interchange of axes as one of $(rw_3)$ for the directing process, Fig. 4 as one of $(rw_2)$ for the parent process, and finally Fig. 5 as a

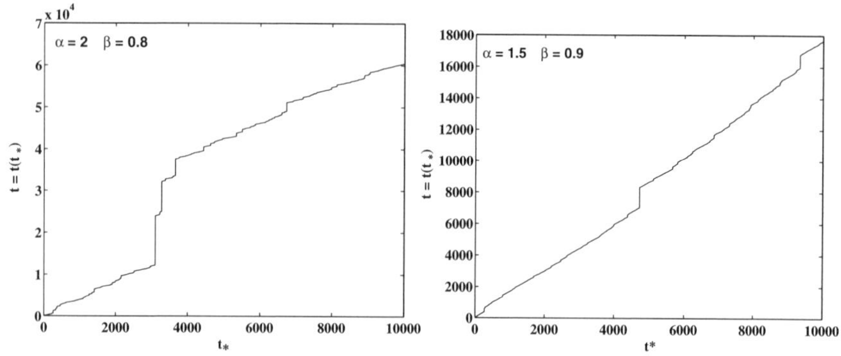


Fig. 3. A sample path for $(rw_1)$, the leading process $t = t(t_*)$.
Left: $\{\alpha = 2, \beta = 0.80\}$, Right: $\{\alpha = 1.5, \beta = 0.90\}$.

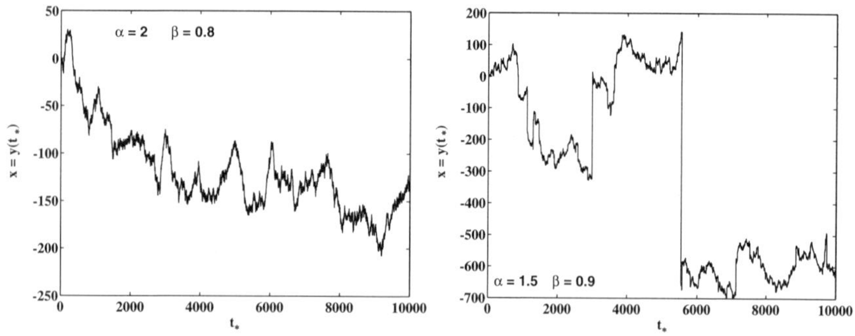


Fig. 4. A sample path for $(rw_2)$, the parent process $x = y(t_*)$.
Left: $\{\alpha = 2, \beta = 0.80\}$, Right: $\{\alpha = 1.5, \beta = 0.90\}$.

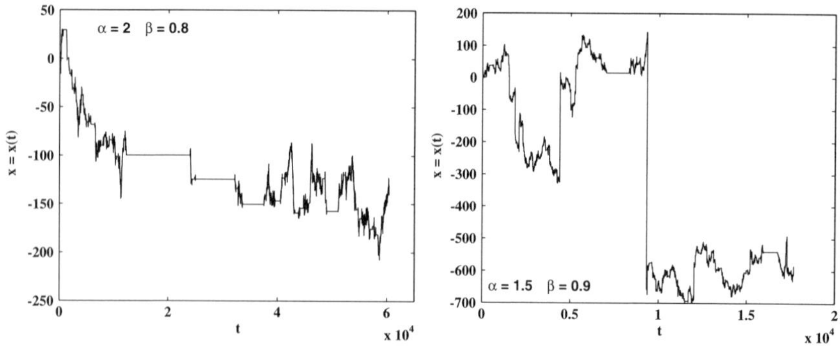


Fig. 5. A sample path for $(rw)$, the subordinated process $x = x(t)$.
Left: $\{\alpha = 2, \beta = 0.80\}$, Right: $\{\alpha = 1.5, \beta = 0.90\}$.

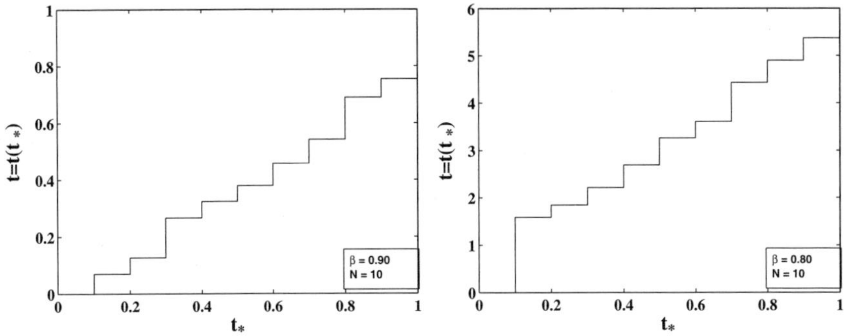


Fig. 6. A sample path for the leading process $t = t(t_*)$.
Left: $\{\beta = 0.9, N = 10^1\}$, Right: $\{\beta = 0.8, N = 10^1\}$.

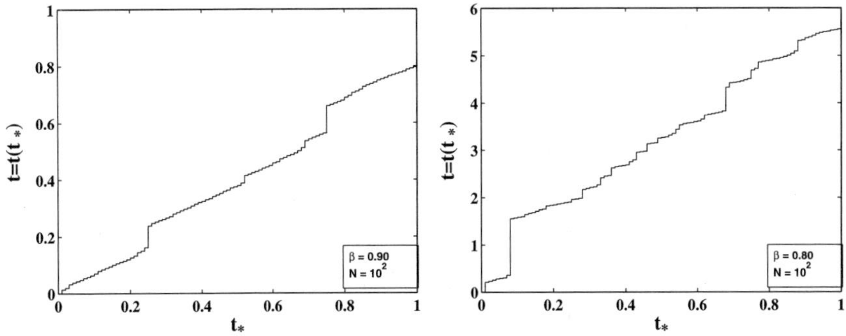


Fig. 7. A sample path for the leading process $t = t(t_*)$.
Left: $\{\beta = 0.9, N = 10^2\}$, Right: $\{\beta = 0.8, N = 10^2\}$.

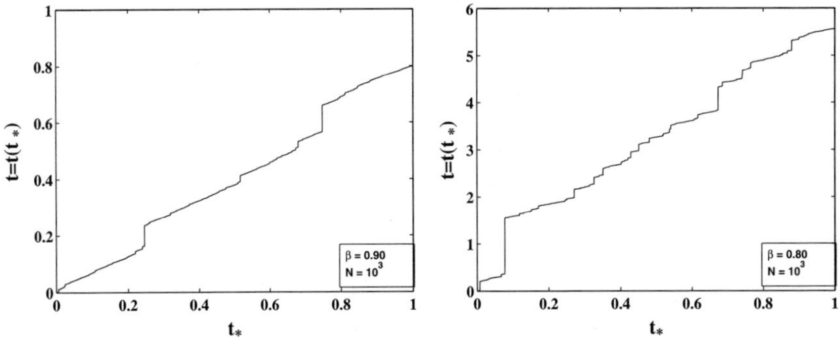


Fig. 8. A sample path for the leading process $x = t(t_*)$.
Left: $\{\beta = 0.9, N = 10^3\}$, Right: $\{\beta = 0.8, N = 10^3\}$.

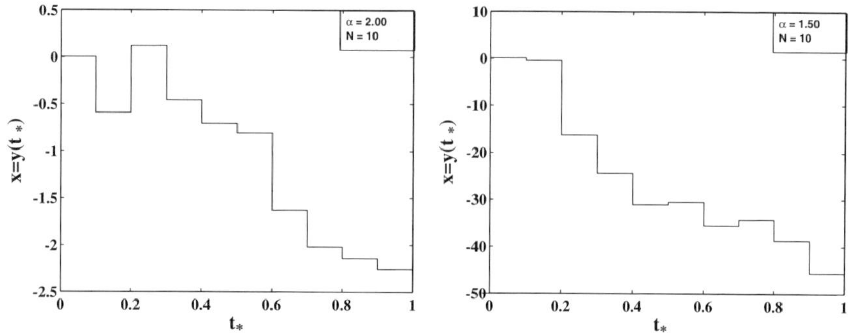


Fig. 9. A sample path for the parent process $x = y(t_*)$.
Left: $\{\alpha = 2, N = 10^1\}$, Right: $\{\alpha = 1.5, N = 10^1\}$.

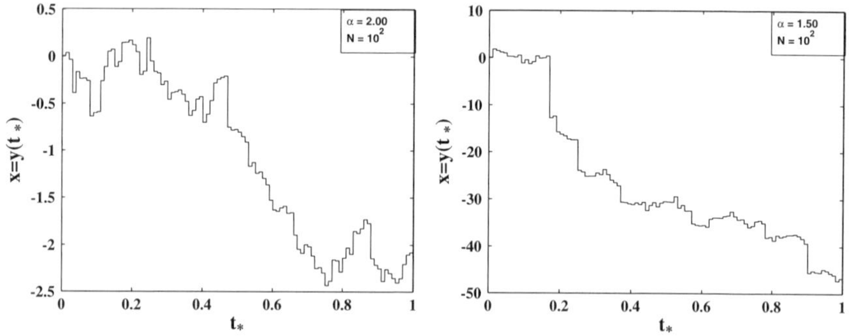


Fig. 10. A sample path for the parent process $x = y(t_*)$.
Left: $\{\alpha = 2, N = 10^2\}$, Right: $\{\alpha = 1.5, N = 10^2\}$.

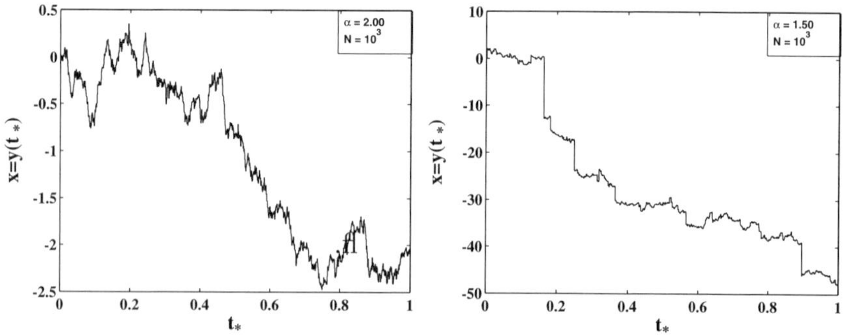


Fig. 11. A sample path for the parent process $x = y(t_*)$.
Left: $\{\alpha = 2, N = 10^3\}$, Right: $\{\alpha = 1.5, N = 10^3\}$.

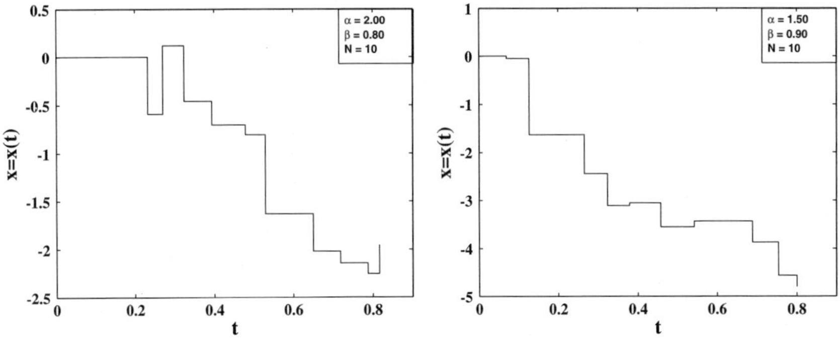


Fig. 12. A sample path for the subordinated process $x = x(t)$.
Left: $\{\alpha = 2, \beta = 0.80, N = 10^1\}$, Right: $\{\alpha = 1.5, \beta = 0.90, N = 10^1\}$.

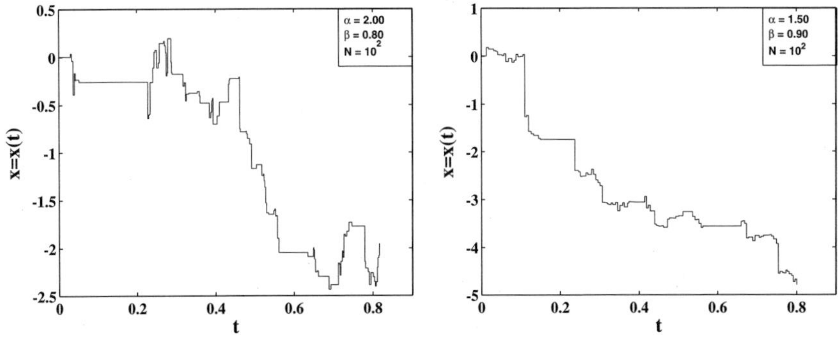


Fig. 13. A sample path for the subordinated process $x = x(t)$.
Left: $\{\alpha = 2, \beta = 0.80, N = 10^2\}$, Right: $\{\alpha = 1.5, \beta = 0.90, N = 10^2\}$.

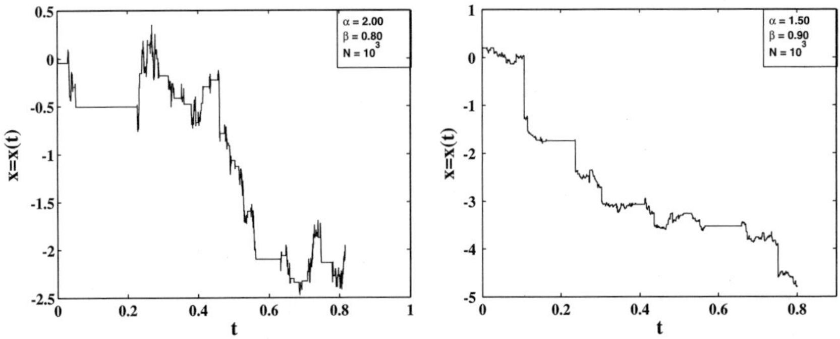


Fig. 14. A sample path for the subordinated process $x = x(t)$.
Left: $\{\alpha = 2, \beta = 0.80, N = 10^3\}$, Right: $\{\alpha = 1.5, \beta = 0.90, N = 10^3\}$.

representation of $(rw)$ for the subordinated process which is our space-time fractional diffusion process.

We conclude by including additional figures, showing the effect of taking smaller step-sizes $\tau_*$, equivalently larger values $N$ of steps following our analysis for the CTRW [18]. Figures 6–8, Figs. 9–11, Figs. 12–14 show the effect of making the operational step-length $\tau_*$ smaller or, equivalently, the number $N$ of operational steps larger for the sample paths of the leading, parent and subordinated processes, respectively. In these pictures $\tau_* = 1/N$, and we have taken $N = 10$, $N = 100$ and $N = 1000$. Finer details will become visible by choosing in the operational time $t_*$ the step-length $\tau_*$ smaller and smaller.

In the figures we can clearly see what happens for finer and finer discretization of the operational time $t_*$, by adopting $10^1$, $10^2$, $10^3$ of the number of steps. As a matter of fact there is no visible difference in the transition for the successive decades $10^4$, $10^5$, $10^6$ of number of steps as the great majority of spatial jumps and waiting times even for very small steps $\tau_*$ of the operational time. This property also explains the visible persistence of large jumps and waiting times.

## Acknowledgments

We acknowledge the valuable assistance of A. Mura and A. Vivoli in producing the figures. We are grateful to the editors of this volume for giving us an opportunity to present our view of subordination in fractional diffusion and we have appreciated their suggestions.

## References

1. E. Barkai, *Phys. Rev. E* **63**, 046118 (2001).
2. E. Barkai, *Chem. Phys.* **284**, 13 (2002).
3. P. Butzer and U. Westphal, Introduction to fractional calculus, in *Fractional Calculus, Applications in Physics*, ed. H. Hilfer (World Scientific, 2000), pp. 1–85.
4. K. Diethelm, *The Analysis of Fractional Differential Equations. An Application Oriented Exposition Using Differential Operators of Caputo Type* (Springer, 2010).
5. A. Erdélyi, W. Magnus, F. Oberhettinger and F. G. Tricomi, *Higher Transcendental Functions* (McGraw-Hill, 1955), Vol. 3, Chap. 18.
6. W. Feller, *An Introduction to Probability Theory and its Applications*, Vol. II (Wiley, 1971).
7. H. C. Fogedby, *Phys. Rev. E* **50**, 1657 (1994).

8. D. Fulger, E. Scalas and G. Germano, *Phys. Rev. E* **77**, 021122 (2008).
9. W. Gawronski, *Ann. Probab.* **12**, 230 (1984).
10. G. Germano, M. Politi, E. Scalas and R. L. Schilling, *Phys. Rev. E* **79**, 066102 (2009).
11. R. Gorenflo, Mittag-Leffler waiting time, power laws, rarefaction, continuous time random walk, diffusion limit, in *Proc of the National Workshop on Fractional Calculus and Statistical Distributions*, eds. S. S. Pai, N. Sebastian, S. S. Nair, D. P. Joseph and D. Kumar, as xiv: CMS Pala Campus, India (2010), pp. 1–22 arxiv:1004.4413.
12. R. Gorenflo and E. A. Abdel-Rehim, *Vietnam J. Math.* **32**, 65–75 (2004).
13. R. Gorenflo and F. Mainardi, Fractional calculus: integral and differential equations of fractional order, in *Fractals and Fractional Calculus in Continuum Mechanics*, eds. A. Carpinteri and F. Mainardi (Springer 1997), pp. 223–276.
14. R. Gorenflo and F. Mainardi, *Fractional Calculus Appl. Anal.* **1**, 167 (1998).
15. R. Gorenflo and F. Mainardi, Continuous time random walk, Mittag-Leffler waiting time and fractional diffusion: mathematical aspects, in *Anomalous Transport, Foundations and Applications*, eds. R. Klages, G. Radons and I. M. Sokolov (Wiley-VCH, 2008), pp. 93–127.
16. R. Gorenflo and F. Mainardi, *J. Comput. Appl. Math.* **229**, 400 (2009).
17. R. Gorenflo and F. Mainardi, *Eur. J. Phys., Special Topics* **193**, 119 (2011).
18. R. Gorenflo, F. Mainardi and A. Vivoli, *Chaos, Solitons and Fractals* **34**, 87 (2007).
19. R. Hilfer, Threefold introduction to fractional calculus, in, *Anomalous Transport, Foundations and Applications*, eds. R. Klages, G. Radons and I. M. Sokolov (Wiley-VCH, 2008), pp. 17–73.
20. R. Hilfer, *Fractals* **3**, 211 (1995).
21. R. Hilfer and L. Anton, *Phys. Rev. E* **51**, R848 (1995).
22. A. Janicki, *Numerical and Statistical Approximation of Stochastic Differential Equations with Non-Gaussian Measures*, Monograph No. 1, H. Steinhaus Center for Stochastic Methods in Science and Technology, Technical University, Wroclaw, Poland (1996).
23. A. Janicki and A. Weron, *Simulation and Chaotic Behavior of $\alpha$–Stable Stochastic Processes* (Marcel Dekker, 1994).
24. A. A. Kilbas, H. M. Srivastava and J. J. Trujillo, *Theory and Applications of Fractional Differential Equations* (Elsevier, 2006).
25. D. Kleinhans and R. Friedrich, *Phys. Rev E* **76**, 061102 (2007).
26. F. Mainardi, *Fractional Calculus and Waves in Linear Viscoelasticity* (Imperial College Press, 2010).
27. F. Mainardi, Yu. Luchko and G. Pagnini, *Fract. Calculus Appl. Anal.* **4**, 153 (2001).
28. F. Mainardi, A. Mura and G. Pagnini, *Int. J. Differential Equations* **2010**, Article ID 104505, p. 29 (2010).
29. F. Mainardi, G. Pagnini and R. K. Saxena, *J. Comput. Appl. Math.* **178**, 321 (2005).
30. F. Mainardi and M. Tomirotti, *Ann. Geofisica* **40**, 1311 (1997).

31. O. I. Marichev, *Handbook of Integral Transforms of Higher Transcendental Functions, Theory and Algorithmic Tables* (Ellis Horwood, 1983).
32. A. M Mathai, R. K. Saxena and H. J Haubold, *The H-function, Theory and Applications* (Springer, 2010).
33. M. M. Meerschaert, D. A. Benson, H. P. Scheffler and B. Baeumer, *Phys. Rev. E* **65**, 041103 (2002).
34. M. M. Meerschaert, D. A. Benson, H. P. Scheffler and P. Becker-Kern, *Phys. Rev. E* **66**, 060102 (2002).
35. M. M. Meerschaert, E. Nane and P. Vellaisamy, The fractional Poisson process and the inverse stable subordinator, arxiv:1007.505.
36. M. M. Meerschaert and H. P. Scheffler, Limit theorems for continuous-time random walks with infinite mean waiting times, *J. Appl. Probab.* **41**, 623 (2004).
37. M. M. Meerschaert, Y. Zhang and B. Baeumer, *Comput. Math. Appl.* **59**, 1078 (2010).
38. R. Metzler, E. Barkai and J. Klafter, *Europhys. Lett.* **46**, 431 (1999).
39. R. Metzler and J. Klafter, *J. Phys. Chem. B* **104**, 3851 (2000).
40. R. Metzler, J. Klafter and I. M. Sokolov, *Phys. Rev. E* **58**, 1621 (1998).
41. R. Metzler and J. Klafter, *Phys. Rep.* **339**, (2000).
42. R. Metzler and J. Klafter, *J. Phys. A. Math. Gen.* **37**, R161 (2004).
43. G. M. Mittag-Leffler, *C. R. Acad. Sci. Paris* **137**, 554 (1903).
44. G. M. Mittag-Leffler, Sopra la funzione $E_\alpha(x)$, *Rendi. Accad. Lincei* (*Ser. V*) **13**, 3 (1904).
45. G. M. Mittag-Leffler, *Acta Math.* **29**, 101 (1905).
46. A. Piryatinska, A. I. Saichev and W. A. Woyczynski, *Physica A* **349**, 375 (2005).
47. I. Podlubny, *Fractional Differential Equations* (Academic Press, 1999).
48. A. Saichev and G. Zaslavsky, *Chaos* **7**, 753 (1997).
49. S. G. Samko, A. A. Kilbas and O. I. Marichev, *Fractional Integrals and Derivatives: Theory and Applications* (Gordon and Breach, 1993).
50. G. Samorodnitsky and M. S. Taqqu, *Stable non-Gaussian Random Processes* (Chapman & Hall, 1994).
51. K.-I. Sato, *Lévy Processes and Infinitely Divisible Distributions* (Cambridge Univ. Press, 1999).
52. E. Scalas, R. Gorenflo and F. Mainardi, *Phys. Rev. E* **69**, 011107 (2004).
53. W. R. Schneider, Stable distributions: Fox function representation and generalization, in *Stochastic Processes in Classical and Quantum Systems*, eds. S. Albeverio, G. Casati and D. Merlini (Springer-Verlag, 1986), pp. 497–511.
54. H. Takayasu, *Fractals in the Physical Sciences* (Manchester Univ. Press, 1990).
55. Z. Tomovski, R. Hilfer and H. M. Srivastava, *Integral Transforms Spec. Funct.* **21**, 797 (2010).
56. V. V. Uchaikin and V. M. Zolotarev, *Chance and Stability. Stable Distributions and their Applications* (VSP, 1999).
57. G. H. Weiss, *Aspects and Applications of Random Walks* (North-Holland, 1994).
58. A. Wiman, *Acta Math.* **29**, 191 (1905).

59. E. M. Wright, *J. London Math. Soc.* **8**, 71 (1933).
60. E. M. Wright, *Proc. London Math. Soc. Ser. II* **38**, 257 (1935).
61. E. M. Wright, *J. London Math. Soc.* **10**, 287 (1935).
62. E. M. Wright, The generalized Bessel function of order greater than one, *Quart. J. Math., Oxford Ser.* **11**, 36 (1940).
63. Y. Zhang, M. M. Meerschaert and B. Baeumer, *Phys. Rev. E.* **78**, 036705 (2008).
64. V. M. Zolotarev, *One-dimensional Stable Distributions* (Amer. Math. Soc. 1986).

# Chapter 11

## Fractional Calculus, Anomalous Diffusion, and Probability

Mark M. Meerschaert
*Department of Statistics and Probability,*
*Michigan State University,*
*East Lansing, MI 48824, USA*

Ideas from probability can be very useful to understand and motivate fractional calculus models for anomalous diffusion. Fractional derivatives in space are related to long particle jumps. Fractional time derivatives code particle sticking and trapping. This probabilistic point of view also leads to some interesting extensions, including vector fractional derivatives, and tempered fractional derivatives. This paper reviews the basic ideas along with some practical applications.

1. Introduction . . . . . . . . . . . . . . . . . . . . . . . . 265
2. Fractional Derivatives and Probability . . . . . . . . . . . 267
3. Fractional Derivatives in Time . . . . . . . . . . . . . . . 273
4. Vector Fractional Calculus . . . . . . . . . . . . . . . . . 275
5. Multi-Scaling Fractional Derivatives . . . . . . . . . . . . 278
6. Simulation . . . . . . . . . . . . . . . . . . . . . . . . . 279
7. Tempered Fractional Derivatives . . . . . . . . . . . . . . 281

### 1. Introduction

The connection between the deterministic diffusion equation, and probabilistic Brownian motion, is a powerful and useful idea that has been exploited in many forms. The basic idea is that $p(x,t)$, the probability density function (PDF) of a Brownian motion stochastic process $B(t)$, solves the partial differential equation $\partial_t p = \partial_x^2 p$. On one hand, this means that solutions to a deterministic partial differential equation provide valuable information about random evolution. On the other hand, simulations of a random process can be used to generate numerical solutions to a deterministic model, a method called particle tracking. More fundamental is that the random path of a particle, described by the Brownian motion process, provides a physical explanation for diffusion. Even in a completely deterministic derivation of the diffusion equation, in terms of flux and conservation of mass, random particle motions are the basic driving force.

Anomalous diffusion occurs when a cloud of particles spreads in a different manner than the traditional diffusion equation predicts. Fractional diffusion equations have become popular as the most reasonable and tractable models for anomalous diffusion. The traditional diffusion equation governs a Brownian motion, the long-time limit of a simple random walk with independent and identically distributed (IID) particle jumps $(X_n)$. The approximation is a result of the central limit theorem of probability, and assumes finite first and second moments for the particle jumps.

If the jumps have power law probability tails, $\mathbb{P}(|X_n| > r) \approx r^{-\alpha}$ with $0 < \alpha < 2$, then the moment conditions are violated, and the random walk behavior is anomalous. In this case, the random walk limit is a stable Lévy motion $A(t)$ with index $\alpha$, a natural mathematical extension of Brownian motion [45]. The PDF of the stable limit solves a space-fractional diffusion equation $\partial_t p = \partial_x^\alpha p$ that reduces to the traditional form when $\alpha = 2$. The underlying probability model gives a specific physical meaning for the fractional derivative in space: It codes large particle jumps, that lead to anomalous super-diffusion.

Note that the power law index $\alpha$ in the probability of long jumps equals the order of the fractional space derivative. A third interpretation for the index $\alpha$ comes from considerations of fractals and self-similarity. The path of a Brownian particle in space traces out a random fractal of dimension two, while a Lévy stable particle draws a fractal of dimension $\alpha$, see [51]. This is closely connected to the idea of a self-similar stochastic process, the relation $B(ct) \approx c^{1/2}B(t)$ or $A(ct) \approx c^{1/\alpha}A(t)$ between processes rescaled in space and time. The self-similarity index $H = 1/\alpha$, also called the Hurst index, provides a useful way to categorize diffusion models. In traditional diffusion, with $H = 1/2$, a plume of diffusing particles spreads away from their center of mass at the rate $t^H$, which is evident from the scaling. The case $H > 1/2$ is called super-diffusion since particles spread at a faster rate.

Anomalous sub-diffusion is a model for particle sticking or trapping. Suppose that each particle jump $X_n$ occurs at the end of a random waiting time $J_n$, with $\mathbb{P}(J_n > t) \approx t^{-\beta}$ for some $0 < \beta < 1$. This is called a continuous time random walk (CTRW), but it is really just a simple random walk in spacetime. The CTRW has a long-time limit density that solves a time-fractional diffusion equation $\partial_t^\beta p = \partial_x^2 p$ that reduces to the traditional form when $\beta = 1$. This shows that the time-fractional derivative models time delays between particle motion. Again, the order of the fractional derivative is the same number that controls the probability model. The limit process is a time-changed Brownian motion $B(E_t)$ where the fractal time $E_{ct} \approx c^\beta E_t$, which leads to $B(E_{ct}) \approx c^{\beta/2} B(E_t)$. Since the Hurst index $H = \beta/2 < 1/2$, a plume of particles spreads slower than traditional diffusion. When long waiting times are combined with long particle jumps, a spacetime fractional diffusion equation $\partial_t^\beta p = \partial_x^\alpha p$ governs the CTRW limit, a process with $A(E_{ct}) \approx c^{\beta/\alpha} A(E_t)$. This can be sub-diffusive, super-diffusive, or even represent an anomalous diffusion that spreads with the same rate $H = 1/2$ as a Brownian motion.

The remainder of this chapter reviews some of the mathematics behind the fractional diffusion equation and its underlying stochastic process. We will also introduce some useful extensions and variations, including fractional vector calculus for vector-valued diffusions, and tempered models that smoothly interpolate between traditional and fractional diffusions. A more complete and detailed development of the ideas presented here can be found in the forthcoming book [36].

## 2. Fractional Derivatives and Probability

In this section, we introduce fractional derivatives from two different points of view: Differential equations, and probability. Then we will show that both points of view are really just two aspects of the same idea. Recall that the first derivative $\partial_x f(x) = \lim_{h \to 0} h^{-1} \Delta f(x)$ where the difference

$$\Delta f(x) = f(x) - f(x-h).$$

For positive integers $\alpha$, $\partial_x^\alpha f(x) = \lim_{h \to 0} h^{-\alpha} \Delta^\alpha f(x)$, where

$$\begin{aligned}
\Delta^2 f(x) &= (f(x) - f(x-h)) - (f(x-h) - f(x-2h)) \\
&= f(x) - 2f(x-h) + f(x-2h), \\
\Delta^3 f(x) &= f(x) - 3f(x-h) + 3f(x-2h) - f(x-3h) \\
&\ \vdots \\
\Delta^\alpha f(x) &= \sum_{m=0}^{\alpha} (-1)^m \binom{\alpha}{m} f(x-mh),
\end{aligned}$$

where the Binomial coefficients

$$g_m = \binom{\alpha}{m} = \frac{\alpha!}{m!(\alpha-m)!} = \frac{\Gamma(\alpha+1)}{\Gamma(m+1)\Gamma(\alpha-m+1)}.$$

For $\alpha > 0$, the fractional derivative $\partial_x^\alpha f(x) = \lim_{h\to 0} h^{-\alpha}\Delta^\alpha f(x)$ where

$$\Delta^\alpha f(x) = \sum_{m=0}^{\infty} \binom{\alpha}{m} (-1)^m f(x - mh). \tag{1}$$

This is actually the same formula as before, since $g_m = 0$ for $m > \alpha$ when $\alpha$ is an integer. Note that the integer derivative is a local operator, since it only depends on values of $f$ near $x$, while the nonlocal fractional derivative depends on values of $f$ in $(-\infty, x]$. Numerical analysis of fractional differential equations is based on (1).

The traditional diffusion equation is the result of two basic ideas: The concentration of particles $p(x,t)$, at location $x$ and time $t$, must obey conservation of mass $\partial_t p = -\partial_x q$, where the particle flux $q(x,t)$ follows Fick's Law $q = -\mathcal{D}\partial_x p$. Fick's Law $q\Delta x \approx -\mathcal{D}\Delta p$ formalizes the empirical observation that particles cross a boundary between regions of differing concentration at a rate proportional to the difference in concentrations. The combination of these two laws gives the diffusion equation $\partial_t p = \mathcal{D}\partial_x^2 p$, where now we explicitly show the diffusivity $\mathcal{D}$. Implicit in Fick's law is the idea that all particles move at more or less the same velocity, so that we can account for flux over an interval of length $h = \Delta x$ in terms of the local difference in concentration $\Delta p = p(x,t) - p(x-h,t)$.

In fractional diffusion, a fractional Fick's law $q\Delta x \approx -\mathcal{D}\Delta^{\alpha-1} p$ for $1 < \alpha < 2$ recognizes the possibility that, when particle velocities are sufficiently heterogeneous, the flux can also depend on concentrations far upstream. The relative contributions of those concentrations depend on the weights in (1). Using Stirling's formula $\Gamma(x+1) \sim \sqrt{2\pi x}\, x^x e^{-x}$ as $x \to \infty$ you can check that

$$w_m = (-1)^m \binom{\alpha}{m} \sim \frac{-\alpha}{\Gamma(1-\alpha)} m^{-1-\alpha} \quad \text{as } m \to \infty. \tag{2}$$

In fact, $w_0 = 1$, $w_1 = -\alpha$, $w_2 = \alpha(\alpha - 1)$ and so forth. The binomial formula states that

$$(1+z)^\alpha = \sum_{m=0}^{\infty} \binom{\alpha}{m} z^m \tag{3}$$

for any complex $|z| \le 1$ and any $\alpha > 0$. Set $z = -1$ to check that the weights $w_m$ in (1) sum to zero, which ensures that all the mass leaving any given location arrives at some point downstream. The proportion of particles transported $m$ steps downstream falls off like a power law. Now the space-fractional diffusion equation

$$\partial_t p(x,t) = \mathcal{D}\partial_x^\alpha p(x,t) \tag{4}$$

comes from combining the fractional Fick's law and conservation of mass.

The Fourier transform (FT)

$$\hat{f}(k) = \int_{-\infty}^{\infty} e^{-ikx} f(x)dx$$

converts differential equations to algebra, since $\partial_x^\alpha f(x)$ has FT $(ik)^\alpha \hat{f}(k)$. When $\alpha$ is a real number, this provides another equivalent definition of the fractional derivative (or fractional integral, if $\alpha < 0$). By the binomial formula (3) and the fact that $f(x-h)$ has FT $e^{-ikh}\hat{f}(k)$, $\Delta^\alpha f(x)$ has FT

$$\sum_{m=0}^{\infty} \binom{\alpha}{m} (-1)^m e^{-ikmh} \hat{f}(k) = (1 - e^{-ikh})^\alpha \hat{f}(k)$$

and then the FT of $h^{-\alpha}\Delta^\alpha f(x)$ is

$$h^{-\alpha}(ikh)^\alpha \left(\frac{1 - e^{-ikh}}{ikh}\right)^\alpha \hat{f}(k) \to (ik)^\alpha \hat{f}(k) \quad \text{as } h \to 0$$

by a Taylor series expansion $e^z = 1 + z + z^2/2! + \cdots$. Take FT in the fractional diffusion equation (4) to get

$$\partial_t \hat{p}(k,t) = \mathcal{D}(ik)^\alpha \hat{p}(k,t)$$

and solve to get $\hat{p}(k,t) = \exp(t\mathcal{D}(ik)^\alpha)$. Inverting the FT in the case $\alpha = 2$ gives a normal density

$$p(x,t) = \frac{1}{\sqrt{4\pi\mathcal{D}t}} \exp\left(-\frac{x^2}{4\mathcal{D}t}\right) \tag{5}$$

which is the point source solution of the traditional diffusion equation. In the case $0 < \alpha < 2$, the solution is a Lévy stable PDF with index $\alpha$. Usually, this PDF cannot be written in closed form.

Next we consider the fractional diffusion equation from the point of view of probability. The random particle jump $X_n$ has PDF $f(x)$ with FT

$$\hat{f}(k) = \int_{-\infty}^{\infty} \left(1 - ikx + \frac{1}{2!}(ikx)^2 + \cdots\right) f(x)dx = 1 - ik\mu_1 - \frac{1}{2}k^2\mu_2 + \cdots,$$

where the $p$th moment

$$\mu_p = \int_{-\infty}^{\infty} x^p f(x)dx. \tag{6}$$

The random walk $S(n) = X_1 + \cdots + X_n$ gives the particle location after $n$ IID jumps. If we take centered jumps with $\mu_1 = 0$ and finite variance $\mu_2 = 2\mathcal{D}$, then $\hat{f}(k) = 1 - \mathcal{D}k^2 + \cdots$, the PDF of $S([nt])$ has FT $\hat{f}(k)^{[nt]}$, and the rescaled sum $S([nt])/\sqrt{n}$ has FT

$$\hat{f}(k/\sqrt{n})^{[nt]} = \left(1 - \frac{\mathcal{D}k^2}{n} + \cdots\right)^{[nt]} \to e^{-t\mathcal{D}k^2} \quad \text{as } n \to \infty.$$

Inverting the FT shows that the rescaled random walk converges in the limit to a Brownian motion $B(t)$ with PDF (5), the same formula that solves the traditional diffusion equation. This is the traditional central limit theorem (CLT) of probability. Figure 1 illustrates the random walk convergence, as the number of jumps increases, to a Brownian motion path, a continuous (but not differentiable) random fractal of dimension 3/2.

For particle jumps with heavy tails $\mathbb{P}(X > x) \sim \mathcal{D}x^{-\alpha}/\Gamma(1-\alpha)$, the jump PDF

$$f(x) \sim \mathcal{D}\frac{\alpha}{\Gamma(1-\alpha)}x^{-\alpha-1} \tag{7}$$

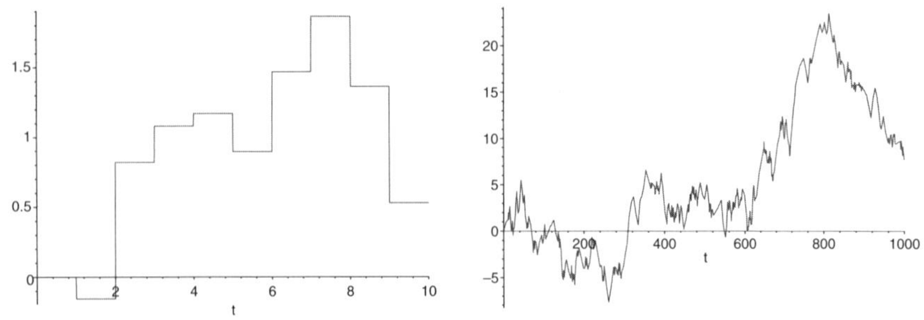


Fig. 1. Random walk simulation, showing convergence to Brownian motion.

and some moments $\mu_p$ are undefined, because the integral (6) does not converge, so the traditional CLT is not valid. If $1 < \alpha < 2$ and $\mu_1 = 0$, then a Tauberian theorem [18] shows that $X$ has FT

$$\hat{f}(k) = 1 + \mathcal{D}(ik)^\alpha + \cdots$$

and the rescaled random walk $n^{-1/\alpha}S([nt])$ has FT

$$\hat{f}(k/n^{1/\alpha})^{[nt]} = \left(1 + \frac{\mathcal{D}(ik)^\alpha}{n} + \cdots\right)^{[nt]} \longrightarrow e^{t\mathcal{D}(ik)^\alpha} \quad \text{as } n \to \infty. \tag{8}$$

Inverting the FT shows that the rescaled random walk with heavy tail jumps converges in the limit to an $\alpha$-stable Lévy motion $A(t)$, whose PDF solves the fractional diffusion equation (4). This is the Kolmogorov–Feller CLT [18]. Figure 2 illustrates the random walk convergence to the stable limit with fractal dimension $2 - 1/\alpha$. Note that the large jumps persist in the limit.

Check using (5) or its FT that Brownian motion scales according to $B(ct) \approx c^{1/2}B(t)$. The well-known PDF is bell-shaped, symmetric, with rapidly decreasing tails, and spreads like $t^{1/2}$ due to this scaling. The FT $\hat{p}(k,t) = \exp(t\mathcal{D}(ik)^\alpha)$ of the Lévy motion makes it evident that $A(ct) \approx c^{1/\alpha}A(t)$. Figure 3 shows the evolution of the stable PDF in the case $\alpha = 1.5$. The skewness is a consequence of long downstream (left to right) particle jumps. The peak shifts left of the mean (center of mass) at zero to balance the heavy tail on the right.

This point source solution to the fractional diffusion equation (4) has been used to model pollution in ground water [9]. The long downstream jumps result from fast velocity channels eroded through the intervening

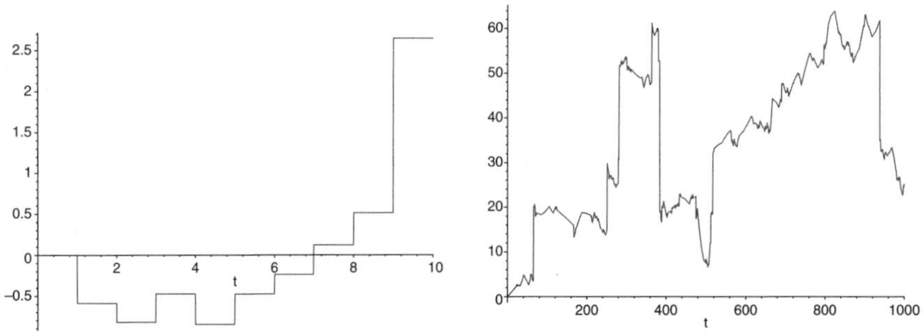


Fig. 2. A random walk with power law jumps converges to stable Lévy motion.

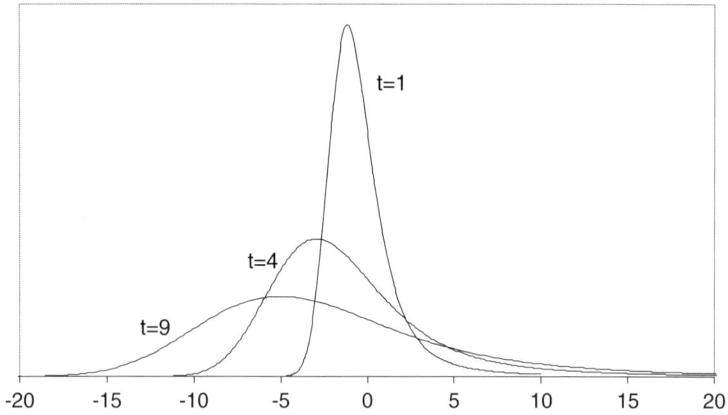


Fig. 3. Evolution of the stable Lévy motion PDF with $\alpha = 1.5$.

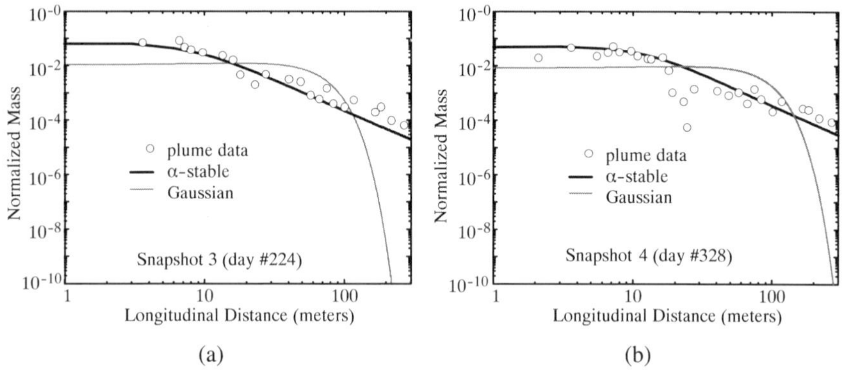


Fig. 4. FADE application to ground water pollution, from [9].

porous medium by historical flows. Figure 4 shows the best fit to measured concentrations of a tracer at the macrodispersion experimental test site (MADE site) near Columbus MS, using $\alpha = 1.1$. The log–log display illustrates the power-law right tail of the stable solution curve. The best fitting normal curve (traditional diffusion equation (4) with $\alpha = 2$) is shown for comparison. The traditional model greatly understates the risk of downstream contamination.

The intimate connection between the deterministic and random points of view is evident, once we compare the power-law jumps in the random walk model (7) to the mass transport in the fractional Fick's law governed by the weights (2). Both assume, as their fundamental premise, that particles travel a long distance downstream, governed by a power law.

## 3. Fractional Derivatives in Time

The (Caputo) fractional derivative in time can be defined, for $0 < \beta \leq 1$, as the function with Laplace transform (LT) $s^\beta \tilde{f}(s) - s^{\beta-1} f(0)$, where

$$\tilde{f}(s) = \int_0^\infty e^{-st} f(t)dt$$

is the LT of $f(t)$. When $\beta = 1$, this is the usual LT relation for the first derivative. The space-time fractional diffusion equation

$$\partial_t^\beta m(x,t) = \mathcal{D}\partial_x^\alpha m(x,t) \tag{9}$$

can be solved using Fourier–Laplace transforms (FLT) $\bar{m}(k,s)$, the FT of the LT of $m$. Take FLT in (9) to get $s^\beta \bar{m}(k,s) - s^{\beta-1} = \mathcal{D}(ik)^\alpha \bar{m}(k,s)$, using $\hat{m}(k,0) = 1$, and solve to get

$$\bar{m}(k,s) = \frac{s^{\beta-1}}{s^\beta - \mathcal{D}(ik)^\alpha} = \int_0^\infty e^{u\mathcal{D}(ik)^\alpha} s^{\beta-1} e^{-us^\beta} du$$

using $\int_0^\infty e^{-au} du = 1/a$. Recognize $\hat{p}(k,u) = e^{u\mathcal{D}(ik)^\alpha}$ as the FT of the solution to the space-fractional diffusion equation $\partial_u p = \mathcal{D}\partial_x^\alpha p$. Invert the FLT to see that

$$m(x,t) = \int_0^\infty p(x,u)h(u,t)du, \tag{10}$$

where $\tilde{h}(u,s) = s^{\beta-1}e^{-us^\beta}$. The effect of the fractional time derivative is to replace the time variable in $p(x,t)$ by an operational time $u$ governed by the function $h(u,t)$. The practical meaning, and inversion of the LT for $h$, will be discussed next.

In a continuous time random walk (CTRW), each particle jump $X_n$ is preceded by a random waiting time $J_n$ [39, 46]. Then the particle arrives at location $S(n) = X_1 + \cdots + X_n$ at time $T_n = J_1 + \cdots + J_n$. The spacetime random vectors $(X_n, J_n)$ are assumed IID, and their running sum $(S(n), T_n)$ is a spacetime random walk. The CTRW is uncoupled if $X_n$ is independent of $J_n$. The number of jumps by time $t$ is $N_t = \max\{n \geq 0 : T_n \leq t\}$, so that $N_t$ is the inverse of $T_n$ (the graph of $N_t$ is the graph of $T_n$ with the axes reversed). The particle position at time $t$ is $S(N_t)$. If $\mathbb{E}J_n = 1$, then the law of large numbers (LLN) guarantees that $N_t \sim t$ as $t \to \infty$, so the long-time behavior of the CTRW is the same as that of a simple random walk. If $\mathbb{P}(J_n > t) \approx t^{-\beta}/\Gamma(1-\beta)$ for some $0 < \beta < 1$ then $\mathbb{E}J_n = \infty$, and neither the CLT nor the LLN applies. A Tauberian theorem [18] shows that the PDF $w(t)$ of $J$ has LT

$$\tilde{w}(s) = 1 - s^\beta + \cdots$$

and the rescaled random walk $n^{-1/\beta}T_{[nt]}$ has FT

$$\tilde{w}(s/n^{1/\beta})^{[nt]} = \left(1 - \frac{s^\beta}{n} + \cdots\right)^{[nt]} \longrightarrow e^{-ts^\beta} \quad \text{as } n \to \infty. \tag{11}$$

Inverting the LT shows that the limit is a $\beta$-stable Lévy motion $D_t$.

Take inverses and apply the continuous mapping theorem of probability to get $c^{-\beta}N_{ct} \approx E_t$, where the inverse process $E_t = \inf\{x > 0 \colon D_x > t\}$ (see [27] for complete details). Then we have

$$c^{-\beta/\alpha}S(N_{ct}) = c^{-\beta/\alpha}S(c^\beta \cdot c^{-\beta}N_{ct}) \approx (c^\beta)^{-1/\alpha}S(c^\beta E_t) \approx A(E_t)$$

as the time scale $c \to \infty$, for the uncoupled CTRW.

Now we will show that the operational time $h(u,t)$ in (10) is the PDF of $E_t$. Note that $\{E_t \le u\} = \{D_u \ge t\}$, as these are inverse processes. Then the PDF of $E_t$ is

$$h(u,t) = \partial_u \mathbb{P}(E_t \le u) = \partial_u \mathbb{P}(D_u \ge t) = \partial_u \left[1 - \int_0^t g(y,u)\,dy\right],$$

where $g(y,u)$ is the PDF of $D_u$. Use $\tilde{g}(s,u) = \exp(-us^\beta)$ from (11), and the fact that integration corresponds to dividing the LT by $s$, to check that

$$\tilde{h}(u,s) = \partial_u \left[1 - s^{-1}e^{-us^\beta}\right] = s^{\beta-1}e^{-us^\beta}$$

so the solution (10) to the spacetime fractional diffusion equation (9) is the PDF of $A(E_t)$, the long-time limit of a CTRW with power-law jumps $\mathbb{P}(X > x) \approx x^{-\alpha}$ and power-law waiting times $\mathbb{P}(J > t) \approx t^{-\beta}$. The form (10) comes from a conditioning argument

$$\mathbb{P}(A(E_t) = x) = \sum_u \mathbb{P}(A(u) = x | E_t = u)\mathbb{P}(E_t = u)$$

also called the law of total probability. The fractal activity time scales according to $E_{ct} \approx c^\beta E_t$ with $0 < \beta < 1$, so that operational time is slower than clock time, a sub-diffusive effect [43].

## 4. Vector Fractional Calculus

Vector fractional derivatives are associated with power law jumps in $d$-dimensional space. To understand this connection, it is useful to view the space-fractional diffusion equation as a Cauchy problem $\partial_t p = Lp$ where the space derivative operator $L$ acts on the $x$ variable. The form $L = \mathcal{D}\partial_x^\alpha$ is connected with power law jumps $\mathbb{P}(X > x) \approx x^{-\alpha}$. The CTRW with these jumps, and exponential waiting times $\mathbb{P}(J > t) = e^{-\lambda t}$, is also called a compound Poisson process. Now $\mathbb{P}(N_t = n) = e^{-\lambda t}(\lambda t)^n/n!$ and

$$P(x,t) = \mathbb{P}(S(N_t) \le x) = \sum_{n=0}^{\infty} \mathbb{P}(S(n) \le x | N_t = n)\mathbb{P}(N_t = n)$$

by the law of total probability. Take FT to get

$$\hat{P}(k,t) = \sum_{n=0}^{\infty} \hat{f}(k)^n e^{-\lambda t}\frac{(\lambda t)^n}{n!} = e^{-\lambda t(1-\hat{f}(k))}$$

by the Taylor series for $e^z$, where $f$ is the PDF of $X$. Clearly this solves

$$\partial_t \hat{P}(k,t) = -\lambda(1-\hat{f}(k))\hat{P}(k,t)$$

which inverts to the Cauchy problem

$$\partial_t P(x,t) = -\lambda P(x,t) + \lambda \int P(x-y,t)f(y)dy \tag{12}$$

using the convolution property of FT. Use the fact that $\int f(y)dy = 1$ to rewrite this in the form

$$\partial_t P(x,t) = \int (P(x-y,t) - P(x,t))\lambda f(y)dy.$$

To arrive at the stable limit process, let $\lambda \to \infty$ and rescale the jumps: Let $X_\lambda = \lambda^{-1/\alpha}X$, with PDF $f_\lambda(y) = \lambda^{1/\alpha}f(\lambda^{1/\alpha}y)$. Using $f(y) \approx \alpha y^{-\alpha-1}/\Gamma(1-\alpha)$ we get $\lambda f_\lambda(y) \to \alpha y^{-\alpha-1}/\Gamma(1-\alpha)$ for all $y > 0$, so the CDF of the limit $A(t)$ solves

$$\partial_t P(x,t) = \frac{\alpha}{\Gamma(1-\alpha)}\int_0^\infty (P(x-y,t) - P(x,t))y^{-\alpha-1}dy. \tag{13}$$

The formula on the right-hand side is another form of the fractional derivative $\partial_x^\alpha P$. To check this, compute the FT

$$\frac{\alpha}{\Gamma(1-\alpha)}\int_0^\infty (e^{-iky} - 1)y^{-\alpha-1}dy = (ik)^\alpha$$

for $0 < \alpha < 1$, and use the fact that $e^{-iky}\hat{P}(k,t)$ is the FT of $P(x-y,t)$. Apply $\partial_x$ to both sides of (13) to recover the fractional diffusion equation (4) with $\mathcal{D}=1$. The Poisson limit is an alternative to the FT argument (11), see [25] for complete details.

The balance equation (12) writes the change in probability in terms of the rate $\lambda$ at which particles jump away from location $x$, and the rate at which particles from location $x-y$ jump to location $x$. This is completely analogous to the fractional Fick's law, as the proportion of particle that travels a distance $y$ falls off like $y^{-\alpha-1}$ in both models.

More general notions of anomalous diffusion are given by the Cauchy problem $\partial_t P = LP$ with

$$Lf(x) = -v\cdot\nabla f(x) + \int_0^\infty [f(x-y) - f(x) + y\cdot\nabla f(x)]\phi(dy), \qquad (14)$$

where $\phi(dy)$ is the Poisson jump intensity, and $\nabla = \partial_{x_1} + \cdots + \partial_{x_d}$ is the gradient [1]. The first term adds a drift at velocity $v$. Taking $v=0$ and

$$\phi(dy) = \begin{cases} a\mathcal{D}\dfrac{\alpha}{\Gamma(1-\alpha)}y^{-1-\alpha}dy & \text{for } y>0 \\ b\mathcal{D}\dfrac{\alpha}{\Gamma(1-\alpha)}|y|^{-1-\alpha}dy & \text{for } y<0 \end{cases}$$

in one dimension leads to $L = a\mathcal{D}\partial_x^\alpha + b\mathcal{D}\partial_{-x}^\alpha$. The negative fractional derivative $\partial_{-x}^\alpha$, equivalent to multiplying the FT by $(-ik)^\alpha$, models right-to-left particle jumps. To verify $L = a\mathcal{D}\partial_x^\alpha + b\mathcal{D}\partial_{-x}^\alpha$, compute the FT

$$\frac{\alpha}{\Gamma(1-\alpha)}\int_0^\infty \left(e^{-iky} - 1 + iky\right) y^{-\alpha-1}\, dy = (ik)^\alpha$$

for $1<\alpha<2$, see [25, p. 265]. The corresponding Cauchy problem governs the CTRW limit with $\mathbb{P}(X>x)\approx a\,x^{-\alpha}$ and $\mathbb{P}(X<-x)\approx bx^{-\alpha}$ so that the weights $a,b$ balance the jumps. This version of (4) has been applied to model contamination in ground water and river flows, see [8, 12, 17]. The stable limit $A(t)$ has PDF with FT $\hat{p}(k,t) = \exp(a\mathcal{D}t(ik)^\alpha + b\mathcal{D}t(-ik)^\alpha)$, see [25, p. 456]. If $a=b$, this gives a model for symmetric anomalous diffusion, the case $\alpha=1$ being the familiar Cauchy distribution.

In the vector case with $\phi(dy) = C\|y\|^{-\alpha-d}dy$ we get $L = -(-\Delta)^{\alpha/2}$ the fractional power of the Laplacian $\Delta = \nabla\cdot\nabla$, which has an interesting history [19]. The corresponding stable process in $d$ dimensions is the limit of a random walk with power-law jumps, whose orientation is uniformly distributed over the unit sphere. If we take jumps of the form $X = R\Theta$

where $\mathbb{P}(R > r) \sim r^{-\alpha}/\Gamma(1-\alpha)$ and $\Theta$ has an arbitrary distribution $M(d\theta)$ on the unit sphere, a Poisson limit argument yields a Cauchy problem with jump intensity $\phi(dr, d\theta) = \alpha r^{-\alpha-1} dr M(d\theta)/\Gamma(1-\alpha)$ in polar coordinates $y = r\theta$. Compute

$$Lf(x) = -v \cdot \nabla f(x) + \int_{|\theta|=1} \mathbb{D}_\theta^\alpha f(x) M(d\theta), \tag{15}$$

where $\mathbb{D}_\theta^\alpha f(x)$ is the fractional directional derivative, equal to $\partial_r^\alpha f(x + r\theta)$ at $r = 0$, whose FT is $(ik \cdot \theta)^\alpha \hat{f}(k)$. The Cauchy problem $\partial_t p = Lp$ using (15) is called the fractional advection-dispersion equation (FADE) [23]. The FADE has found many applications in ground water hydrology [8, 9, 16], biology [3, 40], and physics [37, 38]. For example, the FADE

$$\partial_t p = v_x \partial_x p - v_y \partial_y p + \mathcal{D}_x \partial_x^\alpha + \mathcal{D}_y \partial_y^\alpha \tag{16}$$

governs anomalous diffusion with mean velocity $v = (v_x, v_y)$, the long-time limit of a vector random walk with power-law jumps in the positive $x$ and $y$ directions, with a probability of jumps longer than $r$ falling off like $r^{-\alpha}$, and the proportion of $x, y$ jumps governed by the ratio $\mathcal{D}_x/\mathcal{D}_y$.

A vector fractional calculus was developed in [30], see also [50]. Start with the vector flux $q = vp - Q\nabla p$, and write the dispersion tensor $Q$ in terms of the distribution $M(d\theta)$ that controls jump directions:

$$Q\nabla f(x) = \int_{|\theta|=1} \theta\,(\theta \cdot \nabla f(x))\, M(d\theta), \tag{17}$$

a mixture of directional derivatives $\mathbb{D}_\theta f(x) = \theta \cdot \nabla f(x)$ laid out in each radial direction $\theta$ according to the weights $M(d\theta)$. Together with conservation of mass $\partial_t p = \nabla \cdot q$ this leads to the traditional advection-dispersion equation (ADE) $\partial_t p = -v \cdot \nabla p + \nabla \cdot Q\nabla p$ used to model ground water contaminants [7]. Apply the FT $\hat{p}(k,t) = \int e^{-ik\cdot x} p(x,t)\, dx$ to get

$$\partial_t \hat{p}(k,t) = -v(ik)\hat{p}(k,t) + (ik) \cdot Q(ik)\hat{p}(k,t)$$

whose point source solution $\hat{p}(k,t) = \exp(-tv(ik) + (ik) \cdot tQ(ik))$ inverts to a multivariable Gaussian PDF with mean $tv$ and covariance matrix $2tQ$. The FADE comes from a fractional vector flux $q = vp - \nabla_M^{\alpha-1} p$ where the fractional gradient

$$\nabla_M^{\alpha-1} f(x) = \int_{|\theta|=1} \theta \mathbb{D}_\theta^{\alpha-1} f(x) M(d\theta)$$

is a mixture of fractional directional derivatives. Use FT to check that

$$\nabla \cdot \nabla_M^{\alpha-1} f(x) = \int_{|\theta|=1} \mathbb{D}_\theta^\alpha f(x) M(d\theta).$$

These same ideas can be used to extend the divergence, curl, and basic theorems of vector calculus (divergence theorem, Stokes theorem) to a fractional form [30]. Time-fractional Cauchy problems are considered in [1, 5, 21, 35].

## 5. Multi-Scaling Fractional Derivatives

There is no reason why the order of the space-fractional derivative in the FADE (16) should be the same in both coordinates. An extended model uses matrix scaling. Consider a random walk with jumps $X = R^E\Theta$ where $\mathbb{P}(R > r) \approx r^{-1}$ and $\Theta$ has an arbitrary distribution $M(d\theta)$ on the unit sphere. The matrix power $R^E = \exp(E \log R)$ where the matrix exponential $\exp(A) = I + A + A^2/2 + \cdots$ as usual. If $E = \mathrm{diag}(1/\alpha_1, \ldots, 1/\alpha_d)$ a diagonal matrix, then $R^E = \mathrm{diag}(R_1, \ldots, R_d)$ with $\mathbb{P}(R_i > r) \approx r^{-\alpha_i}$, allowing a different tail index in each coordinate. For a CTRW with these jumps, and exponential waiting times, a Poisson limit argument shows that the CTRW converges to a vector-valued process $A(t)$ with operator scaling $A(ct) \approx c^E A(t)$ [24, 47]. The limit process $A(t)$ is called an operator stable Lévy motion [25]. Its density $p(x,t)$ solves the Cauchy problem $\partial_t P = LP$ with generator $L$ given by (14), and jump intensity $\phi(dy) = r^{-2}dr\, M(d\theta)$ in the multi-scaling polar coordinates $y = r^E\theta$ [20, 25, 47]. For example, the multi-scaling FADE

$$\partial_t p = -v_x\partial_x p - v_y\partial_y p + \mathcal{D}_x\partial_x^{\alpha_x} + \mathcal{D}_y\partial_y^{\alpha_y} \tag{18}$$

governs the long-time limit of a vector random walk with power-law jumps $R^E\Theta = (X, Y)$, with exponent $E = \mathrm{diag}(1/\alpha_x, 1/\alpha_y)$, and $\Theta$ points in the $x, y$ directions with probability proportional to $\mathcal{D}_x, \mathcal{D}_y$, respectively. Then $\mathbb{P}(X > r) \approx r^{-\alpha_x}$ and $\mathbb{P}(Y > r) \approx r^{-\alpha_y}$ so that each fractional derivative codes power law jumps in the corresponding coordinate direction.

By varying the matrix exponent $E$ and the distribution of the jump angle $\Theta$, a wide variety of models can be constructed. Eigenvectors of $E$ determine the coordinate system, the corresponding eigenvalues code the power law tails (order of the fractional derivative) in each coordinate, and $\Theta$ directs the jumps. Practical details are laid out in [26]. The multi-scaling FADE has been applied to ground water pollution in granular aquifers [41, 47, 52] and fractured rock [44], tick-by-tick stock data [29], and chaotic dynamics of microbes in anisotropic porous media [42].

## 6. Simulation

Numerical solutions to (4) are developed in [28, 31, 48] based on a shifted version of the finite difference $\Delta^\alpha f(x) = \sum_{m=0}^{\infty} w_m f(x - mh + h)$. The shift is necessary for numerical stability, even in the case $\alpha = 2$. Numerical methods for the vector equation (18) are developed in [32, 49] based on operator splitting: In one step of the iteration, the $\partial_x^{\alpha_x}$ term is applied while $y$ is held constant, and vice versa. An application in [3] considers a fractional reaction-dispersion equation that models an invasive species crossing a barrier:

$$\partial_t p = C\partial_x^\alpha p + D\partial_y^2 p + rp\left(1 - \frac{p}{K}\right),$$

where population density $p(x, y, t)$, $r$ is the intrinsic growth rate of the species, and $K$ is the environmental carrying capacity [33]. Figure 5 shows the numerical solution, via operator splitting of the reaction and dispersion terms. In the classical model $\alpha = 2$ (top), the invading species leaks slowly

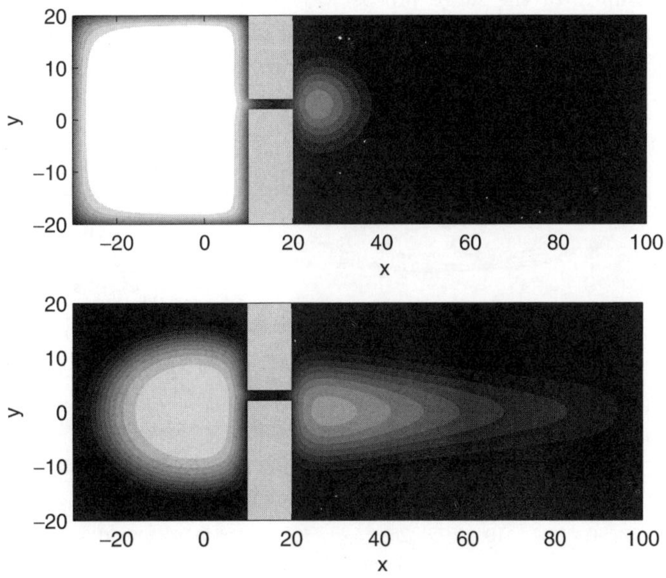


Fig. 5. Fractional model for invasive species crossing a barrier, from [3].

through the slit barrier. In the fractional model $\alpha = 1.7$ (bottom) the invaders jump the barrier, and the slit is irrelevant. This demonstrates the effect of a nonlocal fractional derivative as a model for long jumps, and its implications for applications in ecology. The fractional derivative is relevant in such problems, since many ecological studies have documented heavy tail dispersal kernels (distance between parent and offspring) in ecological applications [3].

A finite difference approximation to the multi-scaling fractional derivative was considered in [2], but for most practical applications, it seems easier to use particle tracking [22]. A random walk with jumps $X = R^E\Theta$ faithfully approximates the operator stable process $A(t)$, and a histogram of particles estimates the PDF $p(x,t)$ that solves the multi-scaling FADE, see [52]. Figure 6 illustrates the particle tracking solution in the case $v_x = 10$, $v_y = 0$ (left-to-right flow), $\alpha_x = 1.5$, and $\alpha_y = 1.9$ (more anomalous dispersion in the direction of flow). The jump directions and respective weights are illustrated in the upper right inset. Dots are individual particles, and continuous curves are the corresponding solution via inverse fast Fourier transform (FFT) of $\hat{p}(k,t)$. The FFT method is only viable for constant coefficients.

A more sophisticated simulation method that preserves the exact location and timing of the large jumps was recently developed [14] based on a shot noise representation $X = R^E\Theta$ for the large jumps, and a Brownian motion approximation for the small jumps (there are infinitely many of those). The same idea was used in [13] for the vector stable process

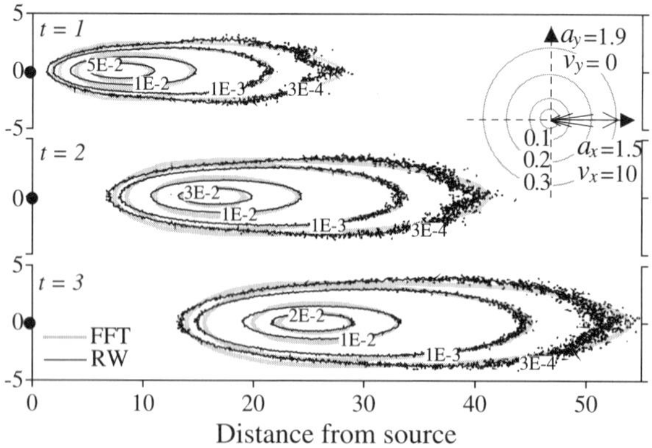


Fig. 6. Particle tracking solution of the multi-scaling FADE (18) from [52].

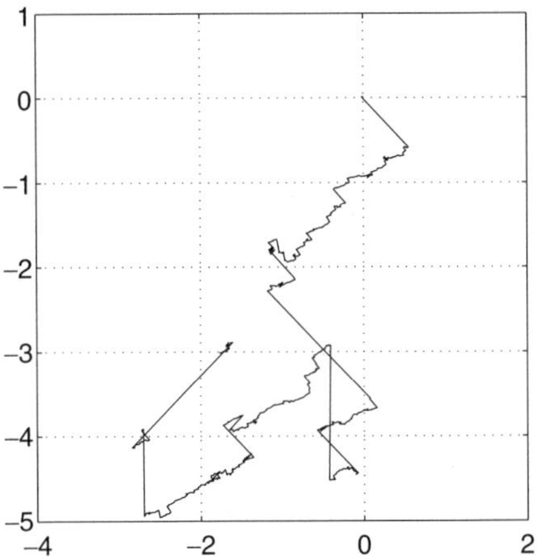


Fig. 7. Particle motion behind the multi-FADE (18) for flow in fractured rock.

that underlies the FADE. Figure 7 shows the resulting particle motion for an application to flow in fractured rock. In this example, transport model number 22 from [44], $E$ has eigenvectors at $+45°$ and $-45°$ with eigenvalues $b_1 = 1/1.1$ and $b_2 = 1/1.2$ respectively. $\Theta$ points to $\pm 45°$ with probability 0.4, and 90° with probability 0.2. The graph shows particle location in a moving coordinate system with origin at the plume center of mass. The $\Theta$ directions model fracture orientation, the $\Theta$ weights determine the proportion of transport events in each fracture direction, and the eigenvalues of $E$ determine the length of power-law particle jumps. The $\Theta = 90°$ jumps are a blend of two power laws according to $X = R^E \Theta$.

## 7. Tempered Fractional Derivatives

Tempering cools the longest jumps in a power law PDF, so that all moments exist. Tempered diffusion models transition from fractional behavior at early time to traditional diffusion at late time, a kind of transient anomalous diffusion. This transition is widely observed in practice, for example, as a basic "stylized fact" in mathematical finance [15]. For applications to geophysics, see [34]. The stable PDF $f(t)$ behind the time-fractional diffusion equation (9) has LT $\tilde{f}(s) = e^{-s^\beta}$. The exponentially tempered version $f(t)e^{-\lambda t}$ integrates to $e^{-\lambda^\beta}$ by the LT formula, so that $f_\lambda(t) = f(t)e^{-\lambda t}e^{\lambda^\beta}$ is a valid PDF on $t > 0$, called the tempered stable. For small $\lambda > 0$ it behaves like the stable PDF until $t$ is large, after which exponential tempering takes over. Compute

$$\tilde{f}_\lambda(s) = \int_0^\infty e^{-st} f(t) e^{-\lambda t} e^{\lambda^\beta} dt = e^{-[(\lambda+s)^\beta - \lambda^\beta]}$$

which reduces to the stable LT as $\lambda \to 0$. Similarly, the stable PDF with FT $\hat{p}(k,t) = \exp(t\mathcal{D}(ik)^\alpha)$ from the space-fractional diffusion equation (4) has a tempered version with FT $\hat{p}_\lambda(k,t) = \exp(t\mathcal{D}[(\lambda+ik)^\alpha - \lambda^\alpha])$. The tempered derivative $\partial_x^{\beta,\lambda} g(x)$ is the inverse FT of $[(\lambda+ik)^\alpha - \lambda^\alpha]\hat{g}(k)$. Invert this FT, using the fact that $\tilde{g}(s-\lambda)$ is the FT of $e^{\lambda t} g(t)$ (twice), to see that

$$\partial_x^{\alpha,\lambda} g(x) = e^{-\lambda x} \partial_x^\alpha (e^{\lambda x} g(x)) - \lambda^\alpha g(x).$$

The PDF $p_\lambda(x,t)$ of the tempered stable process $A_\lambda(x)$ solves the tempered fractional diffusion equation [6, 10]

$$\partial_t p(x,t) = \mathcal{D} \partial_x^{\alpha,\lambda} p(x,t) \tag{19}$$

and transitions from stable PDF to Gaussian PDF as $t \to \infty$. The random walk model behind this tempered anomalous diffusion is laid out in [11]: For each particle jump, an independent exponential with rate $\lambda$ is drawn, and the smaller of the two applies. See [34] for the tempered time-fractional diffusion model.

## Acknowledgments

M.M. Meerschaert was partially supported by NSF grants DMS-1025486, DMS-0803360, EAR-0823965 and NIH grant R01-EB012079-01.

## References

1. B. Baeumer and M. M. Meerschaert, *Frac. Calc. Appl. Anal.* **4**, 481 (2001).
2. B. Baeumer, M. M. Meerschaert and J. Mortensen, *Proc. Amer. Math. Soc.* **133**, 2273 (2005).
3. B. Baeumer, M. Kovács and M. M. Meerschaert, *Bull. Math. Biol.* **69**, 2281 (2007).
4. B. Baeumer, M. Kovács and M. M. Meerschaert, *Comput. Math. Appl.* **55**, 2212 (2008).
5. B. Baeumer, M. M. Meerschaert and E. Nane, *Trans. Amer. Math. Soc.* **361**, 3915 (2009).
6. B. Baeumer and M. M. Meerschaert, *J. Comput. Appl. Math.* **233**, 2438 (2010).
7. J. Bear, *Dynamics of Fluids in Porous Media* (Dover, 1972).

8. D. A. Benson, S. W. Wheatcraft and M. M. Meerschaert, *Water Resour. Res.* **36**, 1403 (2000).
9. D. A. Benson, R. Schumer, M. M. Meerschaert and S. W. Wheatcraft, *Transp. Porous Media* **42**, 211 (2001).
10. Á. Cartea and D. del-Castillo-Negrete, *Phys. Rev. E* **76**, 041105 (2007).
11. A. Chakrabarty and M. M. Meerschaert, *Stat. Probab. Lett.* **81**, 989 (2011).
12. P. Chakraborty, M. M. Meerschaert and C. Y. Lim, *Water Resour. Res.* **45**, W10415 (2009).
13. S. Cohen and J. Rosiński, *Bernoulli* **13**, 195 (2007).
14. S. Cohen, M. M. Meerschaert and J. Rosiński, *Stoch. Proc. Appl.* **120**, 2390 (2010).
15. R. Cont, *Quant. Finance* **1**, 223 (2001).
16. J. H. Cushman and T. R. Ginn, *Water Resour. Res.* **36**, 3763 (2000).
17. Z.-Q. Deng, L. Bengtsson and V. P. Singh, *Environ. Fluid Mech.* **6**, 451 (2006).
18. W. Feller, *An Introduction to Probability Theory and Its Applications*, Vol. II, 2nd edn. (Wiley, 1971).
19. E. Hille and R. S. Phillips, *Functional Analysis and Semi-Groups* (Amer. Math. Soc., 1957).
20. Z. Jurek and J. D. Mason, *Operator-Limit Distributions in Probability Theory* (Wiley, 1993).
21. A. N. Kochubei, *Diff. Eq.* **25**, 967 (1989).
22. M. Magdziarz, A. Weron and K. Weron, *Phys. Rev. E* **75**, 016708 (2007).
23. M. M. Meerschaert, D. A. Benson and B. Baeumer, *Phys. Rev. E* **59**, 5026 (1999).
24. M. M. Meerschaert, D. A. Benson and B. Baeumer, *Phys. Rev. E* **63**, 1112 (2001).
25. M. M. Meerschaert and H. P. Scheffler, *Limit Distributions for Sums of Independent Random Vectors: Heavy Tails in Theory and Practice* (Wiley, 2001).
26. M. M. Meerschaert and H. P. Scheffler, Portfolio modeling with heavy tailed random vectors, in *Handbook of Heavy-Tailed Distributions in Finance*, ed. S. T. Rachev (Elsevier, 2003), pp. 595–640.
27. M. M. Meerschaert and H. P. Scheffler, *J. Appl. Probab.* **41**, 623 (2004).
28. M. M. Meerschaert and C. Tadjeran, *J. Comput. Appl. Math.* **172**, 65 (2004).
29. M. M. Meerschaert and E. Scalas, *Physica A* **370**, 114 (2006).
30. M. M. Meerschaert, J. Mortensen and S. W. Wheatcraft, *Physica A* **367**, 181 (2006).
31. M. M. Meerschaert and C. Tadjeran, *Appl. Numer. Math.* **56**, 80 (2006).
32. M. M. Meerschaert, H. P. Scheffler and C. Tadjeran, *J. Comput. Phys.* **211**, 249 (2006).
33. M. M. Meerschaert, *Mathematical Modeling*, 3rd edn. (Academic Press, 2007).
34. M. M. Meerschaert, Y. Zhang and B. Baeumer, *Geophys. Res. Lett.* **35**, L17403 (2008).
35. M. M. Meerschaert, E. Nane and P. Vellaisamy, *Ann. Probab.* **37** 979 (2009).
36. M. M. Meerschaert and A. Sikorskii, *Fractional Diffusion* (de Gruyter, 2012).
37. R. Metzler and J. Klafter, *Phys. Rep.* **339**, 1 (2000).
38. R. Metzler and J. Klafter, *J. Physics A* **37**, R161 (2004).

39. E. W. Montroll and G. H. Weiss, *J. Math. Phys.* **6**, 167 (1965).
40. R. Parashar and J. H. Cushman, *J. Comput. Phys.* **227**, 6598 (2008).
41. M. Park and J. H. Cushman, *J. Comput. Phys.* **217**, 159 (2006).
42. M. Park and J. H. Cushman, *J. Statist. Mech.* **2009**, P02039 (2009).
43. A. Piryatinska, A. I. Saichev and W. A. Woyczynski, *Physica A* **349**, 375 (2005).
44. D. M. Reeves, D. A. Benson, M. M. Meerschaert and H. P. Scheffler, *Water Resour. Res.* **44**, W05410 (2008).
45. G. Samorodnitsky and M. Taqqu, *Stable non-Gaussian Random Processes.* (Chapman and Hall, 1994).
46. H. Scher and M. Lax, *Phys. Rev. B* **7**, 4491 (1973).
47. R. Schumer, D. A. Benson, M. M. Meerschaert and B. Baeumer, *Water Resour. Res.* **39**, 1022 (2003).
48. C. Tadjeran, M. M. Meerschaert and H. P. Scheffler, *J. Comput. Phys.* **213**, 205 (2006).
49. C. Tadjeran and M. M. Meerschaert, *J. Comput. Phys.* **220**, 813 (2007).
50. V. E. Tarasov, *Ann. Phys.* **323**, 2756 (2008).
51. S. J. Taylor, *Math. Proc. Cambridge Philos. Soc.* **100**, 383 (1986).
52. Y. Zhang, D. A. Benson, M. M. Meerschaert, E. M. LaBolle and H. P. Scheffler, *Phys. Rev. E* **74**, 6706 (2006).

# Chapter 12

## Fractional Langevin Equation

Eric Lutz
*Department of Physics, University of Augsburg,*
*D-86135 Augsburg, Germany*

The fractional Langevin equation is an essential tool for the modeling of anomalous diffusion. We here summarize its main physical properties, focusing on long-range temporal correlations, power-law relaxation and aging. We discuss in addition its critical exponents and ergodic behavior.

1. Introduction . . . . . . . . . . . . . . . . . . . . . . . . 285
2. Langevin Equation . . . . . . . . . . . . . . . . . . . . . 287
   2.1. Solution for the free particle . . . . . . . . . . . . . 288
   2.2. Velocity correlation function . . . . . . . . . . . . . 290
   2.3. Mean-square displacement . . . . . . . . . . . . . . 290
3. Fractional Langevin Equation . . . . . . . . . . . . . . . . 291
   3.1. Memory effects . . . . . . . . . . . . . . . . . . . 292
   3.2. Solution for the free particle . . . . . . . . . . . . . 294
   3.3. Velocity correlation function and mean-square displacement . . . . . . . . . . . . . . 295
4. Experimental Observation of Fractional Brownian Motion . . . . . . . . . . . . . . . . . . . . . . . 296
5. Aging . . . . . . . . . . . . . . . . . . . . . . . . . . . 298
6. Overdamped Regime and Critical Exponents . . . . . . . 299
7. Ergodicity . . . . . . . . . . . . . . . . . . . . . . . . . 302

### 1. Introduction

Fluctuation phenomena are ubiquitous in nature. Consider a small particle moving in a viscous fluid, for example a micron-size latex bead in water or a silica particle in a rarefied gas. According to the laws of mechanics, one would expect that the particle comes to a rest, after some relaxation time, due to the action of frictional forces. Yet, experiments show that the particle keeps jiggling around in an erratic manner and exhibits diffusion

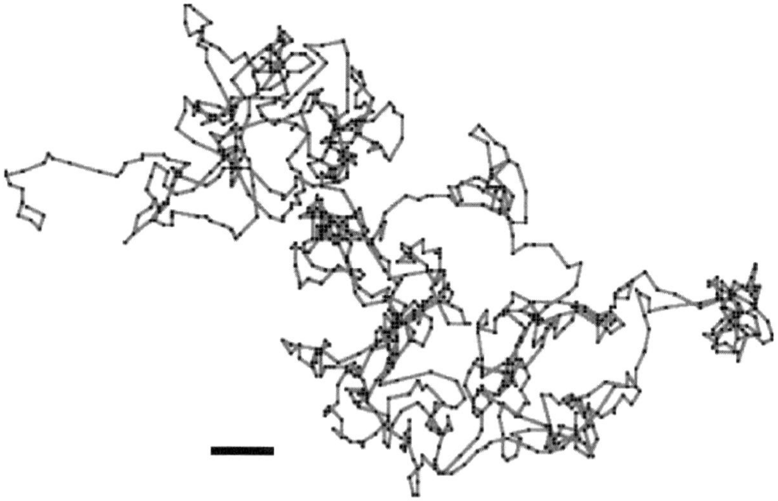

Fig. 1. Brownian trajectory of a single $SiO_2$ sphere of radius 0.8 $\mu$m suspended within rarefied air under microgravity conditions. The trajectory consists of 1024 measurements (black points) and spans a total of 2.2 s. The scale bar indicates 10 $\mu$m (reprinted figure 4 with permission from J. Blum, S. Bruns, D. Rademacher, A. Voss, B. Willenberg and M. Krause, *Phys. Rev. Lett.* **97**, 230601 (2006). © 2006 by the American Physical Society).

(see Fig. 1). While the physical origin of these position fluctuations was mysterious at the time of their first detailed investigation by Robert Brown in 1827 [1], we now know that they are induced by the irregular collisions of the many molecules of the surrounding fluid. As surmised by Albert Einstein in 1905, the observation of the irregular (Brownian) motion of the particle reveals the molecular nature of the fluid and its thermal agitation [2]. While extremely small, and therefore negligible, for macroscopic objects, fluctuation effects dominate at the microscopic scale, and are hence particularly relevant at the level of a biological cell. Mathematically, the dynamics of the small (Brownian) particle can be described, following Einstein, by a diffusion (or more generally a Fokker–Planck) equation for the probability density of the position of the particle [3]. An alternative approach, put forward by Paul Langevin in 1908, makes use of a generalization of Newton's equation of motion [4]: the Langevin equation [5]. Here, a new stochastic force is introduced that takes explicit account of the thermal fluctuations. The dynamical properties of a Brownian particle are strongly influenced by the presence of temporal memory. For vanishing, or short-range memory, the mean-square displacement of a free particle asymptotically grows linearly in time; the latter corresponds to normal diffusion as originally observed by Brown. By contrast, long-range, power-law memory results in anomalous diffusion with an algebraic mean-square displacement. Anomalous diffusion is frequently observed in crowded fluids, for instance in the cytoplasm of living cells [6]. Fractional calculus is a powerful tool to investigate physical or biological systems where these long-range temporal correlations occur [7]. In this approach, time derivatives in evolution equations, such as the Fokker–Planck or Langevin equations, are replaced by their non-integer, fractional extensions, in order to describe the effects of power-law memory.

Our aim in this chapter is to review the main properties of the fractional Langevin equation. After a short introduction to the physics of the ordinary Langevin equation in Sec. 2, and a general discussion of memory effects, we solve the free particle fractional Langevin equation analytically and determine both the velocity correlation function and the mean-square displacement in Sec. 3. In Sec. 4, we present experimental observations of fractional Brownian motion in living cells and in the dynamics of single proteins. Sections 5 and 6 are devoted to the aging properties of nonstationary correlation functions and to the dynamical phase diagram of the fractional Langevin equation. Section 7 finally concludes with a few words about the ergodic behavior of fractional Brownian motion.

## 2. Langevin Equation

We again consider a small particle in a viscous fluid, which we model as a heat bath at temperature $T$. Following Langevin, we describe the time evolution of the Brownian particle with Newton's equation of motion for its position coordinate $x$ (for simplicity, we restrict ourselves to one dimension),

$$m\ddot{x} + U'(x) = \xi(t). \tag{1}$$

Here $m$ is the mass of the particle, $U(x)$ the external, confining potential and $\xi(t)$ the total force which results from the interaction with the surrounding medium. Langevin's insight was to recognize that $\xi(t)$ can be quite generically written as the sum of an average viscous force proportional to the velocity of the particle, $-\gamma\dot{x}$ ($\gamma$ being the friction coefficient), and a rapidly varying fluctuation force $F(t)$ that accounts for the very frequent impacts of the molecules of the fluid on the particle,

$$\xi(t) = -\gamma\dot{x} + F(t). \tag{2}$$

Due to the random character of the incessant molecular bombardment, $F(t)$ is a stochastic force that takes on probabilistic values at any fixed time $t$.

It is assumed that it can be fully characterized by its vanishing mean value and its $\delta$ correlation function,

$$\langle F(t)\rangle = 0, \tag{3}$$

$$\langle F(t)F(t')\rangle = 2D\,\delta(t-t'). \tag{4}$$

The average $\langle \dots \rangle$ is here taken over a large number of realizations of the same process. The second assumption expresses the physical idea that individual collisions occur very rapidly and are therefore independent of each other: the forces $F(t)$ at two different times are completely uncorrelated (no temporal memory). The parameter $D$ measures the strength of the fluctuating force; we will see below that it is determined by the temperature of the fluid and that it is actually related to the diffusion coefficient. The Langevin equation for the Brownian particle then takes the form [3, 5],

$$m\ddot{x} + \gamma\dot{x} + U'(x) = F(t). \tag{5}$$

It represents a stochastic generalization of Newton's equation of motion, familiar from mechanics. In order to develop an intuition on how to handle the random force $F(t)$, we will next solve the Langevin equation for the simple case of a free particle.

### 2.1. *Solution for the free particle*

In the absence of a confining potential, $U(x) = 0$, the Langevin equation (5) can be rewritten in terms of the velocity $v = \dot{x}$ alone as

$$m\dot{v} + \gamma v = F(t). \tag{6}$$

This is a stochastic differential equation of first order, which can be easily solved formally. The physical significance of the fluctuation force can be best appreciated by first setting it to zero.

*a. No noise* $F(t) = 0$

The solution of the linear equation $m\dot{v} = -\gamma v$, with initial value $v(0)$, is

$$v(t) = e^{-\gamma t/m} v(0). \tag{7}$$

As a consequence, $v(t)$, and hence $v^2(t)$, relax exponentially to zero as time increases. This, however, is in contradiction with our previous assumption that the surrounding fluid is in thermal equilibrium. Indeed, according to statistical mechanics, for a system in thermal equilibrium equipartition holds on average, and accordingly $\langle v^2 \rangle_{\text{eq}} = kT/m$, where $k$ is the Boltzmann constant. This simple exercise leads to the important conclusion that, in order to have a consistent description of our Brownian system, both friction and fluctuations have to be properly included.

*b. With noise* $F(t) \neq 0$

The solution of the inhomogeneous first-order linear differential equation $m\dot{v} + \gamma v = F(t)$ reads

$$v(t) = e^{-\gamma t/m} v(0) + \frac{1}{m} \int_0^t dt' e^{-\gamma(t-t')/m} F(t'). \tag{8}$$

By noting that the stochastic force vanishes on average, $\langle F(t) \rangle = 0$, we find that the mean particle velocity obeys

$$\langle v(t) \rangle = e^{-\gamma t/m} v(0), \tag{9}$$

in agreement with Eq. (7). The mean-square velocity, however, is non-zero. By squaring Eq. (8) and taking the average, we obtain

$$\begin{aligned}\langle v^2(t) \rangle &= e^{-2\gamma t/m} v^2(0) \\ &\quad + \frac{1}{m^2} \int_0^t \int_0^t dt' dt'' e^{-\gamma(t-t')/m} e^{-\gamma(t-t'')/m} \langle F(t') F(t'') \rangle \end{aligned} \tag{10}$$

since the terms linear in $F(t)$ again vanish. The double integral on the R.H.S. can be evaluated explicitly by inserting the $\delta$ correlation function (4), and

$$\langle v^2(t) \rangle = e^{-2\gamma t/m} v^2(0) + \frac{D}{\gamma m} (1 - e^{-2\gamma t/m}). \tag{11}$$

We observe that the mean kinetic energy of the particle, $E_K = m\langle v^2 \rangle/2$, decreases because of friction, as described by the first term on the R.H.S. of Eq. (11). However, the second term shows that it also increases through the action of the fluctuation force. In the long-time limit, the mean-square velocity reaches the stationary value $\langle v^2 \rangle_{\text{eq}} = D/\gamma m$. Equipartition allows us to determine the value of the parameter $D = \gamma kT$; the strength of the stochastic force is thus set by the temperature. This is an example of a fluctuation-dissipation relation which indicates that friction (dissipation, $\gamma$) is necessarily accompanied by noise (fluctuation, $D$). The physical explanation of such a relation is the common origin of both forces in the Langevin equation, namely the interaction with the surrounding fluid.

### 2.2. *Velocity correlation function*

Average quantities, like $\langle v(t)\rangle$ and $\langle v^2(t)\rangle$, provide important statistical information about a stochastic process at a given instant (they are one-time functions), however they contain no information about how much a process influences its time evolution at a later time. This kind of information is provided by correlation functions (these are two-time functions). In this section, we compute the important stationary velocity correlation function $C_v(t) = \langle v(t)v(0)\rangle_{\text{eq}}$. We start with Eq. (8) for the particle velocity which we multiply by $v(0)$. By averaging both over the fluctuating force and the initial velocity, assuming that the particle is initially at thermal equilibrium, that is $\langle v^2(0)\rangle_{\text{eq}} = kT/m$ and $\langle v(0)F(t)\rangle_{\text{eq}} = 0\rangle$, we obtain

$$C_v(t) = \langle v(t)v(0)\rangle_{\text{eq}} = \frac{kT}{m}e^{-\gamma t/m}. \tag{12}$$

Equation (12) shows that velocities are correlated in time, although the stochastic forces $F(t)$ are not. The velocity correlations decay exponentially within a typical correlation time that is inversely proportional to the friction constant, $\tau = m/\gamma$. It is worth mentioning that the equilibrium velocity correlation function can also be calculated by multiplying Eq. (8) by itself at a different time, yielding $\langle v(t)v(t')\rangle_{\text{eq}} = (kT/m)\exp(-\gamma|t-t'|/m)$.

### 2.3. *Mean-square displacement*

A central quantity in the theory of Brownian motion is the mean-square displacement $\langle \Delta x^2(t)\rangle$, already calculated by Einstein in 1905. Its importance stems from the fact that the diffusion coefficient $D_x$ can be directly determined experimentally from it. The mean-square displacement can be computed from the velocity correlation function (12), by noticing that

$$\Delta x(t) = x(t) - x(0) = \int_0^t dt' v(t'). \tag{13}$$

As a result, we find,

$$\begin{aligned}\langle \Delta x^2(t)\rangle &= \int_0^t dt' \int_0^t dt'' \langle v(t')v(t'')\rangle_{\text{eq}} = 2\int_0^t dt' \int_0^{t'} dt'' \langle v(t')v(t'')\rangle_{\text{eq}} \\ &= 2\frac{kT}{\gamma}\left(t - \frac{m}{\gamma} + \frac{m}{\gamma}e^{-\gamma t/m}\right). \end{aligned} \tag{14}$$

At short times, the mean-square displacement grows ballistically, $\langle \Delta x^2(t)\rangle \simeq (kT/m)t^2$. At these short time scales the effect of the molecular collisions

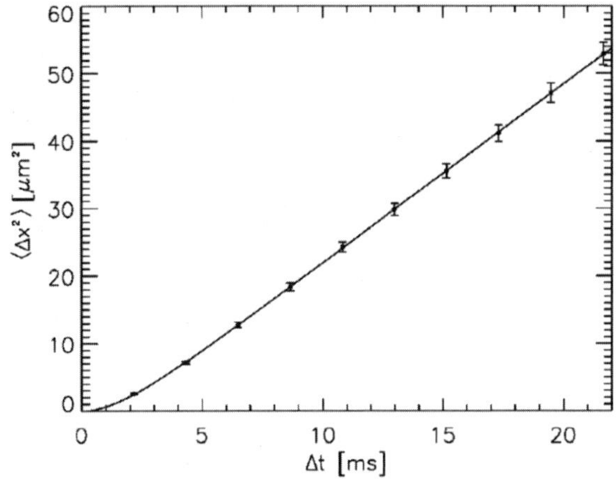


Fig. 2. Mean-square displacement $\langle \Delta x^2(t) \rangle$ of the silica particle of Fig. 1 as a function of the time interval between measurements. The solid line represents the theoretical expression (14) (reprinted figure 2 with permission from J. Blum, S. Bruns, D. Rademacher, A. Voss, B. Willenberg and M. Krause, *Phys. Rev. Lett.* **97**, 230601 (2006). © 2006 by the American Physical Society).

is negligible and the particle behaves as in free motion. By contrast, at long times, when molecular bombardment is significant, the mean-square displacement grows diffusively, $\langle \Delta x^2(t) \rangle \simeq 2D_x t$, where $D_x = kT/\gamma = D/\gamma^2$ is the diffusion coefficient in position (see Fig. 2). This asymptotic linear behavior of the mean-square displacement is the hallmark of normal, memory-free diffusion.

## 3. Fractional Langevin Equation

The phenomenological Langevin equation (5) can be derived microscopically by linearly coupling the Brownian particle to a bath of harmonic oscillators (Kac–Zwanzig model). The resulting generalized Langevin equation is then found to be of the form [9],

$$m\ddot{x}(t) + \int_0^t dt' \gamma(t-t')\dot{x}(t') + U'(x) = F(t), \tag{15}$$

where $\langle F(t) \rangle = 0$ and $\langle F(t)F(t') \rangle = kT\gamma(t-t')$. The latter is a general expression of the fluctuation-dissipation relation between the noise correlation function $\langle F(t)F(t') \rangle$ and the friction kernel $\gamma(t-t')$. In contrast to the memory-free Langevin equation (4), the noise correlation function needs not necessarily be a delta function, allowing for memory effects to play a role. We note, in addition, that the friction term is no longer local in time. The value of the friction force at time $t$ indeed depends on all the past values of the velocity $\dot{x}(t')$, $t' < t$. A general feature of memory effects is thus that the past influences the present. We shall establish in this section that they can profoundly affect the diffusion properties of a Brownian particle. A general measure for the estimation of memory effects in systems described by the generalized Langevin equation (15) has been introduced in [10], it is obtained by comparing the time scales of the decay of the velocity correlation function and its corresponding memory kernel.

### 3.1. *Memory effects*

We begin with an elementary discussion of memory effects [11]. The key function to characterize the latter is the friction kernel $\gamma(t)$ that quantifies how strongly the present depends on the past. The generalized Langevin equation (15) is a stochastic integro-differential equation and is therefore most conveniently analyzed in Laplace space. As in the previous section, we focus on the free particle case $U(x) = 0$. We denote the Laplace transform of a given function $f(t)$ by,

$$\mathcal{L}[f(t)] = \tilde{f}(s) = \int_0^\infty dt\, e^{-st} f(t). \tag{16}$$

The friction term is a convolution, its Laplace transform is thus simply a product of the Laplace transforms of the friction kernel and of the velocity,

$$\mathcal{L}\left[\int_0^t dt' \gamma(t-t') v(t')\right] = \tilde{\gamma}(s)\tilde{v}(s). \tag{17}$$

We may now distinguish three main cases.

*a. No memory*

The friction kernel is a delta function and its Laplace transform a constant,

$$\gamma(t) = \gamma\delta(t) \quad \longrightarrow \quad \tilde{\gamma}(s) = \gamma = \frac{\gamma}{s^0}. \tag{18}$$

The friction force is local in time $\sim\gamma v(t)$.

*b. Full memory*

The friction kernel is a constant and its Laplace transform is given by $1/s$,

$$\gamma(t) = \gamma \quad \longrightarrow \quad \tilde{\gamma}(s) = \frac{\gamma}{s} = \frac{\gamma}{s^1}. \tag{19}$$

The friction force is the integral of all the velocities over time $\sim\gamma \int_0^t dt' v(t')$.

*c. Long-range memory*

The third case is a natural interpolation between the two extreme cases above. The Laplace transform is a power-law $1/s^\mu$, $(0 < \mu < 1)$, and the friction kernel is also a power-law ($\Gamma(x)$ denotes the Gamma function),

$$\gamma(t) = \frac{\gamma}{\Gamma(\mu) t^{\mu-1}} \qquad \longleftarrow \qquad \tilde{\gamma}(s) = \frac{\gamma}{s^\mu}. \tag{20}$$

The distinctive feature of a power-law is the absence of any characteristic time scale. An algebraic memory friction kernel hence results in relaxation without a typical relaxation or memory time. A simple example will illustrate this point. An exponential kernel, $\gamma_{\text{exp}}(t) = \gamma \exp(-t/\tau)$, has a natural time scale set by the relaxation time $\tau$: for times much larger than $\tau$, the kernel relaxes to zero. As a result, its shape is strongly affected by a rescaling of the time variable $\tau \to \kappa\tau$,

$$\gamma_{\text{exp}}(\kappa t) \simeq 0 \quad \text{when } \kappa \gg \tau. \tag{21}$$

By contrast, the shape of a power-law kernel is unmodified by time dilatation since its time dependence is simply rescaled,

$$\gamma_{\text{pow}}(\kappa t) = \frac{\gamma}{\Gamma(\mu)} \frac{1}{\kappa^{\mu-1} t^{\mu-1}} = \frac{1}{\kappa^{\mu-1}} \gamma_{\text{pow}}(t). \tag{22}$$

The absence of a time scale for algebraic friction kernels, and the resulting scale invariance, lead to a temporal fractal structure of the stochastic process. Brownian motion with long-range, power-law memory was therefore coined fractional Brownian motion by Mandelbrot and van Ness [12].

In the following, we consider the generalized Langevin equation (15) with the power-law memory kernel $\gamma(t) = \gamma/\Gamma(1-\alpha)\, t^{-\alpha}$. By using the definition of the Riemann–Liouville fractional derivative $(-1 < \lambda < 0)$,

$$\frac{\partial^\lambda f(t)}{\partial t^\lambda} = \frac{1}{\Gamma(-\lambda)} \int_0^t dt' \frac{f(t')}{(t-t')^{\lambda+1}}, \tag{23}$$

we can rewrite it in the form of a fractional Langevin equation [13],

$$m\ddot{x} + m\gamma_\alpha \frac{\partial^{\alpha-1}}{\partial t^{\alpha-1}} \dot{x}(t) + U'(x) = F(t), \tag{24}$$

where we have defined $\gamma_\alpha = \gamma/(m\Gamma(1-\alpha))$. Fractional Langevin equations have recently been derived from various microscopic models: in the first example, fractional Brownian motion has been shown to arise for a probe particle moving in elastic media, such as membranes and semi-flexible polymers [14]. The second study, on the other hand, has obtained a fractional

Langevin equation for the dynamics of a tracer particle in a one-dimensional many-particle system with a general two-body interaction potential [15].

### 3.2. *Solution for the free particle*

We shall now investigate the properties of the fractional Langevin equation by examining the case of the free particle $U(x) = 0$. Its Laplace transform can be computed by applying the well-known identities,

$$\mathcal{L}[\dot{x}(t)] = s\tilde{x}(s) - x(0), \tag{25}$$

$$\mathcal{L}[\ddot{x}(t)] = s^2\tilde{x}(s) - sx(0) - v(0). \tag{26}$$

We easily obtain,

$$\tilde{x}(s)\left[s^2 + \frac{1}{m}\tilde{\gamma}(s)s\right] = \left[s + \frac{1}{m}\tilde{\gamma}(s)\right]x(0) + v(0) + \frac{1}{m}\tilde{F}(s). \tag{27}$$

The solution of the fractional Langevin equation in time space is then

$$x(t) = x(0) + v(0)\mathcal{B}(t) + \int_0^t dt'\mathcal{B}(t-t')\frac{F(t')}{m}, \tag{28}$$

where we have introduced the function $\tilde{\mathcal{B}}(s) = 1/[s^2 + \tilde{\gamma}(s)s/m] = 1/[s^2 + \gamma_\alpha s^\alpha]$, since $\tilde{\gamma}(s) = m\gamma_\alpha s^{\alpha-1}$. After time derivation, we find that the velocity of the Brownian particle is given by,

$$\dot{x}(t) = v(t) = v(0)\mathcal{C}(t) + \int_0^t dt'\mathcal{C}(t-t')\frac{F(t')}{m}, \tag{29}$$

with $\mathcal{C}(t) = \dot{\mathcal{B}}(t)$ and $\tilde{\mathcal{C}}(s) = 1/[s + \gamma_\alpha s^{\alpha-1}]$. The inverse Laplace transform of $\tilde{\mathcal{C}}(s)$ can be expressed with the help of the Mittag-Leffler function. The Mittag-Leffler function $E_\nu[t]$ is defined by the series expansion [16]

$$E_\nu[t] = \sum_{n=0}^{\infty} \frac{t^n}{\Gamma(\nu n + 1)}. \tag{30}$$

It is a simple generalization of the exponential function to which it reduces when $\nu = 1$. However, the Mittag-Leffler function does not decay exponentially; its asymptotic behavior is given by a stretched exponential at short times and by a power-law at long times,

$$E_\nu[-\lambda_\nu t^\nu] \simeq e^{-\lambda_\nu t^\nu} \qquad \text{for } t \ll 1/\lambda_\nu^{1/\nu}, \tag{31}$$

$$E_\nu[-\lambda_\nu t^\nu] \simeq \frac{t^{-\nu}}{\lambda\Gamma(1-\nu)} \qquad \text{for } t \gg 1/\lambda_\nu^{1/\nu}. \tag{32}$$

Its Laplace transform is given by

$$\mathcal{L}[E_\nu[-\lambda_\nu t^\nu]] = \frac{1}{s + \lambda_\nu s^{1-\nu}}. \tag{33}$$

The functions $\mathcal{C}(t)$ and $\mathcal{B}(t)$ read accordingly,

$$\mathcal{C}(t) = E_{2-\alpha}[-\gamma_\alpha t^{2-\alpha}] \simeq \frac{t^{2-\alpha}}{\gamma_\alpha \Gamma(\alpha - 1)} \quad \text{for large } t, \tag{34}$$

$$\mathcal{B}(t) = tE_{2-\alpha,2}[-\gamma_\alpha t^{2-\alpha}] \simeq \frac{t^{\alpha-1}}{\gamma_\alpha \Gamma(\alpha)} \quad \text{for large } t, \tag{35}$$

where $E_{\nu,\mu} = \sum_n t^n / \Gamma(\nu n + \mu)$ is the generalized Mittag-Leffler function. We stress that the solution (28) of the fractional Mittag-Leffler function is completely specified by the function $\mathcal{B}(t)$.

### 3.3. *Velocity correlation function and mean-square displacement*

Equation (29) has the same form as Eq. (8) derived from the ordinary Langevin equation. The velocity correlation function in the fractional case can be obtained in a similar manner as before by multiplying Eq. (29) by $v(0)$ and averaging over the initial velocity and the fluctuating force:

$$C_v(t) = \langle v(t)v(0)\rangle = \langle v^2(0)\rangle \mathcal{C}(t) = \frac{kT}{m} E_{2-\alpha}[-\gamma_\alpha t^{2-\alpha}]. \tag{36}$$

The mean-square displacement follows immediately as

$$\langle \Delta x^2(t)\rangle = \frac{kT}{m} \int_0^t dt_1 \int_0^t dt_2 \mathcal{C}(t_1 - t_2) \simeq \frac{2kT}{\gamma_\alpha m} \frac{t^\alpha}{\Gamma(1+\alpha)}, \quad t \longrightarrow \infty. \tag{37}$$

The long-range memory of fractional Brownian motion thus expresses itself through slowly decaying, power-law correlations in the velocities. These in turn influence the long-time behavior of the mean-square displacement, which displays the typical anomalous asymptotic behavior $\langle \Delta x^2\rangle \sim t^\alpha$. The following physical picture thus emerges from our analysis [17]:

- For $\alpha = 1$, the noise correlation function is delta correlated (white noise - no memory), the velocity correlation function decays exponentially in time, $C_v(t) \sim \exp(-\gamma t/m)$ and vanishes at long times.

Consequently, the mean-square displacement is asymptotically linear in time, $\langle \Delta x^2(t) \rangle \sim t$, and diffusion is normal.

- For $\alpha < 1$: the noise correlation function is a power-law (colored noise - long range memory), the velocity correlation function displays a *negative* slow decay, $C_v(t) < 0$ at long times, since $\Gamma(\alpha - 1) < 0$. The latter indicates the presence of antipersistent motion (the Brownian particle constantly changes its direction). As a result, the mean-square displacement exhibits an algebraic behavior, $\langle \Delta x^2(t) \rangle \sim t^\alpha$, corresponding to subdiffusion.
- For $\alpha > 1$, the velocity correlation function displays a *positive* slow decay, $C_v(t) > 0$ at long times, since $\Gamma(\alpha - 1) > 0$. The latter indicates the presence of persistent motion (the Brownian particle moves predominantly in the same direction). As a consequence, the mean-square displacement increases asymptotically faster than linear, $\langle \Delta x^2(t) \rangle \sim t^\alpha$, corresponding to superdiffusion.

## 4. Experimental Observation of Fractional Brownian Motion

Long-range memory effects play a crucial role in diffusive processes that take place in crowded fluids, for instance the cytoplasm of a living cell, for which many different experiments have reported the observation of subdiffusion with an algebraic mean-square displacement. Examples include the motion of a polystyrene microsphere in a fibroplast [18], of a lipid granule in a yeast cell [19], and of a telomere in the nucleus of mammallian cell [20]. While the mean-square displacement alone does not suffice to completely characterize the nature of the diffusion process, a recent theoretical study has shown that the motion of mRNA molecules inside live *E. coli* cells investigated by Golding and Cox [21] (see Fig. 3) is likely to be fractional Brownian motion [22]. Moreover, two additional *in vivo* experiments have claimed the direct observation of fractional Brownian motion for particles in the viscoelastic cytoplasm [23, 24].

The harmonic fractional Langevin equation has also been successfully applied in single molecule experiments to describe the distance fluctuations between an electron donor and acceptor pair within the molecule [25, 26]. Here both the power-law friction kernel and the Mittag-Leffler decay of the position correlation function have been measured (see Fig. 4). All these experiments highlight the importance of the fractional Langevin equation (24) for the modeling of anomalous diffusion in biological systems.

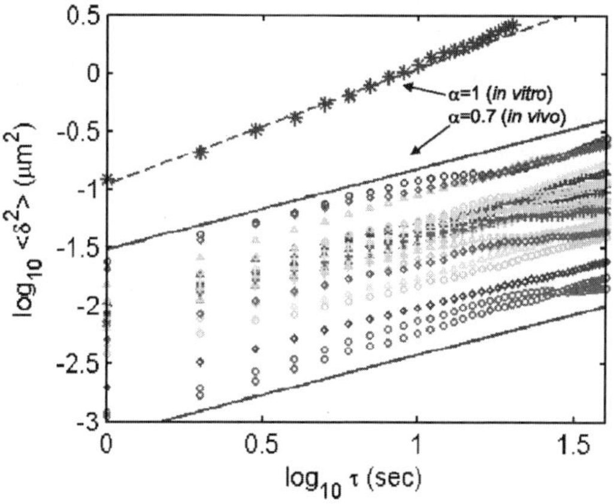


Fig. 3. Mean-square displacement of a mRNA molecule in a live *E. Coli* cell as a function of the time-interval between measurements. Shown are 23 trajectories from three different experiments; they exhibit subdiffusion with $\alpha = 0.7$. Also plotted is the mean-square displacement for the molecule diffusing *in vitro* in 70% glycerol; the latter corresponds to normal diffusion, $\alpha = 1$ (reprinted figure 2 with permission from I. Golding and E. C. Cox, *Phys. Rev. Lett.* **96**, 098102 (2006). © 2006 by the American Physical Society).

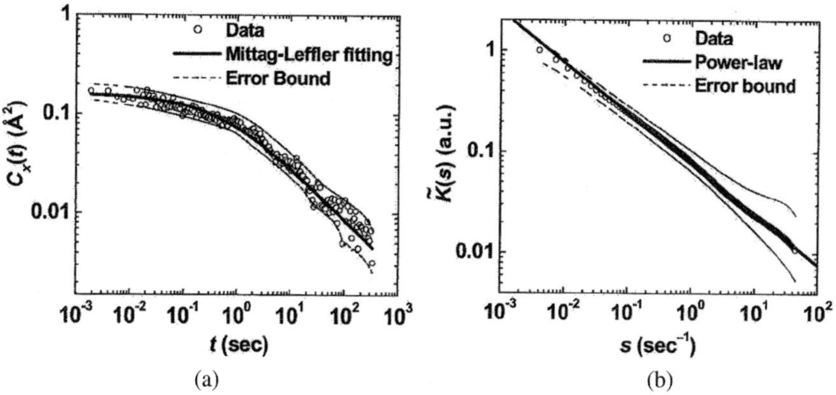


Fig. 4. (a) Mittag-Leffler decay for the correlation function of distance fluctuation $\langle x(t)x(0)\rangle$ between a donor-acceptor pair in a fluorescein/anti-fluorescein molecule complex, averaged over 13 molecules under the same experimental condition, $\alpha \simeq 0.5$. (b) Laplace transform, $\tilde{K}(s) = \tilde{\gamma}(s)$, of the corresponding friction kernel (reprinted figures 4 and 5 with permission from W. Min, G. Lou, B. J. Cherayil, S. C. Kou and X. S. Xie, *Phys. Rev. Lett.* **94**, 198302 (2005). © 2005 by the American Physical Society).

## 5. Aging

In this section, we evaluate the nonstationary two-time correlation function of the velocity and the position of a free fractional Brownian particle. We will show that they generally not only depend on the time difference $t-t'$, but also explicitly on the two times $t$ and $t'$. As a result, the behavior of the correlation functions depends on the particular observation time: the behavior at a later time will be different from the current behavior. The latter property is referred to as aging [27, 28]. We will again work in Laplace space. The double Laplace transform of any stationary correlation function of the form $\langle\Phi(t)\Phi(t')\rangle = Af(|t-t'|)$ can be shown to be given by [27, 28],

$$\langle\tilde{\Phi}(s)\tilde{\Phi}(s')\rangle = A\frac{\tilde{f}(s)+\tilde{f}(s')}{s+s'}. \tag{38}$$

Starting from Eq. (29), one then finds that the double Laplace transform of the two-time velocity correlation function $\langle v(t)v(t')\rangle$ is,

$$\langle\tilde{v}(s)\tilde{v}(s')\rangle = \frac{kT}{m}\frac{\tilde{C}(s)+\tilde{C}(s')}{s+s'} + \left(v^2(0)-\frac{kT}{m}\right)\tilde{C}(s)\tilde{C}(s'). \tag{39}$$

Noting that $\tilde{x}(s) = \tilde{v}(s)/s$, the Laplace transform of the two-time position correlation function $\langle x(t)x(t')\rangle$ is accordingly,

$$\langle\tilde{x}(s)\tilde{x}(s')\rangle = \frac{kT}{m}\frac{1}{s}\frac{1}{s'}\frac{\tilde{C}(s)+\tilde{C}(s')}{s+s'}. \tag{40}$$

Since the correlation functions (39) and (40) are not of the form (38), they will depend separately on $t$ and $t'$ in the time domain, and therefore exhibit aging. They are explicitly given by

$$\begin{aligned}\langle v(t)v(t')\rangle &= \frac{kT}{m}E_{2-\alpha}[-\gamma_\alpha|t-t'|^{2-\alpha}]\\ &\quad+\left(v^2(0)-\frac{kT}{m}\right)E_{2-\alpha}[-\gamma_\alpha t^{2-\alpha}]E_{2-\alpha}[-\gamma_\alpha t'^{2-\alpha}],\end{aligned} \tag{41}$$

$$\begin{aligned}\langle x(t)x(t')\rangle &= \frac{kT}{m}(t^2E_{2-\alpha,3}[-\gamma_\alpha t^{2-\alpha}]+t'^2E_{2-\alpha,3}[-\gamma_\alpha t'^{2-\alpha}]\\ &\quad-(t-t')^2E_{2-\alpha,3}[-\gamma_\alpha(t-t')^{2-\alpha}]),\quad t'<t.\end{aligned} \tag{42}$$

We observe that while the velocity correlation function displays aging only for non-thermal initial conditions, $v^2(0)\neq kT/m$, the position of the fractional Brownian particle always ages in a nonexponential manner.

## 6. Overdamped Regime and Critical Exponents

In many cases of interest, friction is large so that the inertial term in the ordinary Langevin equation, $m\ddot{x}\,(\ll \gamma\dot{x})$ can be neglected for times larger than the velocity relaxation time, $t \gg m/\gamma$. In this limit, velocities have reached their equilibrium values and the Langevin equation (5) reduces to a first-order stochastic differential equation,

$$\gamma\dot{x} + U'(x) = F(t). \tag{43}$$

In a similar way, an overdamped fractional Langevin equation can be written down by setting the inertial term to zero in Eq. (24),

$$m\gamma_\alpha \frac{\partial^{\alpha-1}}{\partial t^{\alpha-1}}\dot{x}(t) + U'(x) = F(t). \tag{44}$$

However, the absence of a well-defined relaxation time in the fractional case suggests that the overdamped approximation (44) should be used with care. It might not always be justified, as we are now going to show.

We consider the fractional Langevin equation (24) for a harmonically bound particle $U(x) = m\omega^2x^2/2$. Before tackling this complicated equation, it is useful to begin with a brief reminder of the properties of the simple damped harmonic oscillator (without thermal fluctuations) with equation:

$$m\ddot{x} + \gamma\dot{x} + m\omega^2 x = 0. \tag{45}$$

The dynamics of the damped oscillator is determined by the solutions of the characteristic equation, $m\lambda^2 + \gamma\lambda + m\omega^2 = 0$. For frequencies smaller than the critical frequency, $\omega < \omega_c = \gamma/(2m)$ (overdamped regime characterized by two real solutions), relaxation is monotonic, while for frequencies larger than the critical frequency, $\omega > \omega_c$ (underdamped regime characterized by two complex solutions), relaxation is oscillatory.

To determine the critical frequency for the fractional oscillator, we proceed in the following way [29, 30]. We consider the normalized position correlation function defined by,

$$C_x(t) = \frac{\langle x(t)x(0)\rangle}{\langle x^2(0)\rangle}. \tag{46}$$

Using the solution of the fractional Langevin equation (28), with the additional contribution coming from the harmonic potential, the Laplace transform $\tilde{C}_x(s)$ can be written, for $\alpha = p/q$ with $p, q$ positive and irreducible, in the form (with $\bar{\gamma} = \gamma/m$),

$$\tilde{C}_x(s) = \frac{s + \bar{\gamma} s^{\alpha-1}}{s^2 + \bar{\gamma} s^{\alpha} + \omega^2} = \frac{(s + \bar{\gamma} s^{(p/q)-1}) + \tilde{Q}(s)}{\tilde{P}(s)}, \tag{47}$$

where we have introduced the two functions,

$$\tilde{Q}(s) = \frac{(s^2+\omega^2)^q + (-1)^{q-1}\bar{\gamma}^q s^p}{s^2 + \bar{\gamma} s^{(p/q)} + \omega^2}, \tag{48}$$

$$\tilde{P}(s) = (s^2+\omega^2)^q + (-1)^{q-1}\bar{\gamma}^q s^p. \tag{49}$$

In the normal case $\alpha = 1$, the critical frequency $\omega_c$ is found, when two zeros of $\tilde{P}(s)$ coincide. The two conditions, $\tilde{P}(s) = 0$ and $d\tilde{P}(s)/ds = 0$, then lead to the expression in the anomalous case $0 < \alpha < 1$,

$$\omega_c = \frac{1}{2^{1/(2-\alpha)}} \sqrt{(2-\alpha)\alpha^{\alpha/(2-\alpha)}\bar{\gamma}^{1/(2-\alpha)}}, \tag{50}$$

for even $q$ or even $(q+p)$; for odd $q$ and even $p$, the critical frequency $\omega_c$ does not exist. However, two additional critical frequencies, $\omega_z$ and $\omega_m$, can be defined in the fractional regime.

By determining the $2q$ zeros of the equation $\tilde{P}(a_k) = 0$ $(k = 1, \ldots, 2q)$, the poles of Eq. (47), one can explicitly compute the inverse Laplace transform of the position correlation function using the Bromwich integral. Defining the constants $A_k = 1/d\tilde{P}(s)/ds|_{s=a_k}$ and $B_{mj}$ by the expansion $\tilde{Q}(s) \times [s + \bar{\gamma} s^{(p/q)-1}] = \sum_{m=0}^{2q-1}\sum_{j=0}^{q-1} B_{mj} s^{m-j/q}$, the solution is,

$$C_x(t) = \sum_{m=0}^{2q-1}\sum_{j=0}^{q-1}\sum_{k=0}^{2q} a_K^m B_{mj} A_k t^{j/q} E_{1,1+(j/q)}[a_k t]. \tag{51}$$

The long time behavior of the position correlation function reads,

$$C_x(t) \simeq \frac{\bar{\gamma}}{\omega^2 \Gamma(1-\alpha)} t^{-\alpha}. \tag{52}$$

For the special value $\alpha = 1/2$, the normalized position correlation function (51) simplifies to,

$$C_x(t) = \sum_{k=1}^{4} \left[ (-\bar{\gamma}^2 + \omega^2 a_k + a_k^3) A_k e^{a_k t} + \bar{\gamma}\omega^2 A_k t^{1/2} E_{1,3/2}[a_k t] \right], \tag{53}$$

where $a_k$ are the solutions of the equation $\tilde{P}(s) = (s^2+\omega^2)^2 - \bar{\gamma}^2 s = 0$. Figure 5 shows the position correlation function (53) for $\alpha = 1/2$ for three different frequencies: (a) for $\omega < \omega_z$, the correlation function decays monotonically, (b) at a critical frequency $\omega = \omega_z$, $C_x(t)$ is nonmonotonic and

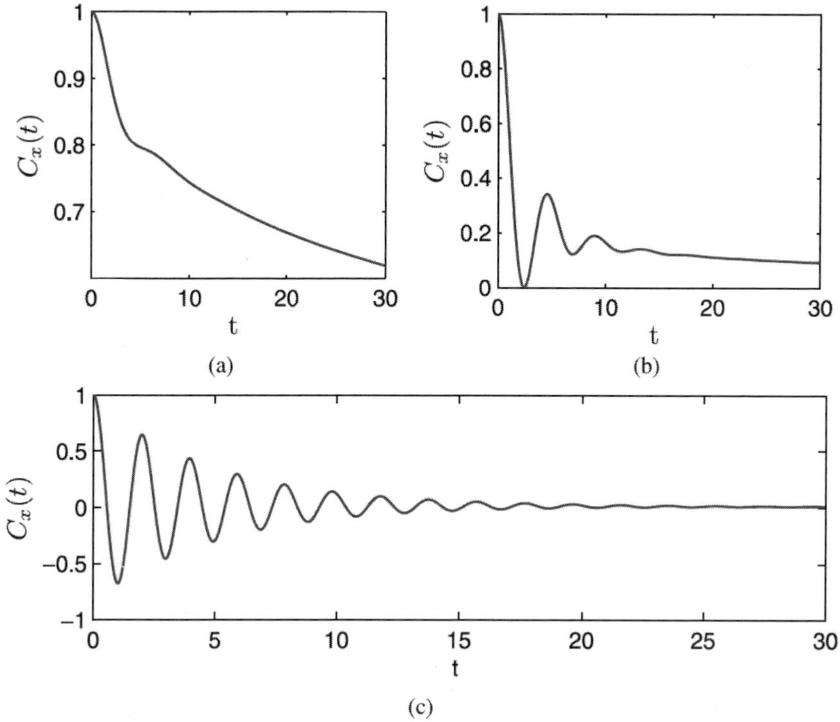


Fig. 5. Position correlation function $C_x(t)$, Eq. (53), as a function of time for $\alpha = 1/2$ and $\bar{\gamma} = 1$. Three types of behaviors are observed: (a) monotonic decay for $\omega = 0.3$, (b) nonmonotonic decay without zero crossing at the crititical frequency $\omega = \omega_z \simeq 1.05$, and (c) nonmonotonic decay with zero crossing for $\omega = 3.0$ (reprinted figure 1 with permission from S. Burov and E. Barkai, *Phys. Rev. Lett.* **100**, 070601 (2008). © 2008 by the American Physical Society).

strictly positive, and finally (c) for $\omega > \omega_z$, $C_x(t)$ is nonmonotonic and can become negative. In view of the different behaviors of the position correlation function, we can define a critical frequency $\omega_z$ as the minimal frequency at which $C_x(t)$ crosses the zero line, and another critical frequency $\omega_m$ as the minimal frequency at which $C_x(t)$ stops being a monotonically decreasing function. A dimensional analysis shows that,

$$\omega_z = \kappa_z(\alpha)\bar{\gamma}^{1/(2-\alpha)}, \quad \omega_m = \kappa_m(\alpha)\bar{\gamma}^{1/(2-\alpha)}, \tag{54}$$

where the functions $\kappa_z(\alpha)$ and $\kappa_m(\alpha)$ depend only on the fractional exponent $\alpha$ and have to be determined numerically. When $\alpha = 1$, all three critical frequencies are identical, $\omega_c = \omega_z = \omega_m = \gamma/(2m)$. The phase diagram of the fractional oscillator is presented in Fig. 6. It reveals the existence of

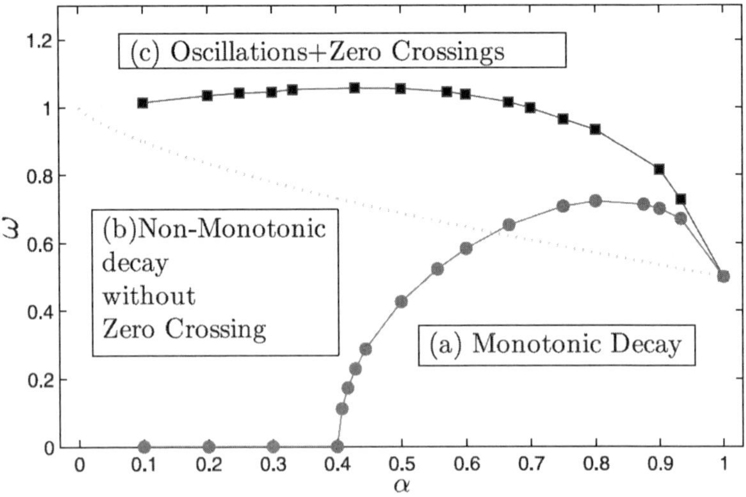


Fig. 6. Phase diagram of the fractional oscillator. The boundary between the nonmonotonic decay without zero-crossing of $C_x(t)$ (b) and the oscillations with zero crossings (c) is given by $\omega_m$; the boundary between the monotonic decay (a) and the nonmonotonic decay (b) by $\omega_z$. For $\alpha < \alpha_c \simeq 0.4$, the phase of monotonic decay disappears. The dotted curve is the critical line $\omega_c$. For all the curves, $\bar{\gamma} = 1$ (reprinted figure 2 with permission from S. Burov and E. Barkai, *Phys. Rev. Lett.* **100**, 070601 (2008). © 2008 by the American Physical Society).

a critical exponent $\alpha_c \simeq 0.4$ below which the position correlation function never relaxes monotonically. Since the overdamped approximation predicts a monotonic decay, one can conclude that the latter fails when $\alpha < \alpha_c$. An analogous study shows the existence of another critical exponent $\alpha_R$. The latter marks the dynamical transition of the response of the oscillator to an external periodic force field, between a regime where a resonance always exists and a regime where it never exists.

## 7. Ergodicity

In this final section, we address the question of ergodicity. Ergodicity is an essential property of stochastic processes. Broadly speaking, a process is ergodic, when time and ensemble average become equal in the long-time limit. Long-range temporal correlations have been shown to lead to ergodicity breaking in some instances [31, 32]. Fractional Brownian motion is ergodic for $\alpha < 2$ in the limit of infinite long times. However, all experiments are performed during a finite time interval, set in the example of *in vivo* experiments by the lifetime of the cell. For finite times, even normal diffusion processes may appear nonergodic, and it is thus important to have a concrete measure of nonergodicity for fractional Brownian motion.

In experiments, single-particle trajectories are recorded in the form of time series of data that are then analyzed by time rather than ensemble averages; in general both need not be equal. The time-averaged mean-square displacement for fractional Brownian motion is

$$\overline{\delta^2(x(t))} = \frac{1}{t-\Delta}\int_0^{t-\Delta}[x(t'+\Delta)-x(t')]^2, \tag{55}$$

where $\Delta$ is the time-lag. On average $\langle\overline{\delta^2(x(t))}\rangle = 2D_\alpha\Delta^\alpha = \langle x^2\rangle$ with $D_\alpha = mkT\gamma_\alpha$. The variance of $\overline{\delta^2(x(t))}$ is given by

$$X_{\text{var}}[\overline{\delta^2(x(t))}] = \langle[\overline{\delta^2(x(t))}]^2\rangle - \langle\overline{\delta^2(x(t))}\rangle^2. \tag{56}$$

A measure of ergodicity breaking is then provided by the parameter

$$E_B(x(t)) = \frac{X_{\text{var}}[\overline{\delta^2(x(t))}]}{\langle\overline{\delta^2(x(t))}\rangle^2}. \tag{57}$$

If the process is ergodic, $E_B(x(t))$ vanishes in the limit $t\to\infty$. Explicit expressions for the ergodicity measure can be derived in the limit of large times [33]. They are given as a function of the Hurst exponent $\alpha = 2H$ by,

$$E_B(x(t)) \simeq \begin{cases} k(H)\dfrac{\Delta}{t}, & 0<H<\dfrac{3}{4}, \\ k(H)\dfrac{\Delta}{t}\ln t, & H=\dfrac{3}{4}, \\ k(H)\left(\dfrac{\Delta}{t}\right)^{4-4H}, & \dfrac{3}{4}<H<1, \end{cases} \tag{58}$$

where the function $k(H)$ reads

$$k(H) = \begin{cases} \displaystyle\int_0^\infty dt'[(t'+1)^{2H}+|t'-1|^{2H}-2t'^{2H}]^2, & 0<H<\dfrac{3}{4}, \\ 4H^2(2H-1)^2 = \dfrac{9}{16}, & H=\dfrac{3}{4}, \\ \left(\dfrac{4}{4H-3}-\dfrac{4}{4H-2}\right)H^2(2H-1)^2, & \dfrac{3}{4}<H<1. \end{cases} \tag{59}$$

Two points are worth emphasizing. First, although fractional Brownian motion is ergodic, the rate of converge is slow and depends on the value of the exponent $H$. Second, the function $k(H)$ diverges when $H$ approaches

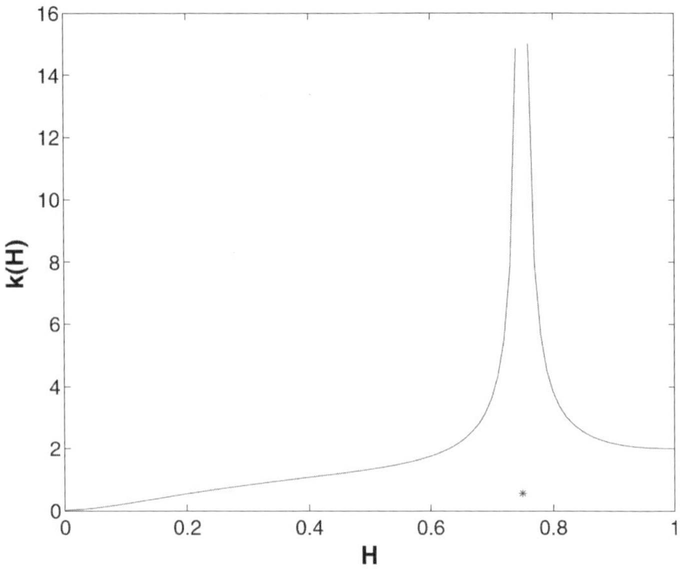


Fig. 7. Function $k(H)$ defined in Eq. (59) marking the nonsmooth transition in the ergodic properties of fractional Brownian motion (reprinted figure 1 with permission from W. Deng and E. Barkai, *Phys. Rev. E.* **79**, 011112 (2009). © 2009 by the American Physical Society).

3/4, indicating a nonsmooth transition (see Fig. 7). More recently, the influence of a confined geometry on ergodicity, important for subdiffusion in the interior of a cell, has been investigated in [34].

## References

1. R. Brown, A brief account of microscopical observations made in the months of June, July and August 1827, on the particles contained in the pollen plants; and on the general existence of active molecules in organic and inorganic bodies, *Edinburgh New Phil. J.* 358–371, July–September (1828).
2. A. Einstein, *Ann. Phys.* (*Leipzig*) **17**, 549 (1905).
3. H. Risken, *The Fokker–Planck Equation* (Springer, 1984).
4. P. Langevin, *C R Acad. Sci.* (*Paris*), **146**, 530 (1908).
5. W. Coffey, Yu. P. Kalmykov and J. T. Waldron, *The Langevin Equation* (World Scientific, 2004).
6. M. Weiss, M. Elsner, F. Kartberg and T. Nilsson, *Biophys. J.* **87**, 3518 (2004).
7. R. Metzler and J. Klafter, *Phys. Rep.* **339**, 1 (2000).
8. J. Blum, S. Bruns, D. Rademacher, A. Voss, B. Willenberg and M. Krause, *Phys. Rev. Lett.* **97**, 230601 (2006).
9. R. Zwanzig, *Nonequilibrium Statistical Mechanics* (Oxford, Univ. Press 2001).

10. A. V. Mokshin, R. M. Yulmetyev and P. Hänggi, *Phys. Rev. Lett.* **95**, 200601 (2005).
11. A. A. Stanislavsky, *Phys. Rev. E* **61**, 4752 (2000).
12. B. B. Mandelbrot and J. W. van Ness, *SIAM Rev.* **10**, 422 (1968).
13. E. Lutz, *Phys. Rev. E* **64**, 051106 (2000).
14. A. Taloni, A. Chechkin and J. Klafter, *Phys. Rev. Lett.* **104**, 160602 (2010).
15. L. Lizana, T. Ambjörnsson, A. Taloni, E. Barkai and M. A. Lomholt, *Phys. Rev. E* **81**, 051118 (2010).
16. A. Erdélyi, *Higher Transcendantal Functions*, vol. 3 (McGraw-Hill, 1955).
17. E. Lutz, *Euro. Phys. Lett.* **54**, 293 (2001).
18. A. Caspi, R. Granek and M. Elbaum, *Phys. Rev. Lett.* **85**, 5655 (2000).
19. I. M. Tolic-Norrelykke, E. L. Munteanu, G. Thon, L. Oddershede and K. Berg-Sorensen, *Phys. Rev. Lett.* **93**, 078102 (2004).
20. I. Bronstein, Y. Israel, E. Kepten, S. Mai, Y. Shav-Tal, E. Barkai and Y. Garini, *Phys. Rev. Lett.* **103**, 018102 (2009).
21. I. Golding and E. C. Cox, *Phys. Rev. Lett.* **96**, 098102 (2006).
22. M. Magdziarz, A. Weron, K. Burnecki and J. Klafter, *Phys. Rev. Lett.* **103**, 180602 (2009).
23. J. Szymanski and M. Weiss, *Phys. Rev. Lett.* **103**, 038102 (2009).
24. S. C. Weber, A. J. Spakowitz and J. A. Theriot, *Phys. Rev. Lett.* **104**, 238102 (2010).
25. S. C. Kou and X. S. Xie, *Phys. Rev. Lett.* **93**, 180603 (2004).
26. W. Min, G. Lou, B. J. Cherayil, S. C. Kou and X. S. Xie, *Phys. Rev. Lett.* **94**, 198302 (2005).
27. N. Pottier, *Physica A* **317**, 371 (2003).
28. N. Pottier and A. Mauger, *Physica A* **332**, 15 (2004).
29. S. Burov and E. Barkai, *Phys. Rev. Lett.* **100**, 070601 (2008).
30. S. Burov and E. Barkai, *Phys. Rev. E* **78**, 031112 (2008).
31. E. Lutz, *Phys. Rev. Lett.* **93**, 190602 (2004).
32. G. Bel and E. Barkai, *Phys. Rev. Lett.* **94**, 240602 (2005).
33. W. Deng and E. Barkai, *Phys. Rev. E* **79**, 011112 (2009).
34. J. H. Jeon and R. Metzler, *Phys. Rev. E* **81**, 022203 (2009).

# Chapter 13

## Subdiffusive Dynamics in Washboard Potentials: Two Different Approaches and Different Universality Classes

Igor Goychuk and Peter Hänggi
*Institute of Physics, University of Augsburg,*
*Universitätsstr. 1, D-86135 Augsburg, Germany*

We consider and compare two different approaches to the fractional subdiffusion and transport in washboard potentials. One is based on the concept of random fractal time and is associated with the fractional Fokker–Planck equation. Another approach is based on the fractional generalized Langevin dynamics and is associated with anti-persistent fractional Brownian motion and its generalizations. Profound differences between these two different approaches sharing the common adjective "fractional" are explained in spite of some similarities they share in the absence of a nonlinear force. In particular, we show that the asymptotic dynamics in tilted washboard potentials obey two different universality classes independently of the form of potential.

1. Introduction . . . . . . . . . . . . . . . . . . . . . . . . 307
2. Free Subdiffusion and Constant Bias . . . . . . . . . . . . 313
3. Other Similarities . . . . . . . . . . . . . . . . . . . . . 316
4. Diffusion and Transport in Washboard Potentials . . . . . 316
   4.1. FFPE dynamics . . . . . . . . . . . . . . . . . . . . 316
   4.2. GLE dynamics in periodic potentials . . . . . . . . 318
5. Summary and Conclusions . . . . . . . . . . . . . . . . . 324

## 1. Introduction

Anomalous diffusion becomes an increasingly popular subject with the number of papers published per year growing fast over last 20 years after a distinct rise occurred in 1990 (cf. Fig. 1). Since then, it spreads from such typical physical applications as charge transport in disordered solids and hot plasmas to biophysical applications and even quantitative finance [1–9, 11–21].

There are several sufficiently generic physical mechanisms and accompanying theoretical approaches to describe the complexity of anomalous

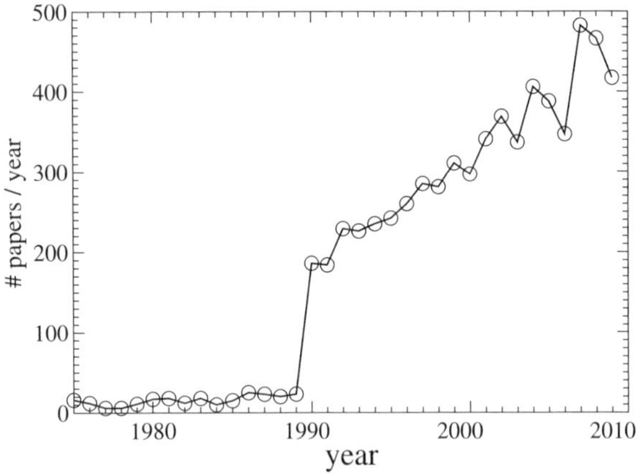


Fig. 1. Number of papers on the subject "anomalous diffusion" published per year according to the ISI Web of Science, Thomson-Reuters. Notice the "phase transition"-like rise during 1990.

transport processes. One approach is intrinsically based on the physical picture of a stochastic time clock [1, 2, 4–6, 10]. It models random sojourns of a traveling particle in trapping domains of a disordered solid, e.g., due to energy disorder. After a random time spent in some spatially located trapping domain, the particle jumps to a neighboring domain, or maybe farther, and such a jumping process continues in time. The jump directions and their lengths are not correlated from jump to jump and the next clock period is not correlated with those passed (semi-Markovian assumption[a]). Such a random clock is completely characterized by the probability density of clock periods $\psi(\tau)$. In a given time interval $t$ there will be a random number of jumps $n$, or stochastic clock periods completed. However, if the mean clock period $\langle\tau\rangle$ exists, the probability distribution $p(n,t)$ of "ticking" $n$ times within the observer time window $t$ becomes for large $n$ a very sharp function [4] around $n^*(t) = t/\langle\tau\rangle$, as characterized by the relative dispersion, $\langle\delta n^2(t)\rangle^{1/2}/\langle n(t)\rangle$ (see Appendix A). In the continuum medium approximation, the trapping domains shrink to points. Then, $\langle\tau\rangle$ can be made arbitrarily small and $n$ becomes quasi-continuous variable for a finite $t$. Correspondingly, the probability density of intrinsic time, $\tau(t) = n(t)\tau_{sc}$, where $\tau_{sc}$ is a time-scaling parameter, an intrinsic clock time unit equal

[a]This does not exclude infinite range memory leading to a weak ergodicity breaking [22].

to the duration of time period for the regular clock, assumes a delta-function, $p(\tau) = \delta(\tau - t)$, and the stochastic clock is not different from the regular one.

The situation changes dramatically if the mean of sojourns does not exist, in other words, it exceeds largely a typical time required to diffuse across the physical medium of a finite size. Then, $p(n,t)$ is not a sharp function around the mean number $\langle n(t) \rangle$. This is the case, for example, when $\psi(\tau)$ possesses a long tail, $\psi(\tau) \propto (\tau/\tau_{\rm sc})^{-1-\alpha}$ for $\tau/\tau_{\rm sc} \to \infty$ with $0 < \alpha < 1$. This implies a divergent mean $\langle \tau \rangle \to \infty$, and diverging higher moments as well. Nevertheless, $\langle n(t) \rangle \propto (t/\tau_{\rm sc})^{\alpha}$ exists for any finite $t$ and it scales sublinearly with the physical time $t$ (see Appendix A). In terms of the one-sided Lévy distribution density $\mathcal{L}_\alpha(z)$, $p(n,t) = (t/\tau_{\rm sc})\mathcal{L}_\alpha(n^{-1/\alpha}t/\tau_{\rm sc})/(\alpha n^{1/\alpha+1})$.

Consider now an ensemble of particles. Until time $t$, each particle has accomplished an *individual* number of intrinsic time periods corresponding to the intrinsic time $\tau(t)$ which becomes a random variable broadly distributed: all the particles have their own history, maintaining individuality and avoiding the fate of self-averaging even in the strict limit $t \to \infty$. Only an additional ensemble averaging smears out this principal randomness [10,23,24]. The *unbiased* diffusion becomes anomalously slow and nonergodic with the spatial variance of a *cloud* of particles growing sublinearly, $\langle \delta x^2(t) \rangle \propto \langle n(t) \rangle \propto t^\alpha$. Such a nonergodic approach to subdiffusion seems appropriate for disordered solids, e.g., thin amorphous films [2,4], with a more recent example provided by $TiO_2$ nanocrystalline electrodes in the Grätzel's photovoltaic cell elements [25]. For charged carriers within such media one can create a potential energy profile $U(x)$ by applying an electrical field of spatially distributed fixed charges and a static external electrical field. Then in the continuum approximation subdiffusion can be described by the fractional Fokker–Planck equation (FFPE) [6,26,27]

$$\frac{\partial}{\partial t}P(x,t) = {}_0\hat{D}_t^{1-\alpha}\left[-\frac{\partial}{\partial x}\frac{f(x)}{\eta_\alpha} + \kappa_\alpha\frac{\partial^2}{\partial x^2}\right]P(x,t), \tag{1}$$

where $f(x) = -dU(x)/dx$ is the force, $\eta_\alpha$ is the fractional friction coefficient related to the fractional subdiffusion coefficient $\kappa_\alpha$ by the generalized Einstein relation, $\eta_\alpha = k_{\rm B}T/\kappa_\alpha$, at temperature $T$ and

$${}_{t_0}\hat{D}_t^{\gamma}\,P(x,t) = \frac{1}{\Gamma(1-\gamma)}\frac{\partial}{\partial t}\int_{t_0}^{t} dt'\frac{P(x,t')}{(t-t')^{\gamma}}, \tag{2}$$

is the Riemann–Liouville operator of the fractional derivative [28], where $0 < \gamma < 1$ and $\Gamma(x)$ is the gamma-function. The FFPE (1) can be derived within the above continuous time random walk (CTRW) framework. It can also be written in the form using the Caputo fractional derivative [28]

$$ {}_{t_0}D_*^{\gamma}\, P(x,t) := \frac{1}{\Gamma(1-\gamma)} \int_{t_0}^{t} dt' \frac{\partial P(x,t')/\partial t'}{(t-t')^{\gamma}} \tag{3} $$

acting on the left-hand side, yielding [29]

$$ {}_0D_*^{\alpha}\, P(x,t) = \kappa_\alpha \frac{\partial}{\partial x} \left( e^{-\beta U(x)} \frac{\partial}{\partial x}\, e^{\beta U(x)} P(x,t) \right) = -\frac{\partial J(x,t)}{\partial x}, \tag{4} $$

in the transport form. Here, $\beta = 1/(k_B T)$ is inverse temperature and

$$ J(x,t) = -\kappa_\alpha e^{-\beta U(x)} \frac{\partial}{\partial x}\, e^{\beta U(x)} P(x,t) \tag{5} $$

is the subdiffusive flux. It should be emphasized that a non-Markovian Fokker–Planck equation never defines the corresponding non-Markovian process completely [31, 32]. It allows to find merely the single-time, conditional, and double-time probability densities, but never the multi-time probability densities. However, the FFPE dynamics can be nicely simulated from the underlying CTRW with the nearest neighbors jumps only [29,30].

A quite different approach to subdiffusion is associated with the fractional Brownian motion [33]. Here, the principal issue is the long-range anti-correlations in the particle displacements, positive increments follow with a greater probability by negative increments and vice versa [34,35], which can reflect e.g., the phenomenon of viscoelasticity in complex glass-forming liquids above, but close to the glass transition. The Brownian particle is temporally trapped in a trap (cage effect) by an elastic force with spring constant $G(t)$ which decays to zero in time releasing the particle. Let us assume that the motion starts at $t_0$, $v(t) = \dot{x}(t) = 0$ for $t < t_0$. In the linear approximation, on the particle acts a viscoelastic force $F_{\mathrm{v-el}}(t) = -\int_{t_0}^{t} G(t-t')\dot{x}(t')dt'$, where $\dot{x}(t)$ is the particle's instant velocity. The first theory of viscoelasticity has been proposed by Maxwell [36] in 1867. It corresponds to $G(t)$ exponentially decaying in time, $G(t) = G_0 \exp(-t/\tau_0)$, with a relaxation time constant $\tau_0$. Departing from the phenomenon of elasticity in solids Maxwell derived the phenomenon of viscosity in liquids in the limit where the decay of elastic modulus is very fast on the time scale of $v(t)$ change, which corresponds to $G(t) = 2\eta_0\delta(t)$, with $\eta_0 = G_0\tau_0$ being the viscous friction coefficient. In the theory of generalized Brownian motion, $G(t)$ is interpreted as the frictional memory kernel $\eta(t)$, rather than a decaying elastic force constant. Here, one departs from the phenomenon of viscosity and viscous Stokes memoryless friction and considers the emerging elasticity in complex fluids or viscoelastic bodies. Both viewpoints are essentially equivalent for a positive $\eta(t) > 0$ departing just from different standing points.[b] The anti-correlations in the particle's displacements are due to the elastic restoring force component.

In complex media, the memory function $G(t)$ is better described by a sum of exponentials reflecting a viscoelastic response with multiple time scales. Moreover, in 1936 Gemant [37] found that some viscoelastic bodies are better described by a $G(t)$ relaxing in accordance with a power law, $G(t) \propto t^{-\alpha}$, rather than a single-exponential and introduced a fractional integro-differential in the viscoelasticity theory. Using the notion of fractional Caputo derivative such a viscoelastic force can be short-handed, written as

$$F_{\text{v-el}}(t) = -\eta_\alpha \, {}_{t_0}D_*^\alpha \, x(t). \tag{6}$$

Indeed, such and similar viscoelastic responses are measured [16, 20, 38] using the microrheology methods [39].

The Brownian motion never stops and the frictional loss of energy must be compensated on average by the energy gain provided by a zero-mean random force of environment so that at the thermal equilibrium the equipartition theorem holds, in accordance with the classical fluctuation-dissipation theorem. Within the considered model of a linear memory-friction, such a force must be Gaussian [40] (but not necessarily so beyond the linear friction model). As a result, the Brownian motion of a particle of mass $m$ is described by the Fractional Langevin Equation (FLE) [41–44]

$$m\ddot{x} + \eta_\alpha \, {}_0D_*^\alpha \, x(t) = f(x) + \xi(t), \tag{7}$$

(from now on we fix $t_0 = 0$) which is a particular case of the celebrated Generalized Langevin Equation (GLE) [45–49]

$$m\ddot{x} + \int_0^t \eta(t-t')\dot{x}(t')dt' = f(x) + \xi(t), \tag{8}$$

[b] $\eta(t)$ can also be negative, e.g., accounting for the hydrodynamic memory or in the case of superdiffusion. Therefore, the memory-friction interpretation is, in fact, more general.

with the memory kernel $\eta(t) = \eta_\alpha t^{-\alpha}/\Gamma(1-\alpha)$ and the noise autocorrelation function obeying the fluctuation dissipation relation

$$\langle \xi(t)\xi(t') \rangle = k_{\rm B}T\eta(|t-t'|). \tag{9}$$

Such a GLE can also be derived from a Hamiltonian model for a particle bilinearly coupled with coupling constants $c_i$ to a thermal bath of harmonic oscillators with masses $m_i$ and frequencies $\omega_i$, $H_{B,\rm int}(p_i, q_i, x) = (1/2)\sum_i \{p_i^2/m_i + m_i\omega_i^2[q_i - c_i x/(m_i\omega_i^2)]^2\}$. The total effect of the bath oscillators, which are initially canonically distributed with $H_{B,\rm int}$ at temperature $T$ and fixed $x = x(0)$, is characterized by the bath spectral density

$$J(\omega) = \frac{\pi}{2}\sum_i \frac{c_i^2}{m_i\omega_i}\delta(\omega - \omega_i). \tag{10}$$

The memory kernel is $\eta(t) = (2/\pi)\int_0^\infty J(\omega)\cos(\omega t)d\omega$ in terms of $J(\omega)$ and the subdiffusive FLE corresponds to a sub-Ohmic, or fracton thermal bath with $J(\omega) = \eta_\alpha \sin(\pi\alpha/2)\omega^\alpha$ [49]. Without frequency cutoffs such a model presents a clear idealization. There always exists a highest frequency of the thermal bath and this leads to a small time regularization of the memory kernel, i.e. a short-time cutoff. Physically, this takes into account the medium's granularity beyond the continuum approximation. Moreover, in the case of a finite-size medium there also always exists a smallest frequency of the medium's oscillators corresponding to the inverse size of the medium. These facts become especially clear if one evaluates the spectral density of low-frequency "fracton" oscillators in proteins, see [21]. This also leads to a cutoff at large times in the memory kernel and the dynamics can be subdiffusive on the time scale smaller than the corresponding memory cutoff. The latter can be but prominently large which makes the considered idealization relevant. The result that an overdamped FLE description of subdiffusion can be derived from a broad class of phenomenological continuum elastic models is also important [50].

In the inertialess limit with $m \to 0$, one can conceive the idea that FFPE (1, 4) is the fractional Fokker–Planck equation corresponding to the FLE (7). This idea is but wrong [43]. The non-Markovian Fokker–Planck equation (NMFPE) which corresponds to the GLE (with arbitrary kernel) [51–53] and to the FLE, in particular [43], is a different one. Presently, its explicit form is known only for constant or linear forces $f(x)$ [51–53]. This resulting NMFPE has the form of Fokker–Planck equation with time-dependent kinetic coefficients. This time-dependence is not universal and it heavily depends on the form of potential. In turn, the Langevin equation which corresponds to the above FFPE is known and it has the form of a Langevin equation which is local in the stochastic time $\tau(t)$ and describes thus a doubly stochastic process [54]. Here lies also the profound mathematical difference between these two approaches to subdiffusion. The physical differences are also immense. In particular, the GLE and FLE approaches are asymptotically mostly ergodic (except for the case of asymptotically ballistic superdiffusion) as they are not based on the concept of fractal stochastic time with divergent mean period and the mean residence time in a finite spatial domain remains finite. Before we discuss the striking differences in more detail, let us start from some apparent, but misleading similarities.

## 2. Free Subdiffusion and Constant Bias

Free subdiffusion, as well as diffusion biased by a constant force $F$ can readily be solved in both approaches using the method of Laplace-transform. First one finds the Laplace-transform of the mean ensemble-averaged displacement $\langle \delta x(t) \rangle$, and of the position variance $\langle \delta x^2(t) \rangle = \langle x^2(t) \rangle - \langle x(t) \rangle^2$, starting from a delta-peaked distribution at $x = 0$ and $t_0 = 0$. Then, one transforms back to the time domain. This gives [6]

$$\langle \delta x(t) \rangle = \mu_\alpha F t^\alpha / \Gamma(1+\alpha) \tag{11}$$

and

$$\langle \delta x^2(t) \rangle = 2\kappa_\alpha t^\alpha / \Gamma(1+\alpha) \tag{12}$$

with the generalized mobility $\mu_\alpha = 1/\eta_\alpha$ related to the subdiffusion coefficient at $F = 0$ by the generalized Einstein relation $\mu_\alpha = \kappa_\alpha/(k_B T)$. Within the FLE approach these results are valid in the strict inertialess limit $m \to 0$. Furthermore, Eq. (12) is still valid for *arbitrary* $F \neq 0$. However, within the FFPE approach, Eq. (12) is valid *only* for $F = 0$, which is the first striking difference, see also below. Furthermore, both results are also valid asymptotically, $t \to \infty$, within the FLE for a finite $m \neq 0$.

Generally, the GLE results can be obtained for arbitrary memory kernel $\eta(t)$. Assuming the particles being initially Maxwellian distributed, i.e. thermalized with thermal r.m.s. velocities

$$v_T = \sqrt{k_B T/m}, \tag{13}$$

one can obtain for the Laplace-transformed stationary velocity (fluctuation) autocorrelation function (VACF) $K_v(\tau) = \langle \delta v(t+\tau)\delta v(t)\rangle$, $\delta v(t) = v(t) - \langle v(t)\rangle$,

$$\tilde{K}_v(s) = \frac{k_B T}{ms + \tilde{\eta}(s)}, \tag{14}$$

where $\tilde{\eta}(s)$ is the Laplace-transform of $\eta(t)$. This is a well-known result which was obtained first by Kubo [45,46] in the Fourier space. For the FLE with $\tilde{\eta}(s) = \eta_\alpha s^{\alpha-1}$ it yields by the inversion to the time-domain [41]

$$K_v(\tau) = v_T^2 E_{2-\alpha}[-(\tau/\tau_v)^{2-\alpha}] \tag{15}$$

with $\tau_v = (m/\eta_\alpha)^{1/(2-\alpha)}$ being the anomalous velocity relaxation time constant. In (15), $E_\gamma(z)$ is the Mittag-Leffler function, $E_\gamma(z) = \sum_{n=0}^{\infty} z^n/\Gamma(n\gamma+1)$ [6]. For $0 < \alpha < 1$, $K_v(\tau)$ is initially positive reflecting ballistic persistence due to inertial effects and then becomes negative (anti-persistence due to decaying elastic cage force). In the limit $m \to 0$, the VACF undergoes a jump starting from $v_T^2$ at $\tau = 0$ and then becoming negative, $K_v(\tau) \propto -1/\tau^{2-\alpha}$ for $\tau > 0$, corresponding to purely anti-persistent motion. The position variance is given by the doubly-integrated VACF. Its Laplace-transform therefore reads,

$$\widetilde{\langle \delta x^2(s)\rangle} = \frac{2k_B T}{s^2[ms + \tilde{\eta}(s)]}. \tag{16}$$

Moreover,

$$\widetilde{\langle \delta x(s)\rangle} = \frac{F}{s^2[ms + \tilde{\eta}(s)]}, \tag{17}$$

for arbitrary kernel, which can also be easily shown from the GLE, and therefore[c]

$$\frac{\langle \delta x(t)\rangle}{\langle \delta x^2(t)\rangle} = \frac{F}{2k_B T} \tag{18}$$

for the thermally equilibrium initial preparation. For the FLE with a finite $m$ the inversion of Eq. (16) to the time domain yields [41],

$$\langle \delta x^2(t)\rangle = 2v_T^2 t^2 E_{2-\alpha,3}[-(t/\tau_v)^{2-\alpha}], \tag{19}$$

[c] For nonequilibrium initial preparations this result holds asymptotically in any asymptotic ergodic case, including the FLE dynamics. The relaxation to the asymptotic regime, or aging, can be but very slow [55, 56] which is a general feature of subdiffusive GLE dynamics also in periodic potentials.

where $E_{\gamma,\beta}(z) = \sum_{n=0}^{\infty} z^n/\Gamma(n\gamma+\beta)$ is the generalized Mittag-Leffler function. One recovers Eq. (12) in the limit $m \to 0$.

However, for the subdiffusive CTRW and FFPE dynamics the behavior of the ensemble-averaged variance is very different from Eq. (12) under a nonzero bias $F \neq 0$. Then, Eq. (12) is not valid anymore. This fact is ultimately related to the properties of the stochastic clock. The point is that starting from a CTRW picture it is easy to show (see Appendix A) that the growing ensemble-averaged variance $\langle \delta x^2(t) \rangle$ depends in the asymmetric case (the probabilities to jump left and right are different) not only on the mean number $\langle n(t) \rangle$ of the stochastic clock periods passed, but also on their variance $\langle \delta n^2(t) \rangle$. For $\alpha = 1$ (regular clock), $\langle \delta n^2(t) \rangle = 0$. However, for $0 < \alpha < 1$, $\langle \delta n^2(t) \rangle \propto t^{2\alpha}$ and this dramatically changes the character of anomalous CTRW and FFPE diffusion in the presence of bias. It becomes asymptotically $\langle \delta x^2(t) \rangle \propto F^2 t^{2\alpha}$, while $\langle \delta x(t) \rangle \propto F t^{\alpha}$. Notice that for $1/2 < \alpha < 1$ the subdiffusion at $F = 0$ transforms into superdiffusion for $F \neq 0$, i.e. a cloud of particles spreads out anomalously fast relative to its center of mass. This yields a remarkable scaling for the ensemble-averaged quantities

$$\lim_{t\to\infty} \frac{\langle \delta x^2(t) \rangle}{\langle \delta x(t) \rangle^2} = \lim_{t\to\infty} \frac{\langle \delta n^2(t) \rangle}{\langle n(t) \rangle^2} = \frac{2\Gamma^2(\alpha+1)}{\Gamma(2\alpha+1)} - 1. \tag{20}$$

This scaling, which was observed first in [1,2] for a CTRW subdiffusion in the absence of any additional potential $U(x)$, has been shown to be *universal* within the FFPE description also for arbitrary tilted washboard potentials and temperature [29,30]. Recently, this astounding fact has been related to the universal fluctuations of anomalous mobility and weak ergodicity breaking [57]. Ultimately, this is just the property of the stochastic clock and it reflects the scaling between the variance and the mean number of stochastic periods passed within the external observed time $t$. Surprisingly, the viscoelastic GLE subdiffusion also exhibits a *universal* asymptotical scaling in tilted washboard potentials. In the $t \to \infty$ limit it is the *same* as in Eq. (18). Astonishingly, it works both for a vanishingly small $F$, and for an arbitrary strong bias. Moreover, both the diffusion and drift in the tilted washboard potentials do not depend *asymptotically* on the amplitude and the form of the periodic potential in the case of GLE subdiffusion and are given by Eqs. (12) and (11), correspondingly. This is again very much different from the FFPE case, where Eq. (18) can be used only to calculate the anomalous flux response at a vanishingly small $F$ from the equilibrium $\langle \delta x^2(t) \rangle_{F=0}$ at $F = 0$. Also, given $\langle \delta x(t) \rangle$ at $F \neq 0$ one can calculate

$\langle \delta x^2(t) \rangle_{F=0}$ using Eq. (18) and the corresponding subdiffusion coefficient in periodic potentials in the limit $F \to 0$, for details see [62] and below.

## 3. Other Similarities

One more similarity emerges for the relaxation of mean fluctuation from equilibrium in harmonic potentials, $U(x) = kx^2/2$. Then, both the FFPE approach and the FLE approach (in the limit $m \to 0$) yield the same relaxation law [19, 26, 43], $\langle \delta x(t) \rangle = \langle \delta x(0) \rangle E_\alpha[-(t/\tau_r)^\alpha]$ with the ultra-slow position relaxation time constant $\tau_r = (\eta_\alpha/k)^{1/\alpha}$. Asymptotically, this relaxation follows a power-law, $\langle \delta x(t) \rangle \propto t^{-\alpha}$.

The asymptotic distributions of the residence times within a half-infinite spatial domain (or the first return times to the origin in the infinite domain) in the case of free subdiffusion are also similar, following the same scaling law [14, 50, 58] $\Psi(\tau) \propto 1/\tau^{2-\alpha/2}$. However, here the similarities end. The asymptotics for a finite-size domain cannot be the same. In particular, the mean residence time in any finite-size domain within the subdiffusive GLE description is finite [34], whereas within the FFPE description is not, except for the case of injection of diffusing particles on the normal radiative boundary, where they can be immediately absorbed [14]. Moreover, the GLE (for arbitrary $\eta(t)$, including FLE) describe a Gaussian process for constant and linear forces $f(x)$,[d] whereas the FFPE does not correspond to a Gaussian process in these cases, see [6].

## 4. Diffusion and Transport in Washboard Potentials

Let us proceed with the case of washboard potentials, where the differences between the two discussed approaches to subdiffusion become particularly transparent. We consider the tilted potential $U(x) = V(x) - xF$, where $V(x+L) = V(x)$ is a periodic potential with the spatial period $L$.

### 4.1. *FFPE dynamics*

In this case, one can find exact analytical results for the ensemble-averaged nonlinear mobility $\mu_\alpha(F)$ using Eq. (11) asymptotically also in washboard potentials. First, one finds the exact analytical expression for the ensemble-averaged subvelocity $v_\alpha(F) = \mu_\alpha(F)F$. The FFPE in the form (4) is more

[d]This is just by the linearity of the transformation from the Gaussian noise $\xi(t)$ to the stochastic process $x(t)$ as described by Eqs. (7) and (8).

convenient for this purpose. Indeed, it has the form of a fractional-time continuity equation with the flux $J(x)$. For the sake of generality we consider its further generalization with a spatially-dependent subdiffusion coefficient $\kappa_\alpha(x)$,

$$J(x,t) = -\kappa_\alpha(x)e^{-\beta U(x)}\frac{\partial}{\partial x}\,e^{\beta U(x)}P(x,t) \tag{21}$$

which is assumed to be periodic with the same period $\kappa_\alpha(x+L) = \kappa_\alpha(x)$, and the generalized Einstein relation is fulfilled locally at any $x$, $\kappa_\alpha(x) = k_{\rm B}T/\eta_\alpha(x)$. We proceed similarly to the case of normal diffusion [59–61], $\alpha = 1$. A spatial period averaged density $\hat{P}(x,t) = \sum_{k=-M}^{M} P(x+kL,t)/(2M+1)$ should attain a steady-state regime (corresponding to a nonequilibrium steady state for $F \neq 0$) in the limit $M \to \infty$, $t \to \infty$, where it becomes periodic with the period $L$, $\hat{P}_{\rm st}(x+L) = \hat{P}_{\rm st}(x)$. The corresponding subdiffusive flux $\hat{J}(x)$, defined with $\hat{P}_{\rm st}(x)$, becomes a constant $J_\alpha$ in the steady state:

$$J_\alpha = -\kappa_\alpha(x)e^{-\beta U(x)}\frac{d}{dx}\,e^{\beta U(x)}\hat{P}_{\rm st}(x). \tag{22}$$

Then, the dynamics of the averaged mean displacement follows as

$${}_0D_*^\alpha \langle x(t)\rangle = LJ_\alpha, \tag{23}$$

which can be shown akin to the normal diffusion case [60]. The appearance of the fractional Caputo time derivative on the L.H.S. of Eq. (23) is the only mathematical difference as compared with the normal diffusion case. The solution of (23) yields for the mean displacement

$$\langle x(t)\rangle = v_\alpha^{(\rm wb)}(F)t^\alpha/\Gamma(1+\alpha), \tag{24}$$

with $v_\alpha^{(\rm wb)}(F) = LJ_\alpha$ being the subvelocity in the washboard potential.

One finds $J_\alpha$ and $v_\alpha^{(\rm wb)}(F)$ by multiplying Eq. (22) with $e^{\beta U(x)}/\kappa_\alpha(x)$ and integrating the result within one spatial period. Taking into account the spatial periodicity of $V(x)$ and $\kappa_\alpha(x)$ this yields:

$$\begin{aligned} J_\alpha \int_y^{y+L} \frac{e^{\beta U(x)}}{\kappa_\alpha(x)}dx &= -e^{\beta U(y+L)}\hat{P}_{\rm st}(y+L) + e^{\beta U(y)}\hat{P}_{\rm st}(y) \\ &= (1-e^{-\beta FL})e^{\beta U(y)}\hat{P}_{\rm st}(y). \end{aligned} \tag{25}$$

Next, multiplying (25) with $e^{-\beta U(y)}$, integrating over $y$ within $[0, L]$, and using the normalization $\int_0^L \hat{P}_{\rm st}(y)dy = 1$ one finds the main result

$$v_\alpha^{(\rm wb)}(F) = \frac{(1 - e^{-\beta F L})L}{\int_0^L e^{-\beta U(y)}dy \int_y^{y+L} \frac{e^{\beta U(x)}}{\kappa_\alpha(x)}dx}. \tag{26}$$

Accordingly, the nonlinear anomalous mobility is $\mu_\alpha^{(\rm wb)}(F) = v_\alpha^{(\rm wb)}(F)/F$. This presents a further generalization of the result for subvelocity in [29,30] to a spatially-dependent subdiffusion coefficient $\kappa_\alpha(x)$. The subdiffusion coefficient in the unbiased washboard potential for $F = 0$ can also be found using the generalized Einstein relation $\kappa_\alpha^{(\rm wb)}(F = 0) = k_{\rm B}T\mu_\alpha^{(\rm wb)}(F = 0)$. It reads,

$$\kappa_\alpha^{(\rm wb)}(F = 0) = \frac{L^2}{\int_0^L e^{-\beta V(y)}dy \int_0^L \frac{e^{\beta V(x)}}{\kappa_\alpha(x)}dx}, \tag{27}$$

and for $\kappa_\alpha = \text{const}$ this is the result of the work [62]. For constant $\kappa_\alpha$ and a number of different potentials $V(x)$, temperatures $T$ and biasing forces $F$, these two general results were beautifully confirmed by numerical simulations of the underlying CTRW [29, 30, 62] on a lattice from which the FFPE in the form (4) was derived in the work [29]. These simulations also confirmed the universality of the scaling relation (20) within the FFPE approach. Surprisingly, it remains invariant also in the presence of a driving which is periodic in time, in the biased case $F \neq 0$ [63], featuring thus the universality class of subdiffusion governed by a stochastic clock with divergent mean period and characterized by the only parameter $\alpha$. The above $v_\alpha^{(\rm wb)}$ is the ensemble-averaged result. The subvelocities of individual particles remain randomly distributed in the limit $t \to \infty$ and they follow a universal subvelocity distribution which reflects the distribution of random individual time of travelling particles, as it has been clarified in [57]. Both the weak ergodicity breaking and the universal fluctuations of anomalous mobility within the FFPE approach are ultimately related to this remarkable property of the stochastic time.

### 4.2. *GLE dynamics in periodic potentials*

The GLE subdiffusion distinctly differs in the physical mechanism and this leads to quite different results for washboard potentials [34, 64]. First of all, it is *asymptotically* ergodic and self-averaging over a single trajectory yields a quite definite non-random result [34]. No additional ensemble averaging is required. Moreover, it turns out that both the particle anomalous mobility $\mu_\alpha^{(\mathrm{wb})}$ and the subdiffusion coefficient $\kappa_\alpha^{(\mathrm{wb})}$ do not depend *asymptotically* on the potential $V(x)$ or on the bias $F$ being universal and the same as for biased GLE subdiffusion in the absence of periodic potential, obeying the generalized Einstein relation. The transition to this asymptotic regime is, however, very slow and it strongly depends on the amplitude of the periodic potential $V_0$ and the temperature $T$. Because of this slowness of the transient aging, this asymptotic regime will not necessarily be relevant on a finite time scale for anomalous transport in finite-size systems. This is especially so if the periodic potential amplitude exceeds the thermal energy by many times. However, this remarkable property features the very mechanism of the GLE subdiffusion, which is based on the long-range velocity and displacement correlations and not on diverging mean residence time within a potential well, in clear contrast to the CTRW subdiffusion with independent increments. It outlines a quite different universality class of subdiffusion. This is the long-range anti-persistence which limits asymptotically the GLE subdiffusion and transport processes in the washboard potentials. Since the mean residence time in a potential well is finite [34], a coarse graining over the potential period, which makes the sojourns in the trapping potential wells irrelevant, becomes *asymptotically* possible. In fact, upon increasing the potential height the escape kinetics out of a potential well (being asymptotically stretched-exponential) becomes ever closer to the normal exponential kinetics [34], where it becomes described by the non-Markovian rate theory [48, 65]. This does not mean, however, that the diffusion spreading over many spatial periods becomes normal. As a matter of fact, in the unbiased periodic potentials the diffusion cannot become faster than the free subdiffusion and this is a reason why the asymptotic limit of free subdiffusion is attained. A signature of this universality has been revealed theoretically for quantum transport in sinusoidal potentials for the case of sub-Ohmic thermal bath which classically corresponds to the considered case of fractional sub-diffusive friction. Technically this was done by using two different approaches, one perturbative [66] and one nonperturbative based on a quantum duality transformation between the quantum dissipative washboard dynamics coupled to a sub-Ohmic bath and a quantum dissipative tight-binding dynamics coupled to a super-Ohmic bath [49]. In the quantum case, there are also tunneling processes which are accounted for. Our numerical results for the classical Brownian dynamics indicate, however, that this feature is purely classical and, moreover, it is universal, i.e. is beyond the particular case of sinusoidal potentials [64]. It is *not* caused by the quantum-mechanical effects.

Our numerical simulation approach is also insightful and it can be considered as an independent theoretical route to model anomalous diffusion and transport processes. The idea is to approximate the non-Markovian GLE dynamics with a power-law kernel by a *finite-dimensional* Markovian dynamics of a sufficiently high dimensionality $D$ [34, 67–69]. Here, "sufficient" means the following: having subdiffusion extending over $r$ time-decades one finds a $D$-dimensional Markovian dynamics whose projection on the $(x, v)$-plane approximates the GLE dynamics over the required time range within the accuracy of stochastic simulations, as it can be checked for the cases where an exact solution of the GLE dynamics is available (free or biased subdiffusion, subdiffusion in harmonic potentials). Increasing $D$ one can cover larger $r$ of experimental interest and the embedding dimension $D$ turns out to be finite to arrive at the asymptotic results valid for the strict power law kernel. Needless to say that the practically observed cases of anomalous diffusion hardly extend over more than six time decades (typically several only) which underpins the practical value of our approach.

We expand the power law kernel into a sum of exponentials

$$\eta(t) = \frac{\eta_\alpha}{\Gamma(1-\alpha)} C_\alpha(b) \sum_{i=1}^{N} \nu_i^\alpha \exp(-\nu_i t) \tag{28}$$

obeying a fractal scaling with $\nu_i = \nu_0/b^{i-1}$, where $b > 1$ is a scaling parameter, $\nu_0 > 0$ is high-frequency (short-time) cutoff corresponding to the fastest time scale $\tau_0 = 1/\nu_0$ in the hierarchy of the relaxation time constants, $\tau_i = \tau_0 b^{i-1}$, of viscoelastic memory kernel. $C_\alpha(b)$ is a numerical constant to provide a best fit to $\eta(t) = \eta_\alpha t^{-\alpha}/\Gamma(1-\alpha)$ in the interval $[\tau_0, \tau_0 b^{N-1}]$. In the theory of anomalous relaxation similar expansions are well-known [4, 70]. In the present context, the approach corresponds to an approximation of the fractional Gaussian noise by a sum of uncorrelated Ornstein–Uhlenbeck (OU) noises, $\xi(t) = \sum_{i=1}^{N} \zeta_i(t)$, with autocorrelation functions, $\langle \zeta_i(t)\zeta_j(t')\rangle = k_B T \kappa_i \delta_{ij} \exp(-\nu_i|t-t'|)$. This idea is also known in the theory of $1/f$ noise [71]. For $t > \tau_0 b^{N-1}$ the tail of (28) is exponential and the diffusion becomes normal for $t \gg \tau_0 b^{N-1}$. However, by increasing $N$ one can enlarge the corresponding time scale and even make it practically irrelevant. The subdiffusion can be modeled in this way over about $r = N \log_{10} b - 2$ time decades and the corresponding embedding dimension,

$D = N + 2$, can be rather small. Such fits are known to exhibit logarithmic oscillations superimposed on the power law [4]. However, their amplitude can be made negligibly small if to choose $b$ sufficiently small, e.g., for $b = 2$ they become already barely detectable. Nevertheless, even the decade scaling with $b = 10$ suffices to arrive at excellent (within the statistical errors of Monte Carlo simulations) approximation of the FLE dynamics by a *finite-dimensional* Markovian dynamics over a huge range of time scales. Weak logarithmic sensitivity of $r$ to $b$ and linear dependence on $N$ allows one to improve the quality of Markovian embedding at a moderate computational price. The choice of Markovian embedding which corresponds to (28) is not unique [68, 69]. A particular one is the following [34]:

$$\begin{aligned} \dot{x} &= v, \\ m\dot{v} &= f(x,t) + \sum_{i=1}^{N} u_i(t), \\ \dot{u}_i &= -k_i v - \nu_i u_i + \sqrt{2\nu_i k_i k_{\mathrm{B}} T}\,\xi_i(t), \end{aligned} \tag{29}$$

where $k_i = C_\alpha(b)\eta_\alpha \nu_i^\alpha/\Gamma(1-\alpha) > 0$ and $\xi_i(t)$ are independent unbiased white Gaussian noise sources, $\langle \xi_i(t)\xi_j(t')\rangle = \delta_{ij}\delta(t-t')$. Indeed, integrating out the auxiliary force variables $u_i$ in Eq. (29) it follows that the resulting dynamics is equivalent to the GLE (8), (9) with the kernel (28), provided that $u_i(0)$ are unbiased random Gaussian variables with variances $\langle u_i^2(0)\rangle = k_i k_{\mathrm{B}} T$. The latter condition ensures the stationarity of $\xi(t)$ in the GLE (8), as well as validity of the FDR (9) *for all times*. Using different non-thermal preparations of $u_i(0)$ one can study the influence of *initial* non-stationarity of the noise $\xi(t)$ in the GLE on the Brownian dynamics [69]. In this aspect, our approach is even more flexible and more general than the standard GLE approach.

The auxiliary variables $u_i$ can be interpreted as viscoelastic forces, $u_i = -k_i(x - x_i)$, exerted by some overdamped particles with positions $x_i$, which are coupled to the central Brownian particle with elastic spring constants $k_i$ and are subjected to viscous friction with frictional constants $\eta_i = k_i/\nu_i = C_\alpha(b)\eta_\alpha \tau_i^{1-\alpha}/\Gamma(1-\alpha)$ and the thermal random forces of environment. This corresponds to motion of $N+1$ particles in a potential $U(x, \{x_i\}) = U(x,t) + (1/2)\sum_{i=1}^{N} k_i(x - x_i)^2$. The Brownian particle is massive (inertial effects are generally included), all other "particles" are overdamped (massless, $m_i \to 0$). For example, one can imagine that some coordination spheres of the viscoelastic environment stick to the Brownian particle and are co-moving. Their influence can be effectively represented by $N$ "quasi-particles". In this insightful physical interpretation, our embedding scheme is equivalent to:

$$\begin{aligned} m\ddot{x} &= f(x,t) - \sum_{i=1}^{N} k_i(x - x_i), \\ \eta_i \dot{x}_i &= k_i(x - x_i) + \sqrt{2\eta_i k_B T}\xi_i(t). \end{aligned} \tag{30}$$

It is worthwhile to notice that in this approach the mass of the Brownian particle and therefore the inertial effects are important. In order to perform an overdamped limit $m \to 0$, one has to include the viscous frictional force $-\eta_0 \dot{x}$ acting directly on the particle and also the corresponding random force. Then, in the limit $m \to 0$, one obtains

$$\begin{aligned} \eta_0 \dot{x} &= f(x,t) - \sum_{i=1}^{N} k_i(x - x_i) + \sqrt{2\eta_0 k_B T}\xi_0(t), \\ \eta_i \dot{x}_i &= k_i(x - x_i) + \sqrt{2\eta_i k_B T}\xi_i(t), \end{aligned} \tag{31}$$

where $\xi_0(t)$ is a zero-mean Gaussian random force of unit intensity which is not correlated with the set $\{\xi_i(t)\}$. However, it was noticed [44] that the inertial effects can be important for the subdiffusive GLE dynamics and therefore we take them into account. Of course, here emerges one more difference with the alternative description of subdiffusion within the FFPE (1), (4).

A proper fractal scaling of coefficients $k_i$ and $\eta_i$ with $i$ (see above) allows one to model viscoelastic subdiffusion over arbitrary time scales of the experimental interest. One can numerically solve these stochastic differential equations (29) e.g., with a standard stochastic Heun method [72] (second-order Runge–Kutta method) as done in [34, 64]. An example of such simulations is given in Figs. 2 and 3 for $\alpha = 0.5$, $\nu_0 = 100$, $b = 10$, $C_\alpha(b) = 1.3$, $N = 12$ and $k_B T = 0.1$. The following scaling is used: time in the units of $\tau_v$,[e] distance in the units of $L$. All the energy units are then scaled in $\Delta E = m(L/\tau_v)^2$ and the force units in $mL/\tau_v^2$. Stochastic Heun method is used to integrate Eq. (29) with a time step $\Delta t = (1-5) \cdot 10^{-3}$ until $t_{\max} = 2 \cdot 10^5$ and $n = 10^4$ trajectories are used for the ensemble averaging. Stochastic numerics are compared against the exact results for the free subdiffusion and for the mean displacement under a constant biasing

[e]This is a natural scaling of the velocity autocorrelation function in time. Other scalings can be also possible [34, 64]. They are more suitable to consider dynamical regimes close to overdamped.

force. The agreement is excellent. The considered particular embedding still works as an approximation to the FLE dynamics until $t = 10^8$. If one needs to describe subdiffusion on an even longer time scale one can increase $N$. If one needs a better precision of approximation one can make $b$ smaller. Initially all the particles are localized at the origin, $x = 0$, with the velocities thermally distributed at the temperature $T$. For the time span $t \lesssim \tau_v$ the motion is always ballistically persistent (superdiffusion). This reflects the inertia of the Brownian particle. It assumes the subdiffusive character for $t \gg \tau_v$, when the VACF is negative. The presence of a periodic potential $V(x) = -V_0 \sin(2\pi x/L)$ dramatically changes both subdiffusion, $\langle x^2(t)\rangle - \langle x(t)\rangle^2$, as well as subdiffusive transport, $\langle x(t)\rangle$, on intermediate time scale. However, the long-time asymptotics of free or biased subdiffusion are gradually attained. The initial behavior still within one potential well remains ballistic. One can conclude that both subdiffusion and subdiffusive transport are indeed asymptotically insensitive to the presence of periodic potential within the GLE approach. This finding is in a striking contrast with the FFPE approach. However, the transient to this asymptotic regime can be very slow, depending on the amplitude of the periodic potential and temperature.

An interesting phenomenon is also accelerated subdiffusion occurring on an intermediate time scale in tilted washboard potentials, as compared with the free subdiffusion. It can be detected in Fig. 2 for a strong yet subcritical

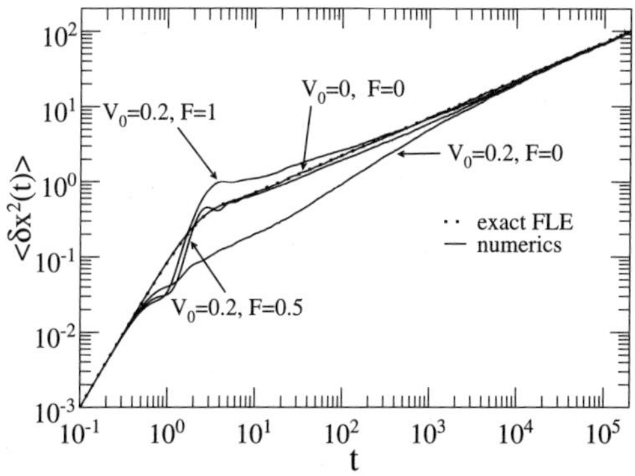


Fig. 2. Anomalous diffusion in the potential $U(x) = -V_0 \sin(2\pi x/L) - Fx$ for various $V_0$ and $F$ at $T = 0.1$ for $\alpha = 0.5$. Notice an excellent agreement (differences practically cannot be detected in this plot) of simulations with the exact FLE result, $\langle \delta x^2(t)\rangle = 2v_T^2 t^2 E_{2-\alpha,3}[-(t/\tau_v)^{2-\alpha}]$, in the absence of periodic potential $V_0 = 0$. Scaling: time in $\tau_v = (m/\eta_\alpha)^{1/(2-\alpha)}$, distance in $L$, energy in $m(L/\tau_v)^2$, force in $mL/\tau_v^2$ and temperature in $mL^2/(\tau_v^2 k_B)$.

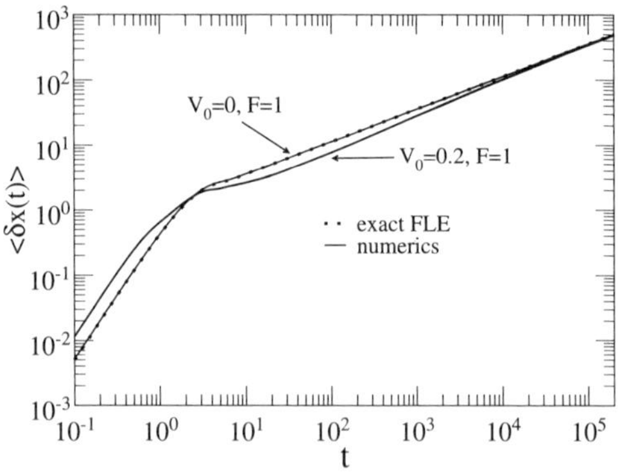


Fig. 3. Anomalous transport for $F > 0$, see Fig. 2 for details.

bias $F = 1 < F_{\rm cr} = 2\pi V_0/L \approx 1.2566\ldots$. This calls to mind the acceleration of normal diffusion in tilted washboard potentials [73]. However, this accelerated subdiffusive phenomenon occurs only on an intermediate time scale because asymptotically the GLE subdiffusion is not sensitive to the presence of the potential. One more interesting effect occurs for the initially ballistic transport. It first seems paradoxical that in the trapping potential the initial transport becomes faster than in the absence of potential and not vice versa, see in Fig. 3. The result can be understood in view of the fact that the minimum of the potential under the strong bias $F$ is essentially displaced in the direction of biasing force and the particles are *initially* accelerated by the additional to $F$ force stemming from the periodic potential.

## 5. Summary and Conclusions

With this Chapter, we reviewed and scrutinized two different approaches, the FFPE approach and the FLE approach, to anomalously slow diffusion and transport in nonlinear force fields with a focus on applications in tilted periodic potentials. In spite of some similarities in the case of constant or linear forcings it was shown that the nonlinear dynamics radically differ, obeying asymptotically two different universality classes. A first one reflects the universal fluctuations of intrinsic time clock and is closely tight to a weak ergodicity breaking. In contrast, within the GLE and FLE approaches the long-range antipersistence of the velocity and position fluctuations renders the asymptotic dynamics ergodic. One approach seems more appropriate for the disordered solids, or glass-forming liquids below the glass-forming transition, as characterized by the nonergodic glass phase. Another one seems more appropriate for the regime above but close to the glass transition, or for crowded viscoelastic environments like cytosols in biological cells. We have left out further pronounced differences between the FFPE and FLE approaches in the case of time-dependent fields [8, 63, 74–76]. We are confident that our results not only shed light on the origin of profound differences, but also will stimulate a further development of both approaches to subdiffusion, and possibly other interrelationships emerging in random potentials.

## Appendix A. Continuous Time Random Walk and Random Clock

Consider a lattice with period $a$ and a particle jumping with probabilities $q_+$ and $q_-$, $q_+ + q_- = 1$, to the neighboring sites after a random clock characterized by the residence time distribution (RTD) $\psi(\tau)$ "ticked" on the next jump. Within the physical time interval $t$ there will be a variable random number of intrinsic time periods $n(t)$. The probability to make $m$ steps forward and $n - m$ steps backward after $n$ periods is given by the binomial distribution, $P(m,n) = n!/[m!(n-m)!]q_+^m q_-^{n-m}$. Using it one can calculate the first two moments, $\langle x^k \rangle = a^k \sum_{m=0}^{n}(2m-n)^k P(m,n), k = 1, 2,$ of the particle displacement after $n$ periods:

$$\begin{aligned} \langle x(t) \rangle &= a(q_+ - q_-)n(t) \\ \langle x^2(t) \rangle &= a^2 \left[ n^2(t)(q_+ - q_-)^2 + 4n(t)q_+q_- \right]. \end{aligned} \tag{A.1}$$

They are still random quantities because of the randomness of $n(t)$. Each particle has an individual number of periods completed until $t$. For the additional ensemble average one obtains

$$\begin{aligned} \langle\langle x(t) \rangle\rangle_{\text{ens}} &= a(q_+ - q_-)\langle n(t) \rangle_{\text{ens}}. \\ \langle\langle x^2(t) \rangle\rangle_{\text{ens}} &= a^2 \left[ \langle n^2(t) \rangle_{\text{ens}}(q_+ - q_-)^2 + 4\langle n(t) \rangle_{\text{ens}} q_+q_- \right] \end{aligned} \tag{A.2}$$

and for the ensemble-averaged variance $\langle\langle [x - \langle\langle x \rangle\rangle_{\text{ens}}]^2 \rangle\rangle_{\text{ens}} = \langle\langle x^2 \rangle\rangle_{\text{ens}} - \langle\langle x \rangle\rangle_{\text{ens}}^2$

$$\langle\langle \delta x^2(t) \rangle\rangle_{\text{ens}} = a^2 \left[ \langle \delta n^2(t) \rangle_{\text{ens}}(q_+ - q_-)^2 + 4\langle n(t) \rangle_{\text{ens}} q_+q_- \right], \tag{A.3}$$

where $\langle \delta n^2(t) \rangle_{\text{ens}} = \langle n^2(t) \rangle_{\text{ens}} - \langle n(t) \rangle_{\text{ens}}^2$ is the variance of random periods passed. Notice that $\langle\langle x^2 \rangle\rangle_{\text{ens}} - \langle\langle x \rangle\rangle_{\text{ens}}^2 \neq \langle\langle [x - \langle x \rangle]^2 \rangle\rangle_{\text{ens}}$. Clearly, for a regular clock, $\langle \delta n^2(t) \rangle_{\text{ens}} = 0$ and the corresponding contribution to the ensemble-averaged position variance is absent. To simplify the notations, we further denote the ensemble averages as $\langle \ldots \rangle$ rather than $\langle\langle \ldots \rangle\rangle_{\text{ens}}$.

The physical time $t$ can be measured by the sum of *independent* stochastic periods $\tau_k$ already completed and one not yet completed period $\tau^*_{n+1}$, $t = \sum_{k=1}^{n(t)} \tau_k + \tau^*_{n+1}$, with $n = 0, 1, 2, \ldots, \infty$. Therefore, the probability distribution $p(n,t)$ to have $n$ time periods within $t$, $\sum_{n=0}^{\infty} p(n,t) = 1$, is the $(n+1)$-time convolution of the RTDs $\psi(\tau)$ ($n$ times) and of the survival probability $\Phi(\tau) = \int_{\tau}^{\infty} \psi(\tau) d\tau$. Its Laplace-transform reads

$$\tilde{p}(n,s) = \frac{1 - \tilde{\psi}(s)}{s} [\tilde{\psi}(s)]^n \tag{A.4}$$

in terms of the Laplace-transformed $\psi(\tau)$. Let us consider $\tilde{\psi}(s) \approx 1 - (s\tau_{\text{sc}})^{\alpha}$ for $s\tau_{\text{sc}} \to 0$, where $\tau_{\text{sc}}$ is a time unit of measurements. The continuous diffusion spatial limit is achieved when $a \to 0$, $\tau_{\text{sc}} \to 0$ with $\kappa_{\alpha} = a^2/\tau_{\text{sc}}^{\alpha}$ being a constant. For a finite $\tau_{\text{sc}}$, considering the scaling limit $n \to \infty$, $s\tau_{\text{sc}} \to 0$ with $n(s\tau_{\text{sc}})^{\alpha}$ being finite, one obtains

$$\tilde{p}(n,s) = \tau_{\text{sc}} (s\tau_{\text{sc}})^{\alpha-1} \exp[-n(s\tau_{\text{sc}})^{\alpha}], \tag{A.5}$$

where $n$ is considered as a continuous variable and

$$\tau(t) = n(t)\tau_{\text{sc}} \tag{A.6}$$

is the intrinsic random time. Notice that for $\alpha = 1$ one finds, $p(n,t) = \delta(n - t/\tau_{\text{sc}})$ by inversion to the time domain. That means to say that $\tau(t) = t$ is not random. For $0 < \alpha < 1$, $p(n,t)$ can be expressed via the one-sided Lévy distribution density $\mathcal{L}_{\alpha}(t)$ whose Laplace transform reads $\tilde{\mathcal{L}}_{\alpha}(s) = \exp(-s^{\alpha})$. Then, all the moments $\langle n^k(t) \rangle$ can be easily found from (A.5) to read

$$\langle n^k(t) \rangle = \frac{\Gamma(1+k)}{\Gamma(1+k\alpha)} (t/\tau_{\text{sc}})^{k\alpha}. \tag{A.7}$$

In spite of the fact that the mean time interval $\langle \tau \rangle$ does not exist all the moments of the intrinsic time $\tau(t)$ are finite. This might seem paradoxical. However, the intrinsic time scales with the number of stochastic periods passed and if the mean period does not exist the moments of $n(t)$ are nevertheless finite for any finite $t$. This is because a frequent occurrence of very long stochastic time periods within some fixed $t$ implies a *smaller* value of

$n(t)$. In particular,

$$\langle n(t)\rangle = (t/\tau_{\rm sc})^\alpha/\Gamma(1+\alpha),$$
$$\langle n^2(t)\rangle = 2(t/\tau_{\rm sc})^{2\alpha}/\Gamma(1+2\alpha) \tag{A.8}$$

and

$$\frac{\langle \delta n^2(t)\rangle}{\langle n(t)\rangle^2} = \frac{\langle \delta\tau^2(t)\rangle}{\langle \tau(t)\rangle^2} = \frac{2\Gamma^2(1+\alpha)}{\Gamma(1+2\alpha)} - 1 \tag{A.9}$$

is the most important property of the stochastic clock. It is primarily responsible for the discussed universality class of the CTRW-based subdiffusion associated with the universal fluctuations, and the weak ergodicity breaking.

## Acknowledgments

We would like to thank E. Heinsalu, M. Patriarca, G. Schmid, and P. Siegle for a very fruitful collaboration on anomalous transport in washboard potentials. This work was supported by the Deutsche Forschungsgemeinschaft, grant No. GO 2052/1-1 (I.G.) and through Nanosystems Initiative Munich (P.H.).

## References

1. M. F. Shlesinger, *J. Stat. Phys.* **10**, 421 (1974).
2. H. Scher and E. M. Montroll, *Phys. Rev. B* **12**, 2455 (1975).
3. J.-P. Bouchaud and A. Georges, *Phys. Rep.* **195**, 127 (1990).
4. B. D. Hughes, *Random Walks and Random Environments*, Vols. 1 and 2 (Clarendon Press, 1995).
5. M. F. Shlesinger, Random processes, in *Encyclopedia of Applied Physics*, Vol. 16 (VCH Publishers, 1996), pp. 45–70.
6. R. Metzler and J. Klafter, *Phys. Rep.* **339**, 1 (2000).
7. B. J. West and W. Deering, *Phys. Rep.* **246**, 2 (1994).
8. B. J. West, E. L. Geneston and P. Grigolini, *Phys. Rep.* **468**, 1 (2008).
9. G. M. Zaslavsky, *Phys. Rep.* **371**, 461 (2002).
10. I. M. Sokolov, *Physics* **1**, 8 (2008).
11. M. J. Saxton and K. Jacobson, *Annu. Rev. Biophys. Biomol. Struct.* **26**, 373 (1997).
12. G. Seisenberger *et al.*, *Science* **294**, 1929 (2001).
13. I. M. Tolić-Nørrelykke *et al.*, *Phys. Rev. Lett.* **93**, 078102 (2004).
14. I. Goychuk and P. Hänggi, *Phys. Rev. E* **70**, 051915 (2004).
15. I. Golding and E. C. Cox, *Phys. Rev. Lett.* **96**, 098102 (2006).
16. C. Wilhelm, *Phys. Rev. Lett.* **101**, 028101 (2008).
17. F. Amblard *et al.*, *Phys. Rev. Lett.* **77**, 4470 (1996).

18. J. Szymanski and M. Weiss, *Phys. Rev. Lett.* **103**, 038102 (2009).
19. S. C. Kou and X. S. Xie, *Phys. Rev. Lett.* **93**, 180603 (2004); W. Min *et al.*, *Phys. Rev. Lett.* **94**, 198302 (2005).
20. D. Mizuno, C. Tardin, C. F. Schmidt and F. C. MacKintosh, *Science* **315**, 370 (2007).
21. R. Granek and J. Klafter, *Phys. Rev. Lett.* **95**, 098106 (2005); S. Reuveni, R. Granek and J. Klafter, *ibid.* **100**, 208101 (2008); M. de Leeuw *et al.*, *PLoS One* **4**, e7296 (2009).
22. G. Bel and E. Barkai, *Phys. Rev. Lett.* **94**, 240602 (2005).
23. Y. He and S. Burov, *Phys. Rev. Lett.* **101**, 058101 (2008).
24. A. Lubelski, I. M. Sokolov and J. Klafter, *Phys. Rev. Lett.* **100**, 250602 (2008).
25. J. Nelson, *Phys. Rev. B* **59**, 15374 (1999).
26. R. Metzler, E. Barkai and J. Klafter, *Phys. Rev. Lett.* **82**, 3563 (1999).
27. E. Barkai, *Phys. Rev. E* **63**, 046118 (2001).
28. R. Gorenflo and F. Mainardi, in *Fractal and Fractal Calculus in Continuum Mechanics*, eds. A. Carpinteri and F. Mainardi (Springer, 1997), pp. 223–276.
29. I. Goychuk, E. Heinsalu, M. Patriarca, G. Schmid and P. Hänggi, *Phys. Rev. E* **73**, 020101 (2006).
30. E. Heinsalu, M. Patriarca, I. Goychuk, G. Schmid and P. Hänggi, *Phys. Rev. E* **73**, 046133 (2006).
31. P. Hänggi and H. Thomas, *Z. Phys. B* **26**, 85 (1977).
32. H. Grabert, P. Hänggi and P. Talkner, *J. Stat. Phys.* **22**, 537 (1980).
33. B. B. Mandelbrot and J. W. van Ness, *SIAM Rev.* **10**, 422 (1968).
34. I. Goychuk, *Phys. Rev. E* **80**, 046125 (2009).
35. J. H. Jeon and R. Metzler, *Phys. Rev. E* **81**, 021103 (2010).
36. J. C. Maxwell, *Phil. Trans. R. Soc. Lond.* **157**, 49 (1867).
37. A. Gemant, *Physics* **7**, 311 (1936).
38. T. G. Mason and D. A. Weitz, *Phys. Rev. Lett.* **74**, 1250 (1995).
39. F. C. MacKintosh and C. F. Schmidt, *Curr. Opin. Coll. Interface Sci.* **4**, 300 (1999).
40. P. Reimann, *Chem. Phys.* **268**, 337 (2001).
41. E. Lutz, *Phys. Rev. E* **64**, 051106 (2001).
42. W. T. Coffey, Y. P. Kalmykov and J. T. Waldron, *The Langevin Equation: With Applications to Stochastic Problems in Physics, Chemistry and Electrical Engineering*, 2nd edn. (World Scientific, 2004).
43. I. Goychuk and P. Hänggi, *Phys. Rev. Lett.* **99**, 200601 (2007).
44. S. Burov and E. Barkai, *Phys. Rev. Lett.* **100**, 070601 (2008).
45. R. Kubo, *Rep. Prog. Phys.* **29**, 255 (1966).
46. R. Kubo, M. Toda and M. Hashitsume, *Nonequilibrium Statistical Mechanics*, 2nd edn. (Springer, 1991).
47. R. Zwanzig, *Nonequilibrium Statistical Mechanics* (Oxford Univ. Press, 2001).
48. P. Hänggi, P. Talkner and B. Borkovec, *Rev. Mod. Phys.* **62**, 251 (1990).
49. U. Weiss, *Quantum Dissipative Systems*, 2nd edn. (World Scientific, 1999).
50. A. Taloni, A. Chechkin and J. Klafter, *Phys. Rev. Lett.* **104**, 160602 (2010).
51. S. A. Adelman, *J. Chem. Phys.* **64**, 124 (1976).

52. P. Hänggi, H. Thomas, H. Grabert and P. Talkner, *J. Stat. Phys.* **18**, 155 (1978).
53. J. T. Hynes, *J. Phys. Chem.* **90**, 3701 (1986).
54. H. C. Fogedby, *Phys. Rev. E* **50**, 1657 (1994); A. A. Stanislavsky, *ibid.* **67**, 021111 (2003); M. Magdziarz, A. Weron and K. Weron, *ibid.* **75**, 016708 (2007).
55. N. Pottier, *Physica A* **317**, 371 (2003).
56. W. H. Deng and E. Barkai, *Phys. Rev. E* **79**, 011112 (2009).
57. I. M. Sokolov, E. Heinsalu, P. Hänggi and I. Goychuk, *Europhys. Lett.* **86**, 30009 (2009).
58. M. Z. Ding and W. M. Yang, *Phys. Rev. E* **52**, 207 (2005).
59. R. L. Stratonovich, *Topics in the Theory of Random Noise*, Vol. II (Gordon and Breach, 1967).
60. P. Reimann, *Phys. Rep.* **361**, 57 (2002).
61. P. Hänggi and F. Marchesoni, *Rev. Mod. Phys.* **81**, 387 (2009).
62. E. Heinsalu, M. Patriarca, I. Goychuk and P. Hänggi, *J. Phys.: Condens. Matt.* **19**, 065114 (2007).
63. E. Heinsalu, M. Patriarca, I. Goychuk and P. Hänggi, *Phys. Rev. E* **79**, 041137 (2009).
64. I. Goychuk, *Chem. Phys.* **375**, 450 (2010).
65. P. Hänggi and F. Mojtabai, *Phys. Rev. A* **26**, 1168 (1982).
66. Y. C. Chen and J. L. Lebowitz, *Phys. Rev. B* **46**, 10743 (1992).
67. M. Ferrario and P. Grigolini, *J. Math. Phys.* **20**, 2567 (1979); F. Marchesoni and P. Grigolini, *J. Chem. Phys.* **78**, 6287 (1983).
68. R. Kupferman, *J. Stat. Phys.* **114**, 291 (2004).
69. P. Siegle, I. Goychuk, P. Talkner and P. Hänggi, *Phys. Rev. E* **81**, 011136 (2010).
70. R. G. Palmer, D. L. Stein, E. Abrahams and P. W. Anderson, *Phys. Rev. Lett.* **53**, 958 (1984).
71. M. B. Weismann, *Rev. Mod. Phys.* **60**, 537 (1988).
72. T. C. Gard, *Introduction to Stochastic Differential Equations* (Dekker, 1988).
73. P. Reimann, C. Van der Broek, H. Linke, P. Hänggi, J. M. Rubi and A. Perez-Madrid, *Phys. Rev. Lett.* **87**, 010602 (2001); *Phys. Rev. E* **65**, 031104 (2002).
74. I. M. Sokolov and J. Klafter, *Phys. Rev. Lett.* **97**, 140602 (2006).
75. E. Heinsalu, M. Patriarca, I. Goychuk and P. Hänggi, *Phys. Rev. Lett.* **99**, 120602 (2007).
76. I. Goychuk, *Phys. Rev. E* **76**, 040102(R) (2007).

# Chapter 14

## Identification and Validation of Fractional Subdiffusion Dynamics

Krzysztof Burnecki, Marcin Magdziarz and Aleksander Weron
*Hugo Steinhaus Center, Institute of Mathematics and Computer Science, Wroclaw University of Technology, Wyspianskiego 27, 50-370 Wroclaw, Poland*

In this chapter we propose a systematic methodology on how to distinguish between three mechanisms leading to single molecule subdiffusion, namely fractional Brownian motion, fractional Lévy stable motion and Fractional Fokker–Planck equation. We illustrate step by step that the methods of sample mean-squared displacement and $p$-variation can be successfully applied for infinite and confined systems. This methodology is based on well-known and not so well-known statistical tools for identification and validation of the above three fractional dynamical systems.

1. Introduction . . . . . . . . . . . . . . . . . . . . . . . . . 331
2. Three Models of Fractional Subdiffusion Dynamics . . . . 333
   2.1. Fractional Brownian motion . . . . . . . . . . . . . 333
   2.2. Fractional Lévy stable motion . . . . . . . . . . . . 333
   2.3. Fractional Fokker–Planck equation . . . . . . . . . 334
3. Selected Identification and Validation Tools . . . . . . . . 335
   3.1. Kolmogorov–Smirnov statistic . . . . . . . . . . . . 335
   3.2. FIRT estimator . . . . . . . . . . . . . . . . . . . . 337
4. Sample MSD . . . . . . . . . . . . . . . . . . . . . . . . 339
5. Sample $p$-variation . . . . . . . . . . . . . . . . . . . . 340
6. Statistical Validation . . . . . . . . . . . . . . . . . . . 343
7. The Case of Confined Systems . . . . . . . . . . . . . . . 346
8. Conclusions . . . . . . . . . . . . . . . . . . . . . . . . 349

## 1. Introduction

The issue of distinguishing between normal and anomalous diffusion, as such, concerns many fields of physics [1–13]. It is usually based on the analysis of the mean-squared displacement (MSD) of the diffusing particles.

In the case of classical diffusion, the second moment is linear in time, whereas anomalous diffusion processes exhibit distinct deviations from this fundamental property: $\langle x^2(t)\rangle \sim t^a$, where for $0 < a < 1$ is subdiffusive and for $a > 1$ is superdiffusive [12]. The origin of anomalous dynamics in a given system is often unknown.

It is not always clear which model applies to a particular system [10, 11, 14], information which is essential when diffusion-controlled processes are considered. Therefore, determining the appropriate model is an important and timely problem; see [10, 11, 15] for discussion on the origins of anomaly in the case of intracellular diffusion.

The MSD can be obtained either by performing an average over an ensemble of particles, or by taking the temporal average over a single trajectory [16, 17]. Recent advances in single molecule spectroscopy enabled single particle tracking experiments following individual particle trajectories [11, 15]. These require temporal moving averages.

Therefore, determining the origins of subdiffusive transport is a vital and timely problem [10, 11, 15, 18, 19]. In Sec. 2, we describe shortly three different models of fractional subdiffusion dynamics. Section 3 contains description of basic identification and validation tools like Kolmogorov–Smirnov statistic and FIRT estimator. We show that they can be used for model identification of stationary or increment stationary data. In Sec. 4, we propose to replace the MSD with sample MSD since the sample (time average) MSD of the fractional Lévy $\alpha$-stable motion (FLSM) behaves very differently from the corresponding ensemble average (second moment). Namely, while the ensemble average MSD diverges for $\alpha < 2$, the sample MSD may exhibit either subdiffusion, normal diffusion or superdiffusion. Traditionally, $\alpha$-stable processes like Lévy flights serve as examples of superdiffusive dynamics since for $\alpha < 2$ their second moment is infinite and the MSD diverges. Nevertheless, $H$-self-similar Lévy stable processes can model either a subdiffusive, diffusive or superdiffusive dynamics in the sense of sample MSD. We also show that the character of the process is controlled by a sign of the memory parameter $d = H - 1/\alpha$. In Sec. 5, we present in detail a new sample $p$-variation test (algorithm) which allows to distinguish between three models of subdiffusive dynamics on simulated data. We already identified [7–9] fractional subdiffusive dynamics on biological data describing the motion of individual fluorescently labeled mRNA molecules inside live *E. coli* cells, but it may concern many other fields of contemporary experimental physics. Finally, in Secs. 6 and 7, we present the main results of this chapter. Namely, statistical validation of the proposed models in the case of freely moving particles and confined systems, respectively.

## 2. Three Models of Fractional Subdiffusion Dynamics

### 2.1. *Fractional Brownian motion*

In the literature, two popular stochastic models have been used to account for anomalous diffusion. The first one is the fractional Brownian motion (FBM) introduced by A. N. Kolmogorov in 1940 [20–22]. The second model of subdiffusion is the continuous time random walk (CTRW) and the corresponding fractional Fokker–Planck equation [12,23]. However, they do not exhaust all possible sources of anomalous diffusion. Another source could be random walks on fractal structures, fractional Langevin equations, generalized Langevin equations, percolation, etc. [12,24].

FBM is a generalization of the classical Brownian motion (BM). Most of its statistical properties are characterized by the Hurst exponent $0 < H < 1$. In particular, the MSD of FBM satisfies $\langle x^2(t)\rangle \sim t^{2H}$, thus for $H < 1/2$ we obtain the subdiffusive dynamics, whereas for $H > 1/2$ the superdiffusive one. For further properties of FBM and its applications to physics see [25–29].

For any $0 < H < 1$, FBM of index $H$ (Hurst exponent) is the mean-zero Gaussian process $B_H(t)$ with the following integral representation [21,26]:

$$B_H(t) = \int_{-\infty}^{\infty} \{(t-u)_+^{H-\frac{1}{2}} - (-u)_+^{H-\frac{1}{2}}\} dB(u), \tag{1}$$

where $B(t)$ is a standard Brownian motion and $(x)_+ = \max(x, 0)$.

FBM is $H$-self-similar, namely for every $c > 0$ we have $B_H(ct) = c^H B_H(t)$ in distribution, and has stationary increments. It is the only Gaussian process satisfying these properties. For $H > 1/2$, the increments of the process are positively correlated and exhibit long-range dependence (long memory, persistence), whereas for $H < 1/2$, the increments of the process are negatively correlated and exhibit short-range dependence (short memory, antipersistence) [26]. For the second moment of the FBM we have $\langle B_H^2(t)\rangle = \sigma^2 t^{2H}$, where $\sigma > 0$, which for $H < 1/2$ gives the subdiffusive dynamics and for $H > 1/2$ the superdiffusive one.

### 2.2. *Fractional Lévy stable motion*

FBM can be generalized to a fractional Lévy stable motion (FLSM) [25,26, 30–32]:

$$L_H^\alpha(t) = \int_{-\infty}^{\infty} \{(t-u)_+^d - (-u)_+^d\} dL_\alpha(u), \tag{2}$$

where $L_\alpha(t)$ is a Lévy $\alpha$-stable motion (LSM), $0 < \alpha \leq 2$, $0 < H < 1$, and $d = H - 1/\alpha$. The process is $\alpha$-stable (for $\alpha = 2$ it becomes a FBM), $H$-self-similar and has stationary increments. Analogously to the FBM case, we say the increments of the process exhibit positive (long-range) dependence if $d > 0$ ($H > 1/\alpha$), and negative dependence when $d < 0$ ($H < 1/\alpha$) [31,32]. This is due to the behavior of the integrand in (2). Therefore, as in the Gaussian case, the parameter $d$ controls sign of dependence.

We show in Sec. 4 that the time average MSD of FLSM behaves very differently from the corresponding ensemble average (second moment). This is a timely subject since single molecule experiments exhibit both anomalous kinetics and a large scatter of the time average MSD. While the ensemble average MSD diverges, the time average MSD may exhibit either subdiffusion, normal diffusion or superdiffusion. Thus in experiment what seems subdiffusive from a single trajectory analysis could in fact be superdiffusive in the ensemble sense.

### 2.3. *Fractional Fokker–Planck equation*

Force-free subdiffusion in the framework of CTRW with heavy-tailed waiting times is conveninty described by the fractional Fokker–Planck equation (FFPE) [12,33]

$$\frac{\partial w(x,t)}{\partial t} = {}_0D_t^{1-\beta}\left[\frac{1}{2}\frac{\partial^2}{\partial x^2}\right] w(x,t) \tag{3}$$

with the initial condition $w(x,0) = \delta(x)$. The operator ${}_0D_t^{1-\beta}$, $0 < \beta < 1$, is the fractional derivative of the Riemann–Liouville type. The MSD corresponding to $w(x,t)$ equals $\frac{t^\beta}{\Gamma(\beta+1)}$, which is characteristic for subdiffusive dynamics.

In Eq. (3), $w(x,t)$ denotes the PDF of some subdiffusive stochastic process $Z_\beta(t)$. The process $Z_\beta(t)$ can be explicitly represented in the following subordination form [34–36]

$$Z_\beta(t) = B(S_\beta(t)), \tag{4}$$

where $B(t)$ is the standard Brownian motion and $S_\beta(t)$ is the so-called inverse $\beta$-stable subordinator defined as

$$S_\beta(t) = \inf\{\tau > 0 \colon U_\beta(\tau) > t\}. \tag{5}$$

Here, $U_\beta(\tau)$ is the $\beta$-stable subordinator [31,32] with the Laplace transform given in the stretched exponential form $E(e^{-uU_\beta(\tau)}) = e^{-\tau u^\beta}$. Moreover, $S_\beta(t)$ is assumed to be independent of $B(t)$.

The Langevin-type process $Z_\beta(t)$ reveals the detailed structure of trajectories corresponding to FFPE (3). Therefore, it allows one to study the statistical properties of the trajectories of subdiffusion in the framework of CTRW with heavy-tailed waiting times.

## 3. Selected Identification and Validation Tools

We now collect a list of identification and validation tools for fractional subdiffusive dynamics used in this Chapter. In Table 1 they are presented along with the related characteristics of the data. Kolmogorov–Smirnov statistic and sample $p$-variation stand for examples of both identification and validation tools, whereas FIRT method and sample MSD are examples of identification tools.

The first tool depicted in Table 1 assumes that the data are stationary. Stationary and nonstationary processes are very different in their properties, and they require different inference procedures. At this point, note that a simple and useful method to tell if a process is stationary in empirical studies is to plot the data. Loosely speaking, if a series does seem to have a trend, or a varying volatility, then very likely, it is not stationary. To make the process stationary it is sometimes enough to calculate its increments.

### 3.1. *Kolmogorov–Smirnov statistic*

Once the distribution class is selected and the parameters are estimated using one of the available methods, the goodness-of-fit has to be tested. A standard approach consists of measuring the distance between the empirical and the fitted analytical distribution function. A group of statistics and tests based on this idea has been discussed [37].

Table 1. List of simple identification and validation tools.

| Tools | Characteristic of the data |
|---|---|
| Kolmogorov–Smirnov statistic | Distribution type |
| FIRT method | Self-similarity |
| Sample mean-squared displacement | Diffusion type, long memory |
| Sample $p$-variation | Diffusion type |

A statistics measuring the difference between the empirical $F_n(x)$ and the fitted $F(x)$ distribution functions, called an empirical distribution function (edf) statistic, is based on the vertical difference between the distributions. This distance is usually measured either by a supremum or a quadratic norm [38].

The most popular supremum statistic:

$$D = \sup_x |F_n(x) - F(x)|, \tag{6}$$

is known as the Kolmogorov or Kolmogorov–Smirnov (KS) statistic. For other classes of measures of discrepancy see, e.g. [37].

Suppose that a sample $x_1, \ldots, x_n$ gives values $z_i = F(x_i), i = 1, \ldots, n$. It can be easily shown that, for values $z$ and $x$ related by $z = F(x)$, the corresponding vertical differences in the edf diagrams for $X$ and $Z$ are equal. Consequently, edf statistics calculated from the empirical distribution function of the $z_i$'s compared with the uniform distribution will take the same values as if they were calculated from the empirical distribution function of the $x_i$'s, compared with $F(x)$. This leads to the following formulas given in terms of the order statistics $z_{(1)} < z_{(2)} < \cdots < z_{(n)}$:

$$D^+ = \max_{1 \le i \le n} \left\{ \frac{i}{n} - z_{(i)} \right\}, \tag{7}$$

$$D^- = \max_{1 \le i \le n} \left\{ z_{(i)} - \frac{(i-1)}{n} \right\}, \tag{8}$$

$$D = \max(D^+, D^-), \tag{9}$$

Kolmogorov–Smirnov statistic can serve as a simple identification tool in the following way. We calculate the statistic for different choices of $F(x)$, e.g., Gaussian and stable distributions. W choose such distribution for further analysis that has smaller value of the statistic. In Table 2 we can see calculated values of the KS statistic for a sample generated from a Lévy stable distribution with $\alpha = 1.85$, $\sigma = 1$, $\beta = \mu = 0$, for two possible choices of $F(x)$, namely Gaussian and Lévy stable. The parameters of the distributions are maximum likelihood estimates [39]. Clearly, KS statistic for the stable distribution is much lower than the one calculated for the Gaussian case.

In order to use Kolmogorov–Smirnov statistic for validation purposes, we have to construct a proper statistical test. The general test of fit is structured as follows. The null hypothesis is that a specific distribution is

Table 2. Parameter estimates obtained via the maximum likelihood method and Kolmogorov–Smirnov test statistic for a sample generated from a Lévy stable distribution with $\alpha = 1.85$, $\sigma = 1$, $\beta = \mu = 0$. The corresponding $p$-values based on 1000 simulated samples are given in parentheses.

| Distributions | Gaussian | Stable |
|---|---|---|
| Parameters | $\mu = 0.0123$ | $\alpha = 1.8410$ |
| | $\sigma = 2.4245$ | $\sigma = 1.0048$ |
| | | $\beta = 0.0920$ |
| | | $\mu = 0.0581$ |
| Kolmogorov–Smirnov test | 0.1239 | 0.005 |
| | (<0.001) | (0.812) |

acceptable, whereas the alternative is that it is not:

$$H_0\colon F_n(x) = F(x;\theta),$$
$$H_1\colon F_n(x) \neq F(x;\theta),$$

where $\theta$ is a vector of known parameters. Small values of the test statistic $D$ are evidence in favor of the null hypothesis, large ones indicate its falsity. To see how unlikely such a large outcome would be if the null hypothesis was true, we calculate the $p$-value by:

$$p\text{-value} = P(D \geq t), \tag{10}$$

where $t$ is the test value for a given sample. It is typical to reject the null hypothesis when a small $p$-value is obtained. The detailed procedure on how to calculate $p$-values for different hypothetical distributions has been described [37].

### 3.2. *FIRT estimator*

Recently introduced [40] estimation procedure of the self-similarity parameter $H$ is the so-called finite impulse response transformation (FIRT). The FIRT estimator involves an array of coefficients. The array is made out of finite impulse response coefficients. The estimator $H_{\text{FIRT}}$ is obtained by performing a log-linear regression on the coefficients and measuring the slope, which is equal to $H$. This is illustrated in Fig. 1(a) for the simulated data from a fractional Lévy stable motion with $H = 0.4$ and $\alpha = 1.85$.

It is important to note that the estimator $H_{\text{FIRT}}$ is considered as one of the most accurate and reliable methods of estimating the Hurst parameter.

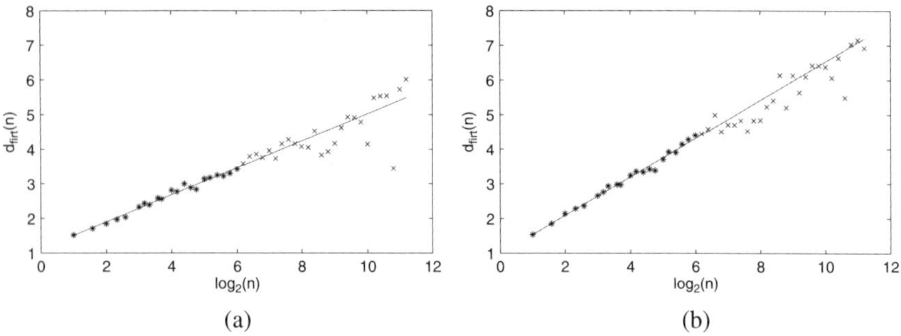


Fig. 1. Finite impulse response transformation (FIRT) for a time series simulated from FLSM with $H = 0.4$ and $\alpha = 1.85$. The estimator $H_{\mathrm{FIRT}}$ is obtained by fitting a least-squares line to the values of the coefficients $d_{\mathrm{FIRT}}(n)$ on a log-linear scale. The points at the high end (marked by "x") are not used to fit the line. (a) The estimated slope of the line for the original data is $H_{\mathrm{FIRT}} = 0.3906$. (b) The estimated slope of the line for the surrogate data is $H_{\mathrm{FIRT}} = 0.5549$, which corresponds to $1/\alpha \sim 0.5405$.

It is unbiased for all $0 < \alpha < 2$. Moreover, for $1 < \alpha < 2$, under certain technical conditions, the estimator is consistent and asymptotically normal.

We note that the estimator gives information on the self-similarity and not on the distribution of the process. However, applying the BMW$^2$ test [26, 41] one can identify the stability index $\alpha$ of the data. The test applies the concept of surrogate data [42], which refers to data that preserve certain linear statistic properties of the experimental time series, without the deterministic component. The surrogate data can be obtained by several different ways. Here, we obtain it by random shuffling of the original data positions.

According to [26] we have he following BMW$^2$ computer test.

- If the data come from FBM, then the values of the applied estimator should change to 1/2 for the surrogate data independently on the initial values.
- If the data come from BM or LSM, then the estimator values should be the same for the original and surrogate data and equal to 1/2 and $1/\alpha$, respectively.
- If the data come from FLSM, then we should observe a change to $1/\alpha$ in the estimators' values.

Therefore, the test can be used to distinguish between diffusion (BM and LSM) and anomalous diffusion (FBM and FLSM) models. This is illustrated in Fig. 1. Comparing Figs. 1(a) and 1(b) we can observe a change of the value of the estimator from 0.3906 to 0.5549. This clearly suggests FLSM as the underlying model, which is true since the data come from FLSM with $H = 0.4$ and $\alpha = 1.85$. Moreover, the FIRT estimator for the surrogate data yields a good approximation of the $1/\alpha$ value.

## 4. Sample MSD

Let $\{X_i,\ i = 0, \ldots, N\}$ be a sample of length $N + 1$. We introduce the sample MSD:

$$M_N(\tau) = \frac{1}{N - \tau + 1} \sum_{k=0}^{N-\tau} (X_{k+\tau} - X_k)^2. \tag{11}$$

The sample MSD is a time average MSD on a finite sample regarded as a function of difference $\tau$ between observations. It is a random variable in contrast to the ensemble average which is deterministic.

We assume now that $N$ is large and $\tau$ small.

- If the sample comes from a fractional Lévy $\alpha$-stable motion with $\alpha < 2$, then

$$M_N(\tau) \overset{d}{\sim} \tau^{2d+1}, \tag{12}$$

  where $d = H - 1/\alpha$ and $\overset{d}{\sim}$ means similarity in distribution [9].
- If the sample comes from a fractional Brownian motion, then

$$M_N(\tau) \overset{d}{\sim} \tau^{2d+1}, \tag{13}$$

  where $d = H - 1/2$.
- If the sample comes from a CTRW, then

$$M_N(\tau) \overset{d}{\sim} \tau, \tag{14}$$

  which corresponds to Eqs. (12) and (13) with $d = 0$, which is the case of the regular diffusion [16].

Therefore, the sample MSD is an identification tool which allows one to distinguish between FFPE and other two models if estimated $d \neq 0$.

Applying the above formula, we estimated memory parameter $d$ for generated subdiffusive FBM, FLSM, and CTRW. The results are presented in Fig. 2 in the form of the so-called box plots [43, 44]. The simulated samples are of length $2^{12}$. The box plot produces a box and whisker plot for each value of $d$. The box has lines at the lower quartile, median, and upper

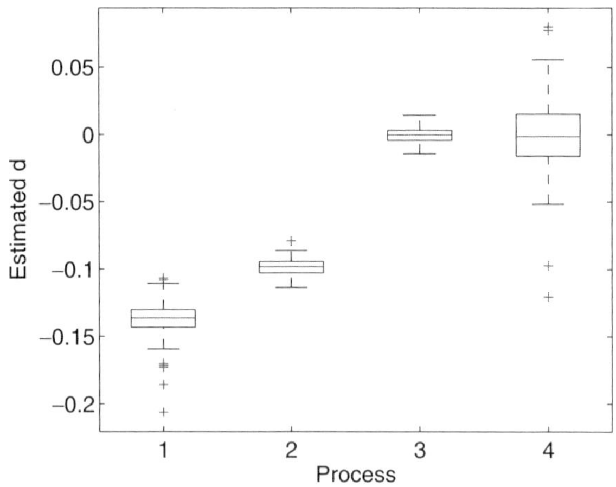


Fig. 2. Box plots of the sample MSD estimates of the memory parameter $d$ obtained via 100 simulated trajectories of FLSM and FPPE of length $2^{12}$ with different $H$'s and $\alpha$'s. (1) FLSM with $H = 0.4$, $\alpha = 1.85$, $d \sim -0.1405$. (2) FLSM with $H = 0.4$, $\alpha = 2$, $d = -0.1$ (FBM). (3) FLSM with $H = 1/1.85$, $\alpha = 1.85$, $d = 0$ (LSM). (4) FFPE with $\beta = 0.8$, $d = 0$.

quartile values. The whiskers are lines extending from each end of the box to show the extent of the rest of the data. Outliers are data with values beyond the ends of the whiskers. We can see that the medians of estimated $d$ coincide well with the theoretical values.

In particular, for LSM and FFPE we arrive at the diffusion case, namely $M_N(\tau) \sim \tau$ since $d = 0$, see also [45].

As a consequence, we see that the memory parameter $d$ controls the type of anomalous diffusion. If $d < 0$ ($H < 1/\alpha$), so in the negative dependence case, the process follows the subdiffusive dynamics, if $d > 0$ ($H > 1/\alpha$), the character of the process changes to superdiffusive. What is even more amazing, it appears that Lévy $\alpha$-stable processes for $\alpha < 2$ can serve both as examples of subdiffusion and superdiffiusion. The subdiffusion pattern arises when the dependence is negative, so possible large positive jumps are quickly compensated by large negative jumps, and on average the process travels shorter distances than the light-tailed Brownian motion.

## 5. Sample $p$-variation

Another method, which can be successfully applied to identify the type of subdiffusion in the experimental data, is the method of $p$-variation

[7–9]. The idea of $p$-variation generalizes the well-established notions of total variation and quadratic variation, which have found applications in various branches of mathematics, physics and engineering, including: optimal control, numerical analysis of differential equations, and calculus of variations [46].

For a stochastic process $X(t)$ observed on time interval $[0, T]$, the corresponding $p$-variation is defined as

$$V^{(p)}(t) = \lim_{n\to\infty} V_n^{(p)}(t), \quad p > 0. \tag{15}$$

Here, $V_n^{(p)}(t)$ is the sum of powers of increments of the process $X(t)$

$$V_n^{(p)}(t) = \sum_{j=0}^{2^n-1} \left| X\left(\frac{(j+1)T}{2^n} \wedge t\right) - X\left(\frac{jT}{2^n} \wedge t\right)\right|^p \tag{16}$$

with $a \wedge b = \min\{a, b\}$. $V_n^{(p)}(t)$ is called the *sample $p$-variation*. We underline that $V_n^{(p)}(t)$ is very easy to calculate numerically. For large enough $n$, the sample $p$-variation is a good approximation of $V^{(p)}(t)$.

It appears that for each subdiffusion model (FBM, FLSM and FFPE) considered here, the $p$-variation displays completely different behavior. This interesting fact plays a crucial role in identifying the proper model of subdiffusion.

The $p$-variation of the FBM $B_H(t)$ satisfies [47]

$$V^{(p)}(t) = \begin{cases} +\infty & \text{if } p < \dfrac{1}{H}, \\ tE(|B_H(1)|^{1/H}) & \text{if } p = \dfrac{1}{H}, \\ 0 & \text{if } p > \dfrac{1}{H}. \end{cases} \tag{17}$$

The expected value in the above formula equals $E(|B_H(1)|^{1/H}) = \frac{\sigma^{1/H} 2^{1/2H}}{\sqrt{\pi}} \Gamma(\frac{1}{2H} + \frac{1}{2})$. Here, $\sigma > 0$ is the scale parameter of $B_H(t)$. Observe that $p = 1/H$ is the critical value in the above formula for FBM.

Since the trajectories of the FLSM in the subdiffusive regime $d < 0$ ($H < 1/\alpha$) are nowhere bounded, the $p$-variation corresponding to $L_H^\alpha(t)$ is equal to infinity

$$V^{(p)}(t) = \infty \tag{18}$$

for any $p > 0$.

The $p$-variation of the Langevin-type process $Z_\beta(t) = B(S_\beta(t))$ corresponding to FFPE (3) satisfies [23]

$$V^{(p)}(t) = \begin{cases} +\infty & \text{if } p < 2, \\ S_\beta(t) & \text{if } p = 2, \\ 0 & \text{if } p > 2. \end{cases} \tag{19}$$

Note that $p = 2$ is the critical value for the $p$-variation of $Z_\beta(t)$.

The above formulas (17)–(19) confirm that the $p$-variation of the discussed subdiffusion models differs considerably. This suggests the following $p$-variation test for distinguishing between FBM, FLSM and FFPE dynamics [7–9]:

- If the underlying model is FBM $B_H(t)$, then by (17) the sample $p$-variation $V_n^{(p)}(t)$, as a function of $n$, should behave in the following way: (a) for $p < 1/H$, $V_n^{(p)}(t)$ should increase with increasing $n$; (b) for $p = 1/H$, $V_n^{(p)}(t)$ should behave as the linear function $tE(|B_H(1)|^{1/H})$; (c) for $p > 1/H$, $V_n^{(p)}(t)$ should decrease with increasing $n$ (see Fig. 3, (a) and (b)).
- If the underlying model is $Z_\beta(t)$ corresponding to FFPE (3), then by (19) the sample $p$-variation $V_n^{(p)}(t)$, as a function of $n$, should behave in the following way: (a) for $p < 2$, $V_n^{(p)}(t)$ should increase with increasing $n$; (b) for $p = 2$, $V_n^{(p)}(t)$ should stabilize around $S_\beta(t)$; (c) for $p > 2$, $V_n^{(p)}(t)$ should decrease with increasing $n$ (see Fig. 3, (c) and (d)).
- If the underlying model is FLSM in the subdiffusive regime $d < 0$, then by (18) the sample $p$-variation $V_n^{(p)}(t)$ should increase with increasing $n$ for any parameter $p > 0$ (see Fig. 3, (e) and (f)).

In Fig. 3 we illustrate the behavior of sample $p$-variation for all three subdiffusion models. The obvious differences in the behavior of $V_n^{(p)}(t)$ corresponding to FBM, FLSM, and FFPE allow to identify the underlying subdiffusion mechanism.

In practice, for a given time series $\{X_i, i = 1, 2, \ldots, 2^N\}$ of experimental observations, one needs to calculate the sample $p$-variation $V_n^{(p)}(t)$ for various parameters $p$ and $n$. Next, one compares the obtained results with the theoretical ones corresponding to FBM, FLSM, and FFPE, and chooses the best fitting model.

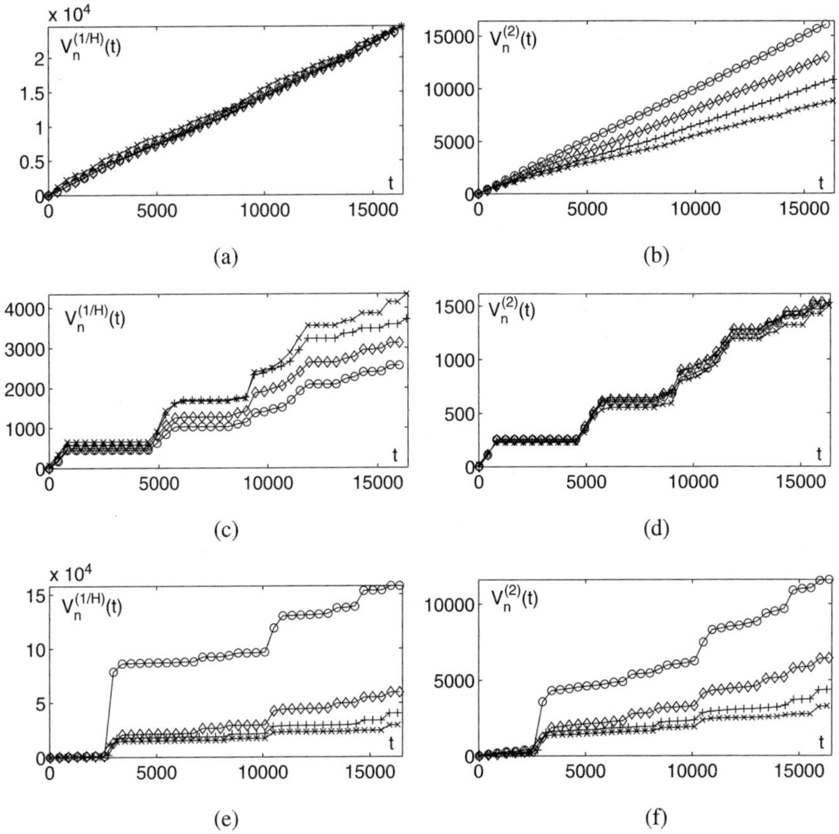


Fig. 3. Behavior of $V_n^{(1/H)}(t)$ and $V_n^{(2)}(t)$ corresponding to FBM ((a) and (b)), FFPE ((c) and (d)) and FLSM ((e) and (f)). The parameters are: $H = \beta/2 = 0.35$, $\alpha = 1.5$. Notation used: x ($n = 11$), +(n=12), ◇ (n=13), o ($n = 14$).

## 6. Statistical Validation

The $p$-variation test described in the previous section, specifically the comparison of empirical data with theoretical predictions and the classification into one of the three underlying scenarios (FBM, FLSM, FFPE), was done in a non-quantitative (visual) manner. Now, we give the $p$-variation method more formal and quantitative shape. To verify the agreement between experimental data and theory, we introduce a proper statistical test (a similar test was introduced in a recent paper [8]), which allows to classify the experimental data into one scenario or another using quantitative confidence level.

First, for a given empirical data, we verify statistically if it originates from the FBM. To be more precise, our null hypothesis is that the data comes from $B_H(t)$ for given $H$. To verify such null hypothesis, we examine the value of the following statistic

$$D_{m,n} = \frac{|\hat{V}_m^{(1/H)}(T) - \hat{V}_n^{(1/H)}(T)|}{\hat{V}_n^{(1/H)}(T)}.$$

Here, $\hat{V}_m^{(1/H)}(T)$ and $\hat{V}_n^{(1/H)}(T)$, $m < n$, are the values of sample $1/H$-variation calculated for the empirical data. Note that by (17), under the null hypothesis, the difference $D_{m,n}$ should be small for large enough $m$ and $n$. Therefore, the null hypothesis is rejected at confidence level $\hat{\alpha}$ (usually $\hat{\alpha} = 0.05$ or $\hat{\alpha} = 0.01$), if

$$D_{m,n} > D_{\hat{\alpha}}.$$

The constant $D_{\hat{\alpha}}$ is found from

$$\mathbb{P}\left(\frac{|V_m^{(1/H)}(T) - V_n^{(1/H)}(T)|}{V_n^{(1/H)}(T)} \leq D_{\hat{\alpha}}\right) = 1 - \hat{\alpha},$$

where $V_m^{(1/H)}(T)$ and $V_n^{(1/H)}(T)$ are the values of sample $1/H$-variation corresponding to FBM, i.e.

$$V_n^{(1/H)}(T) = \sum_{j=0}^{2^n-1} \left| B_H\left(\frac{(j+1)T}{2^n} \wedge T\right) - B_H\left(\frac{jT}{2^n} \wedge T\right)\right|^{1/H},$$

and similarly $V_m^{(1/H)}(T)$. Note that the constant $D_{\hat{\alpha}}$ can be easily found via Monte Carlo techniques. The methods of simulating $B_H(t)$ can be found [48].

We applied the above statistical test to the simulated trajectories of $B_H(t)$, $Z_\beta(t)$ and $L_H^\alpha(t)$. The obtained $p$-values are shown in Fig. 4(a).

Next, let us consider the null hypothesis that the data comes from FFPE. In such case the following statistic should be investigated

$$E_{m,n} = \frac{|\hat{V}_m^{(2)}(T) - \hat{V}_n^{(2)}(T)|}{\hat{V}_n^{(2)}(T)}.$$

Here, $\hat{V}_m^{(2)}(T)$ and $\hat{V}_n^{(2)}(T)$, $m < n$, are the values of sample 2-variation corresponding to the empirical trajectory. Now, by (19), under the null

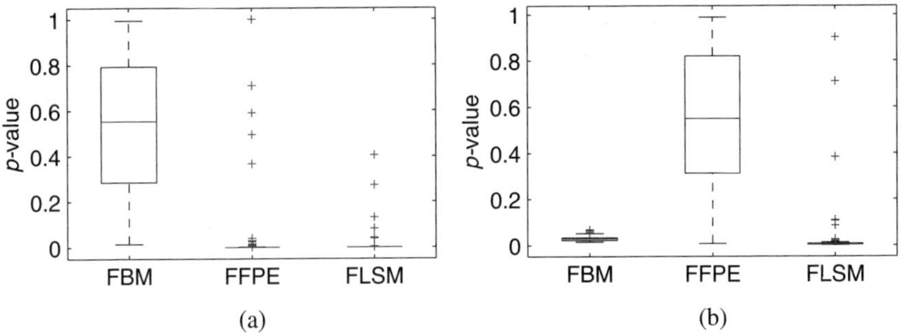


Fig. 4. (a) Results of the introduced statistical test with the null hypothesis that the data comes from FBM. Shown are three boxplots, each with the $p$-values calculated for 100 simulated trajectories of FBM, FFPE and FLSM, respectively. The test accepted almost all the trajectories of FMB, whereas the median of $p$-values obtained for FFPE and FLSM is lower than 0.05. Recall that if the $p$-value is smaller than the confidence level $\hat{\alpha} = 0.05$, the null hypothesis should be rejected. (b) Results of the introduced test for the null hypothesis that the data comes from FFPE model $Z_\beta(t)$. In this case almost all the simulated trajectories of $Z_\beta(t)$ were accepted, while the median of $p$-values obtained for FBM and FLSM is lower than 0.05. The length of trajectories is $2^{12}$, number of trajectories equals 100 for each model, and the parameters are: $H = \beta/2 = 0.4$, $\alpha = 1.5$, $m = 9$ and $n = 12$.

hypothesis the distance $E_{m,n}$ should be small for large enough $m$ and $n$. Consequently, the null hypothesis is rejected at confidence level $\hat{\alpha}$, if

$$E_{m,n} > E_{\hat{\alpha}},$$

where the constant $E_{\hat{\alpha}}$ is derived from

$$\mathbb{P}\left(\frac{|V_m^{(2)}(T) - V_n^{(2)}(T)|}{V_n^{(2)}(T)} \leq E_{\hat{\alpha}}\right) = 1 - \hat{\alpha}.$$

Here $V_m^{(2)}(T)$ and $V_n^{(2)}(T)$ are the values of sample 2-variation corresponding to $Z_\beta(t)$, i.e.

$$V_n^{(2)}(T) = \sum_{j=0}^{2^n-1} \left| Z_\beta\left(\frac{(j+1)T}{2^n} \wedge T\right) - Z_\beta\left(\frac{jT}{2^n} \wedge T\right)\right|^2,$$

and similarly $V_m^{(2)}(T)$. Similarly as before, the constant $E_{\hat{\alpha}}$ should be determined via Monte Carlo techniques. The method of simulating $Z_\beta(t)$ can be found [35, 49].

The above statistical test was applied to the simulated trajectories of $B_H(t)$, $Z_\beta(t)$ and $L_H^\alpha(t)$. The results are presented in Fig. 4(b).

The above tests allow one to validate statistically that the data originates from FBM (or from FFPE). However, the same methodology cannot be applied in a straightforward manner to the case of FLSM. The reason is that for the last process, there is no critical value of $p$, for which the corresponding sample $p$-variation would stabilize. Thus, a different approach to this problem is necessary. Some steps in this direction are in progress.

## 7. The Case of Confined Systems

The results and tests introduced in previous sections were done for the case of freely moving particles. The considered system had no boundaries and no external force was present. Such setting is convenient for a number of physical and biological systems, however, there are some cases for which confinement influences considerably the dynamics of particles. Therefore, it is of great interest to extend the statistical methods of identifying and validating the type of subdiffusion also to the case of confined systems.

FBM in confinement is governed by the fractional Langevin equation of the form

$$dY_H(t) = F(Y_H(t))dt + dB_H(t), \tag{20}$$

with $F(x)$ being the external force.

Similarly, confined subdiffusion in the framework of CTRW with heavy-tailed waiting times is described by the fractional Fokker–Planck equation [12, 33]

$$\frac{\partial w(x,t)}{\partial t} = {}_0D_t^{1-\beta}\left[-\frac{\partial F(x)}{\partial x} + K\frac{\partial^2}{\partial x^2}\right]w(x,t). \tag{21}$$

The force $F(x)$, which is assumed to be continuous, is related to the binding potential $v(x)$ through $F(x) = -v'(x)$. The constant $K$ denotes the anomalous diffusion coefficient. Equation (21) describes the evolution in time of the PDF $w(x,t)$ of some subdiffusion process $W_\beta(t)$. This Langevin-type process has the following representation [34–36]

$$W_\beta(t) = X(S_\beta(t)), \tag{22}$$

where $X(t)$ is the solution of the Itô stochastic differential equation

$$dX(\tau) = F(X(\tau))d\tau + (2K)^{1/2}dB(\tau) \tag{23}$$

driven by the standard Brownian motion $B(\tau)$ and $S_\beta(t)$ is the inverse stable subordinator (5).

In this section we restrict ourselves only to the above two models of subdiffusion ($Y_H(t)$ and $W_\beta(t)$). The case of FLSM in confinement requires further extensive studies.

Now, our goal is to extend the previously introduced methods of identification based on $p$-variation to the case of confined systems described by (20) and (22). The crucial, and somewhat surprising fact is that the presence of external force does not modify the properties of $p$-variation. One can show [8] that the $p$-variation corresponding to $Y_H(t)$ yields

$$V^{(p)}(t) = \begin{cases} +\infty & \text{if } 0 < p < \frac{1}{H}, \\ tE(|B_H(1)|^{1/H}) & \text{if } p = \frac{1}{H}, \\ 0 & \text{if } p > \frac{1}{H}. \end{cases} \tag{24}$$

Analogously, one proves [8] that $V^{(p)}(t)$ corresponding to $W_\beta(t)$ is given by

$$V^{(p)}(t) = \begin{cases} +\infty & \text{if } 0 < p < 2, \\ 2KS_\alpha(t) & \text{if } p = 2, \\ 0 & \text{if } p > 2. \end{cases} \tag{25}$$

Moreover, formulas (24) and (25) are also valid for the case of finite system with reflecting boundaries.

Comparing the above results with (17) and (19), we observe that indeed the confinement has no influence on the behavior of $p$-variation (however, the shape of the potential may influence the speed of convergence of $V_n^{(p)}(t)$ to $V^{(p)}(t)$). Therefore, visual identification of the type of subdiffusion, based on the differences in the behavior of $p$-variation corresponding to FBM and FFPE models, follows exactly the same line as in Sec. 5.

In additional, validation of the model via statistical hypothesis testing for the confined systems can be performed in the analogous manner as in the force-free seetting. Below, for completeness, we present the details of the testing procedures.

Assume that the null hypothesis is that the data comes from $Y_H(t)$. To verify such hypothesis one examines the statistic

$$D_{m,n} = \frac{|\hat{V}_m^{(1/H)}(T) - \hat{V}_n^{(1/H)}(T)|}{\hat{V}_n^{(1/H)}(T)}.$$

Recall that $\hat{V}_m^{(1/H)}(T)$ and $\hat{V}_n^{(1/H)}(T)$, $m < n$, are the values of sample $1/H$-variation calculated for the empirical data. By (24), under the null hypothesis the difference $D_{m,n}$ should be small for large enough $m$ and $n$. Therefore, the null hypothesis is rejected at confidence level $\hat{\alpha}$ if

$$D_{m,n} > D_{\hat{\alpha}}.$$

The constant $D_{\hat{\alpha}}$ is found from

$$\mathbb{P}\left(\frac{|V_m^{(1/H)}(T) - V_n^{(1/H)}(T)|}{V_n^{(1/H)}(T)} \le D_{\hat{\alpha}}\right) = 1 - \hat{\alpha},$$

with $V_m^{(1/H)}(T)$ and $V_n^{(1/H)}(T)$ being the values of sample $1/H$-variation corresponding to the process $Y_H(t)$, i.e.

$$V_n^{(1/H)}(T) = \sum_{j=0}^{2^n-1} \left| Y_H\left(\frac{(j+1)T}{2^n} \wedge T\right) - Y_H\left(\frac{jT}{2^n} \wedge T\right)\right|^{1/H},$$

and similarly $V_m^{(1/H)}(T)$. The constant $D_{\hat{\alpha}}$ can be found by the Monte Carlo method.

If the null hypothesis is that the data comes from $W_\beta(t)$, one should investigate the statistic

$$E_{m,n} = \frac{|\hat{V}_m^{(2)}(T) - \hat{V}_n^{(2)}(T)|}{\hat{V}_n^{(2)}(T)}.$$

Once again, $\hat{V}_m^{(2)}(T)$ and $\hat{V}_n^{(2)}(T)$, $m < n$, are the values of sample 2-variation corresponding to the empirical trajectory. Now, by (25), under the null hypothesis the distance $E_{m,n}$ should be small for large enough $m$ and $n$. Thus, the null hypothesis is rejected at confidence level $\hat{\alpha}$, if

$$E_{m,n} > E_{\hat{\alpha}},$$

where the constant $E_{\hat{\alpha}}$ is derived from

$$\mathbb{P}\left(\frac{|V_m^{(2)}(T) - V_n^{(2)}(T)|}{V_n^{(2)}(T)} \le E_{\hat{\alpha}}\right) = 1 - \hat{\alpha}.$$

Here, $V_m^{(2)}(T)$ and $V_n^{(2)}(T)$ are the values of sample 2-variation corresponding to $W_\beta(t)$, i.e.

$$V_n^{(2)}(T) = \sum_{j=0}^{2^n-1} \left| W_\beta\left(\frac{(j+1)T}{2^n} \wedge T\right) - W_\beta\left(\frac{jT}{2^n} \wedge T\right)\right|^{2},$$

and similarly $V_m^{(2)}(T)$. The constant $E_{\hat{\alpha}}$ can be determined via Monte Carlo techniques. The method of simulating $W_\beta(t)$ can be found [35, 49].

The results of the above tests for the simulated trajectories of $Y_H(t)$ and $W_\beta(t)$ are analogous to the ones presented in Sec. 6.

The advantage of $p$-variation test is its universality – we do not need to know any information about the confinement of the system. Moreover, to perform the test only one sufficiently long trajectory is needed.

## 8. Conclusions

We have described here the dynamics of the sample MSD and sample $p$-variation for CTRW process represented by the FFPE and for general Lévy stable processes, in particular, for a FBM and a FLSM. As a consequence, we constructed a new test which allows to identify the dynamics underlying the data and distinguish between three types of subdiffusive dynamics: FFPE, FBM and FLSM. This was done employing various statistical tests (Kolmogorov–Smirnov and FIRT) and introduced in this chapter sample MSD, and sample $p$-variation tests.

We already showed [7–9] that some of the bacterial cytoplasm data [11] can be modeled by a FLSM or FBM with $d < 0$. We have also observed a similar effect for the data describing the epidermal growth factor receptor labeled with quantum dots in the plasma membrane of live cells [5].

We hope that the general statistical methodology proposed in this chapter will be useful in identification and validation of the appropriate fractional stochastic model behind the data.

## Acknowledgment

The research of M.M. has been partially supported by the European Union within the European Social Funds.

## References

1. X. S. Xie and H. P. Lu, *J. Bio. Chem.* **274**, 15967 (1999).
2. S. Weiss, *Nature Struct. Biol.* **7**, 724 (2000).
3. W. Moerner, *J. Phys. Chem. B* **106**, 910 (2002).
4. S. C. Kou, V. S. Xie and J. S. Lin, *J. Roy. Stat. Soc. C* **54**, 469 (2005).
5. A. Serge, N. Bertaux, H. Rigneault and D. Marquet, *Nature Meth.* **5**, 687 (2008).
6. I. Bronstein, Y. Israel, E. Kepten, S. Mai, Y. Shav-Tal, E. Barkai and Y. Garini, *Phys. Rev. Lett.* **103**, 018102 (2009).

7. M. Magdziarz, A. Weron, K. Burnecki and J. Klafter, *Phys. Rev. Lett.* **103**, 180602 (2009).
8. M. Magdziarz and J. Klafter, *Phys. Rev. E* **82**, 011129 (2010).
9. K. Burnecki and A. Weron, *Phys. Rev. E* **82**, 021130 (2010).
10. G. Guigas, C. Kalla and M. Weiss, *Biophys. J.* **93**, 316 (2007).
11. I. Golding and E. C. Cox, *Phys. Rev. Lett.* **96**, 098102 (2006).
12. R. Metzler and J. Klafter, *Phys. Rep.* **339**, 1 (2000).
13. R. Metzler and J. Klafter, *J. Phys. A: Math Gen.* **37**, R161 (2004).
14. M. Kotulska, *Biophys. J.* **92**, 2412 (2007).
15. A. Caspi, R. Granek and M. Elbaum, *Phys. Rev. Lett.* **85**, 5655 (2000).
16. A. Lubelski, I. M. Sokolov and J. Klafter, *Phys. Rev. Lett.* **100**, 250602 (2008).
17. Y. He, S. Burov, R. Metzler and E. Barkai, *Phys. Rev. Lett.* **101**, 058101 (2008).
18. W. Min *et al.*, *Phys. Rev. Lett.* **94**, 198302 (2005).
19. M. J. Saxton, *Biophys. J.* **92**, 1178 (2007).
20. A. N. Kolmogorov and C. R. (Dokl.) *Acad. Sci. URSS* (*N.S.*) **26**, 115 (1940).
21. B. B. Mandelbrot and J. W. Van Ness, *SIAM Rev.* **10**, 422 (1968).
22. Y. S. Mishura, *Stochastic Calculus for Fractional Brownian Motion and Related Processes* (Springer, 2008).
23. M. Magdziarz, *Stoch. Models* **26**, 256 (2010).
24. A. Weron and M. Magdziarz, *Euro. Phys. Lett.* **86**, 60010 (2009).
25. E. Lutz, *Phys. Rev. E* **64**, 051106 (2001).
26. A. Weron, K. Burnecki, Sz. Mercik and K. Weron, *Phys. Rev. E* **71**, 016113 (2005).
27. S. C. Kou and X. S. Xie, *Phys. Rev. Lett.* **93**, 180603 (2004).
28. W. Deng and E. Barkai, *Phys. Rev. E* **79**, 011112 (2009).
29. J. H. Jeon and R. Metzler, *Phys. Rev. E* **81**, 021103 (2010).
30. N. W. Watkins, D. Credgington, R. Sanchez, S. J. Rosenberg and S. C. Chapman, *Phys. Rev. E* **79**, 041124 (2009).
31. A. Janicki and A. Weron, *A Simulation and Chaotic Behavior of $\alpha$–Stable Stochastic Processes* (Marcel Dekker 1994).
32. G. Samorodnitsky and M. S. Taqqu, *Stable Non-Gaussian Random Processes* (Chapman & Hall, 1994).
33. R. Metzler, E. Barkai and J. Klafter, *Phys. Rev. Lett.* **82**, 3563 (1999).
34. M. M. Meerschaert, D. A. Benson, H. P. Scheffler and B. Baeumer, *Phys. Rev. E* **65**, 041103 (2002).
35. M. Magdziarz, A. Weron and K. Weron, *Phys. Rev. E* **75**, 016708 (2007).
36. E. Gudowska-Nowak, B. Dybiec, P. F. Gora and R. Zygadlo, *Acta Phys. Pol. B* **40**, 1263 (2009).
37. K. Burnecki, A. Misiorek and R. Weron, in *Statistical Tools for Finance and Insurance*, ed. P. Cizek *et al.* (Springer, 2005).
38. R. B. D'Agostino and M. A. Stephens, *Goodness-of-Fit Techniques* (Marcel Dekker, 1986).
39. I. A. Koutrouvelis, *JASA* **75**, 918 (1980).
40. S. Stoev, V. Pipiras and M. S. Taqqu, *Signal Process.* **82**, 1873 (2002).

41. Sz. Mercik, K. Weron, K. Burnecki and A. Weron, *Acta Phys. Pol. B* **34**, 3773 (2003).
42. T. Chang, T. Sauer and S. J. Schiff, *Chaos* **5**, 376 (1995).
43. R. McGill, J. W. Tukey and W. A. Larsen, *Amer. Statist.* **32**, 12 (1978).
44. P. F. Velleman and D. C. Hoaglin, *Applications, Basics, and Computing of Exploratory Data Analysis* (Duxbury Press, 1981).
45. B. Dybiec and E. Gudowska-Nowak, *Phys. Rev. E* **80**, 061122 (2009).
46. T. F. Chan and J. Shen, *Image Processing. Variational, PDE, Wavelet and Stochastic Methods* (SIAM, 2005).
47. L. C. G. Rogers, *Math. Fin.* **7**, 95 (1997).
48. M. S. Taqqu and V. Teverovsky, in *A Practical Guide To Heavy Tails: Statistical Techniques and Applications*, ed. R. J. Adler *et al.* (Birkhäuser, 1998).
49. M. Magdziarz, *Stoch. Proc. Appl.* **119**, 3238 (2009).

# Chapter 15

## A Class of CTRWs: Compound Fractional Poisson Processes


Enrico Scalas

*Dipartimento di Scienze e Tecnologie Avanzate,*
*Università del Piemonte Orientale,*
*viale T. Michel 11, 15121 Alessandria, Italy*

*and*

*BCAM - Basque Center for Applied Mathematics*
*Bizkaia Technology Park, Building 500, 48160 Derio, Spain*



This chapter is an attempt to present a mathematical theory of compound fractional Poisson processes. It begins with the characterization of a well-known Lévy process: The compound Poisson process. The semi-Markov extension of the compound Poisson process naturally leads to the compound fractional Poisson process, where the Poisson counting process is replaced by the Mittag-Leffler counting process also known as fractional Poisson process. This process is no longer Markovian and Lévy. However, several analytical results are available and some of them are discussed here.



1. Introductory Notes . . . . . . . . . . . . . . . . . . . . . . . . 353
2. Compound Poisson Process and Generalizations . . . . . 354
3. Compound Fractional Poisson Processes . . . . . . . . . . 363
4. Limit Theorems . . . . . . . . . . . . . . . . . . . . . . . . 370


### 1. Introductory Notes

This chapter is an attempt to present a mathematical theory of compound fractional Poisson processes. It is not completely self-contained. The proofs of some statements can be found in widely available textbooks or papers. In several cases, freely downloadable versions of these papers can be easily retrieved.

The chapter begins with the characterization of a well-known Lévy process: The compound Poisson process. This process is extensively discussed in the classical books by Feller [1] and de Finetti [2].

The semi-Markov extension of the compound Poisson process naturally leads to the compound fractional Poisson process, where the Poisson counting process is replaced by the Mittag-Leffler counting process also called fractional Poisson process [3–6]. This process is no longer Markovian and Lévy. However, several analytical results are available and some of them are discussed below.

The functional limit of the compound Poisson process is an $\alpha$-stable Lévy process, whereas in the case of the compound fractional Poisson process, one gets an $\alpha$-stable Lévy process subordinated to the inverse $\beta$-stable subordinator.

I became interested in these processes as possible models for tick-by-tick financial data. The main results obtained by my co-workers and myself are described in a review paper for physicists [7].

The reader interested in Monte Carlo simulations can consult two recent papers [8,9] where algorithms are presented to simulate the fractional compound Poisson process.

## 2. Compound Poisson Process and Generalizations

Let $\{X_i\}_{i=1}^{\infty}$ be a sequence of independent and identically distributed (i.i.d.) real-valued random variables with cumulative distribution function $F_X(x)$, and let $N(t)$, $t \geq 0$ denote the Poisson process. Further assume that the i.i.d. sequence and the Poisson process are independent. We have the following

**Definition 1.** (compound Poisson process) The stochastic process

$$Y(t) = \sum_{i=1}^{N(t)} X_i \tag{1}$$

is called *compound Poisson process.*

Here, we shall consider the one-dimensional case only. The extension of many results to $\mathbb{R}^d$ is often straightforward. The compound Poisson process can be seen as a random walk subordinated to a Poisson process; in other words, it is a random sum of independent and identically distributed random variables. It turns out that the compound Poisson process is a Lévy process.

**Definition 2.** (Lévy process) A stochastic process $Y(t)$, $t \geq 0$ with $Y(0) = 0$ is a *Lévy process* if the following three conditions are fulfilled

(1) (independent increments) if $t_1 < t_2 < \cdots < t_n$, the increments $Y(t_2) - Y(t_1), \ldots, Y(t_n) - Y(t_{n-1})$ are independent random variables;

(2) (time-homogeneous increments) the law of the increment $Y(t+\Delta t) - Y(t)$ does not depend on $t$;
(3) (stochastic continuity) $\forall a > 0$, one has that $\lim_{\Delta t \to 0} \mathbb{P}(|Y(t+\Delta t) - Y(t)| \geq a) = 0$.

Loosely speaking, one can say that Lévy processes are stochastic processes with stationary and independent increments. Due to Kolmogorov's extension theorem [10], a stochastic process is characterized by its finite dimensional distributions. In the case of a Lévy process, the knowledge of the law of $Y(\Delta t)$ is sufficient to compute any finite dimensional distribution. Let us denote by $f_{Y(\Delta t)}(y, \Delta t)$ the probability density function of $Y(\Delta t)$

$$f_{Y(\Delta t)}(y, \Delta t)\, dy \stackrel{\text{def}}{=} \mathbb{P}(Y(\Delta t) \in dy). \tag{2}$$

As an example, suppose you want to know the joint density function $f_{Y(t_1),Y(t_2)}(y_1, t_1; y_2, t_2)$ defined as

$$f_{Y(t_1),Y(t_2)}(y_1, t_1; y_2, t_2)\, dy_1\, dy_2 \stackrel{\text{def}}{=} \mathbb{P}(Y(t_1) \in dy_1, Y(t_2) \in dy_2). \tag{3}$$

This is given by

$$\begin{aligned} f_{Y(t_1),Y(t_2)}(y_1, t_1; y_2, t_2) dy_1 dy_2 &\stackrel{\text{def}}{=} \mathbb{P}(Y(t_1) \in dy_1, Y(t_2) \in dy_2) \\ &= \mathbb{P}(Y(t_2) \in dy_2 | Y(t_1) \in dy_1)\mathbb{P}(Y(t_1) \in dy_1) \\ &= \mathbb{P}(Y(t_2) - Y(t_1) \in d(y_2 - y_1))\mathbb{P}(Y(t_1) \in dy_1) \\ &= f_{Y(t_2)-Y(t_1)}(y_2 - y_1, t_2 - t_1) f_{Y(t_1)}(y_1, t_1) dy_1 dy_2, \end{aligned} \tag{4}$$

and this procedure can be used for any finite dimensional distribution. The extension theorem shows the existence of a stochastic process given a suitable set of finite dimensional distributions obeying Komogorov's consistency conditions [10], but not the uniqueness.

**Definition 3.** (càdlàg process) A stochastic process $Y(t)$, $t \geq 0$ is *càdlàg* (*continu à droite et limite à gauche*) if its realizations are right-continuous with left limits.

A càdlàg stochastic process has realizations with jumps. Let $\bar{t}$ denote the epoch of a jump. Then, in a càdlàg process, one has $Y(\bar{t}) = Y(\bar{t}^+) \stackrel{\text{def}}{=} \lim_{t \to \bar{t}^+} Y(t)$.

**Definition 4.** (Modification of a process) A *modification* $Z(t)$, $t \geq 0$, of a stochastic process $Y(t)$, $t \geq 0$, is a stochastic process on the same probability space such that $\mathbb{P}(Z(t) = Y(t)) = 1$.

**Theorem 1.** *Every Lévy process has a unique càdlàg modification.*

***Proof.*** For a proof of this theorem one can see the first chapter of the book by Sato [11]. □

The following theorem gives a nice characterization of compound Poisson processes.

**Theorem 2.** *$Y(t)$ is a compound Poisson process if and only if it is a Lèvy process and its realizations are piecewise constant càdlàg functions.*

***Proof.*** An accessible proof of this theorem can be found in the book by Cont and Tankov [12]. □

As a consequence of the above results, the compound Poisson process enjoys all the properties of Lévy processes, including the Markov property. To show that a Lévy process has the Markov property, some further definitions are necessary.

**Definition 5.** (Filtration) A family $\mathcal{F}_t$, $t \geq 0$ of $\sigma$-algebras is a *filtration* if it is non-decreasing, meaning that $\mathcal{F}_s \subseteq \mathcal{F}_t$ for $0 \leq s \leq t$.

**Definition 6.** (Adapted process) A process $Y(t)$, $t \geq 0$ is said to be *adapted* to the filtration $\mathcal{F}_t$, $t \geq 0$ if it is $\mathcal{F}_t$-measurable for $t \geq 0$.

**Definition 7.** (Markov process with respect to a filtration) A process $Y(t)$ is a *Markov process with respect to the filtration* $\mathcal{F}_t$, $t \geq 0$ if it is adapted to $\mathcal{F}_t$, $t \geq 0$ and ($A \subset \mathbb{R}$)

$$\mathbb{P}(Y(t) \in A|\mathcal{F}_s) = \mathbb{P}(Y(t) \in A|Y(s)). \tag{5}$$

**Definition 8.** (Natural filtration) The *natural filtration* for a stochastic process is the family of non-decreasing $\sigma$-algebras generated by the process itself $\{\sigma(X(s)), s \in [0,t]\}$, $t \geq 0$.

**Definition 9.** (Markov process with respect to itself) A process $Y(t)$, $t \geq 0$ is a *Markov process with respect to itself* (or simply *a Markov process*) if it is a Markov process with respect to its natural filtration.

The natural filtration $\{\sigma(X(s)), s \in [0,t]\}$, $t \geq 0$, is a formal way to characterize the history of the process up to time $t$. For a Markov process, the future values do not depend on the whole history, but only on the present value of the process.

**Definition 10.** (Transition probability) Given the Markov process $Y(t)$, $t \geq 0$, its *transition probability* $P(y, A, \Delta t, t)$ is defined as

$$P(y, A, \Delta t, t) = \mathbb{P}(Y(t + \Delta t) \in A | Y(t) = y), \quad (6)$$

where $A \subset \mathbb{R}$.

**Definition 11.** (homogeneous Markov process) A Markov process $Y(t)$, $t \geq 0$ is said to be (*time*)-*homogeneous* if its transition probability $P(y, A, \Delta t, t)$ does not depend on $t$.

**Theorem 3.** *A Lévy process is a time-homogeneous Markov process with transition probability*

$$P(y, A, \Delta t) = \mathbb{P}(Y(\Delta t) \in A - y) = \int_{x \in A} f_{Y(\Delta t)}(x - y, \Delta t)\, dx. \quad (7)$$

***Proof.*** The Markov property is a consequence of the independence of increments. The following chain of equalities holds true

$$\begin{aligned} \mathbb{P}(Y(t + \Delta t) \in A | \mathcal{F}_t) &= \mathbb{P}(Y(t + \Delta t) - Y(t) \in A - Y(t) | \mathcal{F}_t) \\ &= \mathbb{P}(Y(t + \Delta t) - Y(t) \in A - Y(t) | Y(t)). \end{aligned} \quad (8)$$

We further have

$$\begin{aligned} &\mathbb{P}(Y(t + \Delta t) \in A | Y(t) = y) \\ &\quad = \mathbb{P}(Y(\Delta t) \in A - y) = \int_{x \in A} f_{Y(\Delta t)}(x - y, \Delta t)\, dx, \end{aligned} \quad (9)$$

as a consequence of time homogeneity. □

This result fully justifies the passages leading to Eq. (4).

It turns out that $f_{Y(\Delta t)}(y, \Delta t)$ can be explicitly written for a compound Poisson process. Let $F_X(x)$ be the law of the jumps $\{X_i\}$ and let $\lambda$ denote the parameter of the Poisson process, then we have the following:

**Theorem 4.** *The cumulative distribution function of a compound Poisson process is given by*

$$F_{Y(t)}(y, t) = \mathrm{e}^{-\lambda t} \sum_{n=0}^{\infty} \frac{(\lambda t)^n}{n!} F_{Y_n}^{*n}(y), \quad (10)$$

*where $F_{Y_n}^{*n}(y)$ is the $n$-fold convolution of $F_X(x)$.*

***Proof.*** Starting from $Y(0) = 0$, at time $t$, there have been $N(t)$ jumps, with $N(t)$ assuming integer values starting from 0 ($N(t) = 0$ means no jumps up to time $t$). To fix the ideas, suppose that $N(t) = n$. Therefore, one has

$$Y(t) = \sum_{i=1}^{N(t)} X_i = \sum_{i=1}^{n} X_i = Y_n \tag{11}$$

and, in this case,

$$F_{Y_n}(y) = \mathbb{P}(Y_n \leq y) = \mathbb{P}\left(\sum_{i=1}^{n} X_i \leq y\right) = F_{Y_n}^{*n}(y). \tag{12}$$

For the Poisson process, one has

$$P(n,t) \stackrel{\text{def}}{=} \mathbb{P}(N(t) = n) = \mathrm{e}^{-\lambda t}\frac{(\lambda t)^n}{n!}. \tag{13}$$

Given the independence between $N(t)$ and the $X_i$'s, one has that

$$\mathbb{P}(Y_n \leq y, N(t) = n) = P(n,t)F_{Y_n}^{*n}(y) = \mathrm{e}^{-\lambda t}\frac{(\lambda t)^n}{n!}F_{Y_n}^{*n}(y). \tag{14}$$

The events $\{Y_n \leq y, N(t) = n\}$ are mutually exclusive and exhaustive, meaning that

$$\{Y(t) \leq y\} = \cup_{n=0}^{\infty}\{Y_n \leq y, N(t) = n\}, \tag{15}$$

and that, for any $m \neq n$

$$\{Y_m \leq y, N(t) = m\} \cap \{Y_n \leq y, N(t) = n\} = \emptyset. \tag{16}$$

Calculating the probability on both sides in Eq. (15) and using Eq. (14) and the axiom of infinite additivity.

$$\begin{aligned} F_{Y(t)}(y,t) &= \mathbb{P}(Y(t) \leq y) = \mathbb{P}\left(\cup_{n=0}^{\infty}\{Y_n \leq y, N(t) = n\}\right) \\ &= \sum_{n=0}^{\infty} \mathbb{P}(Y_n \leq y, N(t) = n) = \sum_{n=0}^{\infty} P(n,t)F_{Y_n}^{*n}(y) \\ &= \mathrm{e}^{-\lambda t}\sum_{n=0}^{\infty}\frac{(\lambda t)^n}{n!}F_{Y_n}^{*n}(y) \end{aligned} \tag{17}$$

leads to the thesis. □

**Remark 1.** (generality of Theorem 4) Note that the above theorem is valid for any counting process $N(t)$ in the following form

**Theorem 5.** *Let* $\{X\}_{i=1}^{\infty}$ *be a sequence of i.i.d. real-valued random variables with cumulative distribution function* $F_X(x)$ *and let* $N(t)$, $t \geq 0$ *denote a counting process independent of the previous sequence and such that the number of events in the interval* $[0,t]$ *is a finite but arbitrary integer* $n = 0, 1, \dots$. *Let* $Y(t)$ *denote the process*

$$Y(t) = \sum_{i=1}^{N(t)} X_i. \tag{18}$$

*Then if* $P(n,t) \stackrel{def}{=} \mathbb{P}(N(t) = n)$, *the cumulative distribution function of* $Y(t)$ *is given by*

$$F_{Y(t)}(y,t) = \sum_{n=0}^{\infty} P(n,t) F_{Y_n}^{*n}(y), \tag{19}$$

*where* $F_{Y_n}^{*n}(y)$ *is the n-fold convolution of* $F_X(x)$.

***Proof.*** The proof of this theorem is exactly the same as that of Theorem 4 without specifying $P(n,t)$. □

Theorem 5 will be useful in the next section when the Poisson process will be replaced by the fractional Poisson process.

**Remark 2.** (the $n = 0$ term) For $n = 0$, one assumes $F_{Y_0}^{*0}(y) = \theta(y)$ where $\theta(y)$ is the Heaviside function. Note that $P(0,t)$ is nothing but the survival function at $y = 0$ of the counting process. Therefore, Eq. (19) can be equivalently written as

$$F_{Y(t)}(y,t) = P(n,0)\,\theta(y) + \sum_{n=1}^{\infty} P(n,t) F_{Y_n}^{*n}(y). \tag{20}$$

**Remark 3.** (uniform convergence) The series (10) and (19) are uniformly convergent for $y \neq 0$ and for any value of $t \in (0,\infty)$ (this statement can be proved applying Weierstrass $M$ test). For $y = 0$, there is a jump in the cumulative distribution function of amplitude $P(0,t)$.

**Example 1.** (the normal compound Poisson process) As an example of compound Poisson process, consider the case in which $X_i \sim \mathcal{N}(\mu, \sigma^2)$, so that the cumulative distribution function is

$$F_X(x) = \Phi(x|\mu,\sigma^2) = \frac{1}{2}\left(1 + \operatorname{erf}\left(\frac{x-\mu}{\sqrt{2\sigma^2}}\right)\right), \tag{21}$$

where

$$\mathrm{erf}(x) = \frac{2}{\sqrt{\pi}} \int_0^x \mathrm{e}^{-u^2}\, du \tag{22}$$

is the error function. In this case, the convolution $F_{Y_n}^{*n}(y)$ is given by $\Phi(y|n\mu, n\sigma^2)$ and one finds

$$F_{Y(t)}(y,t) = \mathrm{e}^{-\lambda t} \sum_{n=0}^{\infty} \frac{(\lambda t)^n}{n!} \Phi(y|n\mu, n\sigma^2). \tag{23}$$

**Corollary 1.** *In the same hypotheses as in Theorem* 5, *the probability density* $f_{Y(t)}(y,t)$ *of the process* $Y(t)$ *is given by*

$$f_{Y(t)}(y,t) = P(0,t)\,\delta(y) + \sum_{n=1}^{\infty} P(n,t) f_{Y_n}^{*n}(y), \tag{24}$$

*where* $f_{Y_n}^{*n}(y)$ *is the n-fold convolution of the probability density function* $f_{Y_n}(y) = dF_{Y_n}(y)/dy$.

***Proof.*** One has that $f_{Y(t)}(y,t) = dF_{Y(t)}(y,t)/dy$; moreover, Eq. (24) is the formal term-by-term derivative of Eq. (19). If $y \neq 0$, there is no singular term and the series converges uniformly ($f_{Y_n}^{*n}(y)$ is bounded and Weierstrass $M$ test applies), therefore, for any $y$ it converges to the derivative of $F_{Y(t)}(y,t)$. This is true also for $y = 0$ for $n \geq 1$ and the jump in $y = 0$ gives the singular term of weight $P(0,t)$ (see Eq. (20)). □

**Remark 4.** (historical news and applications) The distribution in Eq. (19) is also known as *generalized Poisson law*. This class of distributions was studied by W. Feller in a famous work published in 1943 [13]. It is useful to quote an excerpt from Feller's paper, with notation adapted to this chapter.

> The most frequently encountered application of the generalized Poisson distribution is to problems of the following type. Consider independent random events for which the simple Poisson distribution may be assumed, such as: telephone calls, the occurrence of claims in an insurance company, fire accidents, sickness, and the like. With each event there may be associated a random variable $X$. Thus, in the above examples, $X$ may represent the length of the ensuing conversation, the sum under risk, the damage, the cost (or length) of hospitalization, respectively. To mention an interesting example of a different type, A. Einstein, Jr. [14] and G. Polya [15,16] have studied a problem arising out of engineering practice connected with building of dams, where the events consists of the motions of a stone at the bottom of a river; the variable $X$ is the distance through which the stone moves down the river.

Now, if $F(x)$ is the cumulative distribution function of the variable $X$ associated with a single event, then $F^{*n}(x)$ is the cumulative distribution function of the accumulated variable associated with $n$ events. Hence the following equation

$$G(x) = \mathrm{e}^{-a} \sum_{n=0}^{\infty} \frac{a^n}{n!} F^{*n}(x)$$

is the probability law of the sum of the variables (sum of the conversation times, total sum paid by the company, total damage, total distance travelled by the stone, etc.).

In view of the above examples, it is not surprising that the law, or special cases of it, have been discovered, by various means and sometimes under disguised forms, by many authors.

Indeed, the rediscovery and/or reinterpretation of Eq. (19) went on also after Feller's paper. In physics, $X$ is interpreted as the position of a walker on a lattice and $N(t)$ is the number of walker jumps occurred up to time $t$ [17–21]. In finance, $X$ is the tick-by-tick log-return for a stock and $N(t)$ is the number of transactions up to time $t$ [7].

The application of Fourier and Laplace transforms to Eq. (24) leads to an equation which is known as Montroll–Weiss equation in the physics literature [17]. For reasons which will become clear in the following, it can also be called semi-Markov renewal equation. Let

$$\widehat{f}(\kappa) = \int_{-\infty}^{+\infty} \mathrm{e}^{i\kappa x} f(x)\, dx$$

denote the Fourier transform of $f(x)$ and

$$\widetilde{g}(s) = \int_{0}^{\infty} \mathrm{e}^{-st} g(t)\, dt$$

denote the Laplace transform of $g(t)$, then the following theorem holds true.

**Theorem 6.** (Montroll–Weiss equation) *Let $J$ denote the sojourn time of the Poisson process in $N(t) = 0$, with $F_J(t) = 1 - \mathrm{e}^{-\lambda t}$, $f_J(t) = \lambda \mathrm{e}^{-\lambda t}$ and $P(0,t) = 1 - F_J(t) = \mathrm{e}^{-\lambda t}$. We have that:*

(1) *The Fourier–Laplace transform of the probability density $f_{Y(t)}(y,t)$ is given by*

$$\widetilde{\widehat{f}}_{Y(t)}(\kappa, s) = \frac{1 - \widetilde{f}_J(s)}{s} \frac{1}{1 - \widetilde{f}_J(s)\widehat{f}_X(\kappa)}. \tag{25}$$

(2) *The probability density function $f_{Y(t)}(y,t)$ obeys the following integral equation*

$$f_{Y(t)}(y,t) = P(0,t)\delta(y) + \int_0^t dt' f_J(t-t') \int_{-\infty}^{+\infty} dy' f_X(y-y') f_{Y(t)}(y',t'). \tag{26}$$

In order to prove Theorem 6, we need the following lemma.

**Lemma 1.** *Let $T_1, T_2, \ldots, T_n, \ldots$ denote the epoch of the first, second, ..., the nth, ... event of a Poisson process, respectively. Let $J = J_1 = T_1$ denote the initial sojourn time and, in general, let $J_i = T_i - T_{i-1}$ be the ith sojourn time. Then $\{J_i\}_{i=1}^{\infty}$ is a sequence of i.i.d. random variables.*

***Proof.*** The proof of this lemma can be derived by combining Lemma 2.1 and Proposition 2.12 in the book by Cont and Tankov [12]. □

It is now possible to prove the theorem

***Proof.*** (Theorem 6) Let us start from Eq. (24) and compute its Fourier–Laplace transform. It is given by

$$\widetilde{\widehat{f}}_{Y(t)}(\kappa, s) = \widetilde{P}(0,s) + \sum_{n=1}^{\infty} \widetilde{P}(n,s)[\widehat{f}_X(\kappa)]^n. \tag{27}$$

Now, we have that

$$T_n = \sum_{i=1}^{n} J_i \tag{28}$$

is a sum of i.i.d. positive random variables and $P(n,t) \stackrel{\text{def}}{=} \mathbb{P}(N(t) = n)$ meaning that there are $n$ jumps up to $t = t_n$ and no jumps in $t - t_n$. Therefore, from pure probabilistic considerations, one has that

$$P(n,t) = P(0, t - t_n) * f_{T_n}(t_n) \tag{29}$$

and, as a consequence of Eq. (28), one further has that

$$f_{T_n}(t_n) = f_J^{*n}(t_n). \tag{30}$$

Therefore, one can conclude that

$$\widetilde{P}(n,s) = \widetilde{P}(0,s)[\widetilde{f}_J(s)]^n. \tag{31}$$

Inserting this result in Eq. (27), noting that $\widetilde{P}(0,s) = (1 - \widetilde{f}_J(s))/s$, and summing the geometric series ($f_X$ and $f_J$ are probability densities) leads to

Eq. (25):

$$\widetilde{\widehat{f}}_{Y(t)}(\kappa, s) = \widetilde{P}(0, s) + \widetilde{P}(0, s) \sum_{n=1}^{\infty} [\widetilde{f}_J(s) \widehat{f}_X(\kappa)]^n$$

$$= \widetilde{P}(0, s) \sum_{n=0}^{\infty} [\widetilde{f}_J(s) \widehat{f}_X(\kappa)]^n$$

$$= \frac{1 - \widetilde{f}_J(s)}{s} \frac{1}{1 - \widetilde{f}_J(s) \widehat{f}_X(\kappa)}. \tag{32}$$

Equation (25) can be rewritten as

$$\widetilde{\widehat{f}}_{Y(t)}(\kappa, s) = \frac{1 - \widetilde{f}_J(s)}{s} + \widetilde{f}_J(s) \widehat{f}_X(\kappa) \widetilde{\widehat{f}}_{Y(t)}(\kappa, s); \tag{33}$$

Fourier–Laplace inverting and recalling the behavior of convolutions under Fourier–Laplace transform, immediately leads to Eq. (26). □

**Remark 5.** Theorem 6 was proved in the hypothesis that $N(t)$ is a Poisson process. In this case, one has $\widetilde{P}(0, s) = 1/(\lambda + s)$ and $\widetilde{f}_J(s) = \lambda/(\lambda + s)$ and Eq. (25) becomes

$$\widetilde{\widehat{f}}_{Y(t)}(\kappa, s) = \frac{1}{\lambda - \lambda \widehat{f}_X(\kappa) + s}. \tag{34}$$

The inversion of the Laplace transform yields the characteristic function of the compound Poisson process

$$\widehat{f}_{Y(t)}(\kappa, t) = \mathbb{E}(e^{i\kappa Y(t)}) = \mathrm{e}^{-\lambda(1 - \widehat{f}_X(\kappa))t}. \tag{35}$$

**Remark 6.** The proof of Theorem 6 does not depend on the specific form of $P(0, t)$ and $f_J(t)$, provided that the positive random variables $\{J\}_{i=1}^{\infty}$ are i.i.d. Therefore, Eqs. (25) and (26) are true also in the case of general compound renewal processes starting from $Y(0) = 0$ at time 0.

## 3. Compound Fractional Poisson Processes

**Definition 12.** (Renewal process) Let $\{J\}_{i=1}^{\infty}$ be a sequence of i.i.d. positive random variables interpreted as sojourn times between subsequent events arriving at random time. They define a *renewal process* whose epochs of renewal (time instants at which the events take place) are the random times $\{T\}_{n=0}^{\infty}$ defined by

$$T_0 = 0,$$
$$T_n = \sum_{i=1}^{n} J_i. \tag{36}$$

The name renewal process is due to the fact that at any epoch of renewal, the process starts again from the beginning.

**Definition 13.** (counting process) Associated to any renewal process, there is a *counting process* $N(t)$ defined as

$$N(t) = \max\{n : T_n \leq t\} \tag{37}$$

that counts the number of events up to time $t$.

**Remark 7.** As mentioned in the previous section $N(t)$ is the Poisson process if and only if $J \sim \exp(\lambda)$. Incidentally, this is the only case of Lévy and Markov counting process related to a renewal process (see Çinlar's book [22] for a proof of this statement).

**Remark 8.** In this paper, we shall assume that the counting process has càdlàg sample paths. This means that the realizations are represented by step functions. If $t_k$ is the epoch of the $k$th jump, we have $N(t_k^-) = k - 1$ and $N(t_k^+) = k$.

Let $\{X_i\}_{i=1}^{\infty}$ be a sequence of independent and identically distributed (i.i.d.) real-valued random variables and let $N(t)$, $t \geq 0$ denote the counting process. Further assume that the i.i.d. sequence and the counting process are independent. We have the following

**Definition 14.** (compound renewal process) The stochastic process

$$Y(t) = \sum_{i=1}^{N(t)} X_i \tag{38}$$

is called *compound renewal process*.

**Remark 9.** Again, here, it is assumed that the sample paths are represented by càdlàg step functions. Compound renewal processes generalize compound Poisson processes and they are called *continuous-time random walks* in the physical literature.

**Remark 10.** As compound renewal processes are just Markov chains (actually, random walks) subordinated to a counting process, their existence can be proved as a consequence of the existence of the corresponding discrete-time random walks and counting processes.

In general, compound renewal processes are non-Markovian, but they belong to the wider class of semi-Markov processes [9, 22–25].

**Definition 15.** (Markov renewal process) A *Markov renewal process* is a two-component Markov chain $\{Y_n, T_n\}_{n=0}^{\infty}$, where $Y_n$, $n \geq 0$ is a Markov chain and $T_n$, $n \geq 0$ is the $n$th epoch of a renewal process, homogeneous with respect to the second component and with transition probability defined by

$$\mathbb{P}(Y_{n+1} \in A, J_{n+1} \leq t | Y_0, \dots Y_n, J_1, \dots, J_n) = \mathbb{P}(Y_{n+1} \in A, J_{n+1} \leq t | Y_n), \tag{39}$$

where $A \subset \mathbb{R}$ is a Borel set and, as usual, $J_{n+1} = T_{n+1} - T_n$.

**Remark 11.** In this chapter, homogeneity with respect to the first component will be assumed as well. Namely, if $Y_n = x$, the probability on the right-hand side of Eq. (39) does not explicitly depend on $n$.

**Remark 12.** (semi-Markov kernel) The positive function $Q(x, A, t) = \mathbb{P}(Y_{n+1} = y \in A, J_{n+1} \leq t | Y_n = x)$, with $x \in \mathbb{R}$, $A \subset \mathbb{R}$ a Borel set, and $t \geq 0$ is called semi-Markov kernel.

**Definition 16.** (semi-Markov process) Let $N(t)$ denote the counting process defined as in Eq. (37), the stochastic process $Y(t)$ defined as

$$Y(t) = Y_{N(t)} \tag{40}$$

is the *semi-Markov process* associated to the Markov renewal process $Y_n, T_n$, $n \geq 0$.

**Remark 13.** In Eq. (37), max is used instead of the more general sup as only processes with finite (but arbitrary) number of jumps in $(0, t]$ are considered here.

**Theorem 7.** *Compound renewal processes are semi-Markov processes with semi-Markov kernel given by*

$$Q(x, A, t) = P(x, A) F_J(t), \tag{41}$$

*where* $P(x, A)$ *is the Markov kernel (a.k.a. Markov transition function or transition probability kernel) of the random walk*

$$P(x, A) \stackrel{\text{def}}{=} \mathbb{P}(Y_{n+1} \in A | Y_n = x), \tag{42}$$

*and $F_J(t)$ is the probability distribution function of sojourn times. Moreover, let $f_X(x)$ denote the probability density function of jumps, one has*

$$P(x, A) = \int_{A-x} f_X(u)\, du, \tag{43}$$

*where $A - x$ is the set of values in $A$ translated of $x$ towards left.*

***Proof.*** The compound renewal process is a semi-Markov process by construction, where the couple $Y_n, T_n, n \geq 0$ defining the corresponding Markov renewal process is made up of a random walk $Y_n$, $n \geq 0$ with $Y_0 = 0$ and a renewal process with epochs given by $T_n$, $n \geq 0$ with $T_0 = 0$. Equation (41) is an immediate consequence of the independence between the random walk and the renewal process. Finally, Eq. (43) is the standard Markov kernel of a random walk whose jumps are i.i.d. random variables with probability density function $f_X(x)$. □

**Remark 14.** As a direct consequence of the previous theorem, if the law of the couple $X_n, J_n$ has a joint probability density function $f_{X,J}(x,t) = f_X(x)f_J(t)$, then one has

$$\begin{aligned}\mathbb{P}(Y_{n+1} \in A, J_{n+1} \leq t|Y_n) &= Q(x, A, t) = P(x, A)F_J(t) \\ &= \int_{A-x} f_X(u)\, du \int_0^t f_J(v)\, dv. \end{aligned} \tag{44}$$

**Theorem 8.** (semi-Markov renewal equation) *The probability density function $f_{Y(t)}(y,t)$ of a compound renewal process obeys the semi-Markov renewal equation*

$$f_{Y(t)}(y,t) = P(0,t)\delta(y) + \int_0^t dt' f_J(t-t') \int_{-\infty}^{+\infty} dy' f_X(y-y') f_{Y(t)}(y',t'). \tag{45}$$

***Proof.*** By definition, one has that

$$\mathbb{P}(Y(t) \in dy|Y(0) = 0) = f_{Y(t)}(y,t)dy, \tag{46}$$

and that

$$\begin{aligned}\mathbb{P}(Y(t) \in dy|Y(t') = y') &= \mathbb{P}(Y(t-t') \in dy|Y(0) = y') \\ &= \mathbb{P}(Y(t-t') - y' \in dy|Y(0) = 0) \\ &= f_{Y(t)}(y-y', t-t')dy, \end{aligned} \tag{47}$$

because the increments in time and space are i.i.d. and hence homogeneous. From Eq. (44), one further has

$$\mathbb{P}(Y_1 \in dy, J_1 \in dt | Y_0 = 0) = f_X(y) f_J(t) dy\, dt. \tag{48}$$

Now, the probability in Eq. (46) can be decomposed into two mutually exclusive parts, depending on the behavior of the first interval

$$\begin{aligned}&\mathbb{P}(Y(t) \in dy | Y(0) = 0)\\ &\quad = P(Y(t) \in dy, J_1 > t | Y(0) = 0) + \mathbb{P}(Y(t) \in dy, J_1 \leq t | Y(0) = 0).\end{aligned} \tag{49}$$

The part with no jumps up to time $t$ immediately gives

$$\mathbb{P}(Y(t) \in dy, J_1 > t | Y(0) = 0) = P(0,t)\delta(y) dy, \tag{50}$$

whereas the part with jumps becomes

$$\begin{aligned}&\mathbb{P}(Y(t) \in dy, J_1 \leq t | Y(0) = 0)\\ &\quad = \int_{-\infty}^{+\infty} \int_0^t \mathbb{P}(Y(t) \in dy | Y(t') = y') \mathbb{P}(Y_1 \in dy', J_1 \in dt')\\ &\quad = \int_{-\infty}^{+\infty} \int_0^t f_{Y(t)}(y - y', t - t') dy f_X(y') f_J(t') dy' dt'\\ &\quad = \left[\int_{-\infty}^{+\infty} \int_0^t f_{Y(t)}(y - y', t - t') f_X(y') f_J(t') dy' dt'\right] dy\end{aligned} \tag{51}$$

as a consequence of Bayes' formula and of Eqs. (47) and (48). A replacement of Eqs. (46), (50), (51) into Eq. (49) and a rearrangement of the convolution variables straightforwardly lead to the thesis (45). □

**Remark 15.** Note that the semi-Markov renewal Eq. (45) does coincide with the Montroll-Weiss Eq. (26) as anticipated.

**Definition 17.** (Mittag-Leffler renewal process) The sequence $\{J_i\}_{i=1}^{\infty}$ of positive independent and identically distributed random variables whose complementary cumulative distribution function $P_\beta(0,t)$ is given by

$$P_\beta(0,t) = E_\beta(-t^\beta) \tag{52}$$

defines the so-called *Mittag-Leffler renewal process*.

**Remark 16.** The one-parameter Mittag-Leffler function in (52) is a straightforward generalization of the exponential function. It is given by the following series

$$E_\beta(z) = \sum_{n=0}^{\infty} \frac{z^n}{\Gamma(\beta n + 1)}, \tag{53}$$

where $\Gamma(z)$ is Euler's Gamma function. The Mittag-Leffler function coincides with the exponential function for $\beta = 1$. The function $E_\beta(-t^\beta)$ is completely monotonic and it is 1 for $t = 0$. Therefore, it is a legitimate survival function.

**Remark 17.** The function $E_\beta(-t^\beta)$ is approximated by a stretched exponential for $t \to 0$:

$$E_\beta(-t^\beta) \simeq 1 - \frac{t^\beta}{\Gamma(\beta+1)} \simeq \mathrm{e}^{-t^\beta/\Gamma(\beta+1)}, \quad \text{for } 0 < t \ll 1, \tag{54}$$

and by a power-law for $t \to \infty$:

$$E_\beta(-t^\beta) \simeq \frac{\sin(\beta\pi)}{\pi} \frac{\Gamma(\beta)}{t^\beta}, \quad \text{for } t \gg 1. \tag{55}$$

**Remark 18.** For applications, it is often convenient to include a scale factor in the definition (52), so that one can write

$$P(0,t) = E_\beta(-(t/\gamma_t)^\beta). \tag{56}$$

As the scale factor can be introduced in different ways, the reader is warned to pay attention to its definition. The assumption $\gamma_t = 1$ made in (52) is equivalent to a change of time unit.

**Theorem 9.** (Mittag-Leffler counting process — fractional Poisson process) *The counting process $N_\beta(t)$ associated to the renewal process defined by Eq.* (52) *has the following distribution*

$$P_\beta(n,t) = \mathbb{P}(N_\beta(t) = n) = \frac{t^{\beta n}}{n!} E_\beta^{(n)}(-t^\beta), \tag{57}$$

*where $E_\beta^{(n)}(-t^\beta)$ denotes the nth derivative of $E_\beta(z)$ evaluated at the point $z = -t^\beta$.*

***Proof.*** The Laplace transform of $P(0,t)$ is given by [26]

$$\widetilde{P}_\beta(0,s) = \frac{s^{\beta-1}}{1+s^\beta}, \tag{58}$$

as a consequence, the Laplace transform of the probability density function $f_{J,\beta}(t) = -dP_\beta(0,t)/dt$ is given by

$$\widetilde{f}_{J,\beta}(s) = \frac{1}{1+s^\beta}; \tag{59}$$

recalling Eq. (31), one immediately has

$$\widetilde{P}_\beta(n,s) = \frac{1}{(1+s^\beta)^n}\frac{s^{\beta-1}}{1+s^\beta}. \tag{60}$$

Using Eq. (80) in Podlubny's book [26] for the inversion of the Laplace transform in (60), one gets the thesis (57). □

**Remark 19.** The previous theorem was proved by Scalas *et al.* [27, 28]. Notice that $N_1(t)$ is the Poisson process with parameter $\lambda = 1$. Recently, Meerschaert *et al.* [29] proved that the fractional Poisson process $N_\beta(t)$ coincides with the process defined by $N_1(D_\beta(t))$ where $D_\beta(t)$ is the functional inverse of the standard $\beta$-stable subordinator. The latter process was also known as fractal time Poisson process. This result unifies different approaches to fractional calculus [30, 31].

**Remark 20.** For $0 < \beta < 1$, the fractional Poisson process is semi-Markov, but not Markovian and is not Lévy. The process $N_\beta(t)$ is not Markovian as the only Markovian counting process is the Poisson process [22]. It is not Lévy as its distribution is not infinitely divisible.

**Definition 18.** (compound fractional Poisson process) With the usual hypotheses, the process

$$Y_\beta(t) = Y_{N_\beta(t)} = \sum_{i=1}^{N_\beta(t)} X_i \tag{61}$$

is called *compound fractional Poisson process.*

**Remark 21.** The process $Y_1(t)$ coincides with the compound Poisson process of parameter $\lambda = 1$.

**Theorem 10.** *Let $Y_\beta(t)$ be a compound fractional Poisson process, then*

(1) *its cumulative distribution function $F_{Y_\beta(t)}(y,t)$ is given by*

$$F_{Y_\beta(t)}(y,t) = E_\beta(-t^\beta)\theta(y) + \sum_{n=1}^{\infty} \frac{t^{\beta n}}{n!} E_\beta^{(n)}(-t^\beta) F_{Y_n}^{*n}(y); \tag{62}$$

(2) *its probability density $f_{Y_\beta(t)}(y,t)$ function is given by*

$$f_{Y_\beta(t)}(y,t) = E_\beta(-t^\beta)\delta(y) + \sum_{n=1}^{\infty} \frac{t^{\beta n}}{n!} E_\beta^{(n)}(-t^\beta) f_{Y_n}^{*n}(y); \tag{63}$$

(3) *its characteristic function $\widehat{f}_{Y_\beta(t)}(\kappa,t)$ is given by*

$$\widehat{f}_{Y_\beta(t)}(\kappa,t) = E_\beta[t^\beta(\widehat{f}_X(\kappa) - 1)]. \tag{64}$$

***Proof.*** The first two equations (62) and (63) are a straightforward consequence of Theorem 5, Corollary 1 and Theorem 9. Equation (64) is the straightforward Fourier transform of (63). □

**Remark 22.** For $0 < \beta < 1$, the compound fractional Poisson process is not Markovian and not Lévy (see Remark 20).

## 4. Limit Theorems

**Definition 19.** (space-time fractional diffusion equation) Let $\partial^\alpha/\partial|x|^\alpha$ denote the spatial non-local pseudo-differential operator whose Fourier transform is given by

$$\mathcal{F}\left[\frac{\partial^\alpha f(x)}{\partial|x|^\alpha};\kappa\right] = -|\kappa|^\alpha \widehat{f}(\kappa), \tag{65}$$

for $x \in (-\infty,+\infty)$, $0 < \alpha \leq 2$ and let $\partial^\beta/\partial t^\beta$ denote the time nonlocal pseudo-differential operator whose Laplace transform is given by

$$\mathcal{L}\left[\frac{\partial^\beta g(t)}{\partial t^\beta};s\right] = s^\beta \widetilde{g}(s) - s^{\beta-1} g(0^+), \tag{66}$$

for $t > 0$, $0 < \beta \leq 1$. Then the pseudo-differential equation

$$\frac{\partial^\alpha u(x,t)}{\partial|x|^\alpha} = \frac{\partial^\beta u(x,t)}{\partial t^\beta} \tag{67}$$

is called *space-time fractional differential equation.*

**Remark 23.** The operator $\partial^\alpha/\partial|x|^\alpha$ is called Riesz derivative and is discussed by Saichev and Zaslavsky [32]. The operator $\partial^\beta/\partial t^\beta$ is called Caputo derivative and was introduced by Caputo in 1967 [31] as a regularization of the so-called Riemann–Liouville derivative.

**Theorem 11.** (Cauchy problem for the space-time fractional diffusion equation) *Consider the following Cauchy problem for the space-time fractional diffusion equation* (67)

$$\frac{\partial^\alpha u_{\alpha,\beta}(x,t)}{\partial|x|^\alpha} = \frac{\partial^\beta u_{\alpha,\beta}(x,t)}{\partial t^\beta}$$

$$u_{\alpha,\beta}(x,0^+) = \delta(x), \tag{68}$$

*then the function*

$$u_{\alpha,\beta}(x,t) = \frac{1}{t^{\beta/\alpha}} W_{\alpha,\beta}\left(\frac{x}{t^{\beta/\alpha}}\right), \tag{69}$$

*where*

$$W_{\alpha,\beta}(u) = \frac{1}{2\pi}\int_{-\infty}^{+\infty} d\kappa\, \mathrm{e}^{-i\kappa u} E_\beta(-|\kappa|^\alpha), \tag{70}$$

*solves the Cauchy problem* [35].

***Proof.*** Taking into account the initial condition (68), as a consequence of the operator definition, for nonvanishing $\kappa$ and $s$, the Fourier–Laplace transform of Eq. (67) is given by

$$-|\kappa|^\alpha \widehat{\widetilde{u}}_{\alpha,\beta}(\kappa,s) = s^\beta \widehat{\widetilde{u}}_{\alpha,\beta}(\kappa,s) - s^{\beta-1}, \tag{71}$$

leading to

$$\widehat{\widetilde{u}}_{\alpha,\beta}(\kappa,s) = \frac{s^{\beta-1}}{|\kappa|^\alpha + s^\beta}. \tag{72}$$

Equation (58) can be invoked for the inversion of the Laplace transform yielding

$$\widehat{u}_{\alpha,\beta}(\kappa,t) = E_\beta(-t^\beta|\kappa|^\alpha). \tag{73}$$

Eventually, the inversion of the Fourier transform leads to the thesis. □

**Remark 24.** The function defined by Eqs. (69) and (70) is a probability density function. For $\beta = 1$ and $\alpha = 2$, it coincides with the Green function for the ordinary diffusion equation. The case $\beta = 1$ and $0 < \alpha \le 2$ gives the Green function and the transition probability density for the symmetric and isotropic $\alpha$-stable Lévy process $L_\alpha(t)$ [34].

**Theorem 12.** *Let $Y_{\alpha,\beta}(t)$ be a compound fractional Poisson process and let $h$ and $r$ be two scaling factors such that*

$$Y_n(h) = hX_1 + \cdots + hX_n, \tag{74}$$

$$T_n(r) = rJ_1 + \cdots + rJ_n \tag{75}$$

*and*

$$\lim_{h,r\to 0} \frac{h^\alpha}{r^\beta} = 1, \tag{76}$$

*with $0 < \alpha \leq 2$ and $0 < \beta \leq 1$. To clarify the role of the parameter $\alpha$, further assume that, for $h \to 0$, one has*

$$\widehat{f}_X(h\kappa) \sim 1 - h^\alpha |\kappa|^\alpha, \tag{77}$$

*then, for $h, r \to 0$ with $h^\alpha/r^\beta \to 1$, $f_{hY_{\alpha,\beta}(rt)}(x,t)$ weakly converges to $u_{\alpha,\beta}(x,t)$, the Green function of the fractional diffusion equation.*

***Proof.*** In order to prove weak convergence, it suffices to show the convergence of the characteristic function (64) [1]. Indeed, one has

$$\widehat{f}_{hY_{\alpha,\beta}(rt)}(\kappa,t) = E_\beta\left(-\frac{t^\beta}{r^\beta}(\widehat{f}_X(h\kappa) - 1)\right) \overset{h,k\to 0}{\rightarrow} E_\beta(-t^\beta|\kappa|^\alpha), \tag{78}$$

which completes the proof. □

**Remark 25.** Condition (77) is not void. It is satisfied by all the distributions belonging to the basin of attaction of symmetric $\alpha$-stable laws. Let $f_{\alpha,X}(x)$ denote the probability density function of a symmetric $\alpha$-stable law whose characteristic function is

$$\widehat{f}_{\alpha,X}(\kappa) = \mathrm{e}^{-|\kappa|^\alpha}, \tag{79}$$

then one can immediately see that (77) holds true. As above, let $L_\alpha(t)$ denote the symmetric $\alpha$-stable Lévy process. Then, Eq. (73) is the characteristic function of $L_{\alpha,\beta}(t) = L_\alpha(D_\beta(t))$, that is of the symmetric $\alpha$-stable Lévy process subordinated to the inverse $\beta$-stable subordinator, $D_\beta(t)$, the functional inverse of the $\beta$-stable subordinator. This remark leads to the conjecture that $L_{\alpha,\beta}(t)$ is the functional limit of $Y_{\alpha,\beta}(t)$, the $\alpha$-stable compound fractional Poisson process defined by Eq. (61) with the law of the jumps $X$ belonging to the basin of attraction of or coinciding with an $\alpha$-stable law.

This conjecture can be found in a paper by Magdziarz and Weron [36] and is proved in Meerschaert *et al.* [29] using the methods discussed in the book by Meerschaert and Scheffler [37].

## Acknowledgments

Enrico Scalas is grateful to Universitat Jaume I for the financial support received from their Research Promotion Plan 2010 during his scientific visit in Castellón de la Plana where this paper was completed. In writing this paper, Enrico profited of useful discussions with Rudolf Gorenflo, Francesco Mainardi, and Mark M. Meerschaert.

## References

1. W. Feller, *An Introduction to Probability Theory and its Applications*, Vol. II (Wiley, 1971).
2. B. de Finetti, *Theory of Probability*, Vol. II (Wiley, 1975).
3. R. Hilfer, *Fractals* **3**, 211 (1995).
4. R. Hilfer and L. Anton, *Phys. Rev. E* **51**, R848 (1995).
5. O. N. Repin and A. I. Saichev, *Radiophys. Quantum Electronics* **43**, 738 (2000).
6. N. Laskin, *Commun. Nonlinear Sci. Numer. Simul.* **8**, 201 (2003).
7. E. Scalas, *Physica A* **362**, 225 (2006).
8. D. Fulger, E. Scalas and G. Germano, *Phys. Rev. E* **77**, 021122 (2008).
9. G. Germano, M. Politi, E. Scalas and R. L. Schilling, *Phys. Rev. E* **79**, 066102 (2009).
10. P. Billingsley, *Probability and Measure* (Wiley, 1986).
11. K. Sato, *Lévy Processes and Infinitely Divisible Distributions* (Cambridge Univ. Press, 1999).
12. R. Cont and P. Tankov, *Financial Modelling with Jump Processes* (Chapman & Hall/CRC Press, 2004).
13. W. Feller, *Ann. Math. Statist.* **14**, 389 (1943).
14. A. Einstein, Jr., *Der Geschiebetrieb als Wahrscheinlichkeitsproblem*, Mitteilungen der Versuchsanstalt für Wasserbau an der Eidgenössischen Technischen Hochschule, Zürich, 1937.
15. G. Pólya, *Zur Kinematik der Geschiebebewegung*, Mitteilungen der Versuchsanstalt für Wasserbau an der Eidgenössischen Technischen Hochschule, Zürich, 1937.
16. G. Pólya, *Actualités Scien. Indust.* **734**, 25 (1938).
17. E. W. Montroll and G. H. Weiss, *J. Math. Phys.* **6**, 167 (1965).
18. H. Scher and M. Lax, *Phys. Rev. B* **7**, 4491 (1973); H. Scher and M. Lax, *Phys. Rev. B* **7**, 4502 (1973); E. W. Montroll and H. Scher, *J. Stat. Phys.* **9**, 101 (1973).

19. M. F. Shlesinger, Random processes, in *Encyclopedia of Applied Physics*, ed. G. L. Trigg (VCH, 1996), pp. 45–70.
20. R. Metzler and J. Klafter, *Phys. Rep.* **339**, 1 (2000).
21. R. Metzler, J. Klafter, *J. Phys. A: Math. Gen.* **37**, R161 (2004).
22. E. Çinlar, *Introduction to Stochastic Processes* (Prentice-Hall, 1975).
23. O. Flomenbom and J. Klafter, *Phys. Rev. Lett.* **95**, 098105 (2005).
24. O. Flomenbom and R. J. Silbey, *Phys. Rev. E* **76**, 041101 (2007).
25. J. Janssen and R. Manca, *Semi-Markov Risk Models for Finance, Insurance and Reliability* (Springer, 2007).
26. I. Podlubny, *Fractional Differential Equations* (Academic Press, 1999).
27. E. Scalas, R. Gorenflo and F. Mainardi, *Phys. Rev. E* **69**, 011107 (2004).
28. F. Mainardi, R. Gorenflo and E. Scalas, *Vietnam J. Math.* **32**, 53 (2004).
29. M. M. Meerschaert, E. Nane and P. Vellaisamy, The fractional Poisson process and the inverse stable subordinator, available from http://www.stt.msu.edu/mcubed/FPP.pdf, *Electronic J. Probab.*, in press.
30. L. Beghin and E. Orsingher, *Electronic J. Probab.* **14**, 1790 (2009).
31. M. M. Meerschaert, E. Nane and P. Vellaisamy, *Ann. Probab.* **37**, 979 (2009).
32. A. I. Saichev and G. M. Zaslavsky, *Chaos* **7**, 753 (1997).
33. M. Caputo, *Geophys. J. R. Astr. Soc.* **13**, 529 (1967).
34. N. Jacob, *Pseudo-differential Operators and Markov Processes*, Vol. III: *Markov Processes and Applications* (Imperial College Press, 2005).
35. F. Mainardi, Yu. Luchko and G. Pagnini, *Fractional Calculus Appl. Anal.* **4**, 153 (2001).
36. M. Magdziarz and K. Weron, *Physica A* **367**, 1 (2006).
37. M. M. Meerschaert and H. P. Scheffler, *Limit Distributions for Sums of Independent Random Vectors: Heavy Tails in Theory and Practice* (Wiley, 2001).

# Chapter 16

## Origin of Allometry Hypothesis

Bruce J. West
*Information Sciences Directorate, Army Research Office,*
*Research Triangle Park, NC, USA*

Damien West
*Physics Department, Rensselaer Polytechnic Institute,*
*Troy, NY 12180, USA*

The theoretical allometry relation between two properties $X$ and $Y$ of an organism is of the form $X = aY^b$ and has been known for over a century. The allometry coefficient $a$ and allometry exponent $b$ have been fit by various data sets over that time. Most theories focus on establishing the proper value of $b$ entailed by an appropriate biological model and with few exceptions dismissed $a$ as a normalization constant. Data analysis indicates that the empirical allometry relation is obtained with the replacements $X \to \langle X \rangle$ and $Y \to \langle Y \rangle$ and the brackets denote an average over an ensemble of realizations of the organism. We show that the resulting empirical allometry relation cannot be derived from the theoretical relation by averaging. Consequently we hypothesize that the statistics of living complex networks can be described by fractional diffusion equations along with the assumption of strong anticipation. Furthermore, we demonstrate that taken together these two assumptions suffice to explain the origin of empirical allometry relations.

1. Introduction . . . 376
 1.1. Theoretical allometry . . . 376
 1.2. Empirical allometry . . . 378
2. Random Allometry Coefficients . . . 380
3. Fractional Diffusion Equation . . . 382
4. Support Origin of Allometry Hypothesis . . . 386
5. Discussion . . . 387
6. Conclusion . . . 389

## 1. Introduction

The rabbit being chased by the fox twists and turns to evade capture; the outfielder lunges to intercept the arc of the long ball and the stock broker sweats over what price to sell the erratically moving stock. Whether the rabbit survives, the batter is out, or the stock broker makes a killing are all determined by how well they anticipate the movement of the fox, the ball's trajectory and the fluctuation in the stock price. Historically *anticipation* was interpreted as the prediction made by a living beings' internal model of the external world; the model interpolates the past trajectories into the futures of the fox, ball and stock. However this perspective has a number of limitations not the least of which is computational intensity. The internal model may satisfactorily predict the parabolic trajectory of the baseball, but it probably cannot determine the erratic trajectory of a stock price or the evasion tactics of the rabbit. But even so some stock brokers do "guess" when stock price will drop better than others and some rabbits seem to be able to zig when the fox zags and live to a ripe old age. This adaptability by the living has led to the partitioning of dynamics into *weak anticipation*, based on internal models, and *strong anticipation*, which depends on things other than such internal models. This paper is about those other things in the context of allometry relations.

Dubois [1] defined *strong anticipation* in terms of the embedding of an organism within an environment with strong anticipation emerging as the lawful regularity of the coupled organism-environment network. We do not review the various models of strong anticipation here, but we make use of the observation of Stephen *et al.* [2] that in master-slave dynamic networks the slave system anticipates the future behavior of the master system by synchronizing with a state that lies in the future of the master systems. Herein we adapt the strong anticipation concept to complex phenomena whose intermittent fluctuations are described by inverse power-law probability densities. In this latter perspective we interpret strong anticipation as a manifestation of the long-range influence stochastic fluctuations have on one another with allometry emerging from the scaling of the statistics.

### 1.1. *Theoretical allometry*

Natural Science encapsulates the information contained in vast amounts of biological and botanical data through allometry relations, which were introduced in the 19th century. The allometry laws are often discussed as if they described a single class of phenomena when, in fact, they actually characterize two distinct groups [3]. In Group 1 the allometry law interrelates variables within a specific species such as the total body mass (TBM) to the mass of a specific organ, e.g., TBM to a deer's antlers or to a crab's claws [4]; this is intraspecies allometry. In Group 2 the allometry law interrelates variables across species such as basal metabolic rate (BMR) to TBM for mammals [5, 6]; this is interspecies allometry.

In 1931, Sir Julian Huxley [4]; grandson of the Huxley of Darwin evolution fame, brother of the novelist Aldous (*Brave New World*) and half-brother of the biophysicist Andrew (the Hodgkin–Huxley equations) [3]; proposed that two parts of the same organism have different rates of growth. In his treatment of the dynamics of differential growth the two rates were assumed proportional to one another. In this way, if $X$ is a living subnetwork observable with growth rate $\alpha$ and $Y$ is a living host network observable with growth rate $\beta$, then

$$\frac{1}{X}\frac{dX}{dt} = \frac{\alpha}{\beta}\frac{1}{Y}\frac{dY}{dt}. \tag{1}$$

Equation (1) can be directly integrated to obtain the time-independent allometry growth law involving the host network $Y$ and the subnetwork $X$:

$$X = aY^b, \tag{2}$$

where $a$ and $b$ $(= \alpha/\beta)$ are empirically determined constants. The theoretical form of the allometry relation considered by Sir Julian is the basis of subsequent theoretical discussion in such excellent books as Schmidt–Nielson's *Scaling. Why is Animal Size so Important?* [5] and Calder's *Size, Function and Life History* [6]. Savageau [7] provided a "first principles" derivation of Eq. (2) treated as a generic relation using the properties of the underlying determinants of complex dynamic systems to generalize Huxley's analysis. Thus, the phenomenological relation describing interspecies allometry appeared to be well established by theory.

The interspecies allometry relation reaches across species and interrelates such variables as the BMR $X$ to TBM $Y$ also using Eq. (2). Both theoretical and observational controversies arose in interspecies allometry relations concerning the value of the allometry exponent $b$. In 1932, Kleiber [8] established by means of linear regression of the logarithmically transformed observational data on

$$\ln X = b \ln Y + \ln a, \tag{3}$$

that $b \approx 3/4$. Subsequent observational studies have suggested other values of the phenomenological exponent $b$ [5, 6, 9, 10]. Controversy also arises in the theoretical explanation as to why the allometry exponent $b$ should take on specific values. In the middle of the 19th century Sarrus and Rameaux [11] reasoned that the heat loss from a warm-blooded animal must be approximately proportional to the animal's free surface. This simple dimensional analysis requires $b = 2/3$ as reviewed in a number of excellent sources [5, 6]. On the other hand, the quarter-power allometry scaling law is explained by West *et al.* [12] using geometric scaling arguments from fractal physics to establish the value $b = 3/4$ in fractal network transport and other quarter-power scaling laws in biology. The constraint on the allometry coefficient $a$ is that it intercept the vertical axis when the property being fit vanishes.

In Huxley's book there are literally dozens of pairs $(X, Y)$ shown to satisfy Eq. (3). In such intraspecies allometry there is relatively little variability and the deterministic equation Eq. (3) fits the single species data quite well. However, the power-law index $b$ changes markedly from species to species, see Withers [13]. Similarly when fitting interspecies data the fluctuations in the linear regression to Eq. (3) are substantial. It has been suggested by a number of authors that the parameter $a$ be treated as a random variable [9, 10] to account for the observed allometry variability.

### 1.2. *Empirical allometry*

Careful reading of Huxley's book reveals that Eq. (3) is not being directly fit to the data because the data do not consist of individual measurements of $X$ and $Y$. The empirical data consist of averages over a collection of such measurements on individual specimens and are denoted by $\langle X \rangle$ and $\langle Y \rangle$. This change in variables is not discussed in the text of his book but is explicitly stated in the table captions. This same duality permeates the literature subsequent to Huxley in which the theoretical discussions focus on $(X, Y)$ whereas data fits are concerned with $(\langle X \rangle, \langle Y \rangle)$. Consequently the empirical form for the allometry law is

$$\langle X \rangle = \bar{a} \langle Y \rangle^{\bar{b}} \tag{4}$$

the brackets denote an average over an ensemble of realizations of the observable that varies from specimen to specimen in the ensemble, and the overbars on the parameters denote average values.

In an attempt to reconcile these very different descriptions of allometry, we consider the average of Eq. (2) over an ensemble of realizations of the data

$$\langle X \rangle = a \left\langle Y^b \right\rangle \tag{5}$$

and note the inequality $\left\langle Y^b \right\rangle \neq \langle Y \rangle^b$ with $b \neq 1$. So without knowing the statistical distribution, the averaging in Eq. (5) cannot be carried out. In order to derive the empirical allometry relation from Eq. (2), we introduce the deviations of the variables from their average values $X = \langle X \rangle + \delta X$ and $Y = \langle Y \rangle + \delta Y$ into the equation. After some algebra the allometry coefficient can be expressed as $a = \bar{a} + \delta a$, where $\bar{a}$ is the average allometry coefficient obtained from data and $\delta a$ is the fluctuation from the mean value that to first order is linear in the relative fluctuations $\delta X / \langle X \rangle$ and $\delta Y / \langle Y \rangle$. Thus, we rewrite Eq. (2) as

$$\langle X \rangle = (\bar{a} + \delta a) \langle Y \rangle^{\bar{b}},$$

which when $\delta a / \bar{a} \ll 1$ simplifies to the empirical allometry law Eq. (4).

We emphasize that the empirical allometry relation Eq. (4) only follows from the theoretical allometry relation Eq. (2) when $\delta a / \bar{a} \ll 1$ implying that the distribution of fluctuations must be very narrow. This condition on the fluctuations in the allometry coefficient can be directly tested by data.

We hypothesize that the origin of the allometry law Eq. (4) with $b \neq 1$ is tied to the strong anticipation of the average behavior of a complex host network by a complex subnetwork, both of which have fractal statistics but with different fractal dimensions. Consequently, the "lawful regularity" of the coupled networks in Dubois' definition of strong anticipation is, in conjunction with the fractal statistics, the origin of allometry laws.

Notice that we use the fractal assumption implemented in contemporary arguments regarding the origins of allometry [12]. However, we use statistical fractals, not geometrical fractals, in our analysis. Consequently, the dynamics of the probability density functions that characterize the transfer of information from one living network to another is described by fractional diffusion equations [14].

In Sec. 2, we assume the allometric coefficient is a random variable and using available data determine an empirical Pareto probability density function (*pdf*) for the coefficient's variability. In Sec. 3, we introduce fractional diffusion equations to describe the phase space dynamics of the probability densities describing the statistical behavior of complex networks. The general scaling form of the solution is all that we require for application to allometry relations. In Sec. 4 we support the *origin of allometry hypothesis* by taking advantage of the fact that Eq. (4) entails a scaling relation between the first moments of the probability densities of the two observables. We use the scaling solution obtained in Sec. 3 to formally calculate these averages. In Sec. 5, the implications of this scaling for allometry are discussed and in Sec. 6 some conclusions are drawn.

## 2. Random Allometry Coefficients

The data relating BMR that measures the energy expended by a given species in watts to the TBM of that species in kilograms for 391 species of mammal is plotted in Fig. 1, see also Dodds *et al.* [10] and Heusner [9]. A regression of the logarithmically transformed data on

$$\ln B = \ln a + b \ln M \tag{6}$$

that minimizes the mean-square error is a straight line on log-log graph paper and was found by Heusner [9] to have slope $b = 0.71 \pm 0.008$ that empirically falls between the theoretical values $2/3 \leq \bar{b} \leq 3/4$. The fit also determines the value of the allometry coefficient fitting the smallest available TBM $\bar{a} = 0.02$. The random fluctuations around the allometry relation with this empirical allometry exponent and normalized allometry coefficient are depicted in Fig. 1:

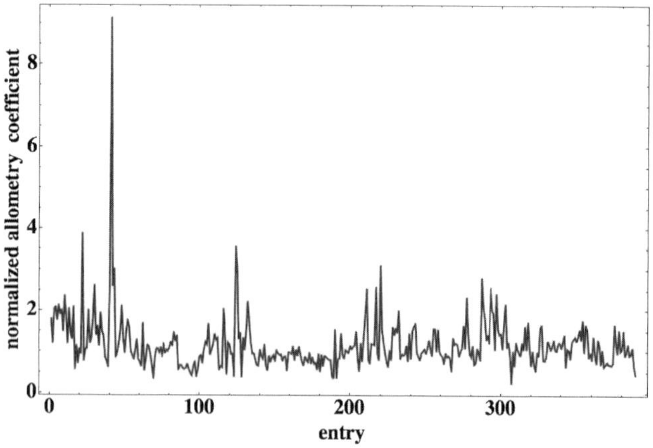


Fig. 1. The fluctuations around the allometry relation between the BMR and TBM in Eq. (6) with $\bar{b} = 0.71$ for the 391 mammalian species tabulated in Huesner [9]. The line segments connect the data points to aid the eye in assessing the variability. Note that this figure depicts the random nature of the allometry coefficient $a$ relative to $\bar{a} = 0.02$.

It is evident from Fig. 1 that the allometry coefficient fluctuates over orders of magnitude indicating that the coupling of TBM and energy use by physiologic processes is not confined to a single scale. The extreme variability suggests that the fluctuations are not additive, but are instead multiplicative as discussed by Packard and Broadman [15] regarding the logarithmic transformation of the data in Eq. (6). What is important is not the individual realization but the *pdf* for the allometry coefficient that we now determine from a histogram of the fluctuations depicted in Fig. 1.

The variability in the allometry coefficient determined by the data in Fig. 1 is depicted in Fig. 2 where these data, using the normalized variable $a' = a/\bar{a} = B/(\bar{a}M^{\bar{b}})$, are partitioned into 20 equal sized bins. A histogram is then constructed by counting the number of data points within each of the bins as indicated by the dots. The solid line segment in this latter figure is the best fit to these 20 numbers with minimum mean-square error.

The functional form for the histogram capturing the allometry coefficient fluctuations is determined by the curve in Fig. 2 and the quality of the fit to the diversity data is determined by the correlation coefficient to be $r^2 = 0.98$. The normalized histogram on the interval $(0, A')$ is interpreted as the *pdf*:

$$P(a') = \frac{\alpha^2 - 1}{2\alpha - (1+\alpha)A'^{1-\alpha}} \begin{cases} a'^{\alpha}; & a' \leq 1 \\ \dfrac{1}{a'^{\alpha}}; & a' \geq 1 \end{cases} \tag{7}$$

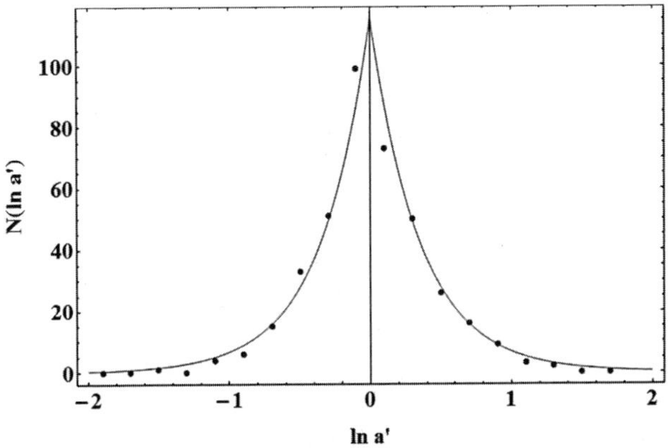


Fig. 2. The histogram of the deviations from the prediction of the allometry relation using the data depicted in Fig. 1 is partitioned into 20 equal sized bins. The solid line segment is the best fit of Eq. (7) to the 20 histogram numbers, which yields the power-law index $\alpha = 2.79$ and the quality of the fit is measured by the correlation coefficient $r^2 = 0.98$.

and empirically $\alpha = 2.79$. The distribution falls off as an inverse power law on either side of $a' = 1$ as depicted in Fig. 2. Consequently, the probability that the allometry coefficient exceeds a given large value of the allometry coefficient is given by a Pareto distribution. The allometry relation describes a trait across multiple species and the Pareto distribution characterizes the variability of that trait. Therefore the variability in allometry is a consequence of the intermittent statistical fluctuations in the allometry coefficient.

Most importantly the variance of the allometry Pareto *pdf* diverges which implies that we cannot use the dynamical variables as the starting point for deriving the empirical allometry relation, as Huxley did. Therefore we adopt a complementary view starting from the perspective of probability theory, but bearing in mind the Pareto *pdf* scales.

## 3. Fractional Diffusion Equation

The origins of the scientific laws and principles that we find so attractive in the physical sciences are the result of years of experiment and data acquisition. Data analysis leads to the identification of patterns and the subsequent casting of these patterns into the mathematical form of laws or rules. These laws stimulate the development of theory to explain them and that predict the outcome of new experiments to further enhance our understanding. This idealized use of the scientific method from the physical sciences encountered an impasse when it came to developing mathematical theories to interpret allometry laws. The theoretical efforts to date have not addressed the proper variables in that they focus on the stochastic variables and not on their average values. Consequently, we are at the very beginning of developing a theory to explain how allometry relations originate and this necessitated making a fundamental hypothesis in the Introduction.

To test the reasonableness of the origin of the allometry hypothesis, we replace the discussion of the dynamics of the observables [4, 7] and/or the traditional geometric scaling of the observables [5, 6] with an examination of the assumed scaling properties of the stochastic variables. Scaling is a nearly ubiquitous property of complex networks indicating that they simultaneously fluctuate over many time scales, coupling the "here and now" with the "there and then." In the physical sciences such phenomena have historically been referred to as $1/f$ noise, but with the more recent identification of such phenomena in the social and life sciences such networks are now said to have $1/f$ variability [14] with the pejorative term noise being suppressed. Mandelbrot [16] was probably the first to recognize the positive significance of this $1/f$ variability with his introduction of fractals into the scientist's lexicon. The $1/f$ variability implies that the stochastic variable describing the dynamics of the underlying network scales algebraically in time as we discuss in the sequel.

Consider the dynamics of a complex process that is heterogeneous in the random variable and whose fluctuations in time contain historical information. In general $P(z,t)dz$ is the probability that the dynamic variable $Z(t)$ has a value in the interval $(z, z+dz)$ at time $t$ given a value $q(t)$ at the origin and the phase space dynamics are formally expressed as

$$G\left(\frac{\partial}{\partial t}, i\frac{\partial}{\partial z}\right) P(z,t) = q(t)\delta(z). \tag{8}$$

The analytic function $G(\cdot,\cdot)$ of the indicated operators determines the dynamics of the probability density function. A phase space equation for a classical diffusion process is determined by the choice of operator and inhomogeneous term

$$G\left(\frac{\partial}{\partial t}, i\frac{\partial}{\partial z}\right) = \frac{\partial}{\partial t} - D\frac{\partial^2}{\partial z^2} \quad \text{and} \quad q(t) = 1. \tag{9}$$

This choice reduces Eq. (8) to the Fokker–Planck equation (FPE) for the probability density with $D$ the diffusion coefficient and the solution to the FPE is the well-known distribution of Gauss.

Correspondingly, as was first observed by West and Seshadri [17], a Lévy diffusive process is determined by the choice of fractional diffusion operator

$$G\left(\frac{\partial}{\partial t}, i\frac{\partial}{\partial z}\right) = \frac{\partial}{\partial t} - K_0 D_z^\alpha \quad \text{and} \quad q(t) = 1, \tag{10}$$

where ${}_0D_z^\alpha$ denotes a Reisz potential operator [18] and $K$ is a constant. In this case, using the Fourier transform of the Reisz potential $-|k|^\alpha$, the solution to the fractional diffusion equation is the symmetric $\alpha$-stable Lévy probability density.

More general dynamics can be simply expressed in terms of the Fourier–Laplace transform $G^*(u,k)$ without directly solving the fractional diffusion equation (FDE). The Fourier transform of an analytic function $f(z)$ scales as

$$f(az) = \int_{-\infty}^{\infty} e^{-ikaz} \tilde{f}(k) \frac{dk}{2\pi} = \frac{1}{a}\int_{-\infty}^{\infty} e^{-ik'z} \tilde{f}\left(\frac{k'}{a}\right) \frac{dk'}{2\pi} \tag{11}$$

and the Laplace transform of the analytic function $g(t)$ scales as

$$g(bt) = \int_0^\infty e^{-ubt}\hat{g}(u)du = \frac{1}{b}\int_0^\infty e^{-u't}\hat{g}\left(\frac{u'}{b}\right)du'. \tag{12}$$

It is clear that the Fourier transform of $f(z)$ is $\tilde{f}(k)$ and the Laplace transform of $g(t)$ is $\hat{g}(u)$ and the double transform is denoted with an asterisk. Consequently the phase space dynamics can be expressed in Fourier–Laplace space as

$$G^*(u,k)\,P^*(k,u) = \hat{q}(u). \tag{13}$$

We have a great deal of lattitude in the choice of dynamics represented by the function $G^*$. In our discussion of interspecies allometry we consider Eq. (10) to be a Fourier–Laplace space representation of a fractional diffusion equation [18, 19]. We choose a fractional derivative in time given by a Riemann–Liouville operator of order $\beta$; the inhomogeneous term to be

$$g(t) = \frac{t^{-\beta}}{\Gamma(1-\beta)} \tag{14}$$

and a Reisz potential in $z$ of order $\alpha$ so that Eq. (13) can be written

$$P^*(k,u) = \frac{u^{\beta-1}}{u^\beta + K\,|k|^\alpha} \tag{15}$$

and $K$ is a constant. The probability density that solves the phase space FDE is given by the inverse Fourier–Laplace transform of Eq. (15). We note that this representation of the FDE in space and time and its solutions for various combinations of parameters are reviewed by Klafter and Metzler [19]. They also show how to derive fractional diffusion equations using the continuous time random walk approach (CTRW) of Montroll and Weiss [20].

It is not an easy task to explicitly carry out the inverse transforms for Eq. (15) for arbitrary $\alpha$ and $\beta$, but it has been done by Uchaikin [21]. However, we do not find these explicit solutions to be particularly illuminating in the present context. For our purposes more insight is provided by adopting a scaling technique using the properties of Fourier and Laplace transforms to obtain the general inverse to Eq. (15) without doing the inverse transforms explicitly.

From the scaling of the Fourier and Laplace transforms in equations Eqs. (11) and (12), we conclude that the scaled probability density $P(az, bt)$

has the Fourier–Laplace transform

$$P(az, bt) \Rightarrow \frac{1}{ab} P^*\left(\frac{k}{a}, \frac{u}{b}\right),$$

which when compared with Eq. (12) yields the equality

$$\frac{1}{ab} P^*\left(\frac{k}{a}, \frac{u}{b}\right) = \frac{1}{a} P^*\left(\frac{b^{\beta/\alpha}}{a} k, u\right). \tag{16}$$

The inverse Fourier–Laplace transform of Eq. (16) is therefore

$$P(z,t) = \frac{1}{b^{\beta/\alpha}} P\left(\frac{a}{b^{\beta/\alpha}} z, t\right). \tag{17}$$

The scaling relations Eqs. (16) and (17) are both satisfied by a probability density function of the scaling form

$$P(az, bt) = \frac{1}{b^{\mu} t^{\mu}} F_z\left(\frac{az}{b^{\mu} t^{\mu}}\right), \tag{18}$$

which when applied to (17) yields

$$\frac{1}{b^{\mu} t^{\mu}} F_z\left(\frac{az}{b^{\mu} t^{\mu}}\right) = \frac{1}{b^{\beta/\alpha} t^{\mu}} F_z\left(\frac{az}{b^{\beta/\alpha} t^{\mu}}\right).$$

The probability density function that solves the fractional diffusion equation therefore satisfies the scaling equation

$$P(z,t) = \frac{1}{t^{\mu}} F_z\left(\frac{z}{t^{\mu}}\right); \quad \mu = \frac{\beta}{\alpha}. \tag{19}$$

The function $F_z(\cdot)$ is left unspecified but it is analytic in the similarity variable $z/t^{\mu}$. Note that in a standard diffusion process $Z(t)$ is the displacement of the diffusing particle from its initial position at time $t$, $\mu = 1/2$ and the functional form of $F_z(\cdot)$ is a Gauss distribution. However, for general complex phenomenon there is a broad class of distributions for which the functional form of $F_z(\cdot)$ is not Gaussian and the scaling index $\mu \neq 1/2$. For example, the $\alpha$-stable Lévy process that we mentioned earlier scales in this way and the Lévy index is in the range $0 < \alpha \leq 2$, with the equality holding for the Gauss distribution and the scaling index is related to the Lévy index by $\mu = 1/\alpha$ when $\beta = 1$, see West *et al.* [18] and Uchaikin [21] for very different discussions. More generally we can write the solution to the FDE as

$$P(z,t) = \int_{-\infty}^{\infty} e^{-ikz} e^{-\gamma t^{\beta}|k|^{\alpha}} \frac{dk}{2\pi};$$

a symmetric Lévy distribution with an algebraic increase in the time dependence of the width.

## 4. Support Origin of Allometry Hypothesis

We assume that it is reasonable for two parts of the same organism to share the same class of probability densities to describe the interacting $X$ and $Y$ observables. Or more generally to assume the same class of ensemble distributions for the property $X$ of an ensemble of realizations of an organism and a second property $Y$ for a second ensemble for the same organism. Consequently, we assume that the host network and the subnetwork have the same form of probability density with suitably indexed parameters. Consider the average of a generic observable using the *pdf* Eq. (19)

$$\langle Z(t)\rangle = \int zP(z,t)dz = \bar{Z}t^{\mu} \tag{20}$$

and $\bar{Z}$ is a finite time-independent constant determined by the form of the distribution in Eq. (19) with the transformation $q = z/t^{\mu_z}$:

$$\bar{Z} \equiv \int qF_z(q)dq. \tag{21}$$

Therefore unlike Huxley who assumed the differential growth equations have the same form with proportional growth rates, we presume the probability densities have the same form and invoke strong anticipation to characterize the interaction between the two networks. This latter assumption implies that the host network and the subnetwork interact strongly with one another and consequently can be viewed as essential parts of one another.

We use strong anticipation to constrain the discussion by assuming the average in Eq. (20) corresponds to the host network; that time is the dependent variable

$$t = f(\langle Y\rangle) = \left(\frac{\langle Y\rangle}{\bar{Y}}\right)^{1/\mu_y}, \tag{22}$$

and the average $\langle Y\rangle$ is the independent variable. Thus, $t$ is the time at which the host network has achieved the average value $\langle Y\rangle$ and we assume that the probability density function for the subnetwork variable most strongly influenced at those times best represent the dynamical response of the subnetwork. Consequently, we assume the subnetwork probability density is expressed in terms of the anticipated average value of the host network variable

$$P(x; \langle Y \rangle) = \int_0^\infty P(x;t)\delta\left(t - f\left(\langle Y \rangle\right)\right) dt$$
$$= \frac{1}{f\left(\langle Y \rangle\right)^{\mu_x}} F_x\left(\frac{x}{f\left(\langle Y \rangle\right)^{\mu_x}}\right). \tag{23}$$

Note that here the subnetwork's statistical variability is dependent on the average behavior of the host network whose broad range of time scales is manifest through the scaling. The $X$-average can therefore be expressed in terms of the $Y$-average as

$$\langle X \rangle = \int x P(x; \langle Y \rangle) dx = \bar{Y} f(\langle Y \rangle)^{\mu_x} \tag{24}$$

so that substituting Eqs. (22) into (24) results in the allometry relation Eq. (4)

$$\langle X \rangle = \bar{X}\left(\frac{\langle Y \rangle}{\bar{Y}}\right)^{\mu_x/\mu_y} = a\,\langle Y \rangle^b$$

with the allometry parameter values related to the parameters in the probability density functions

$$b = \frac{\mu_x}{\mu_y} \quad \text{and} \quad a = \frac{\bar{X}}{\bar{Y}^b}. \tag{25}$$

Note that the allometry exponent $b$ is now the ratio of the power-law indices that appear in the solution to the FDEs and not the ratio of growth rates as assumed by Huxley. This derivation of the allometry relation entails a significant change in the interpretation of the dynamics underlying coupled networks. Here it is the average response or adaptation of the subnetwork to the average behavior of the host network that is captured by the allometry growth law, not the instantaneous growth.

## 5. Discussion

We have demonstrated that the mathematical justification of the allometry relation is not necessarily a consequence of deterministic differential growth as has been assumed since Huxley's seminal book. A modern version of this argument in terms of the dynamics of self-similar deterministic functions is given by Stevens [3]. The derivation of the allometry law given in Sec. 3 is based on the assumed self-similar scaling of the probability density functions that solve the FDE. Consequently allometry depends on the fractal nature of the stochastic variables and the network coupling induced by strong anticipation. Using the average values obtained from Eq. (20), it is straightforward to obtain the deterministic equations for the averages

$$\frac{1}{\langle X\rangle}\frac{d\langle X\rangle}{dt}=\frac{\mu_x}{\mu_y}\frac{1}{\langle Y\rangle}\frac{d\langle Y\rangle}{dt}, \tag{26}$$

that can be directly integrated to obtain Eq. (4). Equation (26) is a direct analogue of Huxley's differential growth equation Eq. (1), but with the dynamic variables replaced with their average values. It might be argued that it was implicitly assumed that Huxley's equations Eq. (1) were actually of the form Eq. (26) even though they were written in terms of $X$ and $Y$. If that were the case, the scaling parameters $\mu_x$ and $\mu_y$would necessarily be misinterpreted as the growth rates $\alpha$ and $\beta$. Consequently, the growth represented by Eq. (1) was intended to represent the instantaneous differential relation between the network variables and not that between their average values.

If Eq. (1) were the correct differential growth equation, then as the growth rate of the host network changes, the growth rate of the subnetwork to which it is coupled changes proportionately. Huxley argued that this proportionality is required to maintain the time-independent form of the allometry relation, which is mathematically correct. However, this matching requirement is not so stringent when applied to Eq. (26) in which case it is the scaling exponents of the averages that retain the proportionality and not the instantaneous rates. In this latter case a burst of growth in the host network is not instantaneously compensated for by an increase in the growth rate of the subnetwork, but requires time over which the subnetwork responds and adjusts its average. Complementary considerations apply for suppressed growth of the host network.

Consequently at any point in time in an ensemble of coupled host networks and subnetworks, the subnetworks may be ordered according to their rates of growth relative to those of the host networks. If these subnetworks are labeled 1 to $N$, starting from the fastest growing and ending with the slowest growing, then at a later point in time the same ordering with respect to growth rates would result in a random reordering of the labels, which is to say, that some subnetworks speed up and other slow down due to their being regulated by their host networks. This regulation is done by the subnetwork strongly anticipating the average change in the host network variable. The statistics of the reordering of the labels in time is characterized by the probability density Eq. (20), which we rewrite, using the dimensionless variables $\chi = X/\bar{X}$ and $\zeta = Y/\bar{Y}$, in the more transparent form

$$p(\zeta; \langle\chi\rangle) = \frac{1}{\langle\chi\rangle^{1/b}} F_y\left(\frac{\zeta}{\langle\chi\rangle^{1/b}}\right). \tag{27}$$

The dependence of the statistics of the subnetwork on the average of the host network variable is explicit in Eq. (27). Note that in terms of these dimensionless variables, the allometry relation is given by

$$\langle\chi\rangle = \langle\zeta\rangle^{b}, \tag{28}$$

which is equivalent to Eq. (4).

## 6. Conclusion

A complex network is defined for our purposes as one whose statistical behavior changes over time as described by a FDE and whose probability density function is consequently of the form Eq. (19). An allometry relation between a property of a complex host network and a property of a complex subnetwork embedded within it arises when the subnetwork fluctuations strongly anticipate the average fluctuations of the host network as manifest in Eq. (23).

We conclude that the implied relation between averages based on fractal scaling of the probability densities in conjunction with strong anticipation supports the origin of allometry hypothesis. Moreover the allometry relation Eq. (28) is the lawful regularity of the coupling between a host complex network and a complex subnetwork.

## Acknowledgments

B.J.W acknowledges the US Army Research Office for partial support of this research.

## References

1. D. M. Dubois, Mathematical foundations of discrete and functional systems with strong and weak anticipations, in *Anticipatory Behavior in Adaptive Learning Systems*, LNAI, Vol. **2683**, eds. M.V. Butz, S. Olivier and P. Gerard (Springer-Verlag, 2003), pp. 110–132.

2. D. G. Stephen, N. Stepp, J. A. Dixon and M. T. Turvey, *Physica A* **387**, 5271 (2008).
3. C. F. Stevens, *J. Biol.* **8**, 14 (2009).
4. J. S. Huxley, *Problems of Relative Growth* (Dial Press, 1931).
5. K. Schmidt-Nielsen, *Scaling, Why is Animal Size so Important?* (Cambridge Univ. Press, 1984).
6. W. W. Calder III, *Size, Function and Life History* (Harvard Univ. Press, 1984).
7. M. A. Savageau, *Proc. Natl. Acad. Sci. USA* **76**, 6023 (1979).
8. M. Kleiber, *Hilgarida* **6**, 315 (1932).
9. A. A. Heusner, *J. Exp. Biol.* **160**, 25 (1991).
10. P. S. Dodds, D. H. Rothman and J. S. Weitz, *J. Theor. Biol.* **209**, 9 (2001).
11. F. Sarrus and J. Rameaux, *Bull. Acad. Roy. Med.* (*Paris*) **3**, 1094 (1838).
12. G. B. West, J. H. Brown and B. J. Enquist, *Science* **276**, 122 (1997).
13. P. C. Withers, *Comparative Animal Physiology* (Saunders, 1992).
14. B. J. West, E. L. Geneston and P. Grigolini, *Phys. Rep.* **468**, 1 (2009).
15. G. C. Packard and T.J. Boardman, *Physiol. Biochem. Zool.* **81**, 496 (2008).
16. B. B. Mandelbrot, *Fractals, Form and Chance* (W. F. Freeman, 1977).
17. V. Seshadri and B. J. West, *Proc. Natl. Acad. Sci. USA* **79**, 4501 (1982).
18. B. J. West, M. Bologna and P. Grigolini, *Physics of Fractal Operators* (Springer, 2003).
19. J. Klafter and R. Metzler, *Phys. Rep.* **339**, 1 (2000).
20. E. W. Montroll and G. Weiss, *J. Math. Phys.* **6**, 167 (1965).
21. V. V. Uchaikin, *Int. J. Theor. Phys.* **39**, 2087 (2000).

# Quantum Systems

# Chapter 17

## Principles of Fractional Quantum Mechanics

Nick Laskin
*TopQuark Inc., Toronto, ON, M6P 2P2, Canada*

A review of fundamentals and physical applications of fractional quantum mechanics has been presented.

Fundamentals cover fractional Schrödinger equation, quantum Riesz fractional derivative, path integral approach to fractional quantum mechanics, hermiticity of Hamiltonian operator, parity conservation law, and current density.

Applications of fractional quantum mechanics cover dynamics of a free particle, a new representation for a free particle quantum mechanical kernel, infinite potential well, bound state in $\delta$-potential well, linear potential, fractional Bohr atom and fractional oscillator.

We also review fundamentals of the Lévy path integral approach to fractional statistical mechanics.

1. Introduction . . . . . . . . . . . . . . . . . . . . . . . . . 394
2. Fractional Schrödinger Equation . . . . . . . . . . . . . . 396
2.1. Quantum Riesz fractional derivative . . . . . . . . 396
2.2. Hermiticity of the fractional Hamiltonian operator 398
2.3. The parity conservation law for fractional quantum mechanics . . . . . . . . . . . . . . . . . 399
2.4. Current density . . . . . . . . . . . . . . . . . . . . 400
2.5. The time-independent fractional Schrödinger equation . . . . . . . . . . . . . . . . . . . . . . . . 402
3. Path Integral . . . . . . . . . . . . . . . . . . . . . . . . 403
3.1. Free particle . . . . . . . . . . . . . . . . . . . . . . 405
3.1.1. Fox $H$-function representation for a free particle kernel . . . . . . . . . . . . . . . . 405
4. Applications of Fractional Quantum Mechanics . . . . . . 409
4.1. A free particle fractional Schrödinger equation . . 409
4.1.1. Scaling properties of the 1D fractional Schrödinger equation . . . . . . . . . . . 409

4.1.2. Exact solution . . . . . . . . . . . . . . . . 410
4.1.3. The 3D generalization . . . . . . . . . . . 411
4.2. The infinite potential well . . . . . . . . . . . . . . . 413
4.3. Fractional Bohr atom . . . . . . . . . . . . . . . . . 415
4.4. Fractional oscillator . . . . . . . . . . . . . . . . . . 417
4.4.1. Quarkonium and fractional oscillator . . . 417
4.4.2. Spectrum of the 1D fractional oscillator in semiclassical approximation . . . . . . 418
5. Some Solvable Models of Fractional Quantum Mechanics 420
5.1. Bound state in $\delta$-potential well . . . . . . . . . . . 420
5.2. Linear potential field . . . . . . . . . . . . . . . . . . 421
6. Fractional Statistical Mechanics . . . . . . . . . . . . . . 422
6.1. Density matrix . . . . . . . . . . . . . . . . . . . . . 422
6.1.1. Motion equation for the density matrix . 424
6.1.2. The 3D generalization of FSM . . . . . . 425
7. Conclusion . . . . . . . . . . . . . . . . . . . . . . . . . 425

## 1. Introduction

Classical mechanics and quantum mechanics are based on the assumption that the Hamiltonian function has the form

$$H(\mathbf{p}, \mathbf{r}) = \frac{\mathbf{p}^2}{2m} + V(\mathbf{r}), \tag{1}$$

where $\mathbf{p}$ and $\mathbf{r}$ are the momentum and space coordinate of a particle with mass $m$, and $V(\mathbf{r})$ is the potential energy. In quantum mechanics, $\mathbf{p}$ and $\mathbf{r}$ should be considered as quantum mechanical operators $\hat{\mathbf{p}}$ and $\hat{\mathbf{r}}$. Then the Hamiltonian function $H(\mathbf{p}, \mathbf{r})$ becomes the Hamiltonian operator $\hat{H}(\hat{\mathbf{p}}, \hat{\mathbf{r}})$,

$$\hat{H}(\hat{\mathbf{p}}, \hat{\mathbf{r}}) = \frac{\hat{\mathbf{p}}^2}{2m} + V(\hat{\mathbf{r}}). \tag{2}$$

The square dependence on the momentum in Eqs. (1) and (2) is empirical physical fact. However, an attempt to get insight on the fundamentals behind this fact posts the question: are there other forms of kinematic term in Eqs. (1) and (2) which do not contradict the fundamental principles of classical mechanics and quantum mechanics? A convenient theoretical physics approach to answer this question is the Feynman path integral approach to quantum mechanics [1], as it was first observed by Laskin [2]. Indeed, the Feynman path integral is the integral over Brownian-like paths.

Brownian motion is a special case of so-called $\alpha$-stable probability distributions developed by Lévy [3] and Khintchine [4]. In mid 1930s they posed the question: Does the sum of $N$ independent identically distributed random quantities $X = X_1 + X_2 + \cdots + X_N$ have the same probability distribution $p_N(X)$ (up to scale factor) as the individual steps $p_i(X_i)$, $i = 1, \ldots, N$? The traditional answer is that each $p_i(X_i)$ should be a Gaussian, because of the central limit theorem. In other words, a sum of $N$ Gaussians is again a Gaussian, but with $N$ times the variance of the original. Lévy and Khintchine proved that there exist the possibility to generalize the central limit theorem. They discovered a class of non-Gaussian $\alpha$-stable (stable under summation) probability distributions. Each $\alpha$-stable distribution has a stability index $\alpha$, often called the Lévy index $0 < \alpha \leq 2$. When $\alpha = 2$, Lévy motion is transformed into Brownian motion.

An option to develop the path integral over Lévy paths was discussed by Kac [5], who pointed out that the Lévy path integral generates a functional measure in the space of left (or right) continuous functions (paths) having only discontinuities of the first kind. The path integral over the Lévy paths has first been introduced and elaborated with applications to fractional quantum mechanics and fractional statistical mechanics by Laskin (see, [2], [16–19]). He followed the framework of the Feynman space-time vision of quantum mechanics, but instead of the Brownian-like quantum mechanical trajectories, Laskin used the Lévy-like ones. If the fractal dimension (for definition of fractal dimension, see [6, 7]) of the Brownian path is $\mathrm{d}_{\mathrm{fractal}}^{(\mathrm{Brownian})} = 2$, then the Lévy path has fractal dimension $\mathrm{d}_{\mathrm{fractal}}^{(\mathrm{Lévy})} = \alpha$, where $\alpha$ is the so-called Lévy index, $1 < \alpha \leq 2$. The Lévy index $\alpha$ becomes a new fundamental parameter in fractional quantum and classical mechanics similar to $\mathrm{d}_{\mathrm{fractal}}^{(\mathrm{Brownian})} = 2$ being a fundamental parameter in standard quantum and classical mechanics. The difference between the fractal dimensions of the Brownian and Lévy paths leads to different physics. In fact, fractional quantum dynamics is generated by the Hamiltonian function $H_\alpha(\mathbf{p}, \mathbf{r})$ of the form [2], [16–19]

$$H_\alpha(\mathbf{p}, \mathbf{r}) = D_\alpha |\mathbf{p}|^\alpha + V(\mathbf{r}), \qquad 1 < \alpha \leq 2, \tag{3}$$

with substitutions $\mathbf{p} \to \hat{\mathbf{p}}$, $\mathbf{r} \to \hat{\mathbf{r}}$, and $D_\alpha$ is the generalized coefficient, the physical dimension of which is $[D_\alpha] = \mathrm{erg}^{1-\alpha} \cdot \mathrm{cm}^\alpha \cdot \mathrm{sec}^{-\alpha}$. One can say that Eq. (3) is a natural generalization of the well-known Eq. (1). When $\alpha = 2$, $D_\alpha = 1/2m$ and Eq. (3) is transformed into Eq. (1) [2]. As a result, the fractional quantum mechanics based on the Lévy path integral generalizes the standard quantum mechanics based on the well-known Feynman path integral. Indeed, if the path integral over Brownian trajectories leads to the well-known Schrödinger equation, then the path integral over Lévy trajectories leads to the fractional Schrödinger equation. The fractional Schrödinger equation is a new fundamental equation of quantum physics and it includes the space derivative of order $\alpha$ instead of the second ($\alpha = 2$) order space derivative in the standard Schrödinger equation. Thus, the fractional Schrödinger equation is the fractional differential equation in accordance with modern terminology (see, for example, [10–15]). This is the main point of the term, *fractional Schrödinger equation*, and for the more general term, *fractional quantum mechanics*, FQM [2,16]. When Lévy index $\alpha = 2$, Lévy motion becomes Brownian motion. Thus, FQM includes standard QM as a particular Gaussian case at $\alpha = 2$. The quantum mechanical path integral over the Lévy paths [2] at $\alpha = 2$ becomes the Feynman path integral [1,8].

In the limit case $\alpha = 2$, the fundamental equations of fractional quantum mechanics are transformed into the well-known equations of standard quantum mechanics [1,8,9].

## 2. Fractional Schrödinger Equation

### 2.1. *Quantum Riesz fractional derivative*

Equation (1) allows us to conclude that the energy $E$ of a particle of mass $m$ under the influence of the potential $V(\mathbf{r})$ is given by

$$E = \frac{\mathbf{p}^2}{2m} + V(\mathbf{r}). \tag{4}$$

To obtain the Schrödinger equation, we introduce the operators following the well-known procedure,

$$E \longrightarrow i\hbar\frac{\partial}{\partial t}, \quad \mathbf{p} \longrightarrow -i\hbar\boldsymbol{\nabla}, \tag{5}$$

where $\boldsymbol{\nabla} = \partial/\partial\mathbf{r}$ and $\hbar$ is the Planck's constant. Further, substituting transformation (5) into Eq. (1) and applying them to the wave function $\psi(\mathbf{r}, t)$ yield

$$i\hbar\frac{\partial\psi(\mathbf{r}, t)}{\partial t} = -\frac{\hbar^2}{2m}\Delta\psi(\mathbf{r}, t) + V(\mathbf{r})\psi(\mathbf{r}, t), \tag{6}$$

here $\Delta = \nabla \cdot \nabla$ is the Laplacian. Thus, we obtain the Schrödinger equation [9].

By repeating the same consideration to Eq. (3) we find the fractional Schrödinger equation [2, 18]

$$i\hbar \frac{\partial \psi(\mathbf{r}, t)}{\partial t} = D_\alpha (-\hbar^2 \Delta)^{\alpha/2} \psi(\mathbf{r}, t) + V(\mathbf{r}) \psi(\mathbf{r}, t), \quad 1 < \alpha \leq 2, \tag{7}$$

with the 3D generalization of the fractional quantum Riesz derivative $(-\hbar^2 \Delta)^{\alpha/2}$ introduced by

$$(-\hbar^2 \Delta)^{\alpha/2} \psi(\mathbf{r}, t) = \frac{1}{(2\pi\hbar)^3} \int d^3 p e^{i \frac{\mathbf{pr}}{\hbar}} |\mathbf{p}|^\alpha \varphi(\mathbf{p}, t), \tag{8}$$

where the wave functions in space $\psi(\mathbf{r}, t)$ and momentum $\varphi(\mathbf{p}, t)$ representations are related to each other by the 3D Fourier transforms

$$\psi(\mathbf{r}, t) = \frac{1}{(2\pi\hbar)^3} \int d^3 p e^{i \frac{\mathbf{pr}}{\hbar}} \varphi(\mathbf{p}, t), \quad \varphi(\mathbf{p}, t) = \int d^3 r e^{-i \frac{\mathbf{pr}}{\hbar}} \psi(\mathbf{r}, t). \tag{9}$$

The 3D fractional Schrödinger equation Eq. (7) has the following operator form

$$i\hbar \frac{\partial \psi(\mathbf{r}, t)}{\partial t} = \hat{H}_\alpha (\hat{\mathbf{p}}, \hat{\mathbf{r}}) \psi(\mathbf{r}, t),$$

where fractional Hamiltonian operator $\hat{H}_\alpha(\hat{\mathbf{p}}, \hat{\mathbf{r}})$ results from Eq. (3) with quantum-mechanical operators $\hat{\mathbf{p}}$ and $\hat{\mathbf{r}}$ substituted instead of $\mathbf{p}$ and $\mathbf{r}$,

$$\hat{H}_\alpha(\hat{\mathbf{p}}, \hat{\mathbf{r}}) = D_\alpha |\hat{\mathbf{p}}|^\alpha + V(\hat{\mathbf{r}}), \quad 1 < \alpha \leq 2. \tag{10}$$

The 1D fractional Schrödinger equation has the form [2], [16–18]

$$i\hbar \frac{\partial \psi(x, t)}{\partial t} = -D_\alpha (\hbar \nabla)^\alpha \psi(x, t) + V(x) \psi(x, t), \quad 1 < \alpha \leq 2, \tag{11}$$

where $(\hbar\nabla)^\alpha$ is the quantum Riesz fractional derivative.[a] The quantum Riesz fractional derivative is defined by the following way [2, 16]

$$(\hbar\nabla)^\alpha \psi(x, t) = -\frac{1}{2\pi\hbar} \int_{-\infty}^{\infty} dp e^{i \frac{px}{\hbar}} |p|^\alpha \varphi(p, t), \tag{12}$$

[a] The Riesz fractional derivative was originally introduced in [20].

where $\varphi(p,t)$ is the Fourier transform of the wave function $\psi(x,t)$ given by

$$\varphi(p,t) = \int_{-\infty}^{\infty} dx e^{-i\frac{px}{\hbar}} \psi(x,t), \tag{13}$$

and reciprocally

$$\psi(x,t) = \frac{1}{2\pi\hbar} \int_{-\infty}^{\infty} dp e^{i\frac{px}{\hbar}} \varphi(p,t).$$

It is easy to see that Eq. (11) can be rewritten in the operator form, namely

$$i\hbar \frac{\partial \psi}{\partial t} = H_\alpha \psi, \tag{14}$$

where $H_\alpha$ is the fractional Hamiltonian operator

$$H_\alpha = -D_\alpha (\hbar \nabla)^\alpha + V(x). \tag{15}$$

For the special case when $\alpha = 2$ and $D_2 = 1/2m$ (see, for details [2,16]), where $m$ is the particle mass, Eqs. (7) and (11) are transformed into the well-known 3D and 1D Schrödinger equations [9].

### 2.2. *Hermiticity of the fractional Hamiltonian operator*

The fractional Hamiltonian $H_\alpha$ given by Eq. (15) is the Hermitian operator in the space with scalar product

$$(\phi, \chi) = \int_{-\infty}^{\infty} dx \phi^*(x,t) \chi(x,t). \tag{16}$$

To prove the hermiticity of $H_\alpha$, let us note that in accordance with the definition of the quantum Riesz fractional derivative given by Eq. (12), there exists the integration by parts formula

$$(\phi, (\hbar\nabla)^\alpha \chi) = ((\hbar\nabla)^\alpha \phi, \chi). \tag{17}$$

The average energy of a fractional quantum system with Hamiltonian $H_\alpha$ is

$$E_\alpha = \int_{-\infty}^{\infty} dx \psi^*(x,t) H_\alpha \psi(x,t). \tag{18}$$

Taking into account Eq. (17), we have

$$E_\alpha = \int_{-\infty}^{\infty} dx \psi^*(x,t) H_\alpha \psi(x,t) = \int_{-\infty}^{\infty} dx (H_\alpha^+ \psi(x,t))^* \psi(x,t) = E_\alpha^*.$$

As a physical consequence, the energy of the system is real. Thus, the fractional Hamiltonian $H_\alpha$ defined by Eq. (15) is the Hermitian or self-adjoint operator in the space with the scalar product defined by Eq. (16) [17, 18]

$$(H_\alpha^+ \phi, \chi) = (\phi, H_\alpha \chi). \tag{19}$$

The generalization of the proof of hermiticity for the 3D case is straightforward. Note that Eq. (11) leads to the important equation

$$\frac{\partial}{\partial t} \int dx \psi^*(x,t)\psi(x,t) = 0, \tag{20}$$

which shows that the wave function remains normalized, if it is normalized once. Multiplying Eq. (11) from the left by $\psi^*(x,t)$ and the conjugate complex of Eq. (11) by $\psi(x,t)$ and then subtracting the two resultant equations finally yield

$$i\hbar \frac{\partial}{\partial t} (\psi^*(x,t)\psi(x,t)) = \psi^*(x,t) H_\alpha \psi(x,t) - \psi(x,t) H_\alpha^* \psi^*(x,t).$$

Integrating this relation over all values of the space variable and using the fact that the operator $H_\alpha$ is self-adjoint, we find Eq. (20). The above consideration can be easy generalized to the 3D case.

### 2.3. *The parity conservation law for fractional quantum mechanics*

From definition (12) of the quantum Riesz fractional derivative it follows that

$$(\hbar \nabla)^\alpha \exp\left\{ i \frac{px}{\hbar} \right\} = -|p|^\alpha \exp\left\{ i \frac{px}{\hbar} \right\}. \tag{21}$$

In other words, the function $\exp\{ipx/\hbar\}$ is the eigenfunction of the quantum Riesz fractional operator $(\hbar\nabla)^\alpha$ with eigenvalue $-|p|^\alpha$.

The 3D generalization is straightforward,

$$(-\hbar^2 \Delta)^{\alpha/2} \exp\left\{ i \frac{\mathbf{px}}{\hbar} \right\} = |\mathbf{p}|^\alpha \exp\left\{ i \frac{\mathbf{px}}{\hbar} \right\}, \tag{22}$$

which means that the function $\exp\{i\mathbf{px}/\hbar\}$ is the eigenfunction of the 3D quantum Riesz fractional operator $(-\hbar^2\Delta)^{\alpha/2}$ with eigenvalue $|\mathbf{p}|^\alpha$.

Thus, the operators $(\hbar\nabla)^\alpha$ and $(-\hbar^2\Delta)^{\alpha/2}$ are the symmetrized fractional derivative, that is

$$(\hbar\nabla_x)^\alpha \ldots = (\hbar\nabla_{-x})^\alpha \ldots, \tag{23}$$

$$(-\hbar^2\Delta_{\mathbf{r}})^{\alpha/2} \ldots = (-\hbar^2\Delta_{-\mathbf{r}})^{\alpha/2} \ldots. \tag{24}$$

Because of properties (17) and (18), the fractional Hamiltonian $H_\alpha$ (see, for example Eqs. (10) or (15)) remains invariant under *inversion* transformation. Inversion, or to be precise, spatial inversion consists of the simultaneous change in the sign of all three spatial coordinates

$$\mathbf{r} \longrightarrow -\mathbf{r}, \quad x \longrightarrow -x, \quad y \longrightarrow -y, \quad z \longrightarrow -z. \tag{25}$$

Let us denote the inversion operator by $\hat{P}$. The inverse symmetry results of the fact that $\hat{P}$ and the fractional Hamiltonian $H_\alpha$ commute,

$$\hat{P}H_\alpha = H_\alpha\hat{P}. \tag{26}$$

We can divide the wave functions of quantum mechanical states with a well-defined eigenvalue of the operator $\hat{P}$ into two classes; (i) functions which are not changed when acted upon by the inversion operator, $\hat{P}\psi_+(\mathbf{r}) = \psi_+(\mathbf{r})$, the corresponding states are called even states; (ii) functions which change sign under action of the inversion operator, $\hat{P}\psi_-(\mathbf{r}) = -\psi_-(\mathbf{r})$, the corresponding states are called odd states. Equation (26) represents the "parity conservation law" for FQM [18]; if the state of a closed fractional quantum mechanical system has a given parity (i.e. if it is even, or odd), then this parity is conserved.

### 2.4. *Current density*

Multiplying Eq. (7) from left by $\psi^*(\mathbf{r},t)$ and the conjugate complex of Eq. (7) by $\psi(\mathbf{r},t)$ and subtracting the two resultant equations yield

$$\frac{\partial}{\partial t}\int d^3r\,(\psi^*(\mathbf{r},t)\psi(\mathbf{r},t)) \\ = \frac{D_\alpha}{i\hbar}\int d^3r(\psi^*(\mathbf{r},t)(-\hbar^2\Delta)^{\alpha/2}\psi(\mathbf{r},t) - \psi(\mathbf{r},t)(-\hbar^2\Delta)^{\alpha/2}\psi^*(\mathbf{r},t)). \tag{27}$$

From this integral relationship we are led to the following well-known differential equation

$$\frac{\partial \rho(\mathbf{r},t)}{\partial t} + \operatorname{div} \mathbf{j}(\mathbf{r},t) = 0, \tag{28}$$

where $\rho(\mathbf{r},t) = \psi^*(\mathbf{r},t)\psi(\mathbf{r},t)$ is the quantum mechanical probability density and the vector $\mathbf{j}(\mathbf{r},t)$ can be called the fractional probability current density vector

$$\mathbf{j}(\mathbf{r},t) = \frac{D_\alpha \hbar}{i}(\psi^*(\mathbf{r},t)(-\hbar^2\Delta)^{\alpha/2-1}\boldsymbol{\nabla}\psi(\mathbf{r},t) \\ - \psi(\mathbf{r},t)(-\hbar^2\Delta)^{\alpha/2-1}\boldsymbol{\nabla}\psi^*(\mathbf{r},t)), \tag{29}$$

where we use the notation $\boldsymbol{\nabla} = \partial/\partial\mathbf{r}$. Introducing the momentum operator $\hat{\mathbf{p}} = \frac{\hbar}{i}\frac{\partial}{\partial \mathbf{r}}$, we can write the vector $\mathbf{j}$ in the form [18]

$$\mathbf{j} = D_\alpha(\psi(\hat{\mathbf{p}}^2)^{\alpha/2-1}\hat{\mathbf{p}}\psi^* + \psi^*(\hat{\mathbf{p}}^{*2})^{\alpha/2-1}\hat{\mathbf{p}}^*\psi). \tag{30}$$

The new fundamental Eqs. (29) and (30) are the fractional generalization of the well-known equations for probability current density vector of standard quantum mechanics [9].

To this end, we express Eq. (30) in terms of the velocity operator, which is defined as follows $\hat{\mathbf{v}} = d\hat{\mathbf{r}}/dt$, where $\hat{\mathbf{r}}$ is the operator of coordinate. Using the general quantum mechanical rule for differentiation of operator

$$\frac{d}{dt}\hat{\mathbf{r}} = \frac{i}{\hbar}[H_\alpha, \mathbf{r}],$$

we have

$$\hat{\mathbf{v}} = \frac{i}{\hbar}(H_\alpha \mathbf{r} - \mathbf{r}H_\alpha),$$

Furthermore, with the help of equation $\mathrm{f}(\hat{\mathbf{p}})\mathbf{r} - \mathbf{r}\mathrm{f}(\hat{\mathbf{p}}) = -i\hbar\partial\mathrm{f}/\partial\mathbf{p}$, which holds for any function $\mathrm{f}(\hat{\mathbf{p}})$ of the momentum operator, and taking into account Eq. (10) for the Hamiltonian operator $\hat{H}_\alpha(\hat{\mathbf{p}},\hat{\mathbf{r}})$ we obtain the equation for the velocity operator

$$\hat{\mathbf{v}} = \alpha D_\alpha |\hat{\mathbf{p}}^2|^{\alpha/2-1}\hat{\mathbf{p}}, \tag{31}$$

here $\hat{\mathbf{p}}$ is the momentum operator. By comparing Eqs. (30) and (31) we finally conclude that

$$\mathbf{j} = \frac{1}{\alpha}\left(\psi\hat{\mathbf{v}}\psi^* + \psi^*\hat{\mathbf{v}}\psi\right), \quad 1 < \alpha \le 2. \tag{32}$$

To get the probability current density to equal to 1 (the current when one particle passes through unit area per unit time), the wave function of a free particle has to be normalized as

$$\psi(\mathbf{r}, t) = \sqrt{\frac{\alpha}{2v}}\exp\left\{\frac{i}{\hbar}\mathbf{pr} - \frac{i}{\hbar}Et\right\}, \quad E = D_\alpha|\mathbf{p}|^\alpha, \quad 1 < \alpha \le 2, \tag{33}$$

where $v$ is the particle velocity, $v = \alpha D_\alpha p^{\alpha-1}$. Then we have

$$\mathbf{j} = \frac{\mathbf{v}}{v}, \quad \mathbf{v} = \alpha D_\alpha|\mathbf{p}^2|^{\frac{\alpha}{2}-1}\mathbf{p}, \tag{34}$$

that is, the vector $\mathbf{j}$ is indeed the unit vector.

Equations (29)–(34) are the fractional generalization of the well-known equations for probability current density vector and velocity vector of standard quantum mechanics [9].

### 2.5. *The time-independent fractional Schrödinger equation*

The special case when Hamiltonian $H_\alpha$ does not depend explicitly on time is of great importance for physical applications. It is easy to see that in this case there exists a special solution of the fractional Schrödinger equation (11) in the form

$$\psi(x, t) = e^{-(i/\hbar)Et}\phi(x), \tag{35}$$

where $\phi(x)$ satisfies

$$H_\alpha\phi(x) = E\phi(x) \tag{36}$$

or

$$-D_\alpha(\hbar\nabla)^\alpha\phi(x) + V(x)\phi(x) = E\phi(x), \quad 1 < \alpha \le 2. \tag{37}$$

We call Eq. (37) the time-independent (or stationary) fractional Schrödinger equation [17, 18]. From Eq. (35) we see that the wave function $\psi(x, t)$ oscillates with a definite frequency. The frequency with which a wave function oscillates corresponds to the energy. Therefore, we say that when the fractional wave function $\psi(x, t)$ is of this special form, the state has a definite energy $E$. The probability to find a particle at $x$ is the absolute square of the wave function $|\psi|^2$. In view of Eq. (35), the probability is equal to $|\phi|^2$ and does not depend upon the time. That is, the probability of finding the particle in any location is independent of time. In other words, the system is in a stationary state–stationary in the sense that there is no variation in the probability as a function of time.

## 3. Path Integral

If a particle at an initial time $t_a$ starts from the point $x_a$ and goes to a final point $x_b$ at time $t_b$, we will simply say that the particle goes from $a$ to $b$ and its trajectory (path)[b] $x(t)$ will have the property that $x(t_a) = x_a$ and $x(t_b) = x_b$. In quantum mechanics to get from the point $a$ to the point $b$, we use a quantum-mechanical amplitude often called a kernel, which may be written as $K(x_b t_b | x_a t_a)$. This will be the sum over all of the trajectories that go between the end points and of a contribution from each [1]. For the one-dimensional version of the Hamiltonian Eq. (3)

$$H_\alpha(p, x) = D_\alpha |p|^\alpha + V(x, t), \tag{38}$$

following consideration provided in [2] and [16], we come to the definition of the kernel $K(x_b t_b | x_a t_a)$ in terms of the path integral in the phase space representation

$$K(x_b t_b | x_a t_a) = \lim_{N \to \infty} \int_{-\infty}^{\infty} dx_1 \cdots dx_{N-1} \frac{1}{(2\pi\hbar)^N} \int_{-\infty}^{\infty} dp_1 \cdots dp_N \times \exp\left\{ \frac{i}{\hbar} \sum_{j=1}^{N} p_j (x_j - x_{j-1}) \right\} \times \exp\left\{ -\frac{i}{\hbar} D_\alpha \varepsilon \sum_{j=1}^{N} |p_j|^\alpha - \frac{i}{\hbar} \varepsilon \sum_{j=1}^{N} V(x_j, j\varepsilon) \right\}. \tag{39}$$

Here $\varepsilon = (t_b - t_a)/N$, $x_j = x(j\varepsilon)$, $p_j = p(j\varepsilon)$ and $x_0 = x_a$, $x_N = x_b$. Then in the continuum limit $N \to \infty$, $\varepsilon \to 0$ we have

$$K_L(x_b t_b | x_a t_a) = \int_{x(t_a)=x_a}^{x(t_b)=x_b} \mathrm{D}x(\tau) \int \mathrm{D}p(\tau) \times \exp\left\{ \frac{i}{\hbar} \int_{t_a}^{t_b} d\tau [p(\tau)\dot{x}(\tau) - H_\alpha(p(\tau), x(\tau), \tau)] \right\}, \tag{40}$$

[b]For simplicity, we consider only one-dimensional motion.

where $\dot{x}$ denotes the time derivative, $H_\alpha(p(\tau), x(\tau), \tau)$ is the fractional Hamiltonian given by Eq. (38) with the replacement $p \to p(\tau)$, $x \to x(\tau)$, and $\{p(\tau), x(\tau)\}$ is the particle trajectory in phase space. Finally, the phase space path integral $\int_{x(t_a)=x_a}^{x(t_b)=x_b} \mathrm{D}x(\tau) \int \mathrm{D}p(\tau) \cdots$ is given by

$$\begin{aligned}
&\int_{x(t_a)=x_a}^{x(t_b)=x_b} \mathrm{D}x(\tau) \int \mathrm{D}p(\tau) \cdots \\
&\quad = \lim_{N\to\infty} \int_{-\infty}^{\infty} dx_1 \cdots dx_{N-1} \frac{1}{(2\pi\hbar)^N} \int_{-\infty}^{\infty} dp_1 \cdots dp_N \\
&\quad \times \exp\left\{ i\frac{p_1(x_1 - x_a)}{\hbar} - i\frac{D_\alpha \varepsilon |p_1|^\alpha}{\hbar} \right\} \\
&\quad \times \cdots \times \exp\left\{ i\frac{p_N(x_b - x_{N-1})}{\hbar} - i\frac{D_\alpha \varepsilon |p_N|^\alpha}{\hbar} \right\} \cdots . \qquad (41)
\end{aligned}$$

The exponential in Eq. (40) can be written as $\exp\{\frac{i}{\hbar} S_\alpha(p, x)\}$ if we introduce the fractional canonical classical mechanical action $S_\alpha(p, x)$ for the trajectory $p(t)$, $x(t)$ in phase space [16]

$$S_\alpha(p, x) = \int_{t_a}^{t_b} d\tau (p(\tau)\dot{x}(\tau) - H_\alpha(p(\tau), x(\tau), \tau)). \qquad (42)$$

Since the coordinates $x_0$, $x_N$ in definition (41) are fixed at their initial and final points, $x_0 = x_a$ and $x_N = x_b$, the all possible trajectories in Eq. (40) satisfy the boundary condition $x(t_b) = x_b$, $x(t_a) = x_a$. We see that the definition given by Eq. (41) includes one more $p_j$-integrals than $x_j$-integrals. Indeed, while $x_0$ and $x_N$ are held fixed and the $x_j$-integrals are done for $j = 1, \ldots, N-1$, each increment $x_j - x_{j-1}$ is accompanied by one $p_j$-integral for $j = 1, \ldots, N$. The above observed asymmetry is a consequence of the particular boundary condition. Namely, the end points are fixed in the position (coordinate) space. There exists the possibility of proceeding in a conjugate way keeping the initial $p_a$ and final $p_b$ momenta fixed. The associated kernel can be derived going through the same steps as before but working in the momentum representation (see, for example, [8]).

The kernel $K(x_b t_b | x_a t_a)$ introduced by Eq. (40) describes the evolution of the quantum mechanical system

$$\psi(x_b, t_b) = \int_{-\infty}^{\infty} dx_a K(x_b t_b | x_a t_a) \psi(x_a, t_a), \qquad (43)$$

where $\psi(x_a, t_a)$ is the wave function of the initial state (at $t = t_a$) and $\psi(x_b, t_b)$ is the wave function of the final state (at $t = t_b$).

### 3.1. *Free particle*

For a free particle when $V(x,t) = 0$, we have $H_\alpha(p) = D_\alpha|p|^\alpha$ and Eq. (39) results in [2]

$$K^{(0)}(x_b t_b|x_a t_a) = \frac{1}{2\pi\hbar}\int_{-\infty}^{\infty} dp \exp\left\{i\frac{p(x_b - x_a)}{\hbar} - i\frac{D_\alpha|p|^\alpha(t_b - t_a)}{\hbar}\right\}, \tag{44}$$

here $K^{(0)}(x_b t_b|x_a t_a)$ stands for a free particle kernel.

Taking into account Eq. (44), it is easy to check the consistency condition

$$K^{(0)}(x_b t_b|x_a t_a) = \int_{-\infty}^{\infty} dx' K^{(0)}(x_b t_b|x' t') \cdot K^{(0)}(x' t'|x_a t_a).$$

This is a special case of the general quantum-mechanical rule: for events occurring in succession in time the amplitudes are multiplied

$$K(x_b t_b|x_a t_a) = \int_{-\infty}^{\infty} dx' K(x_b t_b|x' t') \cdot K(x' t'|x_a t_a). \tag{45}$$

#### 3.1.1. *Fox H-function representation for a free particle kernel*

Let us show how a free particle fractional quantum mechanical kernel $K^{(0)}(x_b t_b|x_a t_a)$ defined by Eq. (44) can be expressed in terms of the Fox $H$-function [21–23]. Following the approach proposed by Laskin [19], we obtain the Mellin transform of the quantum mechanical fractional kernel defined by Eq. (44). Comparison of the inverse Mellin transform with the definition of the Fox function yields the desired expression in terms of "known" function, i.e. Fox $H$-function. Note that the $H$-function bears the name of its discoverer Fox [21] although it has been known at least since 1888, according to [22].

Introducing the notations $x \equiv x_b - x_a$, $\tau \equiv t_b - t_a$, we rewrite Eq. (44)

$$K^{(0)}(x,\tau) = \frac{1}{2\pi\hbar}\int_{-\infty}^{\infty} dp \exp\left\{i\frac{px}{\hbar} - i\frac{D_\alpha|p|^\alpha\tau}{\hbar}\right\}. \tag{46}$$

It is easy to see that the relation $K^{(0)}(x,\tau) = K^{(0)}(-x,\tau)$ holds. Hence, it is sufficient to consider $K_L^{(0)}(x,\tau)$ for $x \geq 0$ only. Further, we will use the following definitions of the Mellin transform

$$\overset{\wedge}{K}^{(0)}(s,\tau) = \int_0^\infty dx\, x^{s-1} K^{(0)}(x,\tau) \tag{47}$$

and inverse Mellin transform

$$K^{(0)}(x,\tau) = \frac{1}{2\pi i}\int_{c-i\infty}^{c+i\infty} ds\, x^{-s} \overset{\wedge}{K}^{(0)}(s,\tau), \tag{48}$$

where the integration path is a straight line from $c - i\infty$ to $c + i\infty$ with $0 < c < 1$.

The Mellin transform of the $K^{(0)}(x,\tau)$ defined in accordance with Eq. (46) is

$$\overset{\wedge}{K}^{(0)}(s,\tau) = \frac{1}{2\pi\hbar}\int_0^\infty dx\, x^{s-1}\int_{-\infty}^{\infty} dp \exp\left\{i\frac{px}{\hbar} - i\frac{D_\alpha |p|^\alpha \tau}{\hbar}\right\}.$$

By changing the variables of integration $p \to (\frac{\hbar}{iD_\alpha \tau})^{1/\alpha}\varsigma$, $x \to (\frac{\hbar}{iD_\alpha \tau})^{1/\alpha}\xi$, one obtains the integrals in the complex $\varsigma$ and $\xi$ planes. Considering the paths of integration in the $\varsigma$ and $\xi$ planes it is easy to represent $\overset{\wedge}{K}^{(0)}(s,\tau)$ as follows:

$$\overset{\wedge}{K}^{(0)}(s,\tau) = \frac{1}{2\pi}\left(\frac{\hbar}{(\hbar/iD_\alpha \tau)^{1/\alpha}}\right)^{s-1}\int_0^\infty d\xi\, \xi^{s-1}\int_{-\infty}^{\infty} d\varsigma \exp\{i\varsigma\xi - |\varsigma|^\alpha\}. \tag{49}$$

The integrals over $d\xi$ and $d\varsigma$ can be evaluated by using the equation

$$\int_0^\infty d\xi\, \xi^{s-1}\int_0^\infty d\varsigma \exp\{i\varsigma\xi - \varsigma^\alpha\} = \frac{4}{s-1}\sin\frac{\pi(s-1)}{2}\Gamma(s)\Gamma\left(1 - \frac{s-1}{\alpha}\right), \tag{50}$$

where $s - 1 < \alpha \leq 2$ and $\Gamma(s)$ is the gamma function.[c]

Inserting Eq. (50) into Eq. (49) and using the functional relations for the gamma function, $\Gamma(1 - z) = -z\Gamma(-z)$ and $\Gamma(z)\Gamma(1 - z) = \pi/\sin \pi z$,

[c] The gamma function $\Gamma(s)$ has familiar integral representation $\Gamma(s) = \int_0^\infty dt\, t^{s-1}e^{-t}$, $\mathrm{Re}\, s > 0$.

yield

$$\overset{\wedge}{K}^{(0)}(s,\tau) = \frac{1}{\alpha}\left(\frac{\hbar}{(\hbar/iD_\alpha\tau)^{1/\alpha}}\right)^{s-1}\frac{\Gamma(s)\Gamma(\frac{1-s}{\alpha})}{\Gamma(\frac{1-s}{2})\Gamma(\frac{1+s}{2})}.$$

The inverse Mellin transform gives a free particle quantum mechanical kernel $K^{(0)}(x,\tau)$

$$\begin{aligned}K^{(0)}(x,\tau) &= \frac{1}{2\pi i}\int_{c-i\infty}^{c+i\infty} ds\, x^{-s}\overset{\wedge}{K}_L^{(0)}(s,\tau)\\ &= \frac{1}{2\pi i}\frac{1}{\alpha}\int_{c-i\infty}^{c+i\infty} ds\left(\frac{\hbar}{(\hbar/iD_\alpha\tau)^{1/\alpha}}\right)^{s-1} x^{-s}\frac{\Gamma(s)\Gamma(\frac{1-s}{\alpha})}{\Gamma(\frac{1-s}{2})\Gamma(\frac{1+s}{2})},\end{aligned}$$

where the integration path is the straight line from $c - i\infty$ to $c + i\infty$ with $0 < c < 1$. By replacing $s$ with $-s$ we obtain

$$\begin{aligned}K^{(0)}(x,\tau) = \frac{1}{\alpha}\left(\frac{\hbar}{(\hbar/iD_\alpha\tau)^{1/\alpha}}\right)^{-1}\frac{1}{2\pi i}\int_{-c-i\infty}^{-c+i\infty} ds\\ \times\left(\frac{1}{\hbar}\left(\frac{\hbar}{iD_\alpha\tau}\right)^{1/\alpha} x\right)^{s}\frac{\Gamma(-s)\Gamma(\frac{1+s}{\alpha})}{\Gamma(\frac{1+s}{2})\Gamma(\frac{1-s}{2})}.\end{aligned} \tag{51}$$

The path of integration may be deformed into one running clockwise around $R_+ - c$. Comparison with the definition of the Fox $H$-function (see, Eqs. (58) and (59), in [19]) leads to

$$\begin{aligned}K^{(0)}(x,\tau) = \frac{1}{\alpha}\left(\frac{\hbar}{(\hbar/iD_\alpha\tau)^{1/\alpha}}\right)^{-1} H_{2,2}^{1,1}\\ \times\left[\frac{1}{\hbar}\left(\frac{\hbar}{iD_\alpha\tau}\right)^{1/\alpha} x\,\middle|\,\begin{matrix}(1-1/\alpha,1/\alpha),(1/2,1/2)\\ (0,1),(1/2,1/2)\end{matrix}\right].\end{aligned} \tag{52}$$

Applying the Property 12.2.5, Ref. [19], of the Fox $H$-function we can express $K^{(0)}(x,\tau)$ as

$$K^{(0)}(x,\tau) = \frac{1}{\alpha x}H_{2,2}^{1,1}\left[\frac{1}{\hbar}\left(\frac{\hbar}{iD_\alpha\tau}\right)^{1/\alpha} x\,\middle|\,\begin{matrix}(1,1/\alpha),(1,1/2)\\ (1,1),(1,1/2)\end{matrix}\right], \quad x \geq 0. \tag{53}$$

Or for any $x$,

$$K^{(0)}(x,\tau)=\frac{1}{\alpha|x|}H_{2,2}^{1,1}\left[\frac{1}{\hbar}\left(\frac{\hbar}{iD_\alpha\tau}\right)^{1/\alpha}|x|\,\middle|\,\begin{matrix}(1,1/\alpha),(1,1/2)\\(1,1),(1,1/2)\end{matrix}\right]. \tag{54}$$

Substituting $x\equiv x_b-x_a$, $\tau\equiv t_b-t_a$, finally yields

$$K^{(0)}(x_bt_b|x_at_a)$$
$$=\frac{1}{\alpha|x_b-x_a|}H_{2,2}^{1,1}\left[\frac{1}{\hbar}\left(\frac{\hbar}{iD_\alpha(t_b-t_a)}\right)^{1/\alpha}|x_b-x_a|\,\middle|\,\begin{matrix}(1,1/\alpha),(1,1/2)\\(1,1),(1,1/2)\end{matrix}\right]. \tag{55}$$

This is the new equation for the 1D free particle fractional quantum mechanical kernel $K^{(0)}(x_bt_b|x_at_a)$.

Let us show that Eq. (55) includes as a particular case at $\alpha=2$ the well-known Feynman quantum mechanical kernel, see Eq. (3-3) in [1]. Putting $\alpha=2$ in Eq. (55), applying the series expansion to the function

$$H_{2,2}^{1,1}\left[\frac{1}{\hbar}\left(\frac{\hbar}{iD_\alpha(t_b-t_a)}\right)^{1/2}|x_b-x_a|\,\middle|\,\begin{matrix}(1,1/2),(1,1/2)\\(1,1),(1,1/2)\end{matrix}\right],$$

and finally, substituting $k\to 2l$ yield

$$K^{(0)}(x,\tau)|_{\alpha=2}=\frac{1}{2\hbar}\left(\frac{\hbar}{iD_2\tau}\right)^{1/2}\sum_{l=0}^{\infty}\left(-\frac{1}{\hbar}\left(\frac{\hbar}{iD_2\tau}\right)^{1/2}\right)^{2l}\frac{|x|^{2l}}{(2l)!}\frac{1}{\Gamma(\frac{1}{2}-l)}. \tag{56}$$

Taking into account identity $\Gamma(\frac{1}{2}+z)\Gamma(\frac{1}{2}-z)=\frac{\pi}{\cos\pi z}$, and applying the Gauss multiplication formula $\Gamma(2l)=\sqrt{\frac{2^{4l-1}}{2\pi}}\Gamma(l)\Gamma(l+\frac{1}{2})$, we find that

$$(2l)!\Gamma\left(\frac{1}{2}-l\right)=\frac{\sqrt{\pi}}{(-1)^l}(2)^{2l}l!. \tag{57}$$

With help of Eq. (57), kernel $K^{(0)}(x,\tau)|_{\alpha=2}$ can be rewritten as

$$K^{(0)}(x,\tau)|_{\alpha=2}=\frac{1}{2\sqrt{\pi}\hbar}\left(\frac{\hbar}{iD_2\tau}\right)^{1/2}\sum_{l=0}^{\infty}\left(-\frac{1}{\hbar}\left(\frac{\hbar}{iD_2\tau}\right)^{1/2}\right)^{2l}\frac{(-1)^l|x|^{2l}}{2^{2l}l!}$$
$$=\frac{1}{2\sqrt{\pi}\hbar}\left(\frac{\hbar}{iD_2\tau}\right)^{1/2}\exp\left\{-\frac{1}{4}\frac{|x|^2}{\hbar iD_2\tau}\right\}. \tag{58}$$

Since $D_2 = 1/2m$, we come to the Feynman kernel (see Eq. (3-3), [1])

$$K^{(0)}(x,\tau)|_{\alpha=2} \equiv K_F^{(0)}(x,\tau) = \sqrt{\frac{m}{2\pi i\hbar\tau}}\exp\left\{\frac{im|x|^2}{2\hbar\tau}\right\}.$$

Thus, it is shown how Feynman's free particle kernel can be derived from the general equation (55) for the fractional quantum mechanical kernel.

## 4. Applications of Fractional Quantum Mechanics

### 4.1. *A free particle fractional Schrödinger equation*

#### 4.1.1. *Scaling properties of the 1D fractional Schrödinger equation*

To make general conclusions regarding solutions of the 1D fractional Schrödinger equation for a free particle, let us study the scaling for the wave function $\psi(x, t; D_\alpha)$, where we keep $D_\alpha$ to remind that besides dependency on $x$ and $t$, the wave function depends on $D_\alpha$ as well.

The scale transformations could be written as

$$t = \lambda t', \qquad x = \lambda^\beta x', \qquad D_\alpha = \lambda^\gamma D'_\alpha, \qquad \psi(x,t;D_\alpha) = \lambda^\delta \psi(x',t';D'_\alpha),$$

where $\beta$, $\gamma$, $\delta$ are exponents of the scale transformations which should leave a free particle 1D fractional Schrödinger equation invariant

$$i\hbar\frac{\partial\psi(x,t;D_\alpha)}{\partial t} = -D_\alpha(\hbar\nabla)^\alpha\psi(x,t;D_\alpha), \tag{59}$$

and save the normalization condition $\int_{-\infty}^{\infty} dx|\psi(x,t;D_\alpha)|^2 = 1$. It results in the relationships between scaling exponents,

$$\alpha\beta - \gamma - 1 = 0, \qquad \delta + \beta/2 = 0, \tag{60}$$

and reduces the number of exponents up to 2.

Hence, we obtain the two-parameter scale transformation group

$$t = \lambda t', \qquad x = \lambda^\beta x', \qquad D_\alpha = \lambda^{\alpha\beta-1}D'_\alpha, \tag{61}$$

$$\psi(\lambda^\beta x, \lambda t; \lambda^{\alpha\beta-1}D_\alpha) = \lambda^{-\beta/2}\psi(x,t;D_\alpha), \tag{62}$$

where $\beta$ and $\lambda$ are arbitrary group parameters.

When the initial condition $\psi(x, t = 0)$ is invariant under the scaling group Eq. (61), the solution of Eq. (59) remains the group invariant.

As an example of the invariant initial condition one may keep in mind $\psi(x, t = 0) = \delta(x)$, which gives us the Green function of the 1D fractional Schrödinger equation.

To get the general scale invariant solutions of the 1D fractional Schrödinger equation, we may use the renormalization group framework. As far as the scale invariant solutions of Eq. (59) should satisfy the identity Eq. (62) for any arbitrary parameters $\beta$ and $\lambda$, the solutions can depend on a combination of $x$ and $t$ to provide the independency of $\beta$ and $\lambda$. Therefore, due to Eq. (60), the solutions are

$$\psi(x,t) = \frac{1}{x}\Phi(x/(D_\alpha t)^{\frac{1}{\alpha}}) = \frac{1}{(D_\alpha t)^{\frac{1}{\alpha}}}\Psi(x/(D_\alpha t)^{\frac{1}{\alpha}}), \tag{63}$$

where arbitrary functions $\Phi$ and $\Psi$ are determined by the conditions, $\Phi(.) = \psi(1, .)$ and $\Psi(.) = \psi(., 1)$.

Thus, we conclude that Eq. (63) brings us a general structure of scale invariant solution of the 1D fractional Schrödinger equation for a free particle.

#### 4.1.2. *Exact solution*

Following [17, 25] let us solve the 1D fractional Schrödinger equation for a free particle (59) with some initial condition $\psi_0(x)$

$$\psi(x, t = 0) = \psi_0(x). \tag{64}$$

Applying the Fourier transforms defined by Eq. (13) and using the quantum Riesz fractional derivative given by Eq. (12) yield for the wave function $\varphi(p, t)$ in the momentum representation,

$$i\hbar\frac{\partial\varphi(p,t)}{\partial t} = D_\alpha|p|^\alpha\varphi(p,t), \tag{65}$$

with the initial condition $\varphi_0(p)$ given by

$$\varphi_0(p) = \varphi(p, t = 0) = \int_{-\infty}^{\infty} dx\, e^{-i\frac{px}{\hbar}}\psi_0(x). \tag{66}$$

The solution of the problem introduced by Eqs. (65) and (66) is

$$\varphi(p,t) = \exp\left\{-i\frac{D_\alpha|p|^\alpha t}{\hbar}\right\}\varphi_0(p). \tag{67}$$

Therefore, the solution of the 1D fractional Schrödinger equation Eq. (59) with initial condition given by Eq. (64) can be presented as

$$\psi(x,t) = \frac{1}{2\pi\hbar}\int_{-\infty}^{\infty} dx' \int_{-\infty}^{\infty} dp \exp\left\{ i\frac{p(x-x')}{\hbar} - i\frac{D_\alpha |p|^\alpha t}{\hbar} \right\} \psi_0(x') \tag{68}$$

or

$$\psi(x,t) = \int_{-\infty}^{\infty} dx' \frac{1}{\alpha|x-x'|} \times H^{1,1}_{2,2}\left[ \frac{1}{\hbar}\left(\frac{\hbar}{iD_\alpha t}\right)^{1/\alpha} |x-x'| \,\middle|\, \begin{matrix} (1,1/\alpha),(1,1/2) \\ (1,1),(1,1/2) \end{matrix} \right] \psi_0(x'). \tag{69}$$

Here we have expressed the integral over $dp$ in Eq. (68) in terms of the $H^{1,1}_{2,2}$-function. If we choose the initial condition $\psi_0(x) = \delta(x)$, then Eq. (69) gives us the quantum mechanical kernel $K^{(0)}(x,t|0,0)$ for the 1D free particle fractional Schrödinger equation

$$K^{(0)}(x,t|0,0) = \frac{1}{\alpha|x|} H^{1,1}_{2,2}\left[ \frac{1}{\hbar}\left(\frac{\hbar}{iD_\alpha t}\right)^{1/\alpha} |x| \,\middle|\, \begin{matrix} (1,1/\alpha),(1,1/2) \\ (1,1),(1,1/2) \end{matrix} \right]. \tag{70}$$

Applying Property 12.2.5 of the Fox $H$-function (see, [19]) we can write for $K^{(0)}(x,t|0,0)$

$$K^{(0)}(x,t|0,0) = \frac{1}{\alpha\hbar}\left(\frac{\hbar}{iD_\alpha t}\right)^{1/\alpha} \times H^{1,1}_{2,2}\left[ \frac{1}{\hbar}\left(\frac{\hbar}{iD_\alpha t}\right)^{1/\alpha} |x| \,\middle|\, \begin{matrix} (1-1/\alpha,1/\alpha),(1/2,1/2) \\ (0,1),(1/2,1/2) \end{matrix} \right]. \tag{71}$$

Thus, we found scale invariant solutions of the 1D fractional Schrödinger equation (see Eq. (59)) for a free particle. It is easy to see that Eqs. (70) and (71) are in agreement with Eq. (63).

### 4.1.3. *The 3D generalization*

Quantum dynamics of a free particle in the 3D space is governed by the following equation,

$$i\hbar\frac{\partial\psi(\mathbf{r},t)}{\partial t} = D_\alpha(-\hbar^2\Delta)^{\alpha/2}\psi(\mathbf{r},t), \qquad \psi(\mathbf{r},t=0) = \psi_0(\mathbf{r}). \tag{72}$$

Using the 3D Fourier transforms defined by Eq. (9) and the definition of the 3D quantum fractional Riesz derivative given by Eq. (8) yield for the wave function $\varphi(\mathbf{p}, t)$ in the momentum representation,

$$i\hbar\frac{\partial\varphi(\mathbf{p},t)}{\partial t} = D_\alpha|\mathbf{p}|^\alpha\varphi(\mathbf{p},t), \tag{73}$$

with the initial condition $\varphi_0(\mathbf{p})$ given by

$$\varphi_0(\mathbf{p}) = \varphi(\mathbf{p}, t=0) = \int d^3r e^{-i\frac{\mathbf{pr}}{\hbar}}\psi_0(\mathbf{r}). \tag{74}$$

Going back to Eq. (72) we can see that the solution $\psi(\mathbf{r}, t)$ has a form

$$\psi(\mathbf{r},t) = \frac{1}{(2\pi\hbar)^3}\int d^3r'\int d^3p \exp\left\{i\frac{\mathbf{p}(\mathbf{r}-\mathbf{r}')}{\hbar} - i\frac{D_\alpha|\mathbf{p}|^\alpha t}{\hbar}\right\}\psi_0(\mathbf{r}').$$

The integral over $d^3p$ can be expressed in terms of $H^{1,2}_{3,3}$-function, see for instance Eqs. (33) and (34) in [19]. Thus, the solution of problem defined by Eq. (72) is

$$\begin{aligned}\psi(\mathbf{r},t) = &-\frac{1}{2\pi\alpha}\int d^3r'\frac{1}{|\mathbf{r}-\mathbf{r}'|^3}H^{1,2}_{3,3}\\ &\times\left[\frac{1}{\hbar}\left(\frac{\hbar}{iD_\alpha t}\right)^{1/\alpha}|\mathbf{r}-\mathbf{r}'|\left|\begin{matrix}(1,1),(1,1/\alpha),(1,1/2)\\(1,1),(1,1/2),(2,1)\end{matrix}\right.\right]\psi_0(\mathbf{r}').\end{aligned} \tag{75}$$

Substituting $\psi_0(\mathbf{r}) = \delta_0(\mathbf{r})$ into Eq. (75), gives us the quantum mechanical kernel $K^{(0)}(\mathbf{r}, t|0, 0)$ for a free particle the 3D fractional Schrödinger equation

$$\begin{aligned}K^{(0)}(\mathbf{r},t|0,0) = &-\frac{1}{2\pi\alpha}\frac{1}{|\mathbf{r}|^3}H^{1,2}_{3,3}\\ &\times\left[\frac{1}{\hbar}\left(\frac{\hbar}{iD_\alpha t}\right)^{1/\alpha}|\mathbf{r}|\left|\begin{matrix}(1,1),(1,1/\alpha),(1,1/2)\\(1,1),(1,1/2),(2,1)\end{matrix}\right.\right].\end{aligned} \tag{76}$$

This is the new equation for a free particle quantum mechanical 3D kernel. We see that in comparison with the 1D case, the 3D quantum kernel is expressed in terms of the $H^{1,2}_{3,3}$ Fox $H$-function. In the case $\alpha = 2$, we come to the well-known equation for the Feynman 3D quantum kernel

$K_F^{(0)}(\mathbf{r}, t|0,0)$,

$$K^{(0)}(\mathbf{r}, t|0,0)|_{\alpha=2} \equiv K_F^{(0)}(\mathbf{r}, t|0,0) = \left(\frac{m}{2\pi i\hbar t}\right)^{3/2} \exp\left\{\frac{im|\mathbf{r}|^2}{2\hbar t}\right\}. \quad (77)$$

### 4.2. *The infinite potential well*

A particle in a one-dimensional well moves in a potential field $V(x)$ which is zero for $-a \leq x \leq a$, and which is infinite elsewhere,

$$V(x) = \begin{cases} \infty, & x < -a \quad \text{(i)} \\ 0, & -a \leq x \leq a \quad \text{(ii)} \\ \infty, & x > a \quad \text{(iii)} \end{cases} \quad (78)$$

It is evident *a priori* that the spectrum will be discrete. We are interested in the solutions of the fractional Schrödinger equation (37) that describe the stationary states with well-defined energies. Such a stationary state with an energy $E$ is described by a wave function $\psi(x,t)$, which can be written as $\psi(x,t) = \exp\{-i\frac{Et}{\hbar}\}\phi(x)$, where $\phi(x)$ is now time independent. In regions (i) and (iii), (see Eq. (78)) we have to substitute $\infty$ for $V(x)$ into Eq. (37), and it is easy to see that here the fractional Schrödinger equation can be satisfied only if we take $\phi(x) = 0$. In the middle region (ii), the time-independent fractional Schrödinger equation is

$$-D_\alpha(\hbar\nabla)^\alpha\phi(x) = E\phi(x). \quad (79)$$

We can treat this as a fractional eigenvalue problem [17]. Within region (ii), the eigenfunctions are determined by Eq. (79). Outside region (ii), $x < -a$ and $x > a$, the eigenfunctions are zero. We want the wave function $\phi(x)$ to be continuous everywhere, thus we impose the boundary conditions $\phi(-a) = \phi(a) = 0$ for the solutions of the fractional differential equation (79). Then the solution of Eq. (79) in region (ii) can be written as

$$\phi^{\text{even}}(x) = A\cos kx \quad \text{or} \quad \phi^{\text{odd}}(x) = A\sin kx,$$

where the following notation is introduced

$$k = \frac{1}{\hbar}\left(\frac{E}{D_\alpha}\right)^{1/\alpha}, \quad 1 < \alpha \leq 2. \quad (80)$$

The even (under reflection $x \to -x$) solution $\phi^{\text{even}}(x)$ satisfies the boundary conditions if

$$k = (2m+1)\frac{\pi}{2a}, \quad m = 0, 1, 2, 3, \ldots.$$

The odd (under reflection $x \to -x$) solution $\phi^{\text{odd}}(x)$ satisfies the boundary conditions if

$$k = \frac{m\pi}{a}, \qquad m = 1, 2, 3, \ldots. \tag{81}$$

It is easy to check that the normalized solutions are

$$\phi_m^{\text{even}}(x) = \frac{1}{\sqrt{a}} \cos\left\{(m+\frac{1}{2})\frac{\pi x}{a}\right\},$$

and

$$\phi_m^{\text{odd}}(x) = \frac{1}{\sqrt{a}} \sin\frac{m\pi x}{a}.$$

Solutions $\phi^{\text{even}}(x)$ and $\phi^{\text{odd}}(x)$ have the property that

$$\int_{-a}^{a} dx \phi_m^{\text{even}}(x)\phi_n^{\text{even}}(x) = \int_{-a}^{a} dx \phi_m^{\text{odd}}(x)\phi_n^{\text{odd}}(x) = \delta_{mn},$$

where $\delta_{mn}$ is the Kronecker symbol and

$$\int_{-a}^{a} dx \phi_m^{\text{even}}(x)\phi_n^{\text{odd}}(x) = 0.$$

The eigenvalues of the particle in a well with help of Eqs. (80) and (81) are [17]

$$E_n = D_\alpha \left(\frac{\pi\hbar}{a}\right)^\alpha n^\alpha, \quad n = 1, 2, 3\ldots, \quad 1 < \alpha \leq 2. \tag{82}$$

It is obvious that in the Gaussian case ($\alpha = 2$), Eq. (82) is transformed into the standard quantum mechanical equation (for example, see Eq. (20.7), [9]) for the energy levels of a particle in a box.

The state of the lowest energy, the ground state, in the infinite potential well is represented by the $\phi_m^{\text{even}}(x)$ at $m = 0$,

$$\phi_{\text{ground}}(x) \equiv \phi_0^{\text{even}}(x) = \frac{1}{\sqrt{a}} \cos\left\{\frac{\pi x}{2a}\right\},$$

and its energy is

$$E_{\text{ground}} = D_\alpha \left(\frac{\pi\hbar}{2a}\right)^\alpha. \tag{83}$$

### 4.3. *Fractional Bohr atom*

When $V(\mathbf{r})$ is the hydrogenlike potential energy

$$V(\mathbf{r}) = -\frac{Ze^2}{|\mathbf{r}|},$$

where $e$ is the electron charge and $Ze$ is the nuclear charge of the hydrogen-like atom, we come to the eigenvalue problem for the fractional hydrogen-like atom,

$$D_\alpha(-\hbar^2\Delta)^{\alpha/2}\phi(\mathbf{r}) - \frac{Ze^2}{|\mathbf{r}|}\phi(\mathbf{r}) = E\phi(\mathbf{r}).$$

This eigenvalue problem has first been solved in [17]. The total energy is $E = E_{\text{kin}} + V$, where $E_{\text{kin}}$ is the kinetic energy $E_{\text{kin}} = D_\alpha|\mathbf{p}|^\alpha$, and $V(\mathbf{r})$ is the potential energy $V(\mathbf{r}) = -\frac{Ze^2}{|\mathbf{r}|}$. It is well-known that if the potential energy is a homogeneous function of the coordinates, and the motion takes place in a finite region of space, there exists a simple relation between the time average values of the kinetic and potential energies, known as the *virial theorem* (see p. 23, [28]). It follows from the virial theorem that between the average kinetic energy and the average potential energy of the system with Hamiltonian (3), there exists the relation

$$\alpha\bar{E}_{\text{kin}} = -\bar{V}, \tag{84}$$

where the average value $\bar{f}$ of any function of time is defined as

$$\bar{f} = \lim_{T\to\infty}\frac{1}{T}\int_0^\infty dt f(t).$$

In order to evaluate the energy spectrum of the fractional hydrogen-like atom, let us remind the *Niels Bohr postulates* [26, 27]:

1. The electron moves in orbits restricted by the requirement that the angular momentum be an integral multiple of $\hbar$, that is, for circular orbits of radius $a_n$, electron's momentum is restricted by

$$pa_n = n\hbar, \qquad n = 1, 2, 3, \ldots, \tag{85}$$

and furthermore, the electrons in these orbits do not radiate in spite of their acceleration. They were said to be in stationary states.

2. Electrons can make discontinuous transitions from one allowed orbit corresponding to $n = n_1$ to another corresponding to $n = n_2$, and the change in energy will appear as radiation with frequency

$$\omega = \frac{E_{n_2} - E_{n_1}}{\hbar}. \tag{86}$$

An atom may absorb radiation by having its electrons making a transition to a higher energy orbit.

Using the first Bohr's postulate and Eq. (84) yields $\alpha D_\alpha \left(\frac{n\hbar}{a_n}\right)^\alpha = \frac{Ze^2}{a_n}$, from which follows the equation for the radius of the fractional Bohr orbits

$$a_n = a_0 n^{\frac{\alpha}{\alpha-1}}. \tag{87}$$

Here $a_0$ is the fractional Bohr radius (the radius of the lowest, $n = 1$, Bohr orbit) defined as,

$$a_0 = \left(\frac{\alpha D_\alpha \hbar^\alpha}{Ze^2}\right)^{\frac{1}{\alpha-1}}. \tag{88}$$

Total average energy $\bar{E}$ as it follows from Eq. (84) is, $\bar{E} = (1-\alpha)\bar{E}_{\text{kin}}$. Thus, for energy levels of the fractional hydrogen-like atom we have

$$E_n = (1-\alpha)E_0 n^{-\frac{\alpha}{\alpha-1}}, \quad 1 < \alpha \le 2, \tag{89}$$

where $E_0$ is the binding energy of the electron in the lowest Bohr orbit, i.e. the energy required to put it in a state with $E = 0$ corresponding to $n = \infty$,

$$E_0 = \left(\frac{Ze^2}{\alpha D_\alpha^{1/\alpha}\hbar}\right)^{\frac{\alpha}{\alpha-1}}. \tag{90}$$

The energy $(\alpha - 1)E_0$ can be considered as a generalization of the Rydberg constant of standard quantum mechanics. It is easy to see that at $\alpha = 2$ the energy $(\alpha - 1)E_0$ is transformed into the well-known expression for the Rydberg constant, $\mathrm{Ry} = me^4/2\hbar^2$.

The frequency of radiation $\omega$ associated with the transition, say, for example from $m$ to $n$, $m \to n$, is, according to the second Bohr postulate,

$$\omega = \frac{(1-\alpha)E_0}{\hbar}\left[\frac{1}{n^{\frac{\alpha}{\alpha-1}}} - \frac{1}{m^{\frac{\alpha}{\alpha-1}}}\right]. \tag{91}$$

The new equations (87)–(91) bring us the fractional generalization of the "Bohr atom" theory. In the special Gaussian case, (standard quantum mechanics) Eqs. (87)–(91) allow us to reproduce the well-known results of the Bohr theory [26,27].

### 4.4. *Fractional oscillator*

#### 4.4.1. *Quarkonium and fractional oscillator*

As another physical application of the developed FQM we propose a new fractional approach to study the quark–antiquark $q\bar{q}$ bound states treated within the nonrelativistic potential picture [2]. Note that only for heavy quark systems (for example, charmonium $c\bar{c}$ or bottonium $b\bar{b}$) the nonrelativistic approach can be justified. The term quarkonium is used to denote any $q\bar{q}$ bound state system [29] in analogy to positronium for the $e^+e^-$ system. The nonrelativistic potential approach remains the most successful and simplest way to calculate and predict energy levels and decay rates.

Thus, from the stand point of "potential" view, we can assume that the confining potential energy of two quarks localized, say, at the space points $\mathbf{r}_i$ and $\mathbf{r}_j$, is given by

$$V(|\mathbf{r}_i - \mathbf{r}_j|) = q_i q_j |\mathbf{r}_i - \mathbf{r}_j|^\beta, \tag{92}$$

where $q_i$ and $q_j$ are the color charges of $i$ and $j$ quarks respectively and the index $\beta > 0$. Equation (92) coincides with the QCD requirements: (i) At short distances the quarks and gluons appear to be weakly coupled; (ii) At large distances the effective coupling becomes strong, resulting in the phenomena of quark confinement.[d]

Considering statistical system of $N$ quarks yields the following equation for the potential energy $U(\mathbf{r}_1, \ldots, \mathbf{r}_N)$ of the system,

$$U(\mathbf{r}_1, \ldots, \mathbf{r}_N) = \sum_{1 \le i < j \le N} q_i q_j |\mathbf{r}_i - \mathbf{r}_j|^\beta. \tag{93}$$

In order to illustrate the main idea, we consider the simplest case, when color $q_i$ charge can only be $q$ or $-q$, and the colorless condition $\sum_{i=1}^{N} q_i = 0$ takes place. Using the general statistical mechanics approach (see Definition 3.2.1 and Proposition 3.2.2, [24]) we can conclude that only for $0 < \beta \le 2$ the many particle system with the potential energy (93) will be thermodynamically stable.

[d]The term quark confinement describes the observation that quarks do not occur isolated in nature, but only in hadronic bound states as the colorless objects such as baryons and mesons.

In order to study the problem of quarkonium it seems reasonable to consider the nonrelativistic FQM model with the fractional Hamiltonian operator $H_{\alpha,\beta}$ defined as

$$H_{\alpha,\beta} = D_\alpha(-\hbar^2\Delta)^{\alpha/2} + q^2|\mathbf{r}|^\beta, \quad 1 < \alpha \leq 2,\ 1 < \beta \leq 2, \tag{94}$$

where $\mathbf{r}$ is the 3D vector, $\Delta = \partial^2/\partial\mathbf{r}^2$ is the Laplacian, and the operator $(-\hbar^2\Delta)^{\alpha/2}$ is defined by Eq. (8).

For the special case, when $\alpha = \beta$, the Hamiltonian operator (94) has a form

$$H_\alpha = D_\alpha(-\hbar^2\Delta)^{\alpha/2} + q^2|\mathbf{r}|^\alpha, \quad 1 < \alpha \leq 2. \tag{95}$$

It is easy to see that Hamiltonian $H_\alpha$ is the fractional generalization of the 3D harmonic oscillator Hamiltonian of standard quantum mechanics. Following [2] and [18], we will call this model by quantum mechanical fractional oscillator.

It would be interesting to calculate the energies of the bound states and the decay rates based on FQM models with the Hamiltonians (94) and (95). These results could be compared with the experimental statistical data of $J/\psi$ decays. The information on decay rates and angular distribution provides an ideal testing ground for the fractional models of $q\bar{q}$ bound states.

### 4.4.2. *Spectrum of the 1D fractional oscillator in semiclassical approximation*

The 1D fractional oscillator with the Hamiltonian function $H_\alpha = D_\alpha|p|^\alpha + q^2|x|^\beta$ poses an interesting problem for semiclassical treatment [18]. We set the total energy equal to $E$, so that

$$E = D_\alpha|p|^\alpha + q^2|x|^\beta, \tag{96}$$

whence $|p| = \left(\frac{1}{D_\alpha}(E - q^2|x|^\beta)\right)^{1/\alpha}$. At the turning points $p = 0$, thus, classical motion is possible in the range $|x| \leq (E/q^2)^{1/\beta}$.

A routine use of the Bohr–Sommerfeld quantization rule [9] yields

$$2\pi\hbar\left(n + \frac{1}{2}\right) = \oint p\,dx = 4\int_0^{x_m} p\,dx = 4\int_0^{x_m} D_\alpha^{-1/\alpha}(E - q^2|x|^\beta)^{1\alpha}dx, \tag{97}$$

where the notation $\oint$ means the integral over one complete period of the classical motion and $x_m = (E/q^2)^{1/\beta}$ is the turning point of classical motion. To evaluate the integral on the right of Eq. (97), we introduce a new variable $y = x(E/q^2)^{-1/\beta}$. Then we have

$$\int_0^{x_m} D_\alpha^{-1/\alpha}(E - q^2|x|^\beta)^{1\alpha} dx = \frac{1}{D_\alpha^{1/\alpha} q^{2/\beta}} E^{\frac{1}{\alpha}+\frac{1}{\beta}} \int_0^1 dy(1 - y^\beta)^{1/\alpha}.$$

The integral over $dy$ can be expressed in terms of the $B$-function. Indeed, substitution $z = y^\beta$ yields[e]

$$\int_0^1 dy(1 - y^\beta)^{1/\alpha} = \frac{1}{\beta}\int_0^1 dz z^{\frac{1}{\beta}-1}(1-z)^{\frac{1}{\alpha}} = \frac{1}{\beta} B\left(\frac{1}{\beta}, \frac{1}{\alpha} + 1\right). \quad (99)$$

With the help of Eq. (99), we rewrite Eq. (97) as

$$2\pi\hbar\left(n + \frac{1}{2}\right) = \frac{4}{D_\alpha^{1/\alpha} q^{2/\beta}} E^{\frac{1}{\alpha}+\frac{1}{\beta}} \frac{1}{\beta} B\left(\frac{1}{\beta}, \frac{1}{\alpha} + 1\right).$$

The above equation gives the values of the energy of stationary states for the 1D fractional oscillator [18],

$$E_n = \left(\frac{\pi\hbar\beta D_\alpha^{1/\alpha} q^{2/\beta}}{2B(\frac{1}{\beta}, \frac{1}{\alpha} + 1)}\right)^{\frac{\alpha\beta}{\alpha+\beta}} \left(n + \frac{1}{2}\right)^{\frac{\alpha\beta}{\alpha+\beta}}. \quad (100)$$

This equation is generalization of the well-known energy spectrum of the standard quantum mechanical oscillator (see, for example, [9]) and is transformed into it at $\alpha = 2$, $\beta = 2$.

We note that at

$$\frac{1}{\alpha} + \frac{1}{\beta} = 1, \quad (101)$$

Eq. (100) gives the equidistant energy spectrum. When $1 < \alpha \leq 2$ and $1 < \beta \leq 2$ the condition given by Eq. (101) takes place for $\alpha = 2$ and $\beta = 2$ only. It means that only the standard quantum mechanical oscillator has an equidistant energy spectrum.

---

[e]The $B$-function is defined by

$$B(u, v) = \int_0^1 dy y^{u-1}(1 - y)^{v-1}. \quad (98)$$

## 5. Some Solvable Models of Fractional Quantum Mechanics

Let us review a few analytically solvable problems of fractional quantum mechanics recently elaborated in [30].

### 5.1. *Bound state in δ-potential well*

For one-dimensional attractive $\delta$-potential well, $V(x) = -\gamma\delta(x)$, $(\gamma > 0)$, where $\delta(x)$ is the Dirac delta function, the fractional Schrödinger equation Eq. (37) becomes

$$-D_\alpha(\hbar\nabla)^\alpha\phi(x) - \gamma\delta(x)\phi(x) = E\phi(x), \quad 1 < \alpha \leq 2. \tag{102}$$

In any one-dimensional attractive potential there will be a bound state that is $E < 0$. In this case, Dong and Xu [30], found the energy and wave function of the bound state. The bound energy has form

$$E = -\left(\frac{\gamma B(1/\alpha, 1-1/\alpha)}{\pi\hbar\alpha D_\alpha^{1/\alpha}}\right)^{\alpha/(\alpha-1)}, \quad 1 < \alpha \leq 2, \tag{103}$$

where $B(1/\alpha, 1-1/\alpha)$ is the $B$-function defined by Eq. (98).

The wave function $\phi(x)$ of the bound state is

$$\phi(x) = C_\alpha H^{2,1}_{2,3}\left[|x|\left(-\frac{D_\alpha\hbar^\alpha}{E}\right)^{-1/\alpha}\middle|\begin{matrix}(1-\frac{1}{\alpha},\frac{1}{\alpha}),(\frac{1}{2},\frac{1}{2})\\(0,1),(1-\frac{1}{\alpha},\frac{1}{\alpha}),(\frac{1}{2},\frac{1}{2})\end{matrix}\right] \tag{104}$$

here, the constant $C_\alpha$ is such that the normalization condition

$$\int_{-\infty}^{\infty} dx|\phi(x)|^2 = 1, \tag{105}$$

holds.

It has been shown in [30], that in the case when $\alpha = 2$ Eqs. (103) and (104) turn into the well-known equations of standard quantum mechanics for bound energy and bound state wave function of a particle with mass $m$ in the attractive $\delta$-potential well,

$$E = -\frac{m\gamma^2}{2\hbar^2} \tag{106}$$

and

$$\phi(x) = C_2\exp(-|x|\sqrt{-2mE}/\hbar), \tag{107}$$

with normalization constant $C_2 = (\sqrt{-2mE}/\hbar)^{1/2}$ satisfying Eq. (105).

### 5.2. *Linear potential field*

For a particle in a linear potential field (for example, see [9], p. 74) the potential function $V(x)$ can be written as:

$$V(x) = \begin{cases} Fx & x \geq 0, \quad (F > 0) \\ \infty & x < 0, \end{cases} \tag{108}$$

and the fractional Schrödinger equation Eq. (37) becomes

$$-D_\alpha(\hbar\nabla)^\alpha \phi(x) + Fx\phi(x) = E\phi(x), \quad 1 < \alpha \leq 2, \quad x \geq 0. \tag{109}$$

The continuity and bounded conditions of the wave function let us conclude that $\phi(x) = 0$, $x < 0$. Besides, $\phi(x)$ must satisfy the boundary conditions

$$\begin{aligned} \phi(x) &= 0, \qquad x = 0, \\ \phi(x) &= 0, \qquad x \to \infty. \end{aligned} \tag{110}$$

Then, wave function $\phi_n(x)$ of the quantum state with energy $E_n$, $n = 1, 2, 3, \ldots$ is [30]

$$\phi_n(x) = \frac{2\pi A}{\alpha+1} H_{2,2}^{1,1}\left[\left(x - \frac{E_n}{F}\right)\frac{1}{\hbar}\left(\frac{D_\alpha}{(\alpha+1)F\hbar}\right)^{-\frac{1}{\alpha+1}} \middle| \begin{matrix} \left(1-\frac{1}{\alpha+1}, \frac{1}{\alpha+1}\right), \left(\frac{\alpha+2}{2(\alpha+1)}, \frac{\alpha}{2(\alpha+1)}\right) \\ (0,1), \left(\frac{\alpha+2}{2(\alpha+1)}, \frac{\alpha}{2(\alpha+1)}\right) \end{matrix}\right], \tag{111}$$

where the constant $A$ is given by

$$A = \frac{1}{2\pi\hbar}\left(\frac{D_\alpha}{(\alpha+1)F\hbar}\right)^{-1/(\alpha+1)}. \tag{112}$$

The energy spectra $E_n$ has the form

$$E_n = \lambda_n F\hbar\left(\frac{D_\alpha}{(\alpha+1)F\hbar}\right)^{1/(\alpha+1)}, \quad 1 < \alpha \leq 2, \; n = 1, 2, 3, \ldots, \tag{113}$$

where $\lambda_n$ are the solutions of the equation [30]

$$H_{2,2}^{1,1}\left[-\lambda_n \middle| \begin{matrix} \left(1-\frac{1}{\alpha+1}, \frac{1}{\alpha+1}\right), \left(\frac{\alpha+2}{2(\alpha+1)}, \frac{\alpha}{2(\alpha+1)}\right) \\ (0,1), \left(\frac{\alpha+2}{2(\alpha+1)}, \frac{\alpha}{2(\alpha+1)}\right) \end{matrix}\right] = 0. \tag{114}$$

When $\alpha = 2$, Eqs. (111) and (113) become the well-known equations of standard quantum mechanics [9, 30].

Other solvable physical models of fractional quantum mechanics include the 1D Coulomb potential [30], a finite square potential well, dynamics in the field of 1D lattice, penetration through a $\delta$-potential barrier, the Dirac comb [31], the bound state problem and penetration through double $\delta$-potential barrier [32].

## 6. Fractional Statistical Mechanics

### 6.1. *Density matrix*

In order to develop *fractional statistical mechanics* (FSM) let us go in Eq. (40) from imaginary time to "inverse temperature" $\beta = 1/k_{\mathrm{B}}T$, where $k_{\mathrm{B}}$ is the Boltzmann's constant and $T$ is the temperature, $it \to \hbar\beta$. Then the partition function $Z$ is expressed as a trace of the density matrix $\rho_L(x, \beta|x_0)$ [2, 16]

$$\begin{aligned} Z &= \int dx \rho_L(x, \beta|x) \\ &= \int dx \int_{x(0)=x(\beta)=x} \mathrm{D}x(\tau) \int \mathrm{D}p(\tau) \\ &\quad \times \exp\left\{ -\frac{1}{\hbar} \int_0^{\hbar\beta} du \{-ip(u)\dot{x}(u) + H_\alpha(p(u), x(u))\} \right\}, \end{aligned} \tag{115}$$

where the fractional Hamiltonian $H_\alpha(p, x)$ has form (38) and $p(u), x(u)$ may be considered as paths running along an "imaginary time axis", $u = it$. The exponential expression of Eq. (115) is very similar to the fractional canonical action (42). Since it governs the fractional quantum-statistical path integrals, it may be called the fractional quantum-statistical action or fractional Euclidean action, indicated by the superscript $(e)$,

$$S_\alpha^{(e)}(p, x) = \int_0^{\hbar\beta} du \{-ip(u)\dot{x}(u) + H_\alpha(p(u), x(u))\}. \tag{116}$$

The parameter $u$ is not the true time in any sense. It is just a parameter in an expression for the density matrix (see, for instance, [1]). Let us call $u$ the "time", leaving the quotation marks to remind us that it is not real time (although $u$ does have the dimension of time). Likewise $x(u)$ will be called the "coordinate" and $p(u)$ the "momentum". Then Eq. (115) may be interpreted in the following way: Consider all the possible paths by which the system can travel between the initial $x(0)$ and final $x(\beta)$ configurations in "time" $\hbar\beta$. The fractional density matrix $\rho_L$ is a path integral over all possible paths, the contribution from a particular path being the "time" integral of the canonical action (116) (considered as the functional of the path $p(u), x(u)$ in the phase space) divided by $\hbar$. The partition function is derived by integrating over only those paths for which initial $x(0)$ and final $x(\beta)$ configurations are the same and after that we integrate over all possible initial (or final) configurations.

The fractional density matrix $\rho_L^{(0)}(x, \beta|x_0)$ of a free particle ($V = 0$) can be written as [2, 16]

$$\rho_L^{(0)}(x, \beta|x_0) = \frac{1}{2\pi\hbar}\int_{-\infty}^{\infty} dp \exp\left\{i\frac{p(x-x_0)}{\hbar} - \beta D_\alpha |p|^\alpha\right\}$$
$$= \frac{1}{\alpha|x-x_0|} H_{2,2}^{1,1}\left[\frac{|x-x_0|}{\hbar(D_\alpha\beta)^{1/\alpha}} \left| \begin{matrix} (1, 1/\alpha), (1, 1/2) \\ (1, 1), (1, 1/2) \end{matrix} \right.\right], \quad (117)$$

where $H_{2,2}^{1,1}$ is the Fox $H$-function (see, [21–23]).

For the 1D system of space scale $\Omega$ the trace of Eq. (117) reads

$$Z = \int_\Omega dx \rho_L^{(0)}(x, \beta|x_0) = \frac{\Omega}{2\pi\hbar}\frac{1}{(\beta D_\alpha)^{1/\alpha}}\Gamma\left(\frac{1}{\alpha}\right).$$

When $\alpha = 2$ and $D_2 = 1/2m$, Eq. (117) gives the well-known density matrix for the 1D free particle (see Eq. (10–46) of [1] or Eq. (2–61) of [33])

$$\rho^{(0)}(x, \beta|x_0) = \left(\frac{m}{2\pi\hbar^2\beta}\right)^{1/2} \exp\left\{-\frac{m}{2\hbar^2\beta}(x-x_0)^2\right\}. \quad (118)$$

The Fourier representation $\rho_L^{(0)}(p, \beta|p')$ of the fractional density matrix $\rho_L^{(0)}(x, \beta|x_0)$ defined by

$$\rho_L^{(0)}(p, \beta|p') = \int_{-\infty}^{\infty} dx\, dx_0 \rho_L^{(0)}(x, \beta|x_0) \exp\left\{-\frac{i}{\hbar}(px - p'x_0)\right\}$$

can be rewritten as

$$\rho_L^{(0)}(p, \beta|p') = 2\pi\hbar\delta(p-p')\; e^{-\beta D_\alpha |p|^\alpha}.$$

In order to obtain a formula for the fractional partition function in the limit of fractional classical mechanics, let us study the case when $\hbar\beta$ is small. It is easy to see that the fractional density matrix $\rho_L(x,\beta|x_0)$ can be written as

$$\rho_L(x,\beta|x_0) = e^{-\beta V(x_0)}\frac{1}{2\pi\hbar}\int_{-\infty}^{\infty} dp\exp\left\{i\frac{p(x-x_0)}{\hbar} - \beta D_\alpha|p|^\alpha\right\}.$$

Then the partition function $Z$ in the limit of classical mechanics becomes

$$Z = \int_{-\infty}^{\infty} dx\rho_L(x,\beta|x) = \frac{\Gamma(1/\alpha)}{2\pi\hbar(\beta D_\alpha)^{1/\alpha}}\int_{-\infty}^{\infty} dxe^{-\beta V(x)}, \qquad (119)$$

where $\Gamma(1/\alpha)$ is the gamma function.

The partition function $Z$ given by Eq. (119) is an approximation valid if the particles of the system cannot wander very far from their initial positions in the "time" $\hbar\beta$. The limit on the distance which the particles can wander before the approximation breaks down can be estimated from Eq. (117). We see that if the final point differs from the initial point by as much as

$$\Delta x \simeq \hbar(\beta D_\alpha)^{1/\alpha} = \hbar\left(\frac{D_\alpha}{kT}\right)^{1/\alpha},$$

the exponential function of Eq. (117) becomes greatly reduced. From this, we can infer that intermediate points only on paths which do not contribute greatly to the path integral of Eq. (117). Thus, we conclude that if the potential $V(x)$ does not alter very much as $x$ moves over this distance, then the fractional classical statistical mechanics is valid.

#### 6.1.1. *Motion equation for the density matrix*

The density matrix $\rho_L(x,\beta|x_0)$ obeys the fractional differential equation [2, 16]

$$-\frac{\partial\rho_L(x,\beta|x_0)}{\partial\beta} = -D_\alpha(\hbar\nabla_x)^\alpha\rho_L(x,\beta|x_0) + V(x)\rho_L(x,\beta|x_0) \qquad (120)$$

or

$$-\frac{\partial\rho_L(x,\beta|x_0)}{\partial\beta} = H_\alpha\rho_L(x,\beta|x_0), \quad \rho_L(x,0|x_0) = \delta(x-x_0), \qquad (121)$$

where the fractional Hamiltonian $H_\alpha$ is defined by Eq. (15). The last equation can be considered as a fractional generalization of the Bloch equation for the density matrix [34].

### 6.1.2. *The 3D generalization of FSM*

The above developments can be generalized to 3D dimension. It is obvious that a free particle density matrix $\rho_L^{(0)}(\mathbf{r},\beta|\mathbf{r}_0)$ for the 3D case has a form

$$\rho_L^{(0)}(\mathbf{r},\beta|\mathbf{r}_0) = \frac{1}{(2\pi\hbar)^3}\int d^3p\,\exp\left\{i\frac{\mathbf{p}(\mathbf{r}-\mathbf{r}_0)}{\hbar} - \beta D_\alpha|\mathbf{p}|^\alpha\right\}, \qquad (122)$$

where $\mathbf{r}$, $\mathbf{r}_0$ and $\mathbf{p}$ are the 3D vectors.

To present the density matrix $\rho_L(\mathbf{r},\beta|\mathbf{r}_0)$ in terms of the Fox $H$-function, we rewrite Eq. (122) as

$$\rho_L^{(0)}(\mathbf{r},\beta|\mathbf{r}_0) = \frac{1}{2\pi^2\hbar^2|\mathbf{r}-\mathbf{r}_0|}\int_0^\infty dp\,p\sin\left(\frac{p|\mathbf{r}_b-\mathbf{r}_a|}{\hbar}\right)\exp\{-\beta D_\alpha|\mathbf{p}|^\alpha\}.$$

With the help of identity $\rho_L^{(0)}(\mathbf{r},\beta|\mathbf{r}_0) = -\frac{1}{2\pi}\frac{\partial}{\partial x}\rho_L^{(0)}(x,\beta|0)|_{x=|\mathbf{r}-\mathbf{r}_0|}$, where $\rho_L^{(0)}(x,\beta|0)$ is the 1D density matrix given by Eq. (117), we find

$$\rho_L^{(0)}(\mathbf{r},\beta|\mathbf{r}_0) = -\frac{1}{2\pi\alpha}\frac{1}{|\mathbf{r}-\mathbf{r}_0|^3}H_{3,3}^{1,2}\left[\frac{|\mathbf{r}-\mathbf{r}_0|}{\hbar(D_\alpha\beta)^{1/\alpha}}\left|\begin{matrix}(1,1),(1,1/\alpha),(1,1/2)\\(1,1),(1,1/2),(2,1)\end{matrix}\right.\right]. \qquad (123)$$

This is the new equation for a free particle fractional density matrix in the 3D space.

The density matrix $\rho_L(\mathbf{r},\beta|\mathbf{r}_0)$ obeys the fractional differential equation

$$-\frac{\partial\rho_L(\mathbf{r},\beta|\mathbf{r}_0)}{\partial\beta} = D_\alpha(-\hbar^2\Delta)^{\alpha/2}\rho_L(\mathbf{r},\beta|\mathbf{r}_0) + V(\mathbf{r})\rho_L(\mathbf{r},\beta|\mathbf{r}_0) \qquad (124)$$

or

$$-\frac{\partial\rho_L(\mathbf{r},\beta|\mathbf{r}_0)}{\partial\beta} = H_\alpha\rho_L(\mathbf{r},\beta|\mathbf{r}_0), \quad \rho_L(\mathbf{r},\beta=0|\mathbf{r}_0) = \delta(\mathbf{r}-\mathbf{r}_0), \qquad (125)$$

where the 3D fractional Hamiltonian $H_\alpha$ is defined by Eq. (3).

Thus, Eqs. (117), (120)–(125) are fundamental equations of fractional statistical mechanics.

## 7. Conclusion

Fractional quantum mechanics has emerged as a field over 10 years ago, attracting the attention of many researchers. The original idea behind fractional quantum mechanics was to develop a path integral over Lévy-like quantum paths instead of the well-known Feynman path integral over Brownian-like paths. The basic outcome of this implementation is an alternative path integral approach, which results in a new fundamental equation — the fractional Schrödinger equation. In other words, if the Feynman path integral approach to quantum mechanics allows one to reproduce the Schrödinger equation, then the path integral over Lévy-like paths leads one to a fractional Schrödinger equation. This is a manifestation of a new non-Gaussian physical paradigm, based on deep relationships between the structure of fundamental physics equations and fractal dimensions of "underlying" quantum paths. The fractional Schrödinger equation includes the space derivative of fractional order $\alpha$ instead of the second-order space derivative in the well-known Schrödinger equation. Thus, the fractional Schrödinger equation is the fractional differential equation in accordance with modern terminology. This is the main point of the term, *fractional Schrödinger equation*, and for the more general term, *fractional quantum mechanics*.

Today, there are two alternative approaches to fractional quantum mechanics — the path integral over Lévy-like quantum paths, and the fractional Schrödinger equation. This paper presents the first review of these fundamentals and some applications of both theoretical frameworks.

Fundamentals cover fractional Schrödinger equation, quantum Riesz fractional derivative, path integral approach to fractional quantum mechanics, hermiticity of Hamiltonian operator, parity conservation law, and current density.

Applications of fractional quantum mechanics cover dynamics of a free particle, a new representation for a free particle quantum mechanical kernel, infinite potential well, bound state in $\delta$-potential well, linear potential, fractional Bohr atom and fractional oscillator. The Lévy path integral approach to *fractional statistical mechanics* has been reviewed as well.

## References

1. R. P. Feynman and A. R. Hibbs, *Quantum Mechanics and Path Integrals* (McGraw-Hill, 1965).
2. N. Laskin, *Phys. Lett. A* **268**, 298 (2000).
3. P. Lévy, *Théorie de l'Addition des Variables Aléatoires* (Gauthier-Villars, 1937).
4. A. Y. Khintchine and P. Lévy, *C. R. Acad. Sci.* (*Paris*) **202**, 374 (1936).
5. M. Kac, in the *Second Berkeley Symposium on Mathematical Statistics and Probability* (University of California Press, 1951).

6. B. B. Mandelbrot, *The Fractal Geometry of Nature* (W. H. Freeman, 1982).
7. J. Feder, *Fractals* (Plenum Press, 1988).
8. H. Kleinert, *Path Integrals in Quantum Mechanics, Statistics, Polymer Physics, and Financial Markets*, 4th edn. (World Scientific, 2006), Chap. 2.
9. L. D. Landau and E. M. Lifshitz, *Quantum Mechanics (Non-relativistic Theory)*, 3rd edn., *Course of Theoretical Physics* (Butterworth-Heinemann, 2003), Vol. 3.
10. K. B. Oldham and J. Spanier, *The Fractional Calculus* (Academic Press, 1974).
11. S. G. Samko, A. A. Kilbas and O. I. Marichev, *Fractional Integrals and Derivatives, Theory and Applications* (Gordon and Breach, 1993).
12. A. I. Saichev and G. M. Zaslavsky, *Chaos* **7**, 753 (1997).
13. K. S. Miller and B. Ross, *An Introduction to the Fractional Calculus and Fractional Differential Equations* (Wiley, 1993).
14. I. Podlubny, *Fractional Differential Equations* (Academic Press, 1999).
15. R. Metzler and J. Klafter, *Phys. Rep.* **339**, 1 (2000).
16. N. Laskin, *Phys. Rev. E* **62**, 3135 (2000).
17. N. Laskin, *Chaos* **10**, 780 (2000).
18. N. Laskin, *Phys. Rev. E* **66**, 056108 (2002).
19. N. Laskin, *Commun. Nonlinear Sci. Numer. Simul.* **12**, 2 (2007).
20. M. Riesz, *Acta Math.* **81**, 1 (1949).
21. C. Fox, *Trans. Amer. Math. Soc.* **98**, 395 (1961).
22. H. M. Srivastava, K. C. Gupta and S. P. Goyal, *The H-function of One and Two Variables with Applications* (South Asian Publishers, 1982).
23. A. M. Mathai and R. K. Saxena, *The H-function with Applications in Statistics and Other Disciplines* (Wiley Eastern, 1978).
24. D. Ruelle, *Statistical Mechanics, Rigorous Results* (W. A. Benjamin, Inc., 1969).
25. X. Y. Guo and M. Y. Xu, *J. Math. Phys.* **47**, 082104 (2006).
26. N. Bohr, *Phil. Mag.* **26**, 1, 476, 857 (1913).
27. N. Bohr, *Collected Works*, ed. J. Rud Nielsen (North-Holland, 1977), Vol. 4.
28. L. D. Landau and E. M. Lifshitz, *Mechanics*, 3rd edn., *Course of Theoretical Physics* (Pergamon, 1976), Vol. 1.
29. K. Berkelman, *Rep. Prog. Phys.* **49**, 1 (1986).
30. J. Dong and M. Xu, *J. Math. Phys.* **48**, 072105 (2007).
31. J. Dong and M. Xu, *J. Math. Phys.* **49**, 052105 (2008).
32. A. Lin, X. Jiang and F. Miao, *J. Shandong Univ. (Eng. Sci.)* **40**, 139 (2010).
33. R. P. Feynman, *Statistical Mechanics* (Benjamin, 1972).
34. F. Bloch, *Z. Phys.* **74**, 295 (1932).

# Chapter 18

## Two Examples of Fractional Quantum Dynamics

Alexander Iomin
*Department of Physics, Technion, Haifa, 32000, Israel*

An application of fractional integro-differentiation in quantum processes is presented. Two examples, related to both fractional time and space derivatives are considered. In particular, it is shown that the quantum dynamics, described by the fractional time Schrödinger equation, is relevant to a quantum comb model. The second example is devoted to the application of the fractional space derivatives to describe the quantum damping dynamics of a fractional kicked rotor.

1. Introduction . . . . . . . . . . . . . . . . . . . . . . . . 429
2. Mathematical Tools . . . . . . . . . . . . . . . . . . . . . 431
3. Fractional Time Schrödinger Equation . . . . . . . . . . . 432
4. Diffusion Comb Model . . . . . . . . . . . . . . . . . . . . 434
5. Quantum Comb Model . . . . . . . . . . . . . . . . . . . . 435
   5.1. Comb FTSE . . . . . . . . . . . . . . . . . . . . . . 436
   5.2. Green's function . . . . . . . . . . . . . . . . . . . 437
6. Fractional Kicked Rotor with Dissipation . . . . . . . . . 438
   6.1. Fractional particle on a circle . . . . . . . . . . . 439
   6.2. Fractional kicked rotor . . . . . . . . . . . . . . . 440
   6.3. Wigner representation . . . . . . . . . . . . . . . . 442
   6.4. Spectrum . . . . . . . . . . . . . . . . . . . . . . . 444
7. Conclusion . . . . . . . . . . . . . . . . . . . . . . . . . 445

### 1. Introduction

Application of the fractional calculus to quantum processes is a new and fast developing part of quantum physics that studies nonlocal quantum phenomena [1–11]. It aims to explore nonlocal effects found for either long-range interactions or time-dependent processes with many scales [3, 12–17]. The concept of differentiation of non-integer orders already arises from the works of Leibniz, Liouville, and Riemann, see e.g. [18, 19]. Its application is related to random processes with power law distributions. This corresponds to the absence of characteristic average values for processes exhibiting many scales [14,20]. A continuously increasing list of applications in many sciences has developed including: material science [13, 21, 22], physical kinetics [15], anomalous transport theory with a variety of applications in solid state physics [12, 14, 20] and in nonlinear dynamics [15, 16], complex phenomena in particle physics [6, 23], and collider phenomenology [24, 25].

In quantum physics, the fractional concept can be introduced by means of the Feynman propagator for nonrelativistic quantum mechanics as for Brownian path integrals [26]. Equivalence between the Wiener and the Feynman path integrals, established by Kac [27], indicates some relation between the classical diffusion equation and the Schrödinger equation. Therefore, the appearance of the space fractional derivatives in the Schrödinger equation is natural, since both the standard Schrödinger equation and that of the space fractional obey the Markov process. As shown in the seminal papers [2, 5], it relates to the path integral approach. As a result of this, the path integral approach for Lévy stable processes, leading to the fractional diffusion equation, can be extended to a quantum Feynman–Lévy measure which leads to the space fractional Schrödinger equation [2, 5].

A fractional time derivative can be introduced in the quantum mechanics by analogy with the fractional Fokker–Planck equation (FFPE), as well, by means of the Wick rotation of time $t \to -it/\hbar$ [7]. However, its physical interpretation is still vague: for example, a phase of the wave function as well as the semiclassical approximation should be understood. The fractional time Schrödinger equation was first considered in [7]. Its generalization to spacetime fractional quantum dynamics [28, 29] was performed and a relation to the fractional uncertainty [30] was studied as well. It is worth noting that, contrary to the space fractional derivative, the fractional time Schrödinger equation (FTSE) describes non-Markovian evolution with a memory effect.

We consider two examples of the application of the fractional integro-differentiation to quantum processes related to both fractional time and space derivatives. The first example concerns the fractional time Schrödinger equation, which is compared with a quantum comb model (QCM). An analogy with the FFPE is used, since the classical counterpart of the comb model is relevant to the FFPE. We show here that, although the FTSE and the QCM are isospectral to each other [10], nevertheless, these systems are different, and the wave function due to the FTSE is insufficient to describe the QCM.

The second example is a fractional kicked rotor, studied in the framework of the fractional Schrödinger equation. The system is described by the non-Hermitian Hamiltonian by virtue of the Weyl fractional derivative, and it has been shown that the Weyl fractional derivative can quantize an open system. Violation of space symmetry leads to acceleration of the orbital momentum. Quantum localization saturates this acceleration, such that the average value of the orbital momentum can be a direct current and the system behaves like a ratchet.

Though these two different examples pertain to special cases, they have it in common that fractional derivatives describe an interaction of a quantum system with its environment by means of the wave function in the framework of the Schrödinger equation.

## 2. Mathematical Tools

Fractional derivation was developed as a generalization of integer order derivatives and is defined as the inverse operation to the fractional integral. Fractional integration of the order of $\alpha$ is defined by the operator (see e.g. [3, 14–16, 18, 19])

$$ {}_aI_x^\alpha f(x) = \frac{1}{\Gamma(\alpha)} \int_a^x f(y)(x-y)^{\alpha-1} dy, $$

where $\alpha > 0$, $x > a$ and $\Gamma(z)$ is the Gamma function. Therefore, the fractional derivative is defined as the inverse operator to ${}_aI_x^\alpha$, namely ${}_aD_x^\alpha f(x) = {}_aI_x^{-\alpha} f(x)$ and ${}_aI_x^\alpha = {}_aD_x^{-\alpha}$. Its explicit form is

$$ {}_aD_x^\alpha f(x) = \frac{1}{\Gamma(-\alpha)} \int_a^x f(y)(x-y)^{-1-\alpha} dy. $$

For arbitrary $\alpha > 0$ this integral diverges, and as a result of this a regularization procedure is introduced with two alternative definitions of ${}_aD_x^\alpha$. For an integer $n$ defined as $n-1 < \alpha < n$, one obtains the Riemann–Liouville fractional derivative of the form

$$ {}_aD_{RL}^\alpha f(x) = \frac{d^n}{dx^n} {}_aI_x^{n-\alpha} f(x), \tag{1} $$

and fractional derivative in the Caputo form (see also [32])

$$ {}_aD_C^\alpha f(x) = {}_aI_x^{n-\alpha} f^{(n)}(x), \quad f^{(n)}(x) \equiv \frac{d^n}{dx^n} f(x). \tag{2} $$

There is no constraint on the lower limit $a$. For example, when $a = 0$, one has ${}_0D^{\alpha}_{RL}x^{\beta} = \frac{x^{\beta-\alpha}\Gamma(\beta+1)}{\Gamma(\beta+1-\alpha)}$. We also have ${}_0D^{\alpha}_{C}f(x) = {}_0D^{\alpha}_{RL}f(x) - \sum_{k=0}^{n-1} f^{(k)}(0^+)\frac{x^{k-\alpha}}{\Gamma(k-\alpha+1)}$ and ${}_aD^{\alpha}_{C}[1] = 0$, while ${}_0D^{\alpha}_{RL}[1] = x^{-\alpha}/\Gamma(1-\alpha)$. When $a = -\infty$, the resulting Weyl derivative is

$$
{}_{-\infty}\mathcal{W}^{\alpha} \equiv {}_{-\infty}D^{\alpha}_{W} = {}_{-\infty}D^{\alpha}_{RL} = {}_{-\infty}D^{\alpha}_{C}. \tag{3}
$$

One also has ${}_{-\infty}D^{\alpha}_{W}e^{x} = e^{x}$. This property is convenient for the Fourier transform

$$
\mathcal{F}\left[{}_{-\infty}\mathcal{W}^{\alpha}f(x)\right] = (ik)^{\alpha}\bar{f}(k), \tag{4}
$$

where $\mathcal{F}[f(x)] = \bar{f}(k)$. This fractional derivation with the fixed low limit is also called the left fractional derivative. However, one can introduce the right fractional derivative, where the upper limit $a$ is fixed and $a > x$. For example, the right Weyl derivative is

$$
\mathcal{W}^{\alpha}_{\infty}f(z) = \frac{1}{\Gamma(-\alpha)}\int_{x}^{\infty}\frac{f(y)dy}{(y-x)^{1+\alpha}}. \tag{5}
$$

The Laplace transform of the Caputo fractional derivative yields

$$
\mathcal{L}[{}_0D^{\alpha}_{C}f(x)] = s^{\alpha}\tilde{f}(s) - \sum_{k=0}^{n-1} f^{(k)}(0^+)s^{\alpha-1-k}, \tag{6}
$$

where $\mathcal{L}[f(x)] = \tilde{f}(s)$, which is convenient for the present analysis, where the initial conditions are imposed in terms of integer derivatives. We also use here a convolution rule for $0 < \alpha < 1$

$$
\mathcal{L}[I^{\alpha}_{x}f(x)] = s^{-\alpha}\tilde{f}(s). \tag{7}
$$

## 3. Fractional Time Schrödinger Equation

The fractional time quantum dynamics with the Hamiltonian $\hat{H}(x)$ is described by the fractional time Schrödinger equation (FTSE)

$$
(i\tilde{h})^{\alpha}\frac{\partial^{\alpha}\psi(\mathbf{x},t)}{\partial t^{\alpha}} = \hat{H}\psi(\mathbf{x},t), \tag{8}
$$

where $\alpha \leq 1$. For agreement of the dimension in Eq. (8) all variables and parameters are considered dimensionless, and $\tilde{h}$ is the dimensionless Planck constant (see also [7,28]). For $\alpha = 1$, Eq. (8) is the "conventional" (standard)

Schrödinger equation. For $\alpha < 1$, the fractional derivative is a formal notation of an integral with a power law memory kernel of the form

$$\frac{\partial^\alpha \psi(t)}{\partial t^\alpha} \equiv I_t^{1-\alpha} \frac{\partial \psi(t)}{\partial t} = \int_0^t \frac{(t-\tau)^{-\alpha}}{\Gamma(1-\alpha)} \frac{\partial \psi(\tau)}{\partial \tau} d\tau, \tag{9}$$

which is the Caputo fractional derivative: $\frac{\partial^\alpha \psi(t)}{\partial t^\alpha} \equiv {}_0D_C^\alpha \psi(t)$, see Eq. (2). This definition makes it possible to carry out the Laplace transform of the fractional derivative. Introducing the Laplace image $\tilde{\psi}(s) = \hat{\mathcal{L}}\psi(t)$, one obtains, according to Eq. (6),

$$\hat{\mathcal{L}}\left[\frac{\partial^\alpha \psi(t)}{\partial t^\alpha}\right] = s^\alpha \tilde{\psi}(s) - s^{\alpha-1}\psi(0). \tag{10}$$

Another interesting property of the FTSE is time evolution in the form of the Mittag-Leffler function. For the time-independent Hamiltonian, the eigenvalue equation (with corresponding boundary conditions) is $H\phi_\lambda = \lambda\phi_\lambda$. Therefore one obtains the Green function in terms of the Mittag-Leffler function [7, 14] $E_{\alpha,1}(z) \equiv E_\alpha(z) = \sum_{j=0}^\infty z^j/\Gamma(j\alpha+1)$. This reads

$$G(\mathbf{x}, t; \mathbf{x}') = \sum_\lambda \phi_\lambda^*(\mathbf{x}')\phi_\lambda(\mathbf{x}) E_\alpha\left(\lambda\left[\frac{t}{i\tilde{h}}\right]^\alpha\right), \tag{11}$$

which is a fractional generalization of Green's function. It is worth noting that this solution for the Green function in the form of the Mittag-Leffler function does not satisfy Stone's theorem on one-parameter unitary groups [31].

In the general case, when the eigenvalue problem cannot be solved rigorously, the analysis of the fractional Green function meets serious deficiencies. For an example, the semiclassical analysis of the FTSE leads to a much more complicated form for the wave function than the physically transparent expression for the local wave function $\psi(x,t) \sim e^{iS(t,x)/\hbar}$. Moreover, the classical action $S(t,x)$ is no longer defined for the FTSE. Another important question could be addressed to both FTSE (8) and Eq. (11): for the fractional Fokker–Planck equation, (FFPE) the fractional time derivative is an asymptotic description of a continuous time random walk (CTRW) and it describes subdiffusion. What, therefore, is the physical meaning of the fractional time derivative in quantum processes when a concept of a multi-scale continuous time random walk is absent?

To shed light on this situation, we consider a case with $\alpha = 1/2$, when fractional quantum dynamics can be modeled by means of the conventional quantum mechanics in the framework of a comb model. The comb model is an analog of a 1D medium where fractional diffusion has been observed [33, 34]. It is a particular example of a non-Markovian phenomenon, explained in the framework of the CTRW [14, 20, 33]. This model is also known as a toy model for a porous medium used for exploration of low dimensional percolation clusters [35].

## 4. Diffusion Comb Model

Anomalous diffusion on the comb structure is described by the 2D distribution function $P = P(t, x, y)$. In some dimensionless variables $x, y$ and $t$ it is governed by the following dimensionless equation

$$\frac{\partial P}{\partial t} + L_{\mathrm{FP}}(x) P \delta(y) - \frac{\partial^2 P}{\partial y^2} = 0, \tag{12}$$

where $\hat{L}_{\mathrm{FP}}$ is the Fokker–Planck operator in the $x$-direction. For example, it is $\hat{L}_{\mathrm{FP}}(x)P = -\frac{\partial^2 P}{\partial x^2} + \frac{\partial v(x)P}{\partial x}$ [34, 36]. The initial condition is $P(0, x, y) = \delta(x)\delta(y)$, and the boundary conditions on the infinities have the form $P(t, \pm\infty, \pm\infty) = 0$ and the same for the first derivatives with respect to $x$ and $y$ $P'_x(t, \pm\infty, \pm\infty) = P'_y(t, \pm\infty, \pm\infty) = 0$. The transport of particles along the structure $x$-axis is described by the function $P(t, x, y = 0) = g(x, t)$. The total number of transporting particles on the structure axis decreases with time due to the comb structure [34, 35]:

$$\langle g \rangle = \int_{-\infty}^{\infty} g(t, x) dx = \frac{1}{2\sqrt{\pi t}}. \tag{13}$$

Therefore, the distribution function $g(x, t)$ describes anomalous diffusion in the $x$-direction when the number of particles $\langle g \rangle$ is not conserved. The formulation of the problem with conservation of the total number of particles is equivalent to the case with a continuous distribution of the delay times [20], where the total number of particles is described by the function $P_1(t, x) = \int_{-\infty}^{\infty} P(t, x, y) dy$, which is the CTRW approach. Carrying out the Laplace transform of Eq. (12), one obtains the solution in the form

$$P(t, x, y) = \mathcal{L}^{-1}\left[\tilde{g}(s, x) e^{-\sqrt{s}|y|}\right], \tag{14}$$

where $\tilde{g}(s, x) = \mathcal{L}[g(x, t)]$. One also obtains the equation for $P_1$ by integrating Eq. (12) with respect to the variable $y$. It reads in the Laplace space for

$\tilde{P}_1(s,x) = \mathcal{L}[P_1(t,x)]$:

$$s\tilde{P}_1(p,x) + \hat{L}_{\mathrm{FP}}(x)\tilde{g}(s,x) = \delta(x). \tag{15}$$

Integration of Eq. (14) with respect to $y$ yields

$$\tilde{g}(s,x) = \frac{1}{2}\sqrt{s}\tilde{P}_1(s,x). \tag{16}$$

Then, after substitution of the last relation in Eq. (15), the CTRW equation in the Laplace space is

$$s\tilde{P}_1 + \frac{1}{2}s^{1/2}\hat{L}_{\mathrm{FP}}(x)\tilde{P}_1 = \delta(x). \tag{17}$$

Now we introduce the Caputo fractional derivatives by means of the Laplace transform (see Eqs. (6) and (10) with $\alpha = 1/2$): $\mathcal{L}[\frac{\partial^{1/2}}{\partial t^{1/2}}P_1)] = s^{1/2}\tilde{P}_1(s) - s^{-1/2}\delta(x)$. Therefore, performing the Laplace inversion, we obtain the FFPE in the following form

$$\frac{\partial^{1/2}P_1}{\partial t^{1/2}} + \hat{L}_{\mathrm{FP}}(x)P_1 = 0. \tag{18}$$

Here the initial condition is $P_1(0,x) = \delta(x)$. Now a relation between the FTSE (8), FFPE (18), and the comb model is established.

## 5. Quantum Comb Model

A special quantum behavior of a particle on the comb can be defined as the quantum motion in the $d+1$ configuration space $(\mathbf{x}, y)$, such that the dynamics in the $d$-dimensional configuration space $\mathbf{x}$ is possible only at $y=0$ and motions in the $\mathbf{x}$ and $y$ directions commute. Therefore the quantum dynamics is described by the following Schrödinger equation

$$i\tilde{h}\frac{\partial\Psi}{\partial t} = \delta(y)\hat{H}(\mathbf{x})\Psi - \frac{\tilde{h}^2}{2}\frac{\partial^2\Psi}{\partial y^2}, \tag{19}$$

where the Hamiltonian $\hat{H} \equiv \hat{H}(\mathbf{x}) = -\frac{\tilde{h}^2}{2}\nabla + V(\mathbf{x})$ is the same as in Eq. (8), and it governs the dynamics with a potential $V(\mathbf{x})$ in the $\mathbf{x}$ space, while the $y$ coordinate corresponds to the 1D free motion. All the parameters and variables are dimensionless. Analogously to the FTSE (8), following [7], one introduces the Planck length $L_{\mathrm{P}} = \sqrt{\hbar G/c^3}$, time $T_{\mathrm{P}} = \sqrt{\hbar G/c^5}$, mass $M_{\mathrm{P}} = \sqrt{\hbar c/G}$, and energy $E_{\mathrm{P}} = M_{\mathrm{P}}c^2$, where $\hbar$, $G$, and $c$ are the Planck constant, the gravitational constant and the speed of light, respectively.

Therefore, quantum mechanics of a particle with mass $m$ is described by the dimensionless units $x/L_{\rm P} \to x$, $y/L_{\rm P} \to y$, $t/T_{\rm P} \to t$, while the dimensionless Planck constant is defined as the inverse dimensionless mass $\tilde{h} = M_{\rm P}/m$. Note that the dimensionless potential is now $V(\mathbf{x}) \to V(\mathbf{x})/M_{\rm P}c^2$.

### 5.1. *Comb FTSE*

By analogy with the diffusion (classical) comb model, we are concerned with the dynamics in the $\mathbf{x}$ space. But simple integration of the wave function over the $y$ coordinate is no longer valid. Therefore, one carries out the Fourier transform in the $y$ space $\mathcal{F}_y\Psi(\mathbf{x}, y, t) = \bar{\Psi}(\mathbf{x}, l, t) \equiv \bar{\Psi}_l$, and as a result of this, (19) reads

$$i\tilde{h}\frac{\partial\bar{\Psi}_l}{\partial t} = \hat{H}(\mathbf{x})\Psi(\mathbf{x}, 0, t) + \frac{l^2}{2}\bar{\Psi}_l. \tag{20}$$

To obtain this equation in a closed form, one needs to express the wave function $\Psi(\mathbf{x}, 0, t)$ by the Fourier image $\bar{\Psi}_l$. To this end the Laplace transform of Eq. (19) is performed $\mathcal{L}[\Psi(\mathbf{x}, y, t)] = \tilde{\Psi}_s(\mathbf{x}, y)$. We have for the solution in the Laplace domain (cf. Eq. (14))

$$\tilde{\Psi}_s(\mathbf{x}, y) = \tilde{\Psi}_s(\mathbf{x}, 0)\exp\left[i(1+i)\sqrt{s/\tilde{h}}|y|\right], \tag{21}$$

where we used $\sqrt{2i} = (1+i)$. Performing the Fourier transform $\bar{\tilde{\Psi}}_{s,l}(\mathbf{x}) = \mathcal{F}[\tilde{\Psi}_s(\mathbf{x}, y)]$ one obtains from Eq. (21)

$$\bar{\tilde{\Psi}}_{s,l}(\mathbf{x}) = \tilde{\Psi}_s(\mathbf{x}, 0)\mathcal{F}_y e^{i(1+i)\sqrt{s/\tilde{h}}|y|} = \frac{2i(1+i)\sqrt{s/\tilde{h}}}{l^2 - is/\tilde{h}}\tilde{\Psi}_s(\mathbf{x}, 0). \tag{22}$$

Finally, the Laplace inversion for $\tilde{\Psi}_s(\mathbf{x}, 0)$ determines the wave function

$$\Psi(\mathbf{x}, 0, t) = \mathcal{L}^{-1}\left[\bar{\tilde{\Psi}}_{s,l}(\mathbf{x})\frac{l^2 - 2is/\tilde{h}}{2i(1+i)\sqrt{s/\tilde{h}}}\right]. \tag{23}$$

Let us, first, consider a simple case with $l = 0$. We have from Eq. (23) $\tilde{\Psi}_s(\mathbf{x}, 0) = -\sqrt{\frac{s}{2\tilde{h}i}}\bar{\tilde{\Psi}}_{s,0}$. Then we define $\bar{\Psi}_0(\mathbf{x}, t) = \psi(x)$, and carrying out the Laplace transform in Eq. (20), we arrive at the definition of the Caputo fractional derivative (9) in the Laplace domain $\mathcal{L}[{}_0D_t^{1/2}\psi(t)] = s^{1/2}\tilde{\psi}(s) - s^{-1/2}\psi(0)$. Finally, carrying out the inverse Laplace transform and redefining $\frac{i\hat{H}}{\sqrt{2}\tilde{h}} \to \hat{H}$, one obtains the FTSE which coincides exactly with Eq. (8) for $\alpha = 1/2$.

Repeating this procedure for an arbitrary $l$, one performs the Laplace transform of the term proportional to $l^2/\sqrt{s}$ in Eq. (23). Performing simple operations of fractional calculus and taking into account Eqs. (1) and (7), one obtains

$$(i\tilde{h})^{1/2}\frac{\partial^{1/2}\bar{\Psi}_l}{\partial t^{1/2}} = -\frac{l^2}{2\sqrt{2}}{}_0I_t^1\hat{H}(\mathbf{x})\bar{\Psi}_l + \frac{i}{\sqrt{2}\tilde{h}}\hat{H}\bar{\Psi}_l + \frac{\tilde{h}^2 l^2}{2}\bar{\Psi}_l. \tag{24}$$

This comb FTSE describes the quantum dynamics in the $\mathbf{x}$ configuration space. The index $l$ corresponds to an effective interaction of a quantum system with an additional degree of freedom, while the fractional time derivatives, with $\alpha = 1/2$, reflect this interaction in a form of non-Markov memory effects.

### 5.2. *Green's function*

The initial value problem with the initial condition $\Psi(t = 0) = \Psi_0(\mathbf{x}, y)$ is described by the Green's function. For the complete analogy with the classical comb model and fractional diffusion, described in Sec. 4, the boundary conditions for the $y$ direction defined at infinities $y = \pm\infty$ are $\Psi(\mathbf{x}, t) = \partial\Psi(\mathbf{x}, t)/\partial y = 0$. Using the eigenvalue problem

$$\hat{H}(\mathbf{x})\psi_\lambda(\mathbf{x}) = \lambda\psi_\lambda(\mathbf{x}), \tag{25}$$

we present the wave function in Eq. (19) as the expansion $\Psi(\mathbf{x}, y, t) = \sum_\lambda \phi_\lambda(y, t)\psi_\lambda(\mathbf{x})$, where $\sum_\lambda$ also assumes integration on $\lambda$ for the continuous spectrum. For fixed $\lambda$, we arrive at the dynamics of a particle in the $\delta$ potential

$$i\tilde{h}\frac{\partial\phi_\lambda}{\partial t} = -\frac{\tilde{h}^2}{2}\frac{\partial^2\phi_\lambda}{\partial y^2} + \lambda\delta(y)\phi_\lambda. \tag{26}$$

Taking into account that the Green function of Eq. (19) has the spectral decomposition $G(\mathbf{x}, y, t; \mathbf{x}'y') = \sum_\lambda G_\lambda(y, t; y')\psi_\lambda^*(\mathbf{x})\psi_\lambda(\mathbf{x}')$, we obtain that the Schrödinger equation for the Green function $G_\lambda(y, t; y')$ reads

$$i\tilde{h}\frac{\partial G_\lambda}{\partial t} = -\frac{\tilde{h}^2}{2}\frac{\partial^2 G_\lambda}{\partial y^2} + \lambda\delta(y)G_\lambda + i\tilde{h}\delta(y)\delta(t). \tag{27}$$

The Green function for this Schrödinger equation has been obtained in [37, 39] for free waves boundary conditions at infinities. For the chosen boundary conditions, it is instructive to employ Eq. (21) in the Laplace domain. Then, replacing the eigenvalues $\lambda$ by the Hamiltonian $\hat{H}(\mathbf{x})$, one obtains the Green's function in the form of the inverse Laplace transform

$$G\left[\hat{H}(\mathbf{x}), y, t)\right] = \mathcal{L}^{-1}\left[\frac{i\tilde{h}e^{i(1+i)\sqrt{s/\tilde{h}}|y|}}{\hat{H}(\mathbf{x}) - i(1+i)\sqrt{\tilde{h}^3 s}}\right]. \tag{28}$$

One performs the inverse Laplace transform, using the following presentation for the denominator

$$\int_0^\infty \exp\left\{-u[\hat{H}(\mathbf{x}) - i(1+i)\sqrt{\tilde{h}^3 s}]\right\} du.$$

Therefore, the Green function reads as

$$G\left[\hat{H}(\mathbf{x}), y, t)\right] = \frac{\sqrt{i\tilde{h}}}{\sqrt{2\pi t^3}} \int_0^\infty (|y| + \tilde{h}^2 u) \exp\left[-u\hat{H}(\mathbf{x}) - \frac{i(|y| + \tilde{h}^2 u)^2}{2\tilde{h}t}\right] du. \tag{29}$$

Using the Fourier transform for the exponential

$$\exp[\frac{i(|y| + \tilde{h}^2 u)^2}{2\tilde{h}t}] = \sqrt{\frac{\tilde{h}t}{2\pi i}} \int_{-\infty}^\infty e^{i\tilde{h}t\xi^2/2} e^{-i\xi(|y|+\tilde{h}^2 u)} d\xi,$$

one presents the Green function in the following convenient form

$$G(x, y, t; x') = \frac{h}{2\pi t} \int_{-\infty}^\infty \left\{ e^{i\tilde{h}t\xi^2/2} \left(i\frac{\partial}{\partial \xi}\right) \right.$$
$$\left. \times \int_0^\infty \exp\left[-u\hat{H}(\mathbf{x}) - i\xi(|y| + \tilde{h}^2 u)\right] du \right\} d\xi. \tag{30}$$

Note that this expression is convenient for further analysis in the framework of the path integral and semiclassical approximation.

## 6. Fractional Kicked Rotor with Dissipation

The second example is devoted to another realization of an interaction of a system with its environment, which is modeled by means of the fractional space derivatives. This task is related to the quantum-to-classical crossover, where the role of dissipation in quantum systems is the subject of extensive research [41–44]. A quantum kicked rotor with dissipation is an example where a quantization of a strange attractor has been considered in the framework of the master equation [44–46]. Fractional space derivatives make it possible to introduce quantum dissipation in such a way that quantum dynamics can still be described by means of a wave function in the framework of the Schrödinger equation. This approach will be demonstrated for the quantum kicked rotor.

### 6.1. *Fractional particle on a circle*

The dynamics of a point particle on a unit circle is described by an angular phase $\phi \in [0, 2\pi]$. If a wave function is the same at $\phi = 0$ and $\phi = 2\pi$, the time evolution of a particle of a unit mass $m = 1$ between two points $\phi_0$ and $\phi = \phi(t)$ in time interval $(0, t)$ is determined by the transition amplitude (see e.g. [11])

$$G(\phi, t; \phi_0) = \sum_{l=-\infty}^{\infty} \int_{-\infty}^{\infty} \frac{dp}{2\pi\tilde{h}} \exp\left\{\frac{i}{\tilde{h}}\left[p(\phi - \phi_0 + 2\pi l) - \frac{p^2}{2m}t\right]\right\}. \quad (31)$$

Here $p$ corresponds to the orbital momentum operator $\hat{p} = -i\tilde{h}\frac{\partial}{\partial\phi}$. A generalization of this expression on the fractional dynamics is straightforward, by substituting $p^2 \to |p|^\alpha$ in the exponential with $1 < \alpha < 2$.

$$G(\phi, t; \phi_0) = \sum_{l=-\infty}^{\infty} \int_{-\infty}^{\infty} \frac{dp}{2\pi\tilde{h}} \exp\left\{\frac{i}{\tilde{h}}\left[p(\phi - \phi_0 + 2\pi l) - \frac{|p|^\alpha}{\alpha}t\right]\right\}. \quad (32)$$

Here modulus $|p|$ ensures the unitary transformation. The dimensionless Planck constant ensures that all variables and parameters are dimensionless, by analogy with the FTSE. One can easily check that $G(\phi, t; \phi_0)$ is the Green function with properties as the initial condition

$$\lim_{t\to 0} G(\phi, t; \phi_0) = \sum_l \delta(\phi - \phi_0 + 2\pi l)$$

and the chain rule $G(\phi, t; \phi_0) = \int_0^{2\pi} G(\phi, t; \phi', t')G(\phi', t'; \phi_0)d\phi'$. The Hamiltonian, which produces the Green function, can be obtained by differentiation of Eq. (32) with respect to time $t > 0$ [5]. This yields

$$i\tilde{h}\partial_t G(\phi, t; \phi_0) = \sum_{l=-\infty}^{\infty} \int_{-\infty}^{\infty} \frac{|p|^\alpha dp}{2\pi\tilde{h}\alpha} \exp\left\{\frac{i}{\tilde{h}}\left[p(\phi - \phi_0 + 2\pi l) - \frac{|p|^\alpha}{\alpha}t\right]\right\}. \quad (33)$$

The term $|p|^{\alpha}e^{\frac{i}{\hbar}pz}$ can be expressed by means of the Weyl fractional integrals (4) and (5). This yields

$$ih\partial_t G = (-i\tilde{h})^{\alpha}\left[{}_{-\infty}\mathcal{W}^{\alpha}G + \mathcal{W}^{\alpha}_{\infty}G\right]. \quad (34)$$

This equation describes the fractional dynamics of a quantum particle on a unit circle. In what follows we consider an example of a perturbed motion on the circle, and we also violate the unitary property of the evolution operator.

### 6.2. *Fractional kicked rotor*

The quantum chaotic dynamics of a fractional kicked rotor (FKR) is described by the Hamiltonian

$$\hat{H} = \hat{\mathcal{T}} + \epsilon\cos x \sum_{n=-\infty}^{\infty}\delta(t-n), \quad (35)$$

where $\epsilon$ is an amplitude of the periodic perturbation which is a train of $\delta$ kicks. The kinetic part of the Hamiltonian is due to Eq. (34)

$$\hat{\mathcal{T}} = (-i\tilde{h})^{\alpha}\mathcal{W}^{\alpha}/\alpha, \quad (36)$$

where $\mathcal{W}^{\alpha} \equiv {}_{-\infty}\mathcal{W}^{\alpha}$ and $\tilde{h}$ is the dimensionless Planck constant. When $\alpha = 2$, Eq. (35) is the quantum kicked rotor [47]. For a periodic function $f(x) = \sum \bar{f}_k e^{-ikx}$, the Fourier transform property of Eq. (4) determines the fractional Weyl derivative $\mathcal{W}^{\alpha}$ in the following simplest way

$$\mathcal{W}^{\alpha}f(x) = \sum_{n=-\infty}^{\infty}(-ik)^{\alpha}\bar{f}_k e^{-ikx}. \quad (37)$$

Thus, the kinetic term in the Hamiltonian (35) is defined on the basis $|k\rangle = e^{ikx}/\sqrt{2\pi}$ as follows:

$$\hat{\mathcal{T}}|k\rangle = \mathcal{T}(k)|k\rangle = \frac{(\tilde{h}k)^{2-\beta}}{2-\beta}|k\rangle, \quad (38)$$

where $\alpha = 2-\beta$ with $0<\beta<1$. This non-Hermitian operator has complex eigenvalues for $k<0$, which are defined on the complex plain with a cut from 0 to $-\infty$, such that $1^{-\beta} = 1$ and $(-1)^{-\beta} = \cos\beta\pi - i\sin\beta\pi$, and therefore, $k^{-\beta} = |k|^{-\beta}e^{-i\pi\beta(k)}$, where $\beta(k) = \beta[1-\mathrm{sgn}(k)]/2$. Note that when $\alpha > 2$, one chooses $(-1) = e^{-i\pi}$, such that $(-1)^{\beta} = e^{-i\pi\beta}$.

A quantum map on the unit period for the wave function $\psi(x,t)$ is

$$\psi(x,t+1) = \hat{U}\psi(x,t), \tag{39}$$

where the evolution operator on the period

$$\hat{U} = \exp\left[\frac{-i\epsilon\cos x}{\tilde{h}}\right]\exp\left[\frac{-i\hat{\mathcal{T}}}{\tilde{h}}\right] \tag{40}$$

describes free dissipative motion and then a kick. Dynamics of the FKR is studied numerically, where Eq. (38) enables one to use the fast Fourier transform as an efficient way to iterate the quantum map (39). A specific property of this Hamiltonian dynamics is quantum dissipation resulting in probability leakage, described by the survival probability

$$P(t) = \langle\psi(t)|\psi(t)\rangle = \sum_{n=-\infty}^{\infty} |f_n|^2, \tag{41}$$

where $|f_n|^2$ is the level occupation distribution at time $t$. The initial occupation is $f_n(t=0) = \delta_{n,0}$. Another specific characteristic is the nonzero mean value of the orbital momentum

$$\langle p(t)\rangle = \tilde{h}\frac{\sum_n n|f_n(t)|^2}{\sum_n |f_n(t)|^2}, \tag{42}$$

due to the asymmetry of the quantum kinetic term $\hat{\mathcal{T}}$. Results of the numerical study of the quantum map (39) are shown in Figs. 1(a),(b). The quantum dissipation leads to the asymmetrical distribution of the level occupation $|f_n(t)|$ (see Fig. 1(a)) that results in the acceleration of the nonzero first moment of the orbital momentum $\langle p\rangle \sim t^{\gamma_1}$, where $\gamma_1 > 0$, which is also accompanied by the power law decay of the survival probability $P(t) \sim t^{-\gamma_2}$ with $0 < \gamma_2 < 1$. Quantum localization leads to the exponential restriction of the initial profile spreading in the orbital momentum space from above. This property results in saturation of acceleration of $\langle p(t)\rangle$; namely, at $t \to \infty$ it follows that $\langle p(t)\rangle \to$ const. Such a behavior is found for a special selection of the parameters $\tilde{h} = 0.76$, $\beta = 0.05$ and $\epsilon = 3$. In Fig. 1(b) one sees a direct current of $\langle p(t)\rangle$ for $5\cdot 10^5$ iterations. This double quantum impact of asymmetric dissipation and localization leads, asymptotically, to a quantum-like ratchet that differs from the quantum one obtained for a classical chaotic attractor [57].

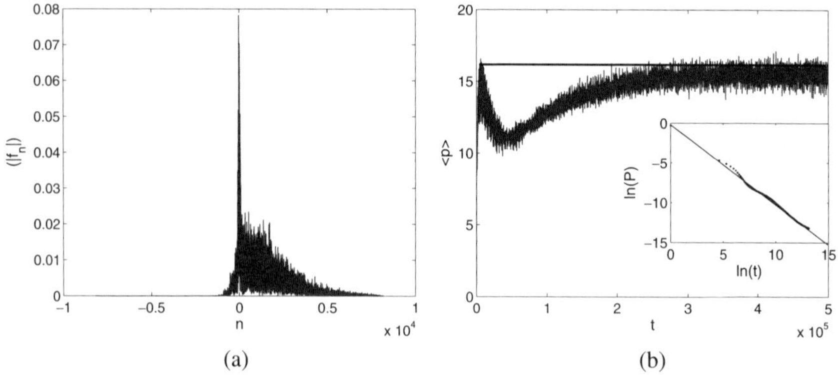


Fig. 1. Asymmetrical properties of quantum map (39) due to dissipation: (a) The level occupation distribution after 250 iterations for $\tilde{h} = 0.2$, $\epsilon = 3$, $\beta = 2 \cdot 10^{-5}$, (b) Ratchet-like behavior of the nonzero quantum current due to Eq. (42).

### 6.3. *Wigner representation*

To understand the physical relevance of the fractional Schrödinger equation with dissipation, the classical limit $\tilde{h} \to 0$ is performed in the Wigner representation. Thus, the system is described by the Wigner function $W(x,p,t)$ which is a $c$-number projection of the density matrix in the Weyl rule of association between $c$-numbers and operators. The Weyl transformation of an arbitrary operator function $G(\hat{x},\hat{p})$ is [48, 49]

$$F(x,p) = \mathrm{Tr}\left[G(\hat{x},\hat{p})\Delta(x-\hat{x},p-\hat{p})\right], \tag{43}$$

where $F(x,p)$ is a $c$-number function and $\Delta(x-\hat{x},p-\hat{p})$ is a projection operator which acts as the two-dimensional Fourier transform. For the cylindrical phase space the projection operator is [50–52]

$$\Delta(x-\hat{x},p-\hat{p}) = \sum_{m=-\infty}^{\infty} \frac{1}{2\pi}\int_{-\pi}^{\pi} d\xi e^{im(x-\hat{x})+i\xi(p-\hat{p})}. \tag{44}$$

This operator determines an inverse transform as well:

$$G(\hat{x},\hat{p}) = \sum_{k=-\infty}^{\infty} \frac{1}{2\pi}\int_{-\pi}^{\pi} F(x,\tilde{h}k)\Delta(x-\hat{x},\tilde{h}k-\hat{p}), \tag{45}$$

where $p = \tilde{h}k$. The quantum map for the density matrix $\hat{\rho}(t)$ is

$$\hat{\rho}(t+1) = \hat{U}^{\dagger}\hat{\rho}(t)\hat{U}. \tag{46}$$

Therefore, evolution of the Wigner function

$$W(t,x,p) = \mathrm{Tr}[\hat{\rho}(t)\Delta(x-\hat{x},p-\hat{p})]\,,$$

for the period determined by the map (46), is

$$\begin{aligned} W(t+1,x,p) &= \mathrm{Tr}\left[\hat{U}^{\dagger}\hat{\rho}(t)\hat{U}\Delta(x-\hat{x},p-\hat{p})\right] \\ &= \sum_{k'=-\infty}^{\infty}\int_0^{2\pi} K_{\tilde{h}}(x,p|x',p')W(t,x',p')dx', \end{aligned} \tag{47}$$

where $K_{\tilde{h}}(x,p|x',p')$ is the Green function for the period

$$\begin{aligned} K_{\tilde{h}}(x,p|x',p') &= \sum_m 1/2\pi \int_{-\pi}^{\pi} e^{im(x-x'+\xi')}e^{i\xi'(p'-p)} \\ &\quad \times \exp\left[\frac{i}{\tilde{h}}\mathcal{T}^*(p+\tilde{h}m/2) - \frac{i}{\tilde{h}}\mathcal{T}(p-\tilde{h}m/2)\right] \\ &\quad \times \exp\left[\frac{i\epsilon}{\tilde{h}}\cos(x'+\tilde{h}\xi'/2) - \frac{i\epsilon}{\tilde{h}}\cos(x'-\tilde{h}\xi'/2)\right]. \end{aligned} \tag{48}$$

The trace is $\mathrm{Tr}[\ldots] = \sum_k \langle k|\ldots|k\rangle$. In the classical limit $\tilde{h} \to 0$, we obtain in Eq. (48) that the difference of the perturbations in the exponential is $-i\epsilon\cos x$, while the difference of the kinetic terms is $imp^{1-\beta} \equiv im\omega(p)$ for $p > 0$ and $-2\sin(\beta\pi)\mathcal{T}(|p|)/\tilde{h}$ for $p < 0$. The last term diverges at $\tilde{h} = 0$ and yields identical zero for Green's function $K_{\tilde{h}=0}(p<0) \equiv 0$. Thus, the classical Green function

$$K_{\tilde{h}=0}(x,p|x'p') = \Theta(p)\delta(x-x'-\omega(p))\delta(p-p'-\epsilon\sin x') \tag{49}$$

corresponds to the classical map $\mathcal{M}$

$$p_{n+1} = p_n + \epsilon\sin x_n, \quad x_{n+1} = x_n + \omega(p_{n+1}) \tag{50}$$

of the nonlinear kicked rotor with the nonlinear frequency $\omega(p)$, and absorbing boundary conditions for $p < 0$, that the Heaviside function $\Theta(p)$ reflects. Therefore, the fractional Hamiltonian (35) corresponds to the open system of Eqs. (49) and (50) in the classical limit. Chaotic dynamics of this open system takes place in the upper half of the cylindrical phase space.

It is worth mentioning that in the class of periodic functions, eigenvalues of the unperturbed Hamiltonian $\mathcal{T}$ coincide with $\hat{H}_0(\hat{p}) = \left(-i\tilde{h}\frac{\partial}{\partial x}\right)^{\alpha}$, and have the same classical limit as Eq. (49). This local derivative has the classical counterpart with the Hamiltonian $H_0(p) = p^\alpha$ which does not coincide with Eq. (49). Namely, the Hamiltonian $H_0(p)$ is the classical system with dissipation for $p < 0$, while the map $\mathcal{M}$ in Eqs. (49) and (50) is the open system where a particle is set apart from the dynamics for $p < 0$.

### 6.4. *Spectrum*

To illuminate the non-Hermitian properties of the Hamiltonian (35) the quasienergy spectrum of the FKR is studied as well. Let $\phi_E \equiv |\phi_E\rangle$ be an eigenfunction of the evolution operator $\hat{U}$ with a corresponding eigenvalue $E$. The eigenvalue problem in the basis of eigenfunctions of the orbital momentum $|k\rangle$ reads

$$\langle k|\hat{U}|\phi_E\rangle = \sum_{k'=-\infty}^{\infty} U_{k,k'}\phi_{k'} = e^{-iE_m}\phi_m, \tag{51}$$

where $\phi_k = \langle k|\phi_E\rangle$, and the matrix elements of the evolution operator are

$$U_{k,k'} = \langle k|\hat{U}|k'\rangle = e^{-i\frac{\mathcal{T}(k)}{\tilde{h}}}(-i)^{k'-k}J_{k'-k}(\kappa), \tag{52}$$

where $J_\nu(z)$ is the Bessel function of the first kind [55].

Note, for $\beta = 0$ and for a special choice of $\tilde{h}/2\pi = M/N$ to be rational, the matrix elements $U_{m,m'}$ obey the following periodicity rule [53] $U_{m+N,m'+N} = (-1)^{MN}U_{m,m'}$. Therefore, the eigenvalue problem of Eq. (51) for the infinite matrix $\hat{U}$ is deduced to the eigenvalue problem for the finite matrix of the form [53, 54]:

$$U_{k,k'} = \frac{1}{N}\sum_{j=1}^{N} \exp\left[-i\frac{\tilde{h}k^2}{2}\right] \exp\left[2\pi i\frac{(k'-k)j}{N} - i\kappa\cos\frac{2\pi j}{N}\right]. \tag{53}$$

This expression for the matrix elements of the $N \times N$ matrix corresponds to the discrete Fourier inversion. Hence, the fast Fourier transform is the efficient way to construct the finite matrix $\hat{U}$ numerically [56].

The same technique of the fast Fourier transform can be applied for the calculation of the spectrum of the fractional evolution operator (40) with $\beta \neq 0$. In this case one has to truncate the matrix $U_{k,k'}$. The size $N$ of the truncated matrix must be large enough to reduce the influence of the truncation on the genuine spectrum and to make it insignificant. Results of numerical calculation of the spectrum for different values of $\beta$ is depicted in Fig. 2. It shows inward migration from the unit circle in the complex

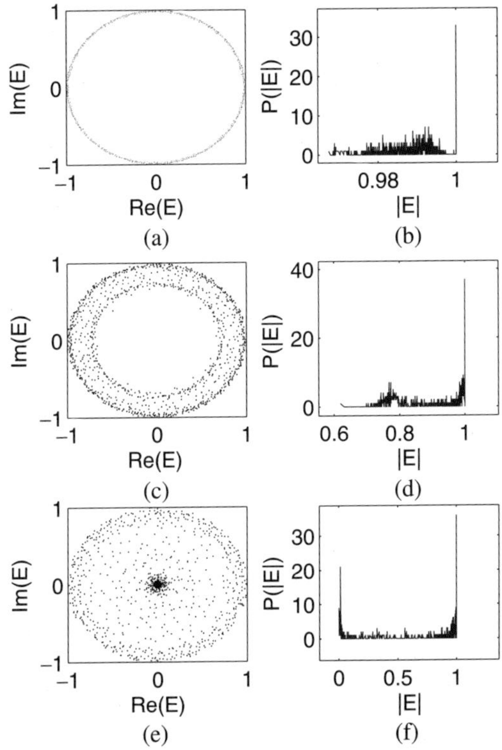


Fig. 2. Complex eigenvalues of $U_{k,k'}$ (left column) and the distribution of eigenvalues $P(|E|)$ (right column) for different values of $\beta$ and $\tilde{h} = 0.02$, $\epsilon = 3$, $N = 1024$: $\beta = 10^{-5}$ for (a),(b), $\beta = 10^{-4}$ for (c),(d), and $\beta = 10^{-3}$ for (e),(f).

plain of the eigenvalues of the dissipative map (39) for different values $\beta$. The degree of this migration inside the complex plain is described by the energy distribution function $P(|E|)$ versus $|E|$.

## 7. Conclusion

We have considered two absolutely different examples of the application of fractional integro-differentiation in quantum mechanics. As already stated, these examples have it in common that fractional derivatives describe an interaction of a quantum system with its environment by means of the wave function.

For the first example, we observed that the fractional time derivative, at least for $\alpha = 1/2$, reflects an effective interaction of a quantum system with an additional degree of freedom. As already mentioned, the $y$ direction in the comb was introduced to model the time fractional derivative $\frac{\partial^{1/2}}{\partial t^{1/2}}$ by analogy with the continuous time random walk, and for $\alpha = 1/2$, the fractional Fokker–Planck equation is identical to the diffusion comb model [34,36]. In the quantum case the situation differs essentially from fractional diffusion. First of all, the quantum comb model and the FTSE are not identical. As shown, the FTSE, being an equation for the zero Fourier component of the wave function, is insufficient to describe the complete dynamics in the $x$ space of the system. Note also that both the quantum motion on the comb and the FTSE introduce new nonlinear phenomena in the semiclassical limit, and this semiclassical approach differs from those described in the framework of the standard Schrödinger equation [10].

An important conclusion from the second example is the following. Fractional Schrödinger equation

$$i\tilde{h}\partial_t\psi = (-i\tilde{h})^\alpha \mathcal{W}^\alpha \psi \tag{54}$$

describes the quantum *dissipative Hamiltonian* dynamics. The classical counterpart is a nonlinear motion with a nonlinear frequency $\omega(p)$ realized on the upper half plane of the phase space with absorption in the lower half plane. It has well-defined physical meaning. Therefore, the fractional Schrödinger equation (54) can be a generalized approach for any functions for which the Fourier transform is valid. In this case, the opposite classical-to-quantum transition can be performed by determining the Heaviside function in Eq. (48)

$$e^{i\omega(p)z}\Theta(p) = \lim_{\tilde{h}\to 0} \exp\left[\frac{i}{\tilde{h}}\mathcal{T}^*\left(p + \frac{\tilde{h}z}{2}\right) - \frac{i}{\tilde{h}}\mathcal{T}\left(p - \frac{\tilde{h}z}{2}\right)\right],$$

where $\mathcal{T}(p)$ is uniquely defined by the condition $\omega(p) = \mathcal{T}'(p)$. Thus, fractional derivatives quantize classical open systems in the framework of the non-Hermitian Hamiltonians.

## Acknowledgments

I would like to thank Elizabeth Yodim for language editing. This work was supported in part by the Israel Science Foundation (ISF) and by the US-Israel Binational Science Foundation (BSF).

## References


1. D. Kusnezov, A. Bulgac and G. D. Dang, *Phys. Rev. Lett.* **82**, 1136 (1999).
2. N. Laskin, *Chaos* **10**, 780 (2000).
3. B. J. West, M. Bologna, and P. Grigolini, *Physics of Fractal Operators* (Springer, 2002).
4. N. Laskin and G. Zaslavsky, *Physica A* **368**, 38 (2006).
5. B. J. West, *J. Phys. Chem. B* **104**, 3830 (2000).
6. R. Hermann, *J. Phys. G: Nucl. Part. Phys.* **34**, 607 (2007).
7. M. Naber, *J. Math. Phys.* **45**, 3339 (2004).
8. A. Iomin, *Phys. Rev. E* **75**, 037201 (2007).
9. V. E. Tarasov, *Phys. Lett. A* **372**, 2984 (2008).
10. A. Iomin, *Phys. Rev. E* **80**, 022103 (2009).
11. See also M. Chaichian and A. Demichev, *Path Integrals in Physics: Stochastic Process and Quantum Mechanics*, Vol. 1 (IOP 2001).
12. J.-P. Bouchaud and A. Georges, *Phys. Rep.* **195**, 127 (1990).
13. Ed. R. Hilfer, *Fractional Calculus in Physics* (World Scientific, 2000).
14. R. Metzler and J. Klafter, *Phys. Rep.* **339**, 1 (2000).
15. G. M. Zaslavsky, *Hamiltonian Chaos and Fractional Dynamics* (Cambridge University Press, 2005).
16. G. M. Zaslavsky, *Phys. Rep.* **371**, 461 (2002).
17. I. M. Sokolov, J. Klafter and A. Blumen, *Phys. Today* **55**, 48 (2002).
18. I. Podlubny, *Fractional Differential Equations* (Academic Press, 1999).
19. K. B. Oldham and J. Spanier, *The Fractional Calculus* (Academic Press, 1974).
20. E. W. Montroll and M. F. Shlesinger, The wonderful wold of random walks, in *Studies in Statistical Mechanics*, Vol. 11, eds. J. Lebowitz and E. W. Montroll, (North-Holland, 1984).
21. R. R. Nigmatullin, *Theor. Math. Phys.* **90**, 242 (1992).
22. R. R. Nigmatullin, *Phys. Status Solidi B* **133**, 425 (1986).
23. R. Herrmann, *Phys. Let. A* **372**, 5515 (2008).
24. E. Goldfain, *Chaos, Solitons and Fractals* **28**, 913 (2006).
25. E. Goldfain, *Commun. Nonlinear Sci. Numer. Simulat.* **14**, 2289 (2009).
26. R. P. Feynman and A. R. Hibbs, *Quantum Mechanics and Path Integrals* (McGraw-Hill, 1965).
27. M. Kac, *Probability and Related Topics in Physical Sciences* (Interscience, 1959).
28. J. Dong and M. Xu, *J. Math. Anal. Appl.* **344**, 1005 (2008).
29. S. Wang and M. Xu, *J. Math. Phys.* **48**, 043502 (2007).
30. M. Bhatti, *Int. J. Contemp. Math. Sci.* **2**, 943 (2007).
31. M. H. Stone, *Ann. Math.* **333**, 643 (1932).
32. F. Mainardi, *Chaos Solitons Fractals* **7**, 1461 (1996).
33. G. H. Weiss and S. Havlin, *Physica A* **134**, 474 (1986).
34. E. Baskin and A. Iomin, *Phys. Rev. Lett.* **93**, 120603 (2004).
35. V. E. Arkhincheev and E. M. Baskin, *Sov. Phys. JETP* **73**, 161 (1991).
36. A. Iomin and E. Baskin, *Phys. Rev. E* **71**, 061101 (2005).

37. L. Schulman, in *Path Integrals from meV to MeV* (World Scientific, 1986).
38. L. Schulman, *Techniques and Applications of Path Integration* (Wiley, 1981).
39. B. Gaveau and L. Schulman, *J. Phys. A — Math. Gen.* **19**, 1833 (1986).
40. H. Bateman and A. Erdélyi, *Higher Transcendental Functions* (McGraw-Hill, 1955), Vol. 3.
41. F. Haake, *Quantum Signature of Chaos* (Springer, 2000).
42. D. Braun, *Dissipative Quantum Chaos and Decoherence* (Springer, 2001).
43. I. Percival, *Quantum State Diffusion* (Cambridge Univ. Press 1998).
44. T. Dittrich, in *Quantum Transport and Dissipation*, eds. T. Dittrich, *et al.* (Wiley–VCH, 1998).
45. R. Graham, *Physica Scripta* **35**, 111 (1987).
46. T. Dittrich and R. Graham, *Physica Scripta* **40**, 409 (1989).
47. G. Casati, B. V. Chirikov, F. M. Izrailev and J. Ford, in *Stochastic Behavior in Classical and Quantum Hamiltonian Systems*, eds. G. Casati and J. Ford (Springer, 1979).
48. E. P. Wigner, *Phys. Rev.* **40**, 749 (1932).
49. G. S. Agarwal and E. Wolf, *Phys. Rev. D* **2**, 2161 (1970).
50. G. P. Berman and A. R. Kolovsky, *Physica D* **17**, 183 (1985).
51. G. P. Berman, F. M. Izrailev, and A. R. Kolovsky, *Physica A* **152**, 273 (1988).
52. G. P. Berman, A. R. Kolovsky, F. M. Izrailev and A. M. Iomin, *Chaos* **1**, 220 (1991).
53. S.-J. Chang and K.-J. Shi, *Phys. Rev. A* **34**, (1986).
54. F. M. Izrailev and D. L. Shepelyansky, *Sov. Phys. Dokl.* **24**, 996 (1979).
55. E. Janke, F. Emde and F. Lösh, *Tables of Higher Functions* (McGraw-Hill, 1960).
56. R. Ketzmerick, K. Kruse and T. Geisel, *Physica D* **131**, 247 (1999).
57. G. G. Carlo, G. Benenti, G. Casati and D. L. Shepelyansky, *Phys. Rev. Lett.* **94**, 164101 (2005).

# Chapter 19

## Fractional Dynamics of Open Quantum Systems

Vasily E. Tarasov
*Skobeltsyn Institute of Nuclear Physics,*
*Moscow State University, Moscow 119991, Russia*

The basic principles of fractional quantum dynamics are described. Fractional power of operators and superoperator are used to describe fractional dynamics of open quantum systems. The theory of open and non-Hamiltonian quantum systems is considered as a fundamental generalization of the quantum Hamiltonian mechanics of pure states. The quantum operations, which describe dynamics of open system, are considered as a real completely positive trace-preserving maps on the operator space. The suggested completely dissipative superoperators are infinitesimal generators of completely positive semigroups for fractional quantum dynamics.

1. Introduction . . . . . . . . . . . . . . . . . . . . . . . . . 449
2. Quantum Operations and Superoperators . . . . . . . . . 451
3. Fractional Powers of Infinitesimal Generators . . . . . . . 452
4. Markovian Dynamics for Quantum Observables . . . . . . 455
5. Fractional Quantum Dynamics for Observables . . . . . . 456
6. Fractional Dynamical Semigroup . . . . . . . . . . . . . . 458
7. Fractional Equation for Quantum States . . . . . . . . . . 459
8. Fractional Hamiltonian Dynamics . . . . . . . . . . . . . 462
9. Fractional Dynamics of Open Quantum Oscillator . . . . 467
10. Quantum Analogs of Derivatives of Integer Orders . . . . 470
11. Quantum Analogs of Fractional Derivatives . . . . . . . . 472
12. Fractional Generalization of Quantum Hamiltonian Systems . . . . . . . . . . . . . . . . . . . . . . . 475
13. Conclusion . . . . . . . . . . . . . . . . . . . . . . . . . 478

### 1. Introduction

Fractional dynamics [1–12] is a field of study in physics, mechanics, chemistry studying the behavior of systems and objects that are described by methods of fractional calculus [13–16], where differentiation and integration of noninteger orders are used. Fractional calculus is a theory of integrals and derivatives of any arbitrary real (or complex) order. The fractional differentiation and fractional integration has a long history [17, 18], which go back to many great mathematicians such as Leibniz, Liouville, Riemann, Abel, Riesz, Weyl. Derivatives and integrals of noninteger orders are used to describe systems that are characterized by power-law nonlocality, power-law long-term memory or fractal properties. We can state that fractional dynamics form a new paradigm in science.

New possibilities in physics appear, when the order $\alpha$ of the differential operator $D^{\alpha}_{x}$ or the integral operator $I^{\alpha}_{x}$ becomes an arbitrary parameter. The fractional calculus is a powerful tool to describe physical systems that have long-time memory and long-range spatial interactions. In general, many usual properties of the ordinary (first-order) derivative $D_x$ are not realized for fractional derivative operators $D^{\alpha}_{x}$. For example, a product rule, chain rule, semigroup property have strongly complicated analogs for the operators $D^{\alpha}_{x}$. The physical processes associated with complex systems have nonlocal dynamics and it can be characterized by long-term memory in time. The integro-differentiation operators of fractional orders allow one to consider some of those characteristics. Using fractional calculus, it is possible to obtain useful dynamical models, where fractional integro-differential operators in the time and space variables describe the long-time memory and nonlocal spatial properties of the complex processes and media.

Fractional quantum dynamics is a field of quantum theory studying the behavior of systems and objects that are described by methods of fractional calculus. Different problems of fractional quantum theory are discussed in [19–52] (see also [60] and [8]).

In this chapter, we consider the fractional dynamics of open quantum systems. The theory of open and non-Hamiltonian quantum systems [53–60] can be considered as a fundamental generalization [53, 60, 67, 72] of the quantum Hamiltonian mechanics. The quantum operations, which describe dynamics of open system, are real completely positive trace-preserving maps on the operator space. These maps form a completely positive semigroup. The infinitesimal generator of this semigroup is completely dissipative.

In quantum theory the states and observables are given by operators. The dynamical description of the quantum system is given by superoperators [60], which are maps from a set of operators into itself. Fractional power of operators and superoperator [45, 60], can be used as a possible approach to describe fractional dynamics of open quantum systems. We consider superoperators that are fractional powers of completely dissipative superoperators [45]. We prove that the suggested superoperators are infinitesimal generators of completely positive semigroups for fractional quantum dynamics. The quantum Markovian equation, which includes an explicit form of completely dissipative superoperator, is the most general type of Markovian master equation describing non-unitary evolution of the density operator that is trace preserving and completely positive for any initial condition. A fractional power of infinitesimal generator can be considered as a parameter to describe a measure of screening of environment [45]. In the limit $\alpha \to 0$, the fractional quantum Markovian master equations for quantum observables and density operators give the Heisenberg and von Neumann equations for Hamiltonian systems. In the case $\alpha = 1$, the suggested fractional quantum Markovian master equations give the usual quantum Markovian equation. For $0 < \alpha < 1$, we have an environmental influence on quantum systems. The physical interpretation of the fractional power of infinitesimal generator can be considered as an existence of a power-like screening of environmental influence.

## 2. Quantum Operations and Superoperators

Quantum theories consist of two parts, a kinematics describing the initial states and observables of the system, and a dynamics describing the change of these states and observables with time. In quantum mechanics, the states and observables are given by operators. The dynamical description of the system is given by superoperators [60], which are maps from a set of operators into itself.

Let us give a brief review of the quantum operations and superoperators to fix notation and provide a convenient reference. If $\mathcal{M}$ is an operator space, then the dual space $\mathcal{M}^*$ is a set of all linear functionals on $\mathcal{M}$. To denote an element of $\mathcal{M}$, we use $|B)$ and $B$. The symbols $(A|$ and $\omega$ denote the elements of $\mathcal{M}^*$. We use the symbol $(A|B)$ for a value of the functional $(A|$ on the operator $|B)$ is a graphic junction of the symbols $(A|$ and $|B)$. If $\mathcal{M}$ is an operator Hilbert space, then $(A|B) = \mathrm{Tr}[A^* B]$. We consider a superoperator as a map $\mathcal{L}$ from an operator space $\mathcal{M}$ into itself. A superoperator $\Lambda = \bar{\mathcal{L}}$ on $\mathcal{M}^*$ is adjoint to superoperator $\mathcal{L}$ on $\mathcal{M}$, if $(\Lambda(A)|B) = (A|\mathcal{L}(B))$ for all $B \in D(\mathcal{L}) \subset \mathcal{M}$ and $A \in D(\Lambda) \subset \mathcal{M}^*$.

The most general state change of a quantum system is called a quantum operation [65, 66, 70, 71] (see also [60, 75, 77]). A quantum operation is a superoperator $\hat{\mathcal{E}}_t$, which maps a density operator $\rho$ to a density operator $\rho_t = \hat{\mathcal{E}}_t(\rho)$. Any density operator $\rho_t = \rho(t)$ is self-adjoint ($\rho_t^\dagger = \rho_t$), positive ($\rho_t > 0$) operator with unit trace ($\mathrm{Tr}[\rho_t] = 1$). As a result, we have the following requirements for a superoperator $\hat{\mathcal{E}}_t$ to be the quantum operation.

(1) The superoperator $\hat{\mathcal{E}}_t$ is a real superoperator. The superoperator $\hat{\mathcal{E}}_t$ on an operator space $\mathcal{M}$ is a real superoperator if $[\hat{\mathcal{E}}_t(A)]^* = \hat{\mathcal{E}}_t(A^*)$ for all $A \in D(\hat{\mathcal{E}}_t) \subset \mathcal{M}$, where $A^* \in D(\hat{\mathcal{E}}_t)$ is adjoint to $A$. The real superoperator $\hat{\mathcal{E}}_t$ maps the self-adjoint operator $\rho$ to the self-adjoint operator $\hat{\mathcal{E}}_t(\rho) = (\hat{\mathcal{E}}_t(\rho))^*$.
(2) The superoperator $\hat{\mathcal{E}}_t$ is a trace-preserving map. The superoperator $\hat{\mathcal{E}}_t$ on an operator space $\mathcal{M}$ is trace-preserving if $\hat{\mathcal{E}}_t^\dagger(I) = I$.
(3) The superoperator $\hat{\mathcal{E}}_t$ is a positive superoperator. A non-negative superoperator is a map $\hat{\mathcal{E}}_t$ from $\mathcal{M}$ into $\mathcal{M}$, such that $\hat{\mathcal{E}}_t(A^2) \geq 0$ for all $A^2 = A^*A \in D(\hat{\mathcal{E}}_t) \subset \mathcal{M}$. A positive superoperator is a map $\hat{\mathcal{E}}_t$ from $\mathcal{M}$ into itself, such that $\hat{\mathcal{E}}_t$ is non-negative and $\hat{\mathcal{E}}_t(A) = 0$ if and only if $A = 0$. A density operator $\rho$ is positive. If $\hat{\mathcal{E}}_t$ is a positive superoperator, then $\rho_t = \hat{\mathcal{E}}_t(\rho)$ is a positive operator. We may assume that the map $\hat{\mathcal{E}}_t$ is not only positive but also completely positive. A map $\hat{\mathcal{E}}_t$ is a completely positive map from an operator space $\mathcal{M}$ into itself if

$$\sum_{k=1}^{n}\sum_{l=1}^{n} A_k^\dagger \hat{\mathcal{E}}_t(B_k^\dagger B_l) A_l \geq 0$$

for all $n \in \mathbb{N}$ and $A_k,\ B_k \in \mathcal{M}$.

Dynamics of open quantum systems is described by the maps $\hat{\mathcal{E}}_t$ that form a completely positive quantum semigroup [56] such that $\hat{\Lambda}$ is an infinite generator of the semigroup [56, 60, 73], and the superoperator $\mathcal{L}$ adjoint to $\hat{\Lambda}$, is completely dissipative

$$\mathcal{L}(A_k A_l) - \mathcal{L}(A_k)A_l - A_k\mathcal{L}(A_l) \geq 0$$

for all $A_k, A_l \in D(\mathcal{L})$. The completely dissipative superoperators are infinitesimal generators of completely positive semigroups $\{\Phi_t | t > 0\}$ that is adjoint of $\{\hat{\mathcal{E}}_t | t > 0\}$. The superoperator $\mathcal{L}$ describes the dynamics of observables of open quantum systems. The evolution of a density operator is described by $\hat{\Lambda}$.

## 3. Fractional Powers of Infinitesimal Generators

Let us give a brief review of the fractional power of operators and superoperators [8, 80–85] to fix notation and provide a convenient reference. To define a fractional power of superoperators $\mathcal{L}$, we assume that $\mathcal{L}$ is a closed linear map on operator space with an everywhere dense domain $D(\mathcal{L})$. If the resolvent $R(-z, \mathcal{L}) = (zL_I + \mathcal{L})^{-1}$, where $z > 0$, satisfies the condition

$$\|R(-z, \mathcal{L})\| \leq C/z, \quad z > 0, \tag{1}$$

then a fractional power of the superoperator $\mathcal{L}$ can be defined [80, 84] by

$$\mathcal{L}^{\alpha} = \frac{\sin \pi\alpha}{\pi} \int_0^{\infty} dz\, z^{\alpha-1} R(-z, \mathcal{L})\, \mathcal{L}, \quad 0 < \alpha < 1. \tag{2}$$

The superoperator $\mathcal{L}^{\alpha}$ allows a closure. If condition (1) holds for a closed superoperator $\mathcal{L}$, then

$$\mathcal{L}^{\alpha} \mathcal{L}^{\beta} = \mathcal{L}^{\alpha+\beta}$$

for $\alpha, \beta > 0$, and $\alpha + \beta < 1$.

If the map $\mathcal{L}$ is a closed infinitesimal generator of the semigroup $\{\Phi_t | \ t \geq 0\}$, i.e.

$$\frac{d}{dt}\Phi_t = \mathcal{L}\Phi_t,$$

then the fractional power $\mathcal{L}^{\alpha}$ of $\mathcal{L}$ is given by the Balakrishnan formula [81]:

$$\mathcal{L}^{\alpha} = \frac{1}{\Gamma(-\alpha)} \int_0^{\infty} dt\, t^{-\alpha-1} (\Phi_t - L_I). \tag{3}$$

The resolvent of the superoperator $\mathcal{L}^{\alpha}$ can be found by the Kato's formula

$$\begin{aligned} R(-z, \mathcal{L}^{\alpha}) &= (zL_I + \mathcal{L}^{\alpha})^{-1} \\ &= \frac{\sin \pi\alpha}{\pi} \int_0^{\infty} dx \, \frac{x^{\alpha}}{z^2 + 2zx^{\alpha} \cos \pi\alpha + x^{2\alpha}} R(-x, \mathcal{L}). \end{aligned} \tag{4}$$

We note that the inequality

$$\|R(-z, \mathcal{L}^{\alpha})\| \leq C/z$$

for $z > 0$ is satisfied with the constant $C$ of inequality (1).

If $\mathcal{L}$ is a closed infinitesimal generator for the semigroup $\{\Phi_t | \ t \geq 0\}$, then the superoperators, which are defined by the Bochner–Phillips formula [80, 84, 86]

$$\Phi_t^{(\alpha)} = \int_0^{\infty} ds f_{\alpha}(t, s)\, \Phi_s, \tag{5}$$

form a semigroup such that $\mathcal{L}^{\alpha}$ is an infinitesimal generator of $\Phi_t^{(\alpha)}$, i.e.

$$\frac{d}{dt}\Phi_t^{(\alpha)} = \mathcal{L}^{\alpha}\Phi_t^{(\alpha)}.$$

In Eq. (5) the function $f_\alpha(t,s)$ is defined by

$$f_\alpha(t,s) = \frac{1}{2\pi i}\int_{a-i\infty}^{a+\imath\infty} dz\, \exp(sz - tz^\alpha), \tag{6}$$

where $a, t > 0$, $s \geq 0$, and $0 < \alpha < 1$. The branch of $z^\alpha$ is chosen such that $\mathrm{Re}(z^\alpha) > 0$ for $\mathrm{Re}(z) > 0$. This branch is a one-valued function in the $z$-plane cut along the negative real axis. Using the factor $\exp(-tz^\alpha)$, it can be proved that integral (6) converges.

Let us note the properties of the function $f_\alpha(t,s)$.

(1) The function $f_\alpha(t,s)$ is non-negative:

$$f_\alpha(t,s) \geq 0.$$

for all $s > 0$.

(2) The function $f_\alpha(t,s)$ satisfies the normalization condition

$$\int_0^\infty ds\, f_\alpha(t,s) = 1.$$

(3) If we denote the path of integration in (6) to the union of two paths $r\, \exp(-i\theta)$, and $r\, \exp(+i\theta)$, where $r \in (0,\infty)$, and $\pi/2 \leq \theta \leq \pi$, then

$$f_\alpha(t,s) = \frac{1}{\pi}\int_0^\infty dr\, F(t,s,r), \tag{7}$$

where

$$F(t,s,r) = \exp(sr\cos\theta - tr^\alpha\cos(\alpha\theta))\, \sin(sr\sin\theta - tr^\alpha \sin(\alpha\theta) + \theta).$$

(4) For example, we can consider $\alpha = 1/2$ and $\theta = \pi$. Then equation (7) gives

$$f_{1/2}(t,s) = \frac{1}{\pi}\int_0^\infty dr\, e^{-sr}\sin(t\sqrt{r}) = \frac{t}{2\sqrt{\pi}s^{3/2}}e^{-t^2/4s}.$$

(5) There is the following relation

$$\int_0^\infty ds\, e^{-sx}\, f_\alpha(t,s) = e^{-tx^\alpha},$$

where $t > 0$ and $x > 0$.

We note that these relations for the function $f_\alpha(t,s)$ is important to describe properties of fractional quantum dynamics.

## 4. Markovian Dynamics for Quantum Observables

The Markovian quantum dynamics of open systems can be described in terms of the infinitesimal change of the system. This change is defined by some form of infinitesimal generator. The most general explicit form of the infinitesimal generators was suggested by Gorini, Kossakowski, Sudarshan in [63,64] and by Lindblad in [73]. There exists a one-to-one correspondence between the completely positive norm continuous semigroups and the completely dissipative generating superoperators. The most general form of a completely dissipative superoperator is defined by the following statement, which was proved in [73].

An infinitesimal generator $\mathcal{L}_V$ of a completely positive unity-preserving semigroup $\{\Phi_t = \exp(-t\mathcal{L}_V)|\ t \geq 0\}$ can be represented by the equation

$$\mathcal{L}_V = L_H^- + \frac{i}{2}\sum_{k=1}^{\infty}\Big(L_{V_k^*}L_{V_k}^- - R_{V_k}L_{V_k^*}^-\Big), \tag{8}$$

where the superoperator $L_H^-$ is a left Lie multiplication by $H^* = H$ such that

$$L_H^- = \frac{1}{i\hbar}\Big(L_H - R_H\Big) \tag{9}$$

and $L_V$ and $R_V$ are superoperators of left and right multiplications

$$L_V A = VA, \quad R_V A = AV.$$

As a result, the quantum observable $A_t = \Phi_t(A)$, where $\Phi_t = \exp(-t\mathcal{L}_V)$, satisfies the equation

$$\frac{d}{dt}A_t = -\frac{1}{i\hbar}[H, A_t] + \frac{1}{2\hbar}\sum_{k=1}^{\infty}\Big(V_k^*[A_t, V_k] + [V_k^*, A_t]V_k\Big). \tag{10}$$

The superoperator $-\mathcal{L}_V$ is an infinitesimal generator of the semigroup $\{\Phi_t|t > 0\}$, i.e.

$$\frac{d}{dt}A_t = -\mathcal{L}_V A_t. \tag{11}$$

Equation (10) defines the Markovian dynamics of quantum observables.

We note that the form of $\mathcal{L}_V$ is not uniquely fixed by (8). The transformations

$$V_k \to V_k + a_k I, \quad H \to H + \frac{1}{2i\hbar}\sum_{k=1}^{\infty}(a_k^* V_k - a_k V_k^*),$$

where $a_k$ are arbitrary complex numbers, preserve the form of Eq. (8). If all operators $V_k$ are equal to zero, then $\mathcal{L}_0 = L_H^-$, and Eq. (11) is the Heisenberg equation for Hamiltonian system

$$\frac{d}{dt}A_t = -\frac{1}{i\hbar}[H, A_t].$$

Equation (10) gives an explicit form of equation for quantum observables, if the following restrictions are satisfied. (1) $\mathcal{L}_V$ and $\Lambda_V$ are bounded superoperators, where $\Lambda_V$ is adjoint to $\mathcal{L}_V$; (2) $\mathcal{L}_V$ and $\Lambda_V$ are completely dissipative superoperators. This result has been extended to a class of quantum dynamical semigroup with unbounded generating superoperators [87].

## 5. Fractional Quantum Dynamics for Observables

Let us obtain a generalization of quantum Markovian equation (10) by using a fractional power of quantum Markovian superoperator (8). Using the superoperator

$$-(\mathcal{L}_V)^{\alpha} = \frac{\sin \pi\alpha}{\pi} \int_0^{\infty} dz\, z^{\alpha-1} R(-z, \mathcal{L}_V)\, \mathcal{L}_V, \tag{12}$$

where $0 < \alpha < 1$, we can define the fractional quantum Markovian equation in the form

$$\frac{d}{dt}A_t = -(\mathcal{L}_V)^{\alpha} A_t, \tag{13}$$

where $t$, $H/\hbar$ and $V_k/\sqrt{\hbar}$ are dimensionless variables. Equation (13) describes the fractional Markovian dynamics of quantum observables.

It is easy to see that Eq. (13) with $V_k = 0$ gives the fractional Heisenberg equation [41]

$$\frac{d}{dt}A_t = -(L_H^-)^{\alpha} A_t. \tag{14}$$

The superoperator $(L_H^-)^{\alpha}$ is a fractional power of left Lie superoperator (9).

Let us consider the interaction representation for quantum Markovian equation (10). Using the operators

$$A_U(t) = U(t)A_t U^*(t), \quad W_k(t) = U(t)V_k U^*(t), \tag{15}$$

where

$$U(t) = \exp\{(1/i\hbar)H\}, \quad U^*(t) = \exp\{-(1/i\hbar)H^*\},$$

Eq. (11) can be written as

$$\frac{d}{dt}A_U(t) = -\tilde{\mathcal{L}}_W A_U(t), \tag{16}$$

where

$$\tilde{\mathcal{L}}_W = \frac{i}{2}\sum_{k=1}^{\infty}\Big(L_{W_k^*}L^-_{W_k} - L^-_{W_k^*}R_{W_k}\Big). \tag{17}$$

Superoperator (17) describes the non-Hamiltonian part of the evolution. Using the fractional power of the superoperator $\tilde{\mathcal{L}}_W$ in the form

$$(\mathcal{L}_W)^\alpha = -\frac{\sin\pi\alpha}{\pi}\int_0^\infty dz\, z^{\alpha-1}R(-z,\mathcal{L}_W)\,\mathcal{L}_W, \quad 0<\alpha<1. \tag{18}$$

As a result, the fractional generalization of Eq. (16) is

$$\frac{d}{dt}A_U(t) = -(\tilde{\mathcal{L}}_W)^\alpha A_U(t). \tag{19}$$

Equation (19) is the fractional quantum Markovian equation in the interaction representation. The parameter $\alpha$ can be considered as a measure of the influence of an environment. For $\alpha = 1$, we have quantum Markovian equation (16). In the limit $\alpha \to 0$, we obtain the Heisenberg equation for observable $A_t$ of a Hamiltonian system. Therefore we can consider the physical interpretation of fractional power of $\tilde{\mathcal{L}}_W$ as an influence of environment. We have the following cases that can be used in quantum dynamics:

(1) absence of the environmental influence ($\alpha = 0$);
(2) complete environmental influence ($\alpha = 1$);
(3) power-like environmental influence ($0 < \alpha < 1$).

As a result, the physical interpretation of Eq. (19) can be connected with an existence of a power-like screening of the environmental influence on the quantum system [45].

## 6. Fractional Dynamical Semigroup

Let us consider the Cauchy problem for Eqs. (11) and (13), where the initial condition is given at the time $t = 0$ by $A_0$. The solution of the Cauchy problem for (11) can be written in the form

$$A_t = \Phi_t A_0.$$

The one-parameter superoperators $\Phi_t$, where $t \geq 0$, form a semigroup

$$\Phi_t \Phi_s = \Phi_{t+s}, \quad (t, s > 0), \quad \Phi_0 = L_I.$$

The superoperator $\mathcal{L}_V$ is a generating superoperator of the semigroup $\{\Phi_t |\ t \geq 0\}$. The Cauchy problem for fractional quantum Markovian equation (13) with the initial condition given by $A_0$, has a solution in the form

$$A_t(\alpha) = \Phi_t^{(\alpha)} A_0.$$

The superoperators $\Phi_t^{(\alpha)}$ for $t > 0$ form a semigroup, i.e.

$$\Phi_t^{(\alpha)} \Phi_s^{(\alpha)} = \Phi_{t+s}^{(\alpha)}, \quad (t, s > 0), \quad \Phi_0^{(\alpha)} = L_I,$$

which is call the fractional dynamical semigroup [8]. The superoperator $(\mathcal{L}_V)^\alpha$ is an infinitesimal generator of the semigroup $\{\Phi_t^{(\alpha)} |\ t \geq 0\}$, i.e.

$$\frac{d}{dt} \Phi_t^{(\alpha)} = -(\mathcal{L}_V)^\alpha \Phi_t^{(\alpha)}.$$

The superoperators $\Phi_t^{(\alpha)}$ can be constructed in terms of $\Phi_t$ by the Bochner–Phillips formula (5), where $f_\alpha(t, s)$ is defined in (6). If $A_t$ is a solution of quantum Markovian equation (11), then formula (5) gives the solution

$$A_t(\alpha) = \int_0^\infty ds\, f_\alpha(t, s)\, A_s, \quad t > 0$$

of fractional quantum Markovian equation (13).

Let us consider some properties of fractional dynamical semigroups $\{\Phi_t^{(\alpha)} | t > 0\}$.

(1) We can prove that the fractional dynamical semigroup $\{\Phi_t^{(\alpha)} |\ t > 0\}$ is completely positive [45]. A linear superoperator $\Phi_t^{(\alpha)}$ is completely positive if the conditions

$$\sum_{i,j} A_i \Phi_t^{(\alpha)} (B_i^* B_j) A_j \geq 0 \tag{20}$$

hold for any $A_i, B_i \in \mathcal{M}$.

Using the Bochner–Phillips formula (5), we get

$$\sum_{i,j} A_i \Phi_t^{(\alpha)}(B_i^* B_j) A_j = \int_0^\infty ds\, f_\alpha(t,s) \sum_{i,j} A_i \Phi_s(B_i^* B_j) A_j$$

for $t > 0$. The property $f_\alpha(t,s) \geq 0$, $s > 0$, and the inequality

$$\sum_{i,j} A_i \Phi_s(B_i^* B_j) A_j \geq 0,$$

gives (20). As a result, we have the following statement. If $\{\Phi_t |\ t > 0\}$ is a completely positive semigroup of superoperator $\Phi_t$, then the fractional superoperators $\Phi_t^{(\alpha)}$, which are defined by Eq. (5), form a completely positive semigroup $\{\Phi_t^{(\alpha)} | t > 0\}$.

(2) The Bochner–Phillips formula and the property $f_\alpha(t,s) \geq 0$, $(s > 0)$, allow us to prove that the superoperator $\Phi_t^{(\alpha)}$ is non-negative, if $\Phi_t$, $t > 0$ is a non-negative superoperator. We note that this statement can also be proved by using $B_1 = I$, $A_1 = A$, and $A_i = B_i = 0$ $(i = 2, ...)$ in the proof of the previous statement. As a result, we have the following statement. If $\Phi_t$, $t > 0$, is a non-negative one-parameter map, i.e. $\Phi_t(A) \geq 0$ for $A \geq 0$, then the map $\Phi_t^{(\alpha)}$ is also non-negative, i.e. $\Phi_t^{(\alpha)}(A) \geq 0$ for $A \geq 0$.

(3) A quantum observable is a self-adjoint operator. If $\Phi_t$ is a real superoperator and $A$ is a self-adjoint operator $(A^* = A)$, then the operator $A_t = \Phi_t A$ is self-adjoint, i.e. $(\Phi_t A)^* = \Phi_t A$. If $\mathcal{M}$ is a set of observables, then superoperators on $\mathcal{M}$ into $\mathcal{M}$ must be real. A real superoperator is a superoperator $\Phi_t$ on $\mathcal{M}$, such that $(\Phi_t A)^* = \Phi_t(A^*)$ for all $A \in D(\Phi_t) \subset \mathcal{M}$, where $A^* \in \mathcal{M}^*$ is adjoint to $A \in \mathcal{M}$. Then the evolution of quantum observables, must be described by real superoperators.

(4) Using the Bochner–Phillips formula, we obtain

$$(\Phi_t^{(\alpha)} A)^* = \int_0^\infty ds f_\alpha^*(t,s)\, (\Phi_s A)^*, \quad t > 0.$$

Equation (7) means that $f_\alpha^*(t,s) = f_\alpha(t,s)$ is a real-valued function. As a result, the condition $(\Phi_t A)^* = \Phi_t A^*$ leads to $(\Phi_t^{(\alpha)} A)^* = \Phi_t^{(\alpha)}(A^*)$ for all $A \in D(\Phi_t^{(\alpha)}) \subset \mathcal{M}$. As a result, if $\Phi_t$ is a real superoperator, then the superoperator $\Phi_t^{(\alpha)}$ is real.

## 7. Fractional Equation for Quantum States

Quantum kinematics descres the observables and states of quantum system. Quantum dynamics describes the change of these states and observables with time. Fractional dynamics of quantum states can be considered as adjoint to fractional equation for quantum observables. The superoperator of $\Phi_t$ that describes dynamics of quantum observable is adjoint to superoperator $\hat{\mathcal{E}}_t$, which describes evolution of quantum states. An adjoint superoperator of $\Phi_t$, which is a map on an operator space $\mathcal{M}$, is a superoperator $\hat{\mathcal{E}}_t$ on $\mathcal{M}^*$, such that

$$(\hat{\mathcal{E}}_t(A)|B) = (A|\Phi_t(B)) \tag{21}$$

for all $B \in D(\Phi_t) \subset \mathcal{M}$ and some $A \in \mathcal{M}^*$.

Let $\hat{\mathcal{E}}_t$ be an adjoint superoperator to $\Phi_t$. Then Eq. (21) is satisfied. As a result, we have

$$\begin{aligned}(\hat{\mathcal{E}}_t^{(\alpha)} A|B) &= \int_0^\infty ds\, f_\alpha(t,s)(\hat{\mathcal{E}}_s A|B) \\ &= \int_0^\infty ds\, f_\alpha(t,s)(A|\Phi_s B) = (A|\Phi_t^{(\alpha)} B).\end{aligned}$$

As a result, we prove the following statement by using the Bochner–Phillips formula: If $\hat{\mathcal{E}}_t$ is a superoperator adjoint to $\Phi_t$, then the superoperator

$$\hat{\mathcal{E}}_t^{(\alpha)} = \int_0^\infty ds\, f_\alpha(t,s)\, \hat{\mathcal{E}}_s, \quad t > 0$$

is adjoint to $\Phi_t^{(\alpha)}$.

It is known that $\hat{\mathcal{E}}_t$ is a real superoperator if $\Phi_t$ is real. Using this statement, the Bochner–Phillips formula and Eq. (7), it is easy to prove that $\hat{\mathcal{E}}_t^{(\alpha)}$ is a real superoperator, if $\Phi_t^{(\alpha)}$ real.

We can define the density operator for $t > 0$ by the equation

$$\rho_t = \hat{\mathcal{E}}_t \rho_0,$$

where $\{\hat{\mathcal{E}}_t | t > 0\}$ is a completely positive semigroup. In infinitesimal form, the dynamics of the density operator $\rho_t = \hat{\mathcal{E}}_t \rho_0$ is described by the equation

$$\frac{d}{dt}\rho_t = -\hat{\Lambda}_V \rho_t, \tag{22}$$

where $\hat{\Lambda}_V$ is adjoint to the quantum Markovian superoperator $\mathcal{L}_V$. The map $\hat{\mathcal{E}}_t$ forms a semigroup

$$\hat{\mathcal{E}}_t \hat{\mathcal{E}}_s = \hat{\mathcal{E}}_{t+s}, \quad \hat{\mathcal{E}}_t = L_I,$$

where $t, s > 0$. The superoperator $\hat{\Lambda}_V$ is an infinitesimal generator of the semigroup, i.e.

$$\frac{d}{dt}\hat{\mathcal{E}}_t = -\hat{\Lambda}_V \hat{\mathcal{E}}_t.$$

The superoperator $\hat{\Lambda}_V$ can be represented in the form

$$\hat{\Lambda}_V = -L^-_H + \frac{1}{\hbar}\sum_{k=1}^{\infty}(L_{V_k}R_{V^*_k} - L_{V^*_k}L_{V_k} - R_{V_k}R_{V^*_k}), \tag{23}$$

where $L_V$ and $R_V$ are superoperators of left and right multiplications

$$L_V A = VA, \quad R_V A = AV,$$

and the superoperator $L^-_H$ is a left Lie multiplication (9). Substitution of (23) into (22) gives

$$\frac{d}{dt}\rho_t = \frac{1}{i\hbar}[H, \rho_t] - \frac{1}{\hbar}\sum_{k=1}^{\infty}(V_k\rho_t V^*_k - \rho_t V^*_k V_k - V^*_k V_k \rho_t). \tag{24}$$

Equation (24) describes quantum Markovian dynamics of state of open systems.

The semigroup $\{\hat{\mathcal{E}}^{(\alpha)}_t | t > 0\}$, which is adjoint to fractional dynamical semigroup $\{\Phi_t | t \geq 0\}$, describes the fractional dynamics of the density operator

$$\rho_t(\alpha) = \hat{\mathcal{E}}^{(\alpha)}_t \rho_0.$$

The fractional Markovian equation for the density operator is

$$\frac{d}{dt}\rho_t(\alpha) = -(\hat{\Lambda}_V)^\alpha \rho_t(\alpha), \tag{25}$$

where the fractional power of the superoperator $\hat{\Lambda}_V$ in the form

$$-(\hat{\Lambda}_V)^\alpha = \frac{\sin \pi\alpha}{\pi}\int_0^\infty dz\, z^{\alpha-1}R(-z, \hat{\Lambda}_V)\,\hat{\Lambda}_V \quad 0 < \alpha < 1. \tag{26}$$

Equation (25) defines fractional Markovian dynamics of quantum states.

We note that Eq. (22) with $V_k = 0$ gives the von Neumann equation

$$\frac{d}{dt}\rho_t = \frac{1}{i\hbar}[H, \rho_t].$$

For $V_k = 0$, Eq. (25) has the form

$$\frac{d}{dt}\rho_t = -(-L_H^-)^\alpha \rho_t. \tag{27}$$

This is the fractional von Neumann equation. Equation (27) cannot be represented in the form

$$\frac{d}{dt}\rho_t(\alpha) = L_{H_{\text{new}}}^- \rho_t(\alpha) = -\frac{i}{\hbar}[H_{\text{new}}, \rho_t(\alpha)]$$

with some operator $H_{\text{new}}$. We also note that the superoperators $\Phi_t^{(\alpha)}$ cannot be represented as

$$\hat{\mathcal{E}}_t^{(\alpha)} = (L_U)^\alpha (R_{U^*})^\alpha,$$

and fractional dynamics cannot be considered as Hamiltonian. Therefore quantum systems described by (27) are not Hamiltonian systems. The systems will be called the fractional Hamiltonian systems. A set of usual Hamiltonian quantum systems is a special case of a set of fractional system.

## 8. Fractional Hamiltonian Dynamics

In the quantum mechanics, the observables are given by self-adjoint operators. The dynamical description of a quantum system is given by a superoperator [60], which is a rule that assigns to each operator exactly one operator. Dynamics of quantum observable is described by the Heisenberg equation. For Hamiltonian systems, the infinitesimal superoperator of the Heisenberg equation is defined by some form of derivation. The infinitesimal generator $(i/\hbar)[H, \,.\,]$, which is used in the Heisenberg equation, is a derivation of observables. A derivation is a linear map $D$, which satisfies the Leibnitz rule $D(AB) = (DA)B + A(DB)$ for all operators $A$ and $B$. Fractional derivative can be defined as a fractional power of derivative (see Sec. 5.7 in [13]). We consider a fractional derivative on a set of observables as a fractional power of derivative $L_H^- = (i/\hbar)[H, \,.\,]$. It allows us to generalize a notion of quantum Hamiltonian systems. In this case, operator equation for quantum observables is a fractional generalization of the Heisenberg equation [41]. The suggested fractional Heisenberg equation can be exactly solved for the Hamiltonians of free particle and harmonic oscillator [41, 60].

Quantum dynamics is described by superoperators [60]. A superoperator $\mathcal{L}$ is a rule that assigns to each operator $A$ exactly one operator $\mathcal{L}(A)$. For Hamiltonian $H$, let $L_H^-$ be a superoperator that is defined by the equation

$$L_H^- A = \frac{1}{i\hbar}(HA - AH).$$

Quantum dynamics of observables of Hamiltonian system is described by the operator differential equation

$$\frac{d}{dt} A_t = -L_H^- A_t. \tag{28}$$

Equation (28) is called the Heisenberg equation for Hamiltonian systems. The time evolution of a Hamiltonian system is induced by the Hamiltonian $H$.

In order to obtain a fractional generalization of Eq. (28), we consider a concept of fractional power for $L_H^-$. If $L_H^-$ is a closed linear superoperator with an everywhere dense domain $D(L_H^-)$, having a resolvent $R(z, L_H^-) = (zL_I - L_H^-)^{-1}$ on the negative half-axis, then there exists the superoperator

$$(L_H^-)^\alpha = -\frac{\sin \pi\alpha}{\pi} \int_0^\infty dz\, z^{\alpha-1} R(-z, L_H^-)\, L_H^-. \tag{29}$$

The superoperator $(L_H^-)^\alpha$ is defined on $D(L_H^-)$ for $0 < \alpha < 1$. It is a fractional power of the Lie left superoperator. Using the superoperator (29), we can describe fractional dynamics of quantum systems by the equation

$$\frac{d}{dt} A_t = -(L_H^-)^\alpha A_t, \tag{30}$$

where $t$ and $H/\hbar$ are dimensionless variables. Equation (30) is called the fractional Heisenberg equation.

We note that Eq. (30) cannot be represented in the form

$$\frac{d}{dt} A_t = -L_{H_{\text{new}}}^- A_t = \frac{i}{\hbar}[H_{\text{new}}, A_t]$$

with some operator $H_{\text{new}}$. Therefore quantum systems described by (30) are not Hamiltonian systems. The systems will be called the fractional Hamiltonian systems. A set of the usual Hamiltonian quantum systems is a special case of a set of fractional system.

The Cauchy problem for operator equation (28) in which the initial condition is given at the time $t = 0$ by $A_0$, can be solved. Solution of this Cauchy problem can be represented in the form $A_t = \Phi_t A_0$, where $\Phi_t$, $t \geq 0$ is one-parameter superoperator. The superoperators $\Phi_t$, $t \geq 0$ have the semigroup property $\Phi_t \Phi_s = \Phi_{t+s}$, for $t, s > 0$, and $\Phi_0 = L_I$, where $L_I$ is unit superoperator ($L_I A = A$). The superoperator $L_H^-$ is a generating superoperator, or infinitesimal generator, of the semigroup $\{\Phi_t, t \geq 0\}$. For conservative quantum Hamiltonian systems, the superoperator $\Phi_t$ can be represented in the form

$$\Phi_t = L_U R_{U^\dagger},$$

where we use the superoperators of left and right multiplication [60]:

$$L_U A = UA, \quad R_U A = AU$$

and

$$U = \exp\Big(\frac{i}{\hbar}H\Big), \quad U^\dagger = \exp\Big(-\frac{i}{\hbar}H\Big).$$

If we consider the Cauchy problem for fractional Heisenberg equation (30) in which the initial condition is given by $A_0$, then its solution can be represented in the form

$$A_t(\alpha) = \Phi_t^{(\alpha)} A_0,$$

where the superoperators $\Phi_t^{(\alpha)}$, $t>0$, have the semigroup property. The semigroup $\{\Phi_t^{(\alpha)}, t \geq 0\}$ is called the fractional dynamical semigroup. The superoperator $(L_H^-)^\alpha$ is infinitesimal generator of the semigroup $\{\Phi_t^{(\alpha)}, t \geq 0\}$. We note that the superoperators $\Phi_t^{(\alpha)}$ cannot be represented as

$$\Phi_t^{(\alpha)} = (L_U)^\alpha (R_{U^\dagger})^\alpha,$$

and fractional dynamics cannot be considered as Hamiltonian.

The superoperators $\Phi_t^{(\alpha)}$, $t > 0$, have the properties

$$\Phi_t^{(\alpha)} \Phi_s^{(\alpha)} = \Phi_{t+s}^{(\alpha)}, \quad (t, s > 0), \quad \Phi_0^{(\alpha)} = L_I.$$

As a result, fractional dynamics that is defined by the equations of form (28) is Markovian. Equation (28) describes processes without a memory.

Let us consider fractional quantum dynamics of a linear harmonic oscillator that is defined by the Hamiltonian

$$H = \frac{1}{2m} P^2 + \frac{m\omega^2}{2} Q^2, \tag{31}$$

where $t$ and $P$ are dimensionless variables. For $A = Q$, and $A = P$, Eq. (28) gives

$$D_t^1 Q_t = \frac{1}{m} P_t, \quad D_t^1 P_t = -m\omega^2 Q_t.$$

The solutions have the form

$$Q_t = Q_0 \cos(\omega t) + \frac{1}{m\omega} P_0 \sin(\omega t),$$
$$P_t = P_0 \cos(\omega t) - m\omega Q_0 \sin(\omega t). \tag{32}$$

The fractional dynamics of linear harmonic oscillator is described by the fractional Heisenberg equations

$$D_t^1 Q_t = -(L_H^-)^{\alpha} Q_t, \quad D_t^1 P_t = -(L_H^-)^{\alpha} P_t, \tag{33}$$

where $H$ is defined by (31). Using the Bochner–Phillips formula, we obtain the solutions of Eq. (33) in the form

$$Q_t(\alpha) = \Phi_t^{(\alpha)} Q_0 = \int_0^{\infty} ds f_{\alpha}(t,s) Q_s,$$
$$P_t(\alpha) = \Phi_t^{(\alpha)} P_0 = \int_0^{\infty} ds f_{\alpha}(t,s) P_s. \tag{34}$$

Substitution of (32) into (34) gives

$$Q_t = Q_0 C_{\alpha}(t) + \frac{1}{m\omega} P_0 S_{\alpha}(t), \tag{35}$$

$$P_t = P_0 C_{\alpha}(t) - m\omega Q_0 S_{\alpha}(t), \tag{36}$$

where

$$C_{\alpha}(t) = \int_0^{\infty} ds\, f_{\alpha}(t,s) \cos(\omega s), \tag{37}$$

$$S_{\alpha}(t) = \int_0^{\infty} ds\, f_{\alpha}(t,s) \sin(\omega s). \tag{38}$$

Equations (35) and (36) describe fractional dynamics of quantum harmonic oscillator.

We can consider $\alpha = 1/2$. In this case, Eqs. (37) and (38) give

$$C_{1/2}(t) = \frac{t}{2\sqrt{\pi}} \int_0^{\infty} ds \, \frac{\cos(\omega s)}{s^{3/2}} \, e^{-t^2/4s},$$

$$S_{1/2}(t) = \frac{t}{2\sqrt{\pi}} \int_0^{\infty} ds \, \frac{\sin(\omega s)}{s^{3/2}} \, e^{-t^2/4s}.$$

We note that these functions can be represented through the Macdonald function (see [98], Sec. 2.5.37.1), which is also called the modified Bessel function of the third kind.

Let us define the average value and dispersion by the equations

$$\langle A_t \rangle = \mathrm{Tr}[|\Psi\rangle\langle\Psi|A_t] = \langle\Psi|A_t|\Psi\rangle,$$
$$D_A(t) = \langle A_t^2 \rangle - \langle A_t \rangle^2 = \langle\Psi|A_t^2|\Psi\rangle - \langle\Psi|A_t|\Psi\rangle^2.$$

Using the coordinate representation and the pure state

$$\Psi(x) = \langle x|\Psi\rangle = (b\sqrt{\pi})^{-1/2} \exp\left( -\frac{(x-x_0)^2}{2b} + \frac{i}{\hbar} p_0 x \right), \tag{39}$$

we can obtain the average values and dispersions of the operators. Using (39), we get the average values

$$\langle Q_t \rangle = x_0 C_\alpha(t) + \frac{1}{m\omega} p_0 S_\alpha(t),$$
$$\langle P_t \rangle = p_0 C_\alpha(t) - m\omega x_0 S_\alpha(t),$$

and the dispersions

$$D_P(t) = \frac{\hbar^2}{2b^2} C_\alpha^2(t) + \frac{b^2 m^2 \omega^2}{2} S_\alpha^2(t),$$

$$D_Q(t) = \frac{b^2}{2} C_\alpha^2(t) + \frac{\hbar^2}{2b^2 m^2 \omega^2} S_\alpha^2(t).$$

It is easy to see that fractional harmonic oscillator is a simple dissipative system. The solutions are characterized by a damping effect for average values of observables of the fractional harmonic oscillator. The fractional dumping is described by the modified Bessel function of the third kind.

Note that solutions of the Cauchy problem for fractional Heisenberg equation are represented by the superoperators $\Phi_t^{(\alpha)}$, $t > 0$, which form a semigroup. Therefore the evolution of observables is Markovian. This means that the suggested fractional derivatives, which are fractional powers of derivative, cannot be connected with long-term memory effects.

Derivatives of non-integer orders can be used as an approach to describe an interaction between the quantum system and an environment. This interpretation is caused by following reasons. Using the properties

$$\int_0^\infty f_\alpha(t,s) = 1, \quad f_\alpha(t,s) \geq 0 \text{ (for all } \ s > 0),$$

we can assume that $f_\alpha(t,s)$ is a density of probability distribution. Then the Bochner–Phillips formula can be considered as a smoothing of Hamiltonian evolution $\Phi_t$ with respect to time $s > 0$. This smoothing can be considered as an influence of the environment on the system. As a result, the parameter alpha can be used to describe an interaction between the system and the environment.

Note that the Schrödinger equation with fractional power of momentum, which can be considered as a fractional derivative in coordinate representation, is discussed in [24, 25, 28]. The fractional quantum dynamics of Hamiltonian systems in pure states [24, 25, 28] can be considered as a special case of the suggested approach [60], when the fractional powers of the superoperator $L_{P^2}$ are used.

## 9. Fractional Dynamics of Open Quantum Oscillator

We can consider an oscillator as an example of fractional quantum dynamics of open systems. The basic assumption is that the general form of a bounded completely dissipative superoperator given by the quantum Markovian equation holds for an unbounded completely dissipative superoperator $\mathcal{L}_V$. We assume that the operators $H$, and $V_k$ are functions of the operators $Q$ and $P$ such that the obtained model is exactly solvable [90, 94] (see also [68, 69]). Therefore we consider $V_k = V_k(Q, P)$ as the first-degree polynomials in $Q$ and $P$, and the Hamiltonian $H = H(Q, P)$ as a second degree polynomial in $Q$ and $P$:

$$H = \frac{1}{2m}P^2 + \frac{m\omega^2}{2}Q^2 + \frac{\mu}{2}(PQ + QP), \quad V_k = a_k P + b_k Q, \tag{40}$$

where $a_k$ and $b_k$, $k = 1, 2$, are complex numbers. These assumptions mean that the friction force is proportional to the velocity. It is easy to obtain

$$\mathcal{L}_V Q = \frac{1}{m}P + \mu Q - \lambda Q,$$
$$\mathcal{L}_V P = -m\omega^2 Q - \mu P - \lambda P,$$

where

$$\lambda = \mathrm{Im}\Big(a_1 b_1^* + a_1 b_1^*\Big).$$

Let us consider a matrix representation of the quantum Markovian equation. We define the matrices

$$A = \begin{pmatrix} Q \\ P \end{pmatrix}, \quad M = \begin{pmatrix} \mu - \lambda & m^{-1} \\ -m\omega^2 & -\mu - \lambda \end{pmatrix}. \tag{41}$$

Then the quantum Markovian equation for observables has the form

$$\frac{d}{dt}A_t = MA_t, \tag{42}$$

where $\mathcal{L}_V A_t = MA_t$. The solution of (42) is $A_t = \Phi_t A_0$, where

$$\Phi_t = \sum_{n=0}^{\infty} \frac{t^n}{n!}\mathcal{L}_V^n = \sum_{n=0}^{\infty} \frac{t^n}{n!} M^n. \tag{43}$$

The matrix $M$ can be represented in the form

$$M = N^{-1} F N, \tag{44}$$

where $F$ is a diagonal matrix. Then we have

$$N = \begin{pmatrix} m\omega^2 & \mu+\nu \\ m\omega^2 & \mu-\nu \end{pmatrix}, \tag{45}$$

$$F = \begin{pmatrix} -(\lambda+\nu) & 0 \\ 0 & -(\lambda-\nu) \end{pmatrix}. \tag{46}$$

Here we use the complex parameter $\nu$, such that $\nu^2 = \mu^2 - \omega^2$.

Using (43), the one-parameter superoperators $\Phi_t$ can be represented by

$$\Phi_t = \sum_{n=0}^{\infty} \frac{t^n}{n!} M^n = N^{-1} \left( \sum_{n=0}^{\infty} \frac{t^n}{n!} F^n \right) N = N^{-1} e^{tF} N. \tag{47}$$

Substitution of (45) and (46) into (47) gives

$$\Phi_t = e^{-\lambda t} \begin{pmatrix} \cosh(\nu t) + (\mu/\nu)\sinh(\nu t) & (1/m\nu)\sinh(\nu t) \\ -(m\omega^2/\nu)\sinh(\nu t) & \cosh(\nu t) - (\mu/\nu)\sinh(\nu t) \end{pmatrix},$$

where sinh and cosh are hyperbolic sine and cosine.

$$\sinh(x) = \frac{1}{2}(e^x - e^{-x}), \quad \cosh(x) = \frac{1}{2}(e^x + e^{-x}).$$

As a result, we obtain $A_t = \Phi_t A_0$ in the form

$$Q_t = e^{-\lambda t}\Big(\cosh(\nu t) + \frac{\mu}{\nu}\sinh(\nu t)\Big) Q_0 + \frac{1}{m\nu} e^{-\lambda t} \sinh(\nu t) P_0, \tag{48}$$

$$P_t = -\frac{m\omega^2}{\nu} e^{-\lambda t} \sinh(\nu t) Q_0 + e^{-\lambda t}\Big(\cosh(\nu t) - \frac{\mu}{\nu}\sinh(\nu t)\Big) P_0. \tag{49}$$

The fractional quantum Markovian equations for $Q_t$ and $P_t$ are

$$\frac{d}{dt}Q_t = -(\mathcal{L}_V)^\alpha Q_t, \quad \frac{d}{dt}P_t = -(\mathcal{L}_V)^\alpha P_t, \tag{50}$$

where $t$ and $V_k/\sqrt{\hbar}$ are dimensionless variables. The solutions of Eq. (50) are given by the Bochner–Phillips formula

$$Q_t(\alpha) = \Phi_t^{(\alpha)} Q_0 = \int_0^\infty ds f_\alpha(t,s) Q_s, \tag{51}$$

$$P_t(\alpha) = \Phi_t^{(\alpha)} P_0 = \int_0^\infty ds f_\alpha(t,s) P_s, \tag{52}$$

where $t > 0$, and the operators $Q_s$ and $P_s$ are given in Eqs. (48) and (49). The function $f_\alpha(t,s)$ is defined in (6). Substitution of (48) and (49) into (51) and (52) gives

$$Q_t(\alpha) = \Big(Ch_\alpha(t) + \frac{\mu}{\nu} Sh_\alpha(t)\Big) Q_0 + \frac{1}{m\nu} Ch_\alpha(t) P_0, \tag{53}$$

$$P_t(\alpha) = -\frac{m\omega^2}{\nu} Sh_\alpha(t) Q_0 + \Big(Ch_\alpha(t) - \frac{\mu}{\nu} Sh_\alpha(t)\Big) P_0, \tag{54}$$

where we use the notations

$$Ch_\alpha(t) = \int_0^\infty ds\, f_\alpha(t,s)\, e^{-\lambda s} \cosh(\nu s), \tag{55}$$

$$Sh_\alpha(t) = \int_0^\infty ds\, f_\alpha(t,s)\, e^{-\lambda s} \sinh(\nu s). \tag{56}$$

As a result, Eqs. (53) and (54) describe fractional dynamics of quantum harmonic oscillator with friction.

We note that (55) and (56) with $\alpha = 1/2$ have the form

$$Ch_{1/2}(t) = \frac{t}{2\sqrt{\pi}} \int_0^\infty ds\, \frac{\cosh(\nu s)}{s^{3/2}}\, e^{-t^2/4s - \lambda s},$$

$$Sh_{1/2}(t) = \frac{t}{2\sqrt{\pi}} \int_0^\infty ds\, \frac{\sinh(\nu s)}{s^{3/2}}\, e^{-t^2/4s - \lambda s}.$$

These functions can be represented by the Macdonald function (see [98], Sec. 2.4.17.2) such that

$$Ch_{1/2}(t) = \frac{t}{2\sqrt{\pi}} (V_+(t) + V_-(t)),$$

$$Sh_{1/2}(t) = -\frac{t}{2\sqrt{\pi}} (V_+(t) - V_-(t)).$$

Here we use the notation

$$V_{\pm}(t) = \left(\frac{t^2 \pm 4\nu}{4\lambda}\right)^{1/4} K_{-1/2}\left(2\sqrt{\frac{\lambda(t^2 \pm 4\nu)}{4}}\right),$$

where $K_{\alpha}(z)$ is the Macdonald function [13] and $\mathrm{Re}(t^2) > \mathrm{Re}(\nu)$, $\mathrm{Re}(\lambda) > 0$.

## 10. Quantum Analogs of Derivatives of Integer Orders

Let us give a brief review of Weyl quantization of derivatives of integer orders with respect to phase-space coordinates [60] to fix notation and provide a convenient reference.

It is well known that the derivatives with respect to coordinates $q_k$ and momenta $p_k$ can be represented as Poisson brackets by the equations

$$D^1_{q_k} A(q,p) = -\{p_k, A(q,p)\},$$
$$D^1_{p_k} A(q,p) = \{q_k, A(q,p)\}$$

for continuously differentiable functions $A(q,p) \in C^1(\mathbb{R}^{2n})$. Quantum analogs of these Poisson brackets are self-adjoint commutators. The Weyl quantization $\pi$ [60–62] gives

$$\pi(\{p_k, A(q,p)\}) = \frac{1}{i\hbar}[\pi(p_k), \pi(A)],$$
$$\pi(\{q_k, A(q,p)\}) = \frac{1}{i\hbar}[\pi(q_k), \pi(A)],$$

where $[A,B] = AB - BA$. As a result, we have that

$$-L^-_{P_k} = -\frac{1}{i\hbar}[P_k,\ .\ ], \tag{57}$$

$$L^-_{Q_k} = \frac{1}{i\hbar}[Q_k,\ .\ ], \tag{58}$$

where $Q_k = \pi(q_k)$ and $P_k = \pi(p_k)$, can be considered as quantum analogs of derivatives $D^1_{q_k}$ and $D^1_{p_k}$. Then quantum analogs of derivatives of integer order $n$ can be defined by the products of $L^-_{P_k}$ and $L^-_{Q_k}$. For example, a quantum analog of $D^1_{q_k} D^1_{p_l}$ has the form $-L^-_{P_k} L^-_{Q_l} = (1/\hbar)^2 [P_k, [Q_l,\ .\ ]]$. For the derivative $D^n_{q_k}$, we have $(-1)^n (L^-_{P_k})^n$.

Let us consider the differential operator

$$\mathcal{L} = \mathcal{L}[q, p, D^1_q, D^1_p] \tag{59}$$

on a phase space $\mathbb{R}^{2n}$, which is a function of the phase-space coordinates $q_k$, $p_k$, $k = 1, ..., n$, and the derivatives $D^1_{q_k}$ and $D^1_{p_k}$, $k = 1, ..., n$. Quantization is usually understood as a procedure, where any real-valued function $A(q,p)$, is associated with a relevant quantum observable, i.e. a self-adjoint operator $A(Q,P)$. Correspondence between operators $\hat{A} = A(Q,P)$ and symbols $A(q,p)$ is completely determined by the formulas that express the symbols of operators $Q\hat{A}$, $\hat{A}Q$, $P\hat{A}$, $\hat{A}P$ in terms of the symbols of the operator $\hat{A}$. It is well known [60] that the Weyl quantization $\pi_W$ is defined by

$$\pi_W(q_k A(q,p)) = \frac{1}{2}(Q_k \hat{A} + \hat{A} Q_k), \tag{60}$$

$$\pi_W(p_k A(q,p)) = \frac{1}{2}(P_k \hat{A} + \hat{A} P_k), \tag{61}$$

$$\pi_W(D^1_{p_k} A(q,p)) = -\frac{1}{i\hbar}(P_k \hat{A} - \hat{A} P_k), \tag{62}$$

$$\pi_W(D^1_{p_k} A(q,p)) = \frac{1}{i\hbar}(Q_k \hat{A} - \hat{A} Q_k), \tag{63}$$

for any $\hat{A} = A(Q,P) = \pi_W(A(q,p))$, where $Q_k = \pi_W(q_k)$ and $P_k = \pi_W(p_k)$.

Let us define the left operators $L^-_A$ and $L^+_A$ acting on phase space functions by the formulas

$$\begin{aligned} L^-_A B(q,p) &= \{A(q,p), B(q,p)\}, \\ L^+_A B(q,p) &= A(q,p) B(q,p). \end{aligned} \tag{64}$$

From these definitions, we obtain

$$L^+_{q_k} A(q,p) = q_k A(q,p), \quad L^+_{p_k} A(q,p) = p_k A(q,p) \tag{65}$$

and

$$L^-_{q_k} A(q,p) = D^1_{p_k} A(q,p), \quad L^-_{p_k} A(q,p) = -D^1_{q_k} A(q,p). \tag{66}$$

Then the operator $\mathcal{L}[q,p,D^1_q,D^1_p]$, will be represented by

$$\mathcal{L}[q,p,D^1_q,D^1_p] = \mathcal{L}[L^+_q, L^+_p, -L^-_p, L^-_q].$$

Using the superoperators $L^\pm_Q$ and $L^\pm_P$, which are defined by

$$L^+_Q \hat{A} = \frac{1}{2}(Q\hat{A} + \hat{A}Q), \quad L^+_P \hat{A} = \frac{1}{2}(P\hat{A} + \hat{A}P), \tag{67}$$

$$L^-_Q \hat{A} = \frac{1}{i\hbar}(Q\hat{A} - \hat{A}Q), \quad L^-_P \hat{A} = \frac{1}{i\hbar}(P\hat{A} - \hat{A}P), \tag{68}$$

we rewrite Eqs. (60)–(63) in the form

$$\pi_W(L^+_{q_k} A) = L^+_{Q_k} \hat{A}, \quad \pi_W(L^-_{q_k} A) = L^-_{Q_k} \hat{A},$$
$$\pi_W(L^+_{p_k} A) = L^+_{P_k} \hat{A}, \quad \pi_W(L^-_{p_k} A) = L^-_{P_k} \hat{A}.$$

Since these relations are valid for any $\hat{A} = \pi_W(A)$, we can define [60–62] the quantization of operators $L^{\pm}_{q_k}$ and $L^{\pm}_{p_k}$.

The Weyl quantization $\pi$ of $q_k$ and $p_k$ gives the operators

$$Q_k = \pi(q_k), \quad P_k = \pi(p_k).$$

The Weyl quantization of the operators $L^{\pm}_{q^k}$ and $L^{\pm}_{p^k}$ is defined by the equation

$$\pi_W(L^+_{q_k}) = L^+_{Q_k}, \quad \pi_W(L^-_{q_k}) = L^-_{Q_k}, \tag{69}$$
$$\pi_W(L^+_{p_k}) = L^+_{P_k}, \quad \pi_W(L^-_{p_k}) = L^-_{P_k}. \tag{70}$$

Equations (69) and (70) define the Weyl quantization of the differential operator $\mathcal{L}[q, p, D^1_q, D^1_p]$. The Weyl quantization $\pi_W$ associates the differential operator $\mathcal{L}[q, p, D^1_q, D^1_p]$ on the function space and the superoperator $\mathcal{L}[L^+_Q, L^+_P, -L^-_P, L^-_Q]$, acting on the operator space. This statement was proved in [60].

The Weyl quantization of polynomial operator (59) is defined by the formula

$$\pi_W\left(\mathcal{L}\left[q, p, D^1_q, D^1_p\right]\right) = \mathcal{L}(L^+_Q, L^+_P, -L^-_P, L^-_Q).$$

We note that the commutation relations for the operators $L^{\pm}_{q_k}$, $L^{\pm}_{p_k}$ and $L^{\pm}_{Q_k}$, $L^{\pm}_{P_k}$ coincide. Then the ordering of $L^{\pm}_{P_k}$ and $L^{\pm}_{Q^k}$ in the superoperator $\mathcal{L}[L^+_Q, L^+_P, -L^-_P, L^-_Q]$ is uniquely determined by ordering in $\mathcal{L}(L^+_q, L^+_p, -L^-_p, L^-_q)$.

## 11. Quantum Analogs of Fractional Derivatives

The fractional derivative has different definitions [13, 14], and exploiting any of them depends on the kind of the problems, initial (boundary) conditions, and the specifics of the considered physical processes. The classical definitions are the so-called Riemann–Liouville derivatives [14]. Note that the Caputo and Riesz derivatives can be represented [13, 14] through the Riemann–Liouville derivatives. Therefore quantization of Riemann–Liouville fractional derivatives allow us to derive quantum analogs for Caputo and Riesz derivatives.

The quantum analogs of the derivatives $D^1_{q_k}$ and $D^1_{p_k}$ are commutators. We can find analogs of Riemann–Liouville fractional derivatives for quantum theory. To obtain these analogs of Riemann–Liouville fractional derivatives, which are defined on a finite interval of $\mathbb{R}$, we can use a representation of these derivatives for analytic functions. In this representation the Riemann–Liouville derivative is a series of derivatives of integer order. It allows us to use the correspondence between the integer derivatives and the self-adjoint commutators [40]. To define a quantum analog of the Liouville fractional derivative, which is defined on the real axis $\mathbb{R}$, we can use the representation of Weyl quantization by the Fourier transformation [60]. Quantum analogs of fractional derivatives give us a notion that allows one to consider quantum processes that are described by fractional differential equations at classical level.

In order to realize Weyl quantization of Riemann–Liouville fractional derivative, we use a fractional analog of the Taylor series. The fractional derivative ${}_0D^\alpha_x$ on $[0, b]$ in the Riemann–Liouville form is defined by the equation

$$ {}_0D^\alpha_x A(x) = \frac{1}{\Gamma(m-\alpha)} \frac{d^m}{dx^m} \int_0^x \frac{A(y)dy}{(x-y)^{\alpha-m+1}}, \tag{71} $$

where $m - 1 < \alpha \leq m$ and $\Gamma(z)$ is a Gamma function. It is known that the derivatives (71) can be represented as a series (see Lemma 15.3 in [13]). If $A(x)$ is an analytic function for $x \in (0, b)$, then the fractional derivative (71) can be represented in the form

$$ {}_0D^\alpha_x A(x) = \sum_{n=0}^{\infty} a(n, \alpha) x^{n-\alpha} D^n_x A(x), \tag{72} $$

where

$$ a(n, \alpha) = \frac{\Gamma(\alpha+1)}{\Gamma(n+1)\Gamma(\alpha-n+1)\Gamma(n-\alpha+1)}. $$

In order to define Weyl quantization of Riemann–Liouville fractional derivative, we consider representation (72) in the phase space. If $A(q, p)$ is an analytic function on the phase-space $\mathbb{R}^{2n}$, then we can use (72) for the Riemann–Liouville fractional derivatives with respect to $q_k$ and $p_k$.

If $A(q,p)$ is an analytic function on the phase-space $\mathbb{R}^{2n}$, then the Riemann–Liouville fractional derivatives with respect to $q_k$ and $p_k$ can be represented (72) in the form

$$ {}_0D^{\alpha}_{q_k} A(q,p) = \sum_{n=0}^{\infty} a(n,\alpha)\, q_k^{n-\alpha}\, D^n_{q_k} A(q,p), \tag{73}$$

$$ {}_0D^{\alpha}_{p_k} A(q,p) = \sum_{n=0}^{\infty} a(n,\alpha)\, p_k^{n-\alpha}\, D^n_{p_k} A(q,p), \tag{74}$$

where $k = 1, \ldots, n$. The Weyl quantization is defined by relations (69) and (70). Therefore fractional derivatives (73) and (74) must be represented through the operators $L^{\pm}_{q_k}$ and $L^{\pm}_{p_k}$, which are defined by (65) and (66). Using the operators $L^{\pm}_{q_k}$ and $L^{\pm}_{p_k}$, Eqs. (73) and (74) are rewritten in the form

$$ {}_0D^{\alpha}_{q_k} A(q,p) = \sum_{n=0}^{\infty} a(n,\alpha)\, (L^{+}_{q_k})^{n-\alpha}\, (-L^{-}_{p_k})^{n} A(q,p),$$

$$ {}_0D^{\alpha}_{p_k} A(q,p) = \sum_{n=0}^{\infty} a(n,\alpha)\, (L^{+}_{p_k})^{n-\alpha}\, (L^{-}_{q_k})^{n} A(q,p).$$

These equations hold for all analytic functions $A(q,p)$ on the phase-space $\mathbb{R}^{2n}$, As a result, the fractional derivatives are defined by

$$ {}_0D^{\alpha}_{q_k} = \sum_{n=0}^{\infty} a(n,\alpha)\, (L^{+}_{q_k})^{n-\alpha}\, (-L^{-}_{p_k})^{n}, \tag{75}$$

$$ {}_0D^{\alpha}_{p_k} = \sum_{n=0}^{\infty} a(n,\alpha)\, (L^{+}_{p_k})^{n-\alpha}\, (L^{-}_{q_k})^{n}. \tag{76}$$

The Weyl quantization of the operators $L^{\pm}_{q^k}$ and $L^{\pm}_{p^k}$ is defined by Eq. (69) and (70). As a result, we obtain the following statement.

The Weyl quantization of the Riemann–Liouville fractional derivatives (75) and (76) gives the superoperators

$$ {}_0\mathcal{D}^{\alpha}_{Q_k} = \pi({}_0D^{\alpha}_{q_k}) = \sum_{n=0}^{\infty} a(n,\alpha)\, (L^{+}_{Q_k})^{n-\alpha}\, (-L^{-}_{P_k})^{n}, \tag{77}$$

$$ {}_0\mathcal{D}^{\alpha}_{P_k} = \pi({}_0D^{\alpha}_{p_k}) = \sum_{n=0}^{\infty} a(n,\alpha)\, (L^{+}_{P_k})^{n-\alpha}\, (L^{-}_{Q_k})^{n}. \tag{78}$$

Equations (77) and (78) can be considered as definitions of the fractional derivation superoperators on an operator space.

For example, it is not hard to prove that

$$ {}_0\mathcal{D}^{\alpha}_{Q} Q^n = \frac{\Gamma(n+1)}{\Gamma(n+1-\alpha)} Q^{n-\alpha}, \quad {}_0\mathcal{D}^{\alpha}_{P} P^n = \frac{\Gamma(n+1)}{\Gamma(n+1-\alpha)} P^{n-\alpha}, $$

where $n \geq 1$, and $\alpha \geq 0$.

Equations (77) and (78) defines the Weyl quantization of fractional Riemann–Liouville derivatives. The Caputo and Riesz derivatives can be represented [13, 14] through the Riemann–Liouville derivatives. Therefore quantum analogs of Riemann–Liouville fractional derivatives allow us to derive correspondent analogs for Caputo and Riesz derivatives. Quantization of fractional derivatives give us a notion that allows one to consider quantum processes that are described by fractional differential equations at classical level.

## 12. Fractional Generalization of Quantum Hamiltonian Systems

Quantum analogs of fractional derivatives [8, 40], allow us to consider a generalization of the notion of fractional Hamiltonian system [8, 78].

Let us consider the phase space $\mathbb{R}^{2n}$ with the canonical coordinates $(q_1, \ldots, q_n, p_1, \ldots, p_n)$. In general, dynamical system is described by the equations

$$ \frac{dq_i}{dt} = G^i(q,p), \quad \frac{dp_i}{dt} = F^i(q,p), \quad i = 1, \ldots, n. \tag{79} $$

For a closed system with potential internal forces, we can describe the motion by using a Hamiltonian function, which is the sum of the kinetic and potential energy in the system

$$ \frac{dq_i}{dt} = \frac{\partial H(q,p)}{\partial p_i}, \quad \frac{dp_i}{dt} = -\frac{\partial H(q,p)}{\partial q_i}, \quad i = 1, \ldots, n. \tag{80} $$

In general, we cannot describe the motion by using a unique function. Only if the right-hand sides of Eq. (79) satisfy the conditions

$$ D^1_{p_j} G^i - D^1_{p_i} G^j = 0, \quad D^1_{q_i} G^j + D^1_{p_j} F^i = 0, \quad D^1_{q_j} F^i - D^1_{q_i} F^j = 0, \tag{81} $$

for all $(q,p) \in W \subset \mathbb{R}^{2n}$, then the dynamical system (79) is a locally Hamiltonian system in the region $W$.

Let us consider the equations of motion

$$\frac{dq_i}{dt} = G^i(q,p), \quad \frac{dp_i}{dt} = F^i(q,p). \tag{82}$$

We can define a fractional generalization of the Helmholtz conditions (81). If the right-hand sides of Eq. (82) satisfy the conditions

$${}_{b_j}^{C}D_{p_j}^{\alpha} G^i - {}_{b_i}^{C}D_{p_i}^{\alpha} G^j = 0, \tag{83}$$

$${}_{a_i}^{C}D_{q_i}^{\alpha} G^j + {}_{b_j}^{C}D_{p_j}^{\alpha} F^i = 0, \tag{84}$$

$${}_{a_j}^{C}D_{q_j}^{\alpha} F^i - {}_{a_i}^{C}D_{q_i}^{\alpha} F^j = 0, \tag{85}$$

then dynamical system (82) is fractional locally Hamiltonian systems. This proposition has been proved in [8, 78].

If we have the relations

$$G^i(q,p) = {}_{b_i}^{C}D_{p_i}^{\alpha} H, \quad F^i(q,p) = -{}_{a_i}^{C}D_{q_i}^{\alpha} H,$$

then Eq. (82) can be written as

$$\frac{dq_i}{dt} = {}_{b_i}^{C}D_{p_i}^{\alpha} H, \quad \frac{dp_i}{dt} = -{}_{a_i}^{C}D_{q_i}^{\alpha} H. \tag{86}$$

These equations describe the motion of fractional globally Hamiltonian systems.

For example, we can consider a fractional dynamical system that is defined by the equations

$$\frac{dq}{dt} = {}_{0}^{C}D_{p}^{\alpha} H, \quad \frac{dp}{dt} = -{}_{0}^{C}D_{q}^{\alpha} H, \tag{87}$$

where $q > 0$, $p > 0$, $0 < \alpha \leq 1$, and the Hamiltonian $H(q,p)$ has the form

$$H(q,p) = \frac{p^2}{2m} + \frac{m\omega^2 q^2}{2}. \tag{88}$$

The equations of motion are

$$\frac{dq}{dt} = \frac{1}{m\Gamma(3-\alpha)}\, p^{2-\alpha}, \quad \frac{dp}{dt} = -\frac{m\omega^2}{\Gamma(3-\alpha)}\, q^{2-\alpha}.$$

These equations describe a non-Hamiltonian system that is a fractional Hamiltonian system. For $\alpha = 1$, Eq. (87) with (88) defines the linear harmonic oscillator.

The Heisenberg's equation for operators of coordinates and momenta of quantum Hamiltonian systems have the form

$$\frac{d}{dt}Q_k = D^1_{P_k} H(Q,P), \quad \frac{d}{dt}P_k = -D^1_{Q_k} H(Q,P). \tag{89}$$

Here

$$D^1_{P_k} H(Q,P) = \frac{1}{i\hbar}[Q_k, H(Q,P)], \quad D^1_{Q_k} H(Q,P) = -\frac{1}{i\hbar}[P_k, H(Q,P)].$$

Using the suggested quantum analogs of fractional derivatives we can define the following equation

$$\frac{d}{dt}Q_k = \ {}_0\mathcal{D}^{\alpha}_{P_k} H_{\alpha}(Q,P), \quad \frac{d}{dt}P_k = \ -{}_0\mathcal{D}^{\alpha}_{Q_k} H_{\alpha}(Q,P). \tag{90}$$

For the Hamiltonian

$$H_{\alpha}(Q,P) = \frac{1}{M\Gamma(\beta+1)} P^{\beta} + \frac{K}{\Gamma(\gamma+1)} Q^{\gamma}, \tag{91}$$

we get

$$\frac{d}{dt}Q_k = \frac{1}{M\Gamma(\beta+1-\alpha)} P_k^{\beta-\alpha}, \quad \frac{d}{dt}P_k = -\frac{K}{\Gamma(\gamma+1-\alpha)} Q_k^{\gamma-\alpha}. \tag{92}$$

If we use the Hamiltonian

$$H_{\alpha}(Q,P) = \frac{1}{M\Gamma(\alpha+2)} P_k^{\alpha+1} + \frac{K}{\Gamma(\alpha+\beta+1)} Q_k^{\alpha+\beta}, \tag{93}$$

then we have

$$\frac{d}{dt}Q_k = \frac{1}{M}P_k, \quad \frac{d}{dt}P_k = -\frac{K}{\Gamma(\beta+1)} Q_k^{\beta}. \tag{94}$$

Using quantum analogs of fractional derivatives, we obtain a generalization of the notion of quantum Hamiltonian system. In this case, a wide class of quantum non-Hamiltonian systems [60] can be considered as fractional Hamiltonian systems. Using this approach, it is possible to study a wide class of quantum analogs of deterministic dynamical systems with regular and strange attaractors in phase-space. Note that quantum analog of the Lorenz system is suggested in [60, 62].

## 13. Conclusion

In quantum dynamics of Hamiltonian systems, the infinitesimal superoperator is defined by some form of derivation. A derivation is a linear map $\mathcal{L}$, which satisfies the Leibnitz rule $\mathcal{L}(AB) = (\mathcal{L}A)B + A(\mathcal{L}B)$ for all operators $A$ and $B$. Fractional derivative can be defined as a fractional power of derivative. It is known that the infinitesimal generator $\mathcal{L} = (1/i\hbar)[H, .]$, which is used in Hamiltonian dynamics, is a derivation of observables. In the general case, quantum systems are non-Hamiltonian and $\mathcal{L}$ is not a derivation. For a wide class of quantum systems, the infinitesimal generator $\mathcal{L}$ is completely dissipative [53, 60, 67, 72]. We consider [45] a fractional generalization of the equation of motion for open quantum systems. A fractional power of completely dissipative superoperator can be used in this generalization.

A generalization of the Markovian equation for quantum observables and states are suggested. In these equations, we use superoperators that are fractional powers of completely dissipative superoperators. The suggested superoperators are infinitesimal generators of completely positive semigroups. The Bochner–Phillips formula allows us to obtain a fractional dynamical description in terms of solution of non-fractional dynamics. Properties of this semigroup are considered. A fractional power of the quantum Markovian superoperator can be considered as a parameter to describe a measure of screening of an environment [45]. Quantum computations by quantum operations with mixed states [75] can be controlled by these parameters. We assume that there exist stationary states of open quantum systems [68, 74, 88, 90–93] that depend on the fractional parameter.

We prove that solutions of the fractional dynamics of quantum observables and states are described by the quantum dynamical semigroup. Therefore the evolutions of observables and states are Markovian. This means that the suggested fractional derivatives, which are fractional powers of operators and superoperator [45,60], cannot be connected with long-term memory effects.

Note that the Schrödinger equation with fractional power of momentum, which can be considered as a fractional derivative in coordinate representation, is discussed in [24,25,28]. The fractional quantum dynamics of Hamiltonian systems in pure states [24, 25, 28] can be considered as a special case of the suggested approach [60], when the fractional powers of the superoperator $L_{P^2}$ are used.

The suggested fractional quantum Markovian equation is exactly solved for the harmonic oscillator with linear friction. We assume that other solutions and properties described in [68, 69, 74, 76, 89, 90, 94–97] can be considered for fractional generalizations of the quantum Markovian equation and the Gorini–Kossakowski–Sudarshan equation [63, 64]. Note that it is important to consider fractional powers of generating superoperators for $N$-level open quantum systems. These systems are described by Gorini–Kossakowski–Sudarshan equation [63,64] (see Sec.15.11 in [60]). Fractional generalizations of the Gorini–Kossakowski–Sudarshan equation to describe the Markovian fractional dynamics can be studied. The solutions of the Cauchy type problem for the fractional equations of finite-dimensional open quantum systems can be obtained. For example, interesting results can be obtained for two-level open quantum systems that is described in Sec. 15.12 of [60].

## References


1. Eds. A. Carpinteri and F. Mainardi, *Fractals and Fractional Calculus in Continuum Mechanics* (Springer, 1997).
2. Ed. R. Hilfer, *Applications of Fractional Calculus in Physics* (World Scientific, 2000).
3. B. West, M. Bologna and P. Grigolini, *Physics of Fractal Operators* (Springer, 2003).
4. G. M. Zaslavsky, *Hamiltonian Chaos and Fractional Dynamics* (Oxford Univ. Press, 2005).
5. Eds. J. Sabatier, O. P. Agrawal and J. A. Tenreiro Machado, *Advances in Fractional Calculus. Theoretical Developments and Applications in Physics and Engineering* (Springer, 2007).
6. Eds. A. C. J. Luo and V. S. Afraimovich, *Long-range Interaction, Stochasticity and Fractional Dynamics* (Springer, 2010).
7. F. Mainardi, *Fractional Calculus and Waves in Linear Viscoelasticity: An Introduction to Mathematical Models* (World Scientific, 2010).
8. V. E. Tarasov, *Fractional Dynamics: Applications of Fractional Calculus to Dynamics of Particles, Fields and Media* (Springer, HEP, 2010).
9. G. M. Zaslavsky, *Phys. Rep.* **371**, 461 (2002).
10. E. W. Montroll and M. F. Shlesinger, The wonderful world of random walks, in: *Studies in Statistical Mechanics*, Vol. 11, eds. J. Lebowitz, E. Montroll, (North-Holland, 1984), pp. 1–121.
11. R. Metzler and J. Klafter, *Phys. Rep.* **339**, 1 (2000).
12. R. Metzler and J. Klafter, *J. Phys. A* **37**, R161 (2004).
13. S. G. Samko, A. A. Kilbas and O. I. Marichev, *Fractional Integrals and Derivatives Theory and Applications* (Gordon and Breach, 1993).

14. A. A. Kilbas, H. M. Srivastava and J. J. Trujillo, *Theory and Application of Fractional Differential Equations* (Elsevier, 2006).
15. K. Miller and B. Ross, *An Introduction to the Fractional Calculus and Fractional Differential Equations* (Wiley, 1993).
16. I. Podlubny, *Fractional Differential Equations* (Academic Press, 1999).
17. B. Ross, A brief history and exposition of the fundamental theory of fractional calculus, in Lecture Notes in Mathematics, Vol. 457 (Springer, 1975), pp. 1–36.
18. J. T. Machado, V. Kiryakova and F. Mainardi, *Commun. Nonlinear Sci. Numer. Simul.* **16**, 1140 (2011).
19. X.-F. He, *Phys. Rev. B.* **42**, 11751 (1990).
20. R. L. P. G. do Amaral and E. C. Marino, *J. Phys. A* **25**, 5183 (1992).
21. V. S. Vladimirov, I. V. Volovich and E. I. Zelenov, *P-Adic Analysis and Mathematical Physics* (Nauka, 1994), in Russian and (World Scientific, 1994) Sec. II.
22. D. G. Barci, L. E. Oxman and M. Rocca, *Int. J. Mod. Phys. A* **11**, 2111 (1996).
23. D. G. Barci, C. G. Bollini, L. E. Oxman and M. C. Rocca, *Int. J. Theor. Phys.* **37**, 3015 (1998).
24. N. Laskin, *Phys. Lett. A* **268**, 298 (2000).
25. N. Laskin, *Phys. Rev. E* **62**, 3135 (2000).
26. A. Matos-Abiague, *J. Phys. A* **34**, 11059 (2001).
27. A. Raspini, *Physica Script.* **64**, 20 (2001).
28. N. Laskin, *Phys. Rev. E.* **66**, 056108 (2002).
29. S. C. Lim and S. V. Muniandy, *Phys. Lett. A* **324**, 396 (2004).
30. M. Naber, *J. Math. Phys.* **45**, 3339 (2004).
31. F. Ben Adda and J. Cresson, *Appl. Math. Comput.* **161**, 323 (2005).
32. W. Chen, arXiv:quant-ph/0505050.
33. S. C. Lim, *Physica. A* **363**, 269 (2006).
34. C. H. Eab and S. C. Lim, *Physica. A* **371**, 303 (2006).
35. X. Guo and M. Xu, *J. Math. Phys.* **47**, 082104 (2006).
36. C. H. Eab, S. C. Lim and L. P. Teo, *J. Math. Phys.* **48**, 082301 (2007).
37. R. Herrmann, *J. Phys. G* **34**, 607 (2007).
38. A. Iomin, *Phys. Rev. E* **75**, 037201 (2007).
39. M. Bhatti, *Int. J. Contemp. Math. Sci.* **2**, 943 (2007).
40. V. E. Tarasov, *J. Math. Phys.* **49**, 102112 (2008).
41. V. E. Tarasov, *Phys. Lett. A* **372**, 2984 (2008).
42. Z. Odibat, S. Momani and A. Alawneh, *J. Phys. Conf. Ser.* **96**, 012066 (2008).
43. A. K. Golmankhaneh, arXiv:0809.3382.
44. A. Iomin, *Phys. Rev. E* **80**, 022103 (2009).
45. V. E. Tarasov, *Theor. Math. Phys.* **158**, 179 (2009).
46. A. Tofighi, *Acta Phys. Pol. A* **116**, 114 (2009).
47. R. Eid, S. I. Muslih, D. Baleanu and E. Rabei, *Nonlinear Anal.* **10**, 1299 (2009).
48. M. Jeng, S.-L.-Y. Xu, E. Hawkins and J. M. Schwarz, *J. Math. Phys.* **51**, 062102 (2010).

49. S. I. Muslih, O. P. Agrawal and D. Baleanu, *Int. J. Theor. Phys.* **49**, 1746 (2010).
50. S. I. Muslih, O. P. Agrawal and D. Baleanu, *J. Phys. A* **43**, 055203 (2010).
51. G. Calcagni, *J. High Energy Phys.* **1003**, 1 (2010).
52. P. Rozmej and B. Bandrowski, *Comput. Methods Sci. Tech.* **16**, 191 (2010).
53. E. B. Davies, *Quantum Theory of Open Systems* (Academic Press, 1976).
54. R. S. Ingarden, A. Kossakowski and M. Ohya, *Information Dynamics and Open Systems: Classical and Quantum Approach* (Springer, 1997).
55. G. Lindblad, *Non-equilibrium Entropy and Irreversibility* (Delta Reidel, 1983).
56. R. Alicki and K. Lendi, *Quantum Dynamical Semigroups and Applications* (Springer-Verlag, 1987).
57. U. Weiss, *Quantum Dissipative Systems* (World Scientific, 1993).
58. L. Accardi, Y. G. Lu and I. V. Volovich, *Quantum Theory and Its Stochastic Limit* (Springer-Verlag, 2002).
59. H.-P. Breuer and F. Petruccione, *Theory of Open Quantum Systems* (Oxford Univ. Press, 2002).
60. V. E. Tarasov, *Quantum Mechanics of Non-Hamiltonian and Dissipative Systems* (Elsevier, 2008).
61. V. E. Tarasov, *Moscow Univ. Phys. Bull.* **56**, 5 (2001).
62. V. E. Tarasov, *Phys. Lett. A* **288**, 173 (2001).
63. V. Gorini, A. Kossakowski and E. C. G. Sudarshan, *J. Math. Phys.* **17**, 821 (1976).
64. V. Gorini, A. Frigerio, M. Verri, A. Kossakowski and E. C. G. Sudarshan, *Rep. Math. Phys.* **13**, 149 (1978).
65. K. E. Hellwing and K. Kraus, *Commun. Math. Phys.* **11**, 214 (1969).
66. K. E. Hellwing and K. Kraus, *Commun. Math. Phys.* **16**, 142 (1970).
67. R. S. Ingarden and A. Kossakowski, *Ann. Phys.* **89**, 451 (1975).
68. A. Isar, A. Sandulescu and W. Scheid, *Int. J. Mod. Phys. B* **10**, 2767 (1996).
69. A. Isar, A. Sandulescu, H. Scutaru, E. Stefanescu and W. Scheid, *Int. J. Mod. Phys. E* **3**, 635 (1994).
70. K. Kraus, *Ann. Phys.* **64**, 311 (1971).
71. K. Kraus, *States, Effects and Operations. Fundamental Notions of Quantum Theory* (Springer, 1983).
72. A. Kossakowski, *Rep. Math. Phys.* **3**, 247 (1972).
73. G. Lindblad, *Commum. Math. Phys.* **48**, 119 (1976).
74. V. E. Tarasov, *Phys. Rev. E* **66**, 056116 (2002).
75. V. E. Tarasov, *J. Phys. A* **35**, 5207 (2002).
76. V. E. Tarasov, *Phys. Lett. A* **299**, 173 (2002).
77. V. E. Tarasov, *J. Phys. A* **37**, 3241 (2004).
78. V. E. Tarasov, *J. Phys. A* **38**, 5929 (2005).
79. V. E. Tarasov, *J. Phys. A* **39**, 8409 (2006).
80. K. Yosida, *Functional Analysis* (Springer, 1965).
81. V. Balakrishnan, *Pacific J. Math.* **10**, 419 (1960).
82. H. Komatsu, *Pacific J. Math.* **19**, 285 (1966).
83. C. Martinez and M. Sanz, *The Theory of Fractional Powers of Operators*, North-Holland Mathematics Studies, Vol. 187 (Elsevier, 2000).

84. E. Hille and R. S. Phillips, *Functional Analysis and Semigroups* (Amer. Math. Soc., 1957).
85. S. G. Krein, *Linear Differential Equations in Banach Space*, Translations of Mathematical Monographs, Vol. 29 (Amer. Math. Soc., 1971).
86. S. Bochner, *Proc. Nat. Acad. Sci. USA* **35**, 369 (1949).
87. E. B. Davies, *Rep. Math. Phys.* **11**, 169 (1977).
88. E. P. Davies, *Commun. Math. Phys.* **19**, 83 (1970).
89. E. B. Davies, *Ann. l'Inst. Poincaré, Sec. A* **35**, 149 (1981).
90. G. Lindblad, *Rep. Math. Phys.* **10**, 393 (1976).
91. H. Spohn, *Rep. Math. Phys.* **10**, 189 (1976).
92. H. Spohn, *Lett. Math. Phys.* **2**, 33 (1977).
93. C. Anastopoulous and J. J. Halliwell, *Phys. Rev. D* **51**, 6870 (1995).
94. A. Sandulescu and H. Scutaru, *Ann. Phys.* **173**, 277 (1987).
95. D. A. Lidar, Z. Bihary and K. B. Whaley, *Chem. Phys.* **268**, 35 (2001).
96. H. Nakazato, Y. Hida, K. Yuasa, B. Militello, A. Napoli and A. Messina, *Phys. Rev. A* **74**, 062113 (2006).
97. K. Dietz, *J. Phys. A* **35**, 10573 (2002).
98. A. P. Prudnikov, Yu. A. Brychkov and O. I. Marichev, *Elementary Functions, Integrals and Series*, Vol. 1 (Gordon and Breach, 1986).

# Chapter 20

## Casimir Effect Associated with Fractional Klein–Gordon Field

S. C. Lim
*Faculty of Engineering, Multimedia University,*
*Cyberjaya, 63100, Selangor, Malaysia*

L. P. Teo
*Faculty of Engineering,*
*University of Nottingham Malaysia Campus,*
*Semenyih, 43500, Selangor, Malaysia*

This chapter gives a brief review on the recent work on fractional Klein–Gordon fields, in particular on the Casimir effect associated to fractional Klein–Gordon fields in various geometries and boundary conditions. New results on Casimir piston due to a fractional Klein–Gordon massive field are given.

1. Introduction . . . . . . . . . . . . . . . . . . . . . . . . . . . 483
2. Casimir Effect Associated with Fractional Klein–Gordon Field . . . . . . . . . . . . . . . . . . . . . . 486
3. Topological Symmetry Breaking of Self-interacting Fractional Klein–Gordon Field . . . . . . . . . . . . . . . . 490
4. Casimir Piston Associated with Massless Fractional Klein–Gordon Field . . . . . . . . . . . . . . . . . . . . . . 493
5. Casimir Piston Associated with Massive Fractional Klein–Gordon Field . . . . . . . . . . . . . . . . . . . . . . 497
6. Conclusion . . . . . . . . . . . . . . . . . . . . . . . . . . . . 502

## 1. Introduction

The possibility of the fractal structure of spacetime has gained ground lately due to the work on quantum theory of gravity. One promising theory known as causal dynamical triangulations proposed by Ambjørn, Jurkiewicz and Loll [1–3] is a nonperturbative approach to quantum gravity which shows how the spacetime has fractal structure, with the number of dimensions changes smoothly from the usual four at large scales to two at the Planck scale. Another approach of quantum gravity, the asymptotically safe quantum Einstein gravity [4,5] also has a similar property such that spacetime evolves from a lower dimensional scale-invariant structure at high energy or micro scales to that of normal four-dimensional geometry at large scales or low energy. In these theories, spacetime geometry cannot be understood in terms of a single metric. There is a need to introduce a different effective metric at each energy scale. Modesto [6] has applied a similar technique used in asymptotically safe quantum gravity to loop quantum gravity and obtained a similar fractal-like spacetime structure. Numerical simulations of the propagation of a scalar particle in a dynamically triangulated spacetime, with a discretized Einstein dynamics have been carried out. It is found that the spectral dimension of the microscopic spacetime is two and it becomes four for large scales [4,7–10]. Another proposal of a Lorentz invariant theory in a fractal spacetime requires the spectral dimension and the ultraviolet Hausdorff dimension of the spacetime to be both equal to two [11]. The fractal nature of quantum spacetime may provide a way out of the ultraviolet problem with the reduction of spacetime dimensions. It is interesting to note that in addition to these different approaches to quantum gravity which indicate the emergence of a ground-scale spacetime with fractal properties, there also exists a recent work [12] which shows that spacetime with quantum group symmetry in general has scale-dependent dimensions, and it allows one to establish a link between noncommutative geometries and quantum group and seemingly unrelated approaches of quantum gravity like causal dynamical triangulations and quantum Einstein gravity. All these different theories have a common property, namely, dimensions of spacetime depend on scale, and they could have fractal properties at small scales. At a more fundamental level, Palmer [13] introduced the postulate of an invariant set which assumes the existence of a state space embedding an invariant fractal subset which represents physical reality. It provides a possible link between fractal geometry and quantum theory and it may help to resolve some basic problems in quantum mechanics.

One useful mathematical tool in dealing with fractals and fractal-related phenomena is fractional calculus [14–18]. An important development that stimulates the application of fractional calculus to physical sciences is the realization of the close connection between fractional calculus and fractal geometries [19–22]. Such a relation makes fractional differential equations a natural tool for describing transport processes with fractal properties, such as anomalous diffusion, non-Debye relaxation processes, and other fractal phenomena [16,17,23,24]. Although the application of fractional differential equations to model various transport phenomena in complex heterogeneous media becomes more and more popular during the past two decades, their use in quantum theory is still quite limited. For many years the trajectories of the well-known fractal process, the Brownian motion, have been extensively used in Feynman path integral approach to quantum mechanics [25]. The paths of the Brownian motion are known to be fractal curves with Hausdorff dimension two [26]. Brownian motion also plays an important role in stochastic mechanics, which provides an alternative formulation to quantum mechanics [27]. In the early application of fractal geometry in quantum field theory, the main focus was on the quantum field models in fractal sets, and quantum field theory of spin systems such as Ising spin model [28,29]. Fractal geometry was later applied to Wilson loops in lattice gauge theory, and in the modeling of random surfaces in quantum gravity [30].

The possible existence of fractal structure of quantum spacetime prompted one to employ fractional calculus and fractional differential equations [14–18] in quantum theory. Just like fractional diffusion equation has replaced diffusion equation to describe anomalous transport systems in fractal media, for quantum theories in fractal spacetime it is natural to consider quantum mechanics and quantum field theory which satisfy fractional generalizations of Schrödinger equation, Klein–Gordon equation and Dirac equation. There are some studies on the formulation of quantum mechanics based on various types of fractional Schrödinger equations and their possible applications [31–38]. By noting that fractional diffusion equations are just the Euclidean Schrödinger equations, results obtained for the former can be directly applied to the latter. In particular, fractional Lévy path integral approach of fractional quantum mechanics has been considered [31, 32, 39, 40]. It will be interesting to see whether it is possible to use the path integral representation of fractional Brownian motion [41, 42] and fractional oscillator processes [43] in the formulation of fractional quantum theory. It has been shown that fractional Brownian motion does appear naturally in the fractional gluon propagator with the temporal gauge condition [44]. Incorporation of supersymmetry into fractional quantum mechanics has also been considered [45].

Fractional Klein–Gordon equation and fractional Dirac equation have been studied by several authors during the past decades [46–48]. Fractional power of D′Alembertian operator was used in the nonlocal kinetic terms Lagrangian field theory in the $(2+1)$-dimensional bosonization. Such fractional power operators also appeared in the effective field theory which has some degrees of freedom integrated out from the underlying local theory [49–51]. Canonical quantization of fractional Klein–Gordon massless and massive fields has been studied by some authors [52, 53]. Quantization of fractional Klein–Gordon field and fractional gauge field based on Nelson's stochastic mechanics and Parisi–Wu stochastic quantization procedure at zero and positive temperature have been considered [44, 54]. There also exists work on axiomatic approach to fractional Klein–Gordon field, where properties of the $n$-point Schwinger or Euclidean Green functions and their analytic continuation to the corresponding $n$-point Wightman functions were given [55, 56]. Fractal fermion propagator has been obtained as a consequence of the QED radiative corrections, which requires the incorporation of the propagator with a fractional exponent connected with the fine structure constant [57]. More recent work is related to the Casimir effect for the massless and massive fractional fields at zero and positive temperature; and work on Casimir effect due to fractional Klein–Gordon field subject to fractional Neumann boundary conditions which interpolate between the usual Dirichlet and Neumann conditions have been carried out [58].

In this chapter, we shall mainly concern with the Casimir effect associated with fractional Klein–Gordon fields. In the next section a brief review on the results obtained for the Casimir energy and Casimir force due to a fractional Klein–Gordon field will be given. We shall also discuss the implication of fractional Neumann boundary conditions on the Casimir force. Calculations of Casimir force related to piston geometry for the massive fractional Klein–Gordon field will be given in detail in Sec. 5. In the concluding section some possible directions for further work will be discussed.

## 2. Casimir Effect Associated with Fractional Klein–Gordon Field

Currently, the biggest challenge in cosmology is the explanation of the accelerated expansion of the universe. This is closely related to the problem of dark energy which has been proposed to explain the cause of this accelerating expansion. Various candidates of dark energy have been considered, among them are positive cosmological constant, quintessence, Casimir energy in extra spacetime dimensions, etc. [59, 60]. As a possible candidate for dark energy, Casimir energy [61–64] has attracted considerable renewed interest. On the other hand, Casimir effect has become significant in nanotechnology. At the nano-scales, the Casimir force cannot be neglected, and therefore scientists working in nanotechnology begin to show interest in this subject. Improvement in the ability to measure small forces near surfaces in nano-devices has allowed the possibility to manipulate Casimir force to drive a nano-device. Experimental advances together with improved computations and simulations of Casimir forces make it possible to tailor a nano-device to eliminate unwanted properties caused by the Casimir force such as stiction or adhesion, and hence facilitate the design of better and more effective nano-devices. Thus, from both large and small scales Casimir effect seems to play an important role, as it is expected from vacuum energy which is all prevalent in the universe [65].

In this section we summarize results on some physical quantities related to Casimir effect due to fractional Klein–Gordon fields. Denote by $\Delta = \frac{\partial^2}{\partial t^2} + \sum_j^D \frac{\partial^2}{\partial x_j^2}$ the $(D+1)$-dimensional Euclidean Laplacian operator. The Riesz fractional derivative $(-\Delta)^\alpha$ is defined as [14, 18]:

$$(-\Delta)^\alpha f(t,x) = F^{-1}((k_0^2 + k^2)^{2\alpha} \hat{f}(k_0, k)),$$

where $\hat{f}(k_0, k) = F(f(t,x))$ is the Fourier transform of $f(t,x)$. Let $\phi(t,x)$, $t \in \mathbb{R}$, $x \in \mathbb{R}^D$, be the real scalar Klein–Gordon field. Consider the Lagrangian density

$$L = -\frac{1}{2}\phi(t,x)\Lambda(\partial)\phi(t,x),$$

where $\Lambda(\partial)$ is a pseudo-differential operator. The resulting Lagrangian equation is then given by $\Lambda(\partial)\phi(t,x) = 0$. For the scalar massive fractional Klein–Gordon field, one can have the following possible choices of fractional Klein–Gordon operator:

$$\Lambda(\partial) = (-\Delta + m^2)^\gamma, \tag{I}$$

$$\Lambda(\partial) = (-\Delta)^\alpha + m^{2\alpha}, \tag{II}$$

$$\Lambda(\partial) = [(-\Delta)^\alpha + m^{2\alpha}]^\gamma. \tag{III}$$

Here we remark that (I) seems to be the simplest choice, and (II) is a special case of (III) when $\gamma = 1$. Simple results on Casimir effect associated with fractional Klein–Gordon field of type (I) have been obtained [54]. Note that the zero temperature fractional Klein–Gordon scalar massive field with Euclidean propagator is given by $S(k) = (k^2 + m^2)^{-\gamma}$; and the Euclidean propagator for the positive temperature case takes the following form: $S(k_n) = (k_n^2 + m^2)^{-\gamma}$, with $k_n^2 = \omega_n^2 + k^2$, $\omega_n = 2n\pi/\beta$, and $1/\beta = T$ is the temperature. The free energy corresponds to the type (I) fractional Klein–Gordon field equals to that for ordinary Klein–Gordon field multiply by a factor $\gamma$. When $D = 3$, the high temperature limit of the free energy is [54]:

$$-\gamma\left(\frac{\pi^2}{90}T^4 + \frac{1}{24}m^2T^2 - \frac{1}{12\pi}m^3T + \cdots\right).$$

In general, it is very difficult to compute the free energy for type (III) fractional Klein–Gordon field. The only case it has been computed is when $D = 0$ [66]. After renormalization, the free energy is given by

$$\mathcal{F} = \frac{\gamma}{2\beta}\left\{\log m^{2\alpha} + \alpha\log\beta^2 - 2\sum_{\substack{l\in\mathbb{N}\\ l\neq\frac{1}{2\alpha}}}\frac{(-1)^l}{l}\left(\frac{\beta m}{2\pi}\right)^{2\alpha l}\zeta_R(2\alpha l)\right.$$
$$\left. + \omega_\alpha(-1)^{\frac{1}{2\alpha}}\frac{\beta m}{\pi}\left[\alpha\left(\log\left(\frac{2\pi}{\beta m}\right)^2 + 2\psi(1)\right) - \psi\left(\frac{1}{2\alpha}\right) + \psi(1)\right]\right\}.$$

Here $\zeta_R(s) = \sum_{n=1}^{\infty} n^{-s}$ is the Riemann zeta function and $\omega_\alpha = 1$ if and only if $1/(2\alpha)$ is an integer. Otherwise $\omega_\alpha = 0$. The dependence of the dimensionless free energy $\mathcal{F}/m$ on $\beta m$ and $\alpha$ are shown in Figs. 1 and 2.

Casimir energy due to massless field between two parallel plates subject to fractional Neumann boundary conditions has been computed based on

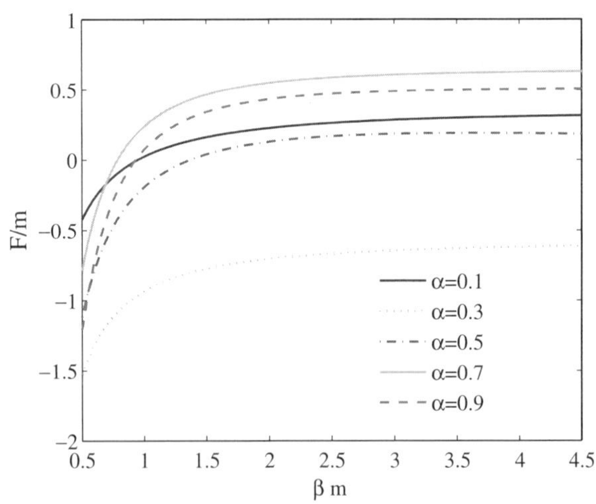


Fig. 1. $\mathcal{F}/m$ as a function of $\beta m$ when $\alpha = 0.1, 0.3, 0.5, 0.7, 0.9$ and $\gamma = 1$.

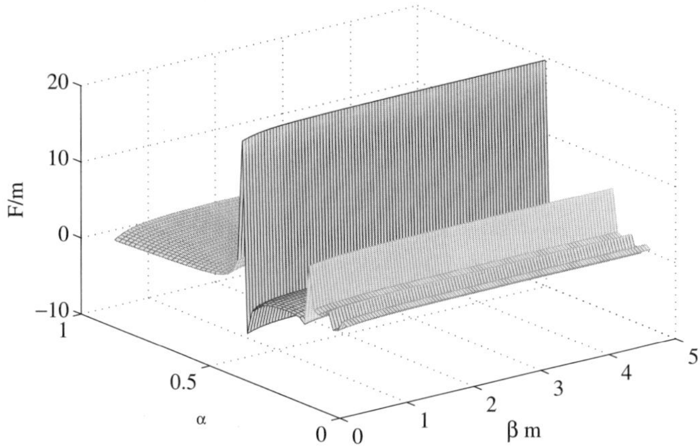


Fig. 2. $\mathcal{F}/m$ as a function of $\beta m$ and $\alpha$. Here $\gamma = 1$.

the zeta function regularization method [85]. Let us recall briefly the basic steps involved in this regularization procedure which can be summarized in three steps. In the case of scalar massless fractional Klein–Gordon field, they are (i) evaluation of eigenvalues $\lambda$ of the fractional D′Alembertian operator $(-\Delta)^\alpha$ with certain boundary conditions, thus leading to spectral zeta function $\zeta_{(-\Delta)^\alpha}(s) = \sum_\lambda \lambda^{-s}$; (ii) analytic continuation of the zeta function to a meromorphic function of the whole complex plane; and (iii) calculation of $\det(-\Delta)^\alpha$ in terms of $\zeta_{(-\Delta)^\alpha}(s)$ using the relation $\det(-\Delta)^\alpha = \exp(-\zeta'_{(-\Delta)^\alpha}(0))$.

Fractional Neumann boundary conditions were introduced to study the Casimir effect due to fractional Klein–Gordon field. On two plates separated by a distance $a$, the fractional Neumann boundary conditions are:

$$\left.\frac{\partial^\mu}{\partial x_D^\mu}\phi(t,\bar{x},x_D)\right|_{x_D=0} = 0,$$

$$\left.\frac{\partial^\nu}{\partial x_D^\nu}\phi(t,\bar{x},x_D)\right|_{x_D=a} = 0,$$

with $\bar{x} \in \mathbb{R}^{D-1}$, $x_D \in \mathbb{R}$ and $\mu, \nu \in [0,1]$, and the fractional derivative is defined in terms of its Fourier transform:

$$\frac{\partial^\mu f(x)}{\partial x^\mu} = \frac{1}{2\pi}\int_{-\infty}^{\infty} dk(ik)^\mu e^{ikx}\hat{f}(-k),$$

where $(\pm ik)^\alpha = |k|^\alpha e^{\pm i\alpha\pi/2}\mathrm{sign}(k)$.

Results on free energy density and pressure due to fractional massless field confined to parallel plates subject to the following fractional Neumann boundary conditions have been obtained [58]. These include (a) general fractional Neumann boundary conditions with $\mu \neq \nu$ and $(\mu, \nu) \neq (0, 1)$ or $(1, 0)$; (b) $\mu = \nu \neq 0, 1$; (c) Dirichlet boundary conditions $\mu = \nu = 0$; (d) Neumann boundary conditions $\mu = \nu = 1$; and (e) Boyer boundary conditions $(\mu, \nu) = (0, 1)$ or $(1, 0)$. Low and high temperature limits of the Casimir energy and pressure have been calculated for the above cases. It is interesting to note that there exists a transition value in the difference in the orders of the fractional Neumann conditions for the two plates, that is $|\mu - \nu|$, for which the Casimir force changes from attractive to repulsive, or vice versa.

## 3. Topological Symmetry Breaking of Self-interacting Fractional Klein–Gordon Field

Quartic self-interacting fractional Klein–Gordon scalar field in a toroidal spacetime and the problem of topological mass generation and symmetry breaking has been investigated [67]. The Lagrangian is

$$L = -\frac{1}{2}\phi(t, x)\Lambda(\partial)\phi(t, x) - \frac{\lambda}{4!}\phi^4(t, x),$$

where $\Lambda(\partial) = (-\Delta + m^2)^\gamma$ is the type (I) fractional Klein–Gordon operator. Based on the Epstein zeta function regularization method, the one-loop effective potential for both the massive and massless $\phi^4$ theory in the toroidal spacetime $T^p \times \mathbb{R}^q$ is derived in terms of power series of $\lambda\tilde{\varphi}^2$, where $\tilde{\varphi}$ is the constant classical background field. Renormalization of these quantities is carried out, and the results for the renormalized mass are obtained explicitly. In the massive case, the renormalized mass $m_{\text{ren}}$ is related to the bare mass $m$ by

$$m_{\text{ren}}^{2\gamma} = m^{2\gamma} + \frac{\lambda\pi^\gamma m^{\frac{d}{2}-\gamma}}{(2\pi)^{\frac{d}{2}+\gamma}\Gamma(\gamma)} \times \sum_{\mathbf{k}\in\mathbb{Z}^p\setminus\{\mathbf{0}\}} \left(\sum_{i=1}^p [L_i k_i]^2\right)^{-\frac{d-2\gamma}{4}} K_{\frac{d-2\gamma}{2}}\left(m\sqrt{\sum_{i=1}^p [L_i k_i]^2}\right).$$

Here $d = D + 1 = p + q$, $L_1, \ldots, L_p$ are the compactification lengths of the torus $T^p$. In the massless $m = 0$ case, the theory is nonrenormalizable when

$\alpha = d/2$. When $\alpha \neq d/2$, the renormalized topologically generated mass $m_{\rm ren}$ is

- If $p = 0$, then $m_{\rm ren}^{2\gamma} = 0$;
- If $p \geq 1$, and
- if $\gamma \neq \frac{q}{2}$, then

$$m_{\rm ren}^{2\gamma} = \frac{\lambda}{\Gamma(\gamma)} \frac{1}{2^{q+1}\pi^{\frac{q}{2}}[\prod_{i=1}^{p} L_i]} \Gamma\left(\gamma - \frac{q}{2}\right) Z_{E,p}\left(\gamma - \frac{q}{2}; \frac{2\pi}{L_1}, \ldots, \frac{2\pi}{L_p}\right)$$

$$= \frac{\lambda}{\Gamma(\gamma)} \frac{1}{2^{2\gamma+1}\pi^{\frac{d}{2}}} \Gamma\left(\frac{d}{2} - \gamma\right) Z_{E,p}\left(\frac{d}{2} - \gamma; L_1, \ldots, L_p\right).$$

- if $\gamma = \frac{q}{2}$, then

$$m_{\rm ren}^{2\gamma} = \frac{\lambda}{\Gamma(\gamma+1)} \frac{1}{2^{q+1}\pi^{\frac{q}{2}}[\prod_{i=1}^{p} L_i]} \times \left\{1 + \gamma\Big[\psi(\gamma) - \psi(1)\Big] + \gamma Z'_{E,p}\left(0; \frac{2\pi}{L_1}, \ldots, \frac{2\pi}{L_p}\right)\right\}.$$

Here $Z_{E,p}(s; a_1, \ldots, a_p)$ is the zeta function

$$Z_{E,p}(s; a_1, \ldots, a_p) = \sum_{\mathbf{k}\in\mathbb{Z}^p} \left(\sum_{i=1}^{p} [a_i k_i]^2\right)^{-s}.$$

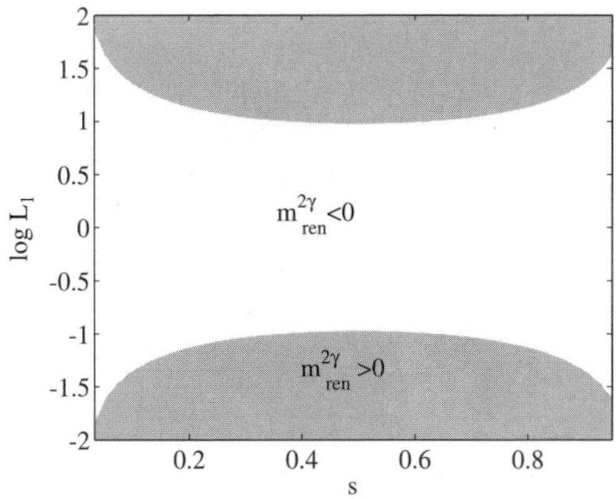


Fig. 3. The regions where $m_{\rm ren}^{2\gamma} > 0$ and $m_{\rm ren}^{2\gamma} < 0$ for $p = 2$ and $L_1 L_2 = 1$. Here $s = \frac{d}{2} - \gamma$.

From these, conditions for symmetry breaking are derived analytically, which show that there is no symmetry breaking in the massive case, however, for the massless case, symmetry breaking occurs provided $0 < d - 2\gamma \leq p$ for a fixed compactified dimension $p \leq 9$. In the case $p \geq 10$, symmetry breaking can only appear for $d - 2\gamma$ in a proper subset of $(0, p]$.

In Figs. 3–7, the regions where the renormalized topologically generated mass $m_{\text{ren}}^{2\gamma}$ are positive and negative respectively are shown.

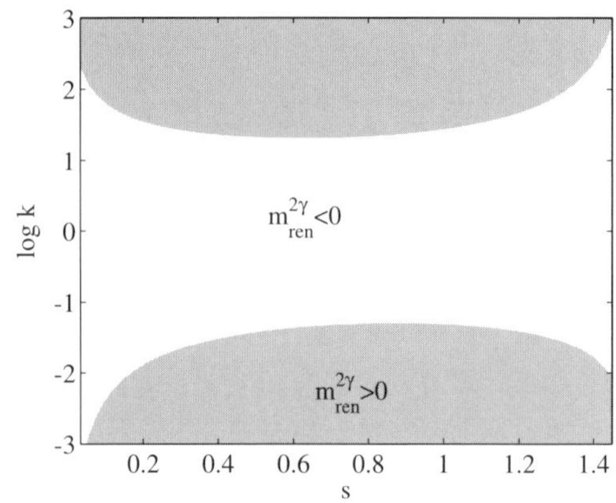


Fig. 4. The regions where $m_{\text{ren}}^{2\gamma} > 0$ and $m_{\text{ren}}^{2\gamma} < 0$ for $p = 3$, $L_1L_2L_3 = 1$, $L_1 : L_2 : L_3 = k : 1 : 1$. Here $s = \frac{d}{2} - \gamma$.

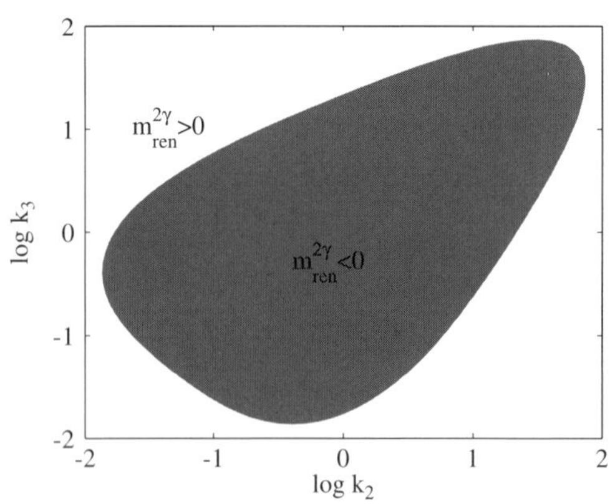


Fig. 5. The region where $m_{\text{ren}}^{2\gamma} > 0$ and $m_{\text{ren}}^{2\gamma} < 0$ for $p = 3$, $L_1L_2L_3 = 1$, $L_1 : L_2 : L_3 = 1 : k_2 : k_3$ and $\frac{d}{2} - \gamma = 0.1$.

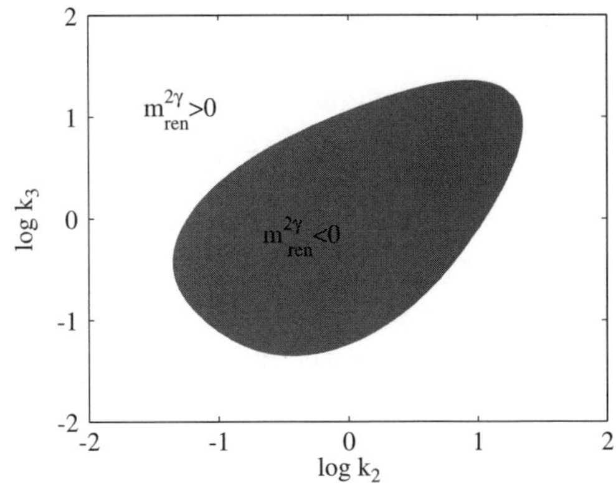


Fig. 6. The regions where $m_{\text{ren}}^{2\gamma} > 0$ and $m_{\text{ren}}^{2\gamma} < 0$ for $p = 3$, $L_1L_2L_3 = 1$, $L_1 : L_2 : L_3 = 1 : k_2 : k_3$ and $\frac{d}{2} - \gamma = 0.3$.

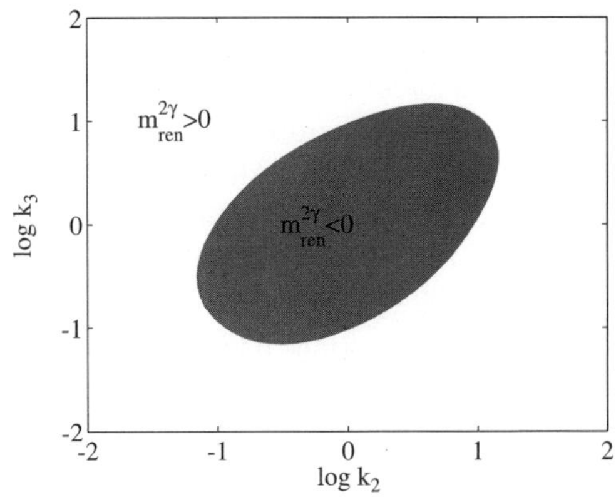


Fig. 7. The regions where $m_{\text{ren}}^{2\gamma} > 0$ and $m_{\text{ren}}^{2\gamma} < 0$ for $p = 3$, $L_1L_2L_3 = 1$, $L_1 : L_2 : L_3 = 1 : k_2 : k_3$ and $\frac{d}{2} - \gamma = 0.6$.

## 4. Casimir Piston Associated with Massless Fractional Klein–Gordon Field

In the calculations of Casimir force based on the conventional methods for a confined geometry such as a rectangular cavity, one usually ignores the nontrivial contribution of vacuum energy outside the cavity. A finite result is obtained by discarding the surface divergent terms which depend on the dimensions and geometry under the pretext of some regularization methods [61–64]. However, there are studies which show that renormalization of physical parameters of the theory fails to remove such surface divergence [68, 69].

In 2004, Cavalcanti [70] introduced a new geometric setup, namely a two-dimensional rectangular piston and showed that for such a geometrical configuration it is possible to obtain an unambiguous finite Casimir force despite the presence of the divergence in the Casimir energy. He demonstrated that in the case of a massless scalar field with Dirichlet boundary conditions, a finite attractive Casimir force without the surface divergence is obtained due to the cancellation of the divergent part of the Casimir force acting in the two regions separated by the piston. This result leads to a surge in the studying of the Casimir force on pistons with various geometric configurations and boundary conditions [71–87]. Casimir force on a rectangular piston in the spacetime with extra compactified dimensions has also been studied [88–93]. It has been shown that the Casimir force between two (non-magnetic) dielectric bodies which are related by reflection is always attractive [94, 95]. The attractive nature of the Casimir force may result in undesirable effects stiction in nano-devices. As a result, it is desirable to search for circumstances under which the Casimir force can become less attractive, or even repulsive. Results obtained by Barton [77] showed that for a thin piston with weakly reflecting dielectrics, the nature of Casimir force is separation-dependent. The force is attractive at small separations, but it becomes repulsive when the separation increases. For Casimir effect on an $n$-dimensional ($n = 1$, 2 or 3) rectangular piston with mixed boundary conditions, namely with one surface having Dirichlet condition, and the other with Neumann condition, the resulting force is repulsive [78]. Other possible ways of obtaining a repulsive Casimir force have also been discussed [79, 83, 85, 86, 92, 96]. In particular, the Casimir force of fractional Klein–Gordon field acting on a piston with fractional boundary conditions can be repulsive [85, 96].

Finite temperature Casimir force acting on a $(D+1)$-dimensional piston in a $(D+1)$-dimensional semi-infinite Dirichlet cylinder due to a type (I) fractional massless Klein–Gordon field at positive temperature has been calculated [85, 96]. For such a system, when the fractional Neumann condition of order $\mu$ is imposed on the piston, the Casimir force acting on the piston is given by

$$F = -2\gamma T \,\mathrm{Re} \sum_j \sum_{l=-\infty}^{\infty} \frac{e^{i\pi\mu}\sqrt{m_j^2 + (2\pi l T)^2}}{\exp(2a\sqrt{m_j^2 + (2\pi l T)^2}) - e^{i\pi\mu}},$$

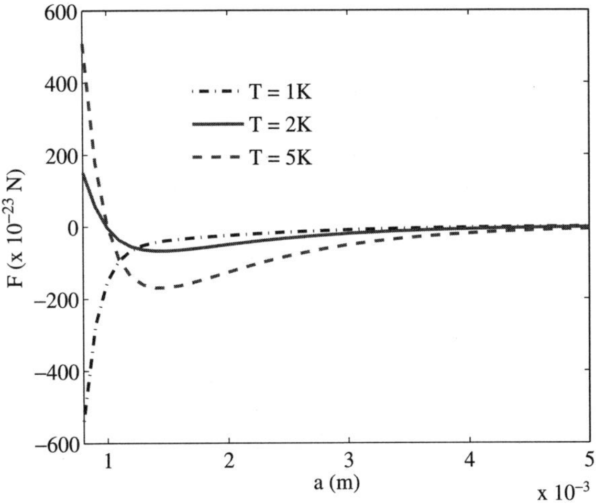


Fig. 8. The Casimir force $F$ as a function of $a$ when $d = 3$, $L_2 = L_3 = 0.01\,\mathrm{m}$, $\mu = 0.47$, and $T = 1\,\mathrm{K}$, $2\,\mathrm{K}$ and $5\,\mathrm{K}$ respectively.

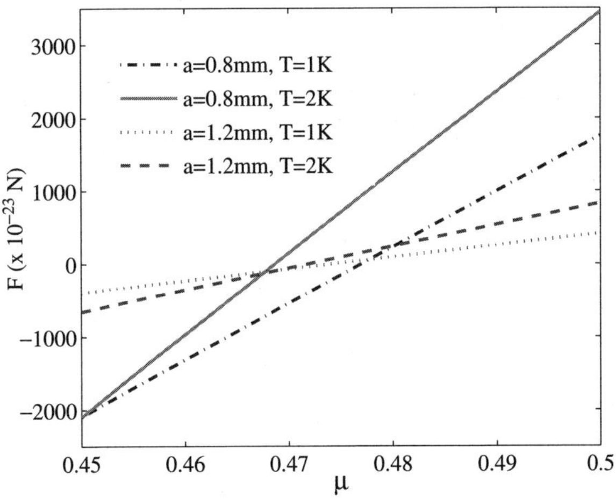


Fig. 9. The Casimir force $F$ as a function of $\mu$ when $d = 3$, $L_2 = L_3 = 0.01\,\mathrm{m}$ for various values of $(a, T)$.

where $a$ is the distance between the piston and the opposite wall; $\{m_j\}$ is the set of eigenvalues of the Laplacian operator with Dirichlet boundary conditions on the cross section of the cylinder.

This result has to be divided by two if $\mu = 0$ (Dirichlet) or $\mu = 1$ (Neumann). From this, we find that for $\mu > 1/2$, the Casimir force acting

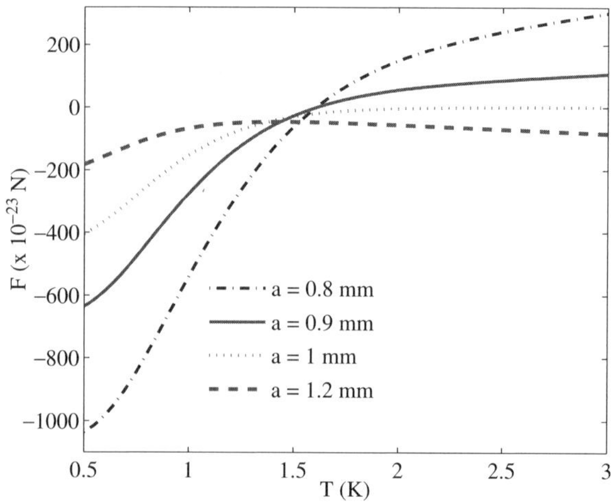


Fig. 10. The Casimir force $F$ as a function of $T$ when $d = 3$, $L_2 = L_3 = 0.01\,\text{m}$, $\mu = 0.47$ and $a = 0.8\,\text{mm}$, $0.9\,\text{mm}$, $1\,\text{mm}$ and $1.2\,\text{mm}$, respectively.

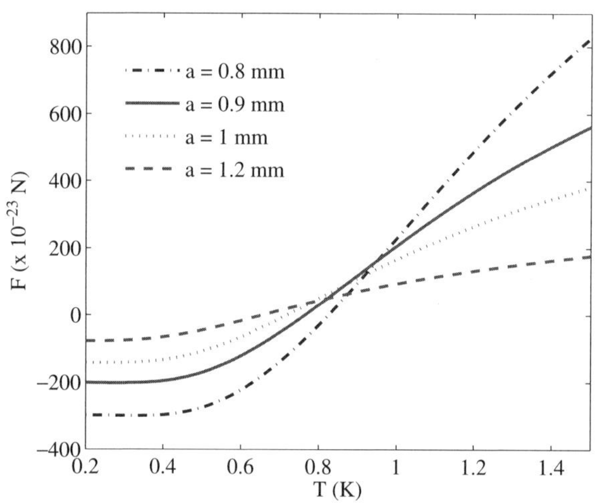


Fig. 11. The Casimir force $F$ as a function of $T$ when $d = 3$, $L_2 = L_3 = 0.01\,\text{m}$, $\mu = 0.48$ and $a = 0.8\,\text{mm}$, $0.9\,\text{mm}$, $1\,\text{mm}$ and $1.2\,\text{mm}$, respectively.

on the piston is always repulsive, and the magnitude of the force decreases with $a$. On the other hand, when $\mu < 1/2$, the Casimir force can change from repulsive to attractive with increasing $a$. For certain values of $\mu$ less than but close to $1/2$, the Casimir force can be attractive or repulsive depending on the aspect ratio of the cavity and the temperature. Increasing temperature can change the Casimir force from attractive to repulsive. In the case $D = 3$ and the cross section of the cylinder is the rectangle box $[0, L_2] \times [0, L_3]$, the dependence of the Casimir force on various quantities are shown in Figs. 8–11.

## 5. Casimir Piston Associated with Massive Fractional Klein–Gordon Field

In this section, we present some new results on the Casimir effect of a massive fractional Klein–Gordon field in the piston scenario. Consider the type (III) fractional Klein–Gordon field $\phi(t, \mathbf{x})$ satisfying the equation:

$$[(-\Delta)^\alpha + m^{2\alpha}]^\gamma \phi(t, \mathbf{x}) = 0.$$

We are interested in the finite temperature Casimir effect of the fractional Klein–Gordon field $\phi(t, \mathbf{x})$ on a piston inside a semi-infinite rectangular cylinder $[0, \infty] \times [0, L_2] \times \cdots \times [0, L_D]$, with Dirichlet boundary conditions on the walls of the cylinder and fractional Neumann condition of order $\mu$ on the piston. The rectangular piston is located at $x_1 = a$. To regularize the Casimir energy, first assume that the system is enclosed in the rectangular box $[0, L_1] \times \cdots \times [0, L_D]$. Take another piston system where the piston is located at $x_1 = L_1/\eta$, for a constant $\eta > 1$. The regularized Casimir energy is defined as the limit (see Fig. 12)

$$E_{\text{Cas}} = \lim_{L_1 \to \infty} (E^{\text{A}}_{\text{Cas}} - E^{\text{B}}_{\text{Cas}}),$$

where $E^{\text{A}}_{\text{Cas}}$ and $E^{\text{B}}_{\text{Cas}}$ are respectively the Casimir energies of system with the piston at $x_1 = a$ and at $x_1 = L_1/\eta$ respectively.

$$E^{\text{A}}_{\text{Cas}} = E^{\text{box}}_{\text{Cas}}(a) + E^{\text{box}}_{\text{Cas}}(L_1 - a), \quad E^{\text{B}}_{\text{Cas}} = E^{\text{box}}_{\text{Cas}}\left(\frac{L_1}{\eta}\right) + E^{\text{box}}_{\text{Cas}}\left(L_1 - \frac{L_1}{\eta}\right).$$

Using zeta regularization, the finite temperature Casimir energy inside the box $[0, a] \times [0, L_2] \times \cdots \times [0, L_D]$ is given by

$$E^{\text{box}}_{\text{Cas}}(a) = -\frac{T}{2}(\zeta_T'(0; a) + \log[\nu^2]\zeta_T(0; a)),$$

where $\nu$ is a normalization constant, and $\zeta_T(s; a)$ is the zeta function

$$\zeta_T(s; a) = \sum_{n=-\infty}^{\infty} \sum_{\mathbf{k}_\perp \in \mathbb{N}^{D-1}} \sum_{l=-\infty}^{\infty} \left\{ \left( \omega^2_{\mathbf{k}_\perp, l} + \left[ \frac{\pi(n - \frac{\mu}{2})}{a} \right]^2 \right)^\alpha + m^{2\alpha} \right\}^{-\gamma s}.$$

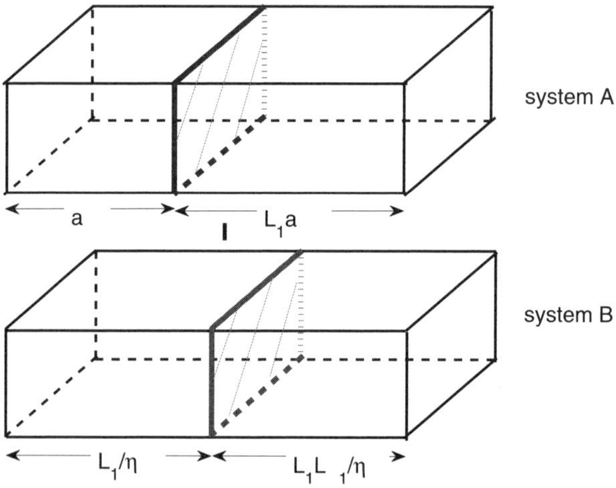


Fig. 12. Regularizing of Casimir energy.

Here $\mathbf{k}_\perp = (k_2, \ldots, k_D)$,

$$\omega_{\mathbf{k}_\perp, l}^2 = \omega_{\mathbf{k}_\perp}^2 + [2\pi l T]^2, \quad \omega_{\mathbf{k}_\perp}^2 = \left(\frac{\pi k_2}{L_2}\right)^2 + \cdots + \left(\frac{\pi k_D}{L_D}\right)^2.$$

When $\mu = 0$ or $\mu = 1$, the summation over $n$ starts from $n = 1$. Using standard techniques,

$$\begin{aligned}\zeta_T(s; a) &= \frac{2}{\Gamma(\gamma s)} \sum_{n=-\infty}^{\infty} \sum_{\mathbf{k}_\perp \in \mathbb{N}^{D-1}} {\sum_{l=0}^{\infty}}' \int_0^\infty t^{\gamma s - 1} \\ &\quad \times \exp\left(-t\left\{\left(\omega_{\mathbf{k}_\perp, l}^2 + \left[\frac{\pi\left(n - \frac{\mu}{2}\right)}{a}\right]^2\right)^\alpha + m^{2\alpha}\right\}\right) dt \\ &= \frac{2}{\Gamma(\gamma s)} \sum_{j=0}^{\infty} \frac{(-1)^j}{j!} m^{2\alpha j} \Gamma(\gamma s + j) Z\left(\alpha(\gamma s + j)\right),\end{aligned}$$

where

$$Z(s) = \sum_{n=-\infty}^{\infty} \sum_{\mathbf{k}_\perp \in \mathbb{N}^{D-1}} {\sum_{l=0}^{\infty}}' \left(\omega_{\mathbf{k}_\perp, l}^2 + \left[\frac{\pi(n - \frac{\mu}{2})}{a}\right]^2\right)^{-s}.$$

Using the fact that

$$\sum_{n=-\infty}^{\infty} \exp\left(-t\left[\frac{\pi(n-\frac{\mu}{2})}{a}\right]^2\right) = \frac{a}{\sqrt{\pi t}} \sum_{n=-\infty}^{\infty} \exp\left(-\frac{n^2a^2}{t}\right) e^{\pi i n\mu},$$

we find that

$$\begin{aligned} Z(s) &= \frac{1}{\Gamma(s)} \int_0^\infty t^{s-1} \sum_{n=-\infty}^{\infty} \sum_{\mathbf{k}_\perp \in \mathbb{N}^{D-1}} \sideset{}{'}\sum_{l=0}^{\infty} \\ &\quad \times \exp\left\{-t\left(\left[\frac{\pi\left(n-\frac{\mu}{2}\right)}{a}\right]^2 + \omega_{\mathbf{k}_\perp,l}^2\right)\right\} dt \\ &= \frac{a}{\sqrt{\pi}\Gamma(s)} \int_0^\infty t^{s-\frac{3}{2}} \sum_{n=-\infty}^{\infty} \sum_{\mathbf{k}_\perp \in \mathbb{N}^{D-1}} \sideset{}{'}\sum_{l=0}^{\infty} e^{\pi i n \mu} \exp\left\{-t\omega_{\mathbf{k}_\perp,l}^2 - \frac{n^2a^2}{t}\right\} dt \\ &= \frac{a}{\sqrt{\pi}\Gamma(s)} \Gamma\left(s-\frac{1}{2}\right) Z_0\left(s-\frac{1}{2}\right) \\ &\quad + \frac{4a}{\sqrt{\pi}\Gamma(s)} \sum_{n=1}^{\infty} \sum_{\mathbf{k}_\perp \in \mathbb{N}^{D-1}} \sideset{}{'}\sum_{l=0}^{\infty} e^{\pi i n \mu} \left(\frac{na}{\omega_{\mathbf{k}_\perp,l}}\right)^{s-\frac{1}{2}} K_{s-\frac{1}{2}}(2na\omega_{\mathbf{k}_\perp,l}), \end{aligned}$$

where

$$Z_0(s) = \sum_{\mathbf{k}_\perp \in \mathbb{N}^{D-1}} \sideset{}{'}\sum_{l=0}^{\infty} \omega_{\mathbf{k}_\perp,l}^{-2s}.$$

Therefore,

$$\begin{aligned} \zeta_T(s;a) &= aY(s) + \frac{8a}{\sqrt{\pi}\Gamma(\gamma s)} \sum_{j=0}^{\infty} \frac{(-1)^j}{j!} m^{2\alpha j} \frac{\Gamma(\gamma s + j)}{\Gamma(\alpha(\gamma s + j))} \sum_{n=1}^{\infty} \sum_{\mathbf{k}_\perp \in \mathbb{N}^{D-1}} \sideset{}{'}\sum_{l=0}^{\infty} \\ &\quad \times e^{\pi i n\mu} \left(\frac{na}{\omega_{\mathbf{k}_\perp,l}}\right)^{\alpha(\gamma s+j)-\frac{1}{2}} K_{\alpha(\gamma s+j)-\frac{1}{2}}(2na\omega_{\mathbf{k}_\perp,l}), \end{aligned}$$

where $Y(s)$ is independent of $a$. One immediately finds that

$$\zeta_T(0;a) = aY_1(0)$$

and

$$\zeta_T'(0;a) = aY_1'(0) + \frac{8\alpha\gamma a}{\sqrt{\pi}}\sum_{j=0}^{\infty}(-1)^j\frac{m^{2\alpha j}}{\Gamma(\alpha j+1)}\sum_{n=1}^{\infty}\sum_{\mathbf{k}_\perp\in\mathbb{N}^{D-1}}{\sum_{l=0}^{\infty}}' e^{\pi i n\mu}$$
$$\times\left(\frac{na}{\omega_{\mathbf{k}_\perp,l}}\right)^{\alpha j-\frac{1}{2}} K_{\alpha j-\frac{1}{2}}(2na\omega_{\mathbf{k}_\perp,l}).$$

From this, we find that the finite temperature Casimir energy in the box $[0,a]\times[0,L_2]\times\cdots\times[0,L_D]$ is given by

$$E_{\text{Cas}}^{\text{box}}(a) = aE_1 - \frac{4\alpha\gamma aT}{\sqrt{\pi}}\sum_{j=0}^{\infty}(-1)^j\frac{m^{2\alpha j}}{\Gamma(\alpha j+1)}\sum_{n=1}^{\infty}\sum_{\mathbf{k}_\perp\in\mathbb{N}^{D-1}}{\sum_{l=0}^{\infty}}' e^{\pi i n\mu}$$
$$\times\left(\frac{na}{\omega_{\mathbf{k}_\perp,l}}\right)^{\alpha j-\frac{1}{2}} K_{\alpha j-\frac{1}{2}}(2na\omega_{\mathbf{k}_\perp,l}),$$

where the term $E_1$ is independent of $a$. This term will vanish when we take the difference of the energies in systems A and B. In the limit $L_1\to\infty$, we find that the finite temperature Casimir energy of the piston system is

$$E_{\text{Cas}} = -\frac{4\alpha\gamma aT}{\sqrt{\pi}}\sum_{j=0}^{\infty}(-1)^j\frac{m^{2\alpha j}}{\Gamma(\alpha j+1)}\sum_{n=1}^{\infty}\sum_{\mathbf{k}_\perp\in\mathbb{N}^{D-1}}{\sum_{l=0}^{\infty}}' e^{\pi i n\mu}$$
$$\times\left(\frac{na}{\omega_{\mathbf{k}_\perp,l}}\right)^{\alpha j-\frac{1}{2}} K_{\alpha j-\frac{1}{2}}(2na\omega_{\mathbf{k}_\perp,l}). \tag{1}$$

For $\mu=0$ or $\mu=1$, this result should be divided by two. Equation (1) is the high temperature expansion of the Casimir energy. It shows that the high temperature leading term is given by the sum of the terms with $l=0$, i.e. when $aT\gg 1$,

$$E_{\text{Cas}} \sim -\frac{2\alpha\gamma aT}{\sqrt{\pi}}\sum_{j=0}^{\infty}(-1)^j\frac{m^{2\alpha j}}{\Gamma(\alpha j+1)}\sum_{n=1}^{\infty}\sum_{\mathbf{k}_\perp\in\mathbb{N}^{D-1}} e^{\pi i n\mu}$$
$$\times\left(\frac{na}{\omega_{\mathbf{k}_\perp}}\right)^{\alpha j-\frac{1}{2}} K_{\alpha j-\frac{1}{2}}(2na\omega_{\mathbf{k}_\perp}).$$

The zero temperature Casimir energy is obtained by taking the limit $T \to 0$ of (1), which gives

$$\begin{aligned} E_{\text{Cas}}^{T=0} &= -\frac{2\alpha\gamma a}{\pi^{\frac{3}{2}}}\sum_{j=0}^{\infty}(-1)^j\frac{m^{2\alpha j}}{\Gamma(\alpha j+1)}\sum_{n=1}^{\infty}\sum_{\mathbf{k}_\perp\in\mathbb{N}^{D-1}} e^{\pi i n\mu} \\ &\quad\times\int_0^\infty\left(\frac{na}{\sqrt{\omega_{\mathbf{k}_\perp}^2+u^2}}\right)^{\alpha j-\frac{1}{2}} K_{\alpha j-\frac{1}{2}}\left(2na\sqrt{\omega_{\mathbf{k}_\perp}^2+u^2}\right)du \\ &= -\frac{\alpha\gamma a}{\pi}\sum_{j=0}^{\infty}(-1)^j\frac{m^{2\alpha j}}{\Gamma(\alpha j+1)}\sum_{n=1}^{\infty}\sum_{\mathbf{k}_\perp\in\mathbb{N}^{D-1}} e^{\pi i n\mu}\left(\frac{na}{\omega_{\mathbf{k}_\perp}}\right)^{\alpha j-1} \\ &\quad\times K_{\alpha j-1}(2na\omega_{\mathbf{k}_\perp}). \end{aligned}$$

The Casimir force acting on the piston is given by

$$\begin{aligned} F_{\text{Cas}}^{\|} &= -\frac{\partial E_{\text{Cas}}^{\|}}{\partial a} \\ &= -\frac{8\alpha\gamma T}{\sqrt{\pi}}\sum_{j=0}^{\infty}(-1)^j\frac{m^{2\alpha j}}{\Gamma(\alpha j+1)}\sum_{n=1}^{\infty}\sum_{\mathbf{k}_\perp\in\mathbb{N}^{D-1}}\sum_{l=0}^{\infty}{}' e^{\pi i n\mu} \\ &\quad\times\frac{(na)^{\alpha j+\frac{1}{2}}}{\omega_{\mathbf{k}_\perp,l}^{\alpha j-\frac{3}{2}}}K_{\alpha j+\frac{1}{2}}(2na\omega_{\mathbf{k}_\perp,l}) \\ &\quad+\frac{8\alpha\gamma T}{\sqrt{\pi}}\sum_{j=1}^{\infty}(-1)^j\frac{m^{2\alpha j}}{\Gamma(\alpha j)}\sum_{n=1}^{\infty}\sum_{\mathbf{k}_\perp\in\mathbb{N}^{D-1}}\sum_{l=0}^{\infty}{}' e^{\pi i n\mu} \\ &\quad\times\left(\frac{na}{\omega_{\mathbf{k}_\perp,l}}\right)^{\alpha j-\frac{1}{2}} K_{\alpha j-\frac{1}{2}}(2na\omega_{\mathbf{k}_\perp,l}). \end{aligned}$$

In the limit $T \to 0$, we find that the zero temperature Casimir force acting on the piston is given by

$$\begin{aligned} F_{\text{Cas}}^{\|,T=0} &= -\frac{2\alpha\gamma}{\pi}\sum_{j=0}^{\infty}(-1)^j\frac{m^{2\alpha j}}{\Gamma(\alpha j+1)}\sum_{n=1}^{\infty}\sum_{\mathbf{k}_\perp\in\mathbb{N}^{D-1}} e^{\pi i n\mu} \\ &\quad\times\frac{(na)^{\alpha j}}{\omega_{\mathbf{k}_\perp}^{\alpha j-2}}K_{\alpha j}(2na\omega_{\mathbf{k}_\perp}) \end{aligned}$$

$$+\frac{\alpha\gamma}{\pi}\sum_{j=0}^{\infty}(-1)^j(2\alpha j-1)\frac{m^{2\alpha j}}{\Gamma(\alpha j+1)}\sum_{n=1}^{\infty}\sum_{\mathbf{k}_\perp\in\mathbb{N}^{D-1}}e^{\pi i n\mu}$$
$$\times\left(\frac{na}{\omega_{\mathbf{k}_\perp}}\right)^{\alpha j-1}K_{\alpha j-1}(2na\omega_{\mathbf{k}_\perp}).$$

For a type (III) massive Klein–Gordon field, we can only expand the Casimir energy and Casimir force as a series in $m^{2\alpha}$. It is not easy to determine the sign of the Casimir force from these series expressions.

## 6. Conclusion

Justifications on the need to study fractional Klein–Gordon fields are given. A brief summary of the results on fractional Klein–Gordon fields mainly by the author and coworkers are given. We also present new results on Casimir piston associated with fractional Klein–Gordon massive field.

Finally, we would like to suggest some other possible directions for further work. So far we have considered fractional Klein–Gordon field with unique fractional exponent. Even though type (III) fractional Klein–Gordon field parametrized by two indices provides more flexibility for modeling, it is still inadequate to represent quantum field in spacetime with dimensions that vary with spacetime scales. In view of the possible variable spacetime dimensions, one may have to deal with fractional quantum fields of variable fractional order, with $\alpha$ replaced by the function $\alpha(x)$. Fractional stochastic processes of variable order such as multifractional Brownian motion [97,98], multifractional Lévy motion [99,100], multifractional Riesz–Bessel process [101] and multifractional Ornstein–Uhlenbeck process [102] have been studied. Variable order one-dimensional Klein–Gordon massive field is given by multifractional Ornstein–Uhlenbeck process, which can be easily extended to higher dimensions to yield multifractional Klein–Gordon field. On the other hand, fractional Klein–Gordon field can also be obtained as a special case of multifractional Riesz–Bessel field. In addition, it may be necessary to impose fractional Neumann boundary conditions of variable fractional order when dealing with multifractional Klein–Gordon field. It should be emphasized that calculations of physical quantities of Casimir effect due to fractional Klein–Gordon field of variable order are expected to be mathematically difficult, and new techniques will be required.

Another direction of generalization is to consider the above results in non-flat space, which requires fractional itegro-differential operators in curved spacetime. One particular case of recent interest will be to study fractional Klein–Gordon field in de Sitter spacetime. Stochastic quantization for ordinary Klein–Gordon field in de Sitter spacetime has been carried out [103]. It will be interesting to extend it to the fractional case. Path integral approach to fractional Klein–Gordon field is another area that has so far not been seriously treated.

## Acknowledgments

S.C.L. would like to thank the Malaysian Ministry of Science, Technology and Innovation for the grant under the Brain Gain Malaysia (Back to Lab) program.

## References

1. J. Ambjørn, J. Jurkiewicz and R. Loll, *Contemp. Phys.* **47**, 103 (2006).
2. J. Ambjørn, A. Grlich, J. Jurkiewicz and R. Loll, *Phys. Rev. Lett.* **100**, 091304 (2008).
3. J. Ambjørn, J. Jurkiewicz and R. Loll, arXiv:1004.0352.
4. O. Lauscher and M. Reuter, *JHEP* **0510**, 050 (2005).
5. M. Niedermaier and M. Reuter, *Liv. Rev. Rel.* **9**, 5 (2006).
6. L. Modesto, *Class. Quantum Grav.* **26**, 242002 (2009).
7. J. Ambjørn, J. Jurkiewicz and R. Loll, *Phys. Rev. Lett.* **95**, 171301 (2005).
8. P. Hofava, *Phys. Rev. Lett.* **102**, 101301 (2009).
9. L. Modesto and P. Nicolini, *Phys. Rev. D* **81**, 104040 (2010).
10. D. Benedetti and J. Henson, *Phys. Rev. D* **80**, 124036 (2009).
11. G. Calcagni, *Phys. Rev. Lett.* **104**, 251301 (2010).
12. D. Benedetti, *Phys. Rev. Lett.* **102**, 111303 (2009).
13. T. Palmer, *Proc. Roy. Soc. Lond. A* **465**, 3187 (2009).
14. S. Samko, A. A. Kilbas and D. I. Maritchev, *Integrals and Derivatives of the Fractional Order and Some of Their Applications* (Gordon and Breach, 1993).
15. I. Podlubny, *Fractional Differential Equations* (Academic Press, 1999).
16. R. Metzler and J. Klafter, *Phys. Rep.* **339**, 1 (2000).
17. B. J. West, M. Bologna and P. Grigolini, *Physics of Fractal Operators* (Springer, 2003).
18. A. A. Kilbas, H. M. Srivastava and J. J. Trujillo, *Theory and Applications of Fractional Differential Equations* (Elsevier, 2006).
19. R. R. Nigmatullin, *Theor. Math. Phys.* **90**, 242 (1992).
20. F. B. Tatom, *Fractals* **3**, 217 (1995).
21. R. S. Rutman, *Theor. Math. Phys.* **105**, 1509 (1995).
22. I. Podlubny, *Fractional Calculus Appl. Anal.* **5**, 367 (2002).
23. G. M. Zaslavsky, *Hamiltonian Chaos and Fractional Dynamics* (Oxford Univ. Press, 2005).

24. R. Klages, G. Randon and I. M. Sokolov, *Anomalous Transport: Foundation and Applications* (Wiley-VCH, 2008).
25. R. P. Feynman and A. R. Hibbs, *Quantum Mechanics and Path Integrals* (McGraw-Hill, 1965).
26. L. F. Abbot and M. B. Wise, *Amer. J. Phys.* **49**, 37 (1981).
27. E. Nelson, *Quantum Fluctuations* (Princeton Univ. Press, 1985).
28. K. Svozil, *J. Phys. A: Math. Gen.* **20**, 3861 (1987).
29. G. Eyink, *Commun. Math. Phys.* **125**, 613; **126**, 85 (1989).
30. H. Kroger, *Phys. Rep.* **323**, 81 (2000).
31. N. Laskin, *Phys. Rev. E* **62**, 3135 (2000).
32. N. Laskin, *Phys. Rev. E* **66**, 056108 (2002).
33. M. Naber, *J. Math. Phys.* **45**, 3339 (2004).
34. X. Guo and M. Xu, *J. Math. Phys.* **47**, 082104 (2006).
35. S. Wong and M. Xu, *J. Math. Phys.* **48**, 043502 (2007).
36. A. Tofighi, *Acta Phys. Pol. A* **116**, 114 (2009).
37. A. Iomin, *Phys. Rev. E* **80**, 022103 (2009).
38. M. Jeng, S.-L.-Y. Xu, E. Hawkins and J. M. Schwarz, *J. Math. Phys.* **51**, 062102 (2010).
39. I. Calvo, R. Sanchez and B. A. Carreras, *J. Phys. A: Math. Theor.* **42**, 055003 (2009).
40. L. Turgeman, S. Carmi and E. Barkai, *Phys. Rev. Lett.* **103**, 190201 (2009).
41. K. L. Sebastian, *J. Phys. A: Math. Gen.* **28**, 4305 (1995).
42. I. Calvo and R. Sanchez, *J. Phys. A: Math. Theor.* **41**, 282002 (2008).
43. C. H. Eab and S. C. Lim, *Physica A* **371**, 303 (2006).
44. S. C. Lim and S. V. Muniandy, *Phys. Lett. A* **324**, 396 (2004).
45. D. Baleanu and S. I. Muslih, *Czech. J. Phys.* **55**, 1063 (2005).
46. M. S. Plyushchay and M. R. de Traubenber, *Phys. Lett. B* **477**, 276 (2000).
47. A. Raspini, *Phys. Script.* **64**, 20 (2001).
48. P. Zavada, *J. Appl. Math.* **2**, 163 (2002).
49. E. C. Marino, *Nucl. Phys. B* **408**, 551 (1993).
50. G. G. Barci, C. D. Fosco and L.E. Oxman, *Phys. Lett. B* **375**, 267 (1996).
51. D. A. R. Dalvit and F. D. Mazzitelli, *Phys. Rev. D* **50**, 1001 (1994).
52. R. L. P. G. do Amaral and E. C. Marino, *J. Phys. A: Math. Gen.* **25**, 5183 (1992).
53. D. G. Barci, L. E. Oxman and M. Rocca, *Int. J. Mod. Phys. A* **11**, 2111 (1996).
54. S. C. Lim, *Physica A* **363**, 269 (2006).
55. S. Albeverio, H. Gottschalk and J.-L. Wu, *Rev. Math. Phys.* **8**, 763 (1996).
56. M. Grothaus and L. Streit, *J. Math. Phys.* **40**, 5387 (1999).
57. S. Gulzari, Y. Srivastava, J. Swain and A. Widom, *Braz. J. Phys.* **37**, 286 (2007).
58. C. H. Eab, S. C. Lim and L. P. Teo, *J. Math. Phys.* **48**, 082301 (2007).
59. J. Frieman, M. Turner and D. Huterer, *Ann. Rev. Astron. Astrophy.* **46**, 385 (2008).
60. P. J. E. Peebles and B. Ratra, *Rev. Mod. Phys.* **75**, 559 (2003).
61. K. A. Milton, *The Casimir Effect* (World-Scientific, 2001).

62. E. Elizalde, S. D. Odintsov, A. Romeo, A. A. Bytsenko and S. Zerbini, *Zeta Regularization Techniques with Applications* (World Scientific, 1994).
63. K. Kirsten, *Spectral Functions in Mathematics and Physics* (CRC Press, 2002).
64. M. Bordag, G. L. Klimchitskaya, U. Mohideen and V. M. Mostepanenko, *Advances in the Casimir Effect* (Oxford Univ. Press, 2009).
65. G. van Dijk and M. Wakayama, *Casimir Force, Casimir Operators and the Riemann Hypothesis: Mathematics for Innovation in Industry and Science* (De Gruyter, 2010).
66. S. C. Lim and L. P. Teo, *J. Phys. A: Math. Theor.* **42**, 065208 (2009).
67. S. C. Lim and L. P. Teo, *J. Phys. A: Math. Theor.* **41**, 145403 (2008).
68. N. Graham *et al.*, *Nucl. Phys. B* **645**, 49 (2002).
69. N. Graham *et al.*, *Nucl. Phys. B* **677**, 379 (2004).
70. R. M. Cavalcanti, *Phys. Rev. D* **69**, 065015 (2004).
71. M. P. Hertzberg, R. L. Jaffe, M. Kardar and A. Scardicchio, *Phys. Rev. Lett.* **95**, 250402 (2005).
72. M. P. Hertzberg, R. L. Jaffe, M. Kardar and A. Scardicchio, *Phys. Rev. D* **76**, 045016 (2007).
73. V. N. Marachevsky, *J. Phys. A: Math. Theor.* **41**, 164007 (2008).
74. A. Edery, *Phys. Rev. D* **75**, 105012 (2007).
75. A. Edery and I. Macdonald, *JHEP* **0709**, 005 (2007).
76. A. Edery and V. N. Marachevsky, *Phys. Rev. D* **78**, 025021 (2008).
77. G. Barton, *Phys. Rev. D* **73**, 065018 (2006).
78. X. H. Zhai and X. Z. Li, *Phys. Rev. D* **76**, 047704 (2007).
79. S. A. Fulling, L. Kaplan and J. H. Wilson, *Phys. Rev. A* **76**, 012118 (2007).
80. S. C. Lim and L. P. Teo, *Ann. Phys.* **324**, 1676 (2009).
81. S. C. Lim and L. P. Teo, *Eur. Phys. J. C* **60**, 323 (2009).
82. X. H. Zhai, Y. Y. Zhang and X. Z. Li, *Mod. Phys. Lett. A* **24**, 393 (2009).
83. S. C. Lim and L. P. Teo, *New J. Phys.* **11**, 013055 (2009).
84. V. K. Oikonomou, *Mod. Phys. Lett. A* **24**, 2405 (2009).
85. S. C. Lim and L. P. Teo, *Phys. Lett. B* **679**, 130 (2009).
86. L. P. Teo, *Phys. Rev. A* **81**, 032502 (2010).
87. K. A. Milton, J. Wagner and K. Kirsten, *Phys. Rev. D* **80**, 125028 (2009).
88. H. B. Cheng, *Phys. Lett. B* **668**, 72 (2008).
89. S. A. Fulling and K. Kirsten, *Phys. Lett. B* **671**, 179 (2009).
90. L. P. Teo, *Phys. Lett. B* **672**, 190 (2009).
91. K. Kirsten and S. A. Fulling, *Phys. Rev. D* **79**, 065019 (2009).
92. E. Elizalde, S. D. Odintsov and A. A. Saharian, *Phys. Rev. D* **79**, 065023 (2009).
93. L. P. Teo, *Nucl. Phys. B* **819**, 431 (2009).
94. O. Kenneth and I. Klich, *Phys. Rev. Lett.* **97**, 160401 (2006).
95. C. P. Bachas, *J. Phys. A* **40**, 9089 (2007).
96. L. P. Teo and K. Kirsten, Finite temperature Casimir effect in the presence of extra dimensions, in *Proc. of the Ninth Conference on Quantum Field Theory Under the Influence of External Conditions* (World Scientific, 2010).

97. R. Peltier and J. Lévy Véhel, Multifractional Brownian motion: Definition and preliminary results, INRIA Report 2645 (1995).
98. A. Benassi, S. Jaffard and D. Roux, *Rev. Mat. Ibro.* **13**, 19 (1997).
99. C. Lacaux, *Ann. Inst. Henri Poincare* **40**, 259 (2004).
100. K. J. Falconer and J. Lévy Véhel, *J. Theor. Probab.* **22**, 375 (2009).
101. S. C. Lim and L. P. Teo, *J. Math. Phys.* **49**, 013509 (2008).
102. S. C. Lim and L. P. Teo, *J. Phys. A: Theor. Gen.* **40**, 6035 (2007).
103. T. C. de Aguiar, G. Menezes and N. F. Svaiter, *Class. Quantum Grav.* **26**, 075003 (2009).
104. J. Ambjørn, R. Loll, W. Westa and S. Zohren, *Phys. Lett. B* **680**, 359 (2009).

# Index

absolute time, 211
absorbing boundaries, 37
advection, 168
ageing, 10, 20, 298
algebra, 213
algebraic decay, 143
allometry, 382
  coefficient, 375, 382
  exponent, 375, 380
  hypothesis, 386, 389
  law, 377
  relation, 375, 387, 389
anomalous diffusion, 3, 34, 37, 41, 54, 55, 62, 63, 65–67, 69, 72, 73, 80, 102, 129, 266
anomalous diffusion constant, 5
anomalous diffusion exponent, 5
anomalous diffusion processes, 332
anomalous transport, 130
anti-Pascal principle, 93
anticipation, 376
arcsine law, 197, 204
Aristotelian instant, 211
Aristotle, 211
asymptotic, 84, 90
attractive, 497
attractive Casimir force, 494
autocorrelation function, 314
average, 387

Balakrishnan formula, 453
bath spectral density, 312
beginning, 212
bi-fractional Smoluchowski–Fokker–Planck equation, 43
bimodality, 16
binding energy, 416
Bochner–Levy diffusion, 223
Bochner–Phillips formula, 453
"Bohr atom" theory, 416
Bohr–Sommerfeld quantization, 418
Boltzmann distribution, 11
Boltzmann's constant, 422
bound
  lower, 92, 93
  upper, 92, 93
bound energy, 420
bound state, 420, 422
boundary condition, 79
  absorbing, 81, 82, 87
  radiation, 82, 99
  Robin, 82
boundary value problem, 16, 81, 83
Bramson-Lebowitz, 92, 94
Bray-Blythe, 94
Brownian motion, 34, 270, 286, 333, 395, 396, 485

C*-algebra, 213
càdlàg process, 355
Casimir effect, 483, 486, 487, 489
Casimir energy, 488, 490, 502
Casimir force, 486, 487, 493–497, 501, 502
Casimir Piston, 493
Cauchy flight, 16
central limit theorem, 7, 237, 270, 395
character, 213–216
  recurrent, 215
characteristic exponent, 136
characteristic function, 134
classification parameter, 218
Cole–Davidson, 222
comb FTSE, 436, 437
comb models, 10

completely dissipative superoperator, 467
completely monotone functions, 233
completely positive map, 452
complex disordered systems, 51
complex network, 389
composite time evolution, 221
compound fractional Poisson process, 353, 354, 369
compound Poisson process, 354, 356, 357, 363
confined systems, 346
confinement, 23
conservative system, 214, 223
Continuous Time Random Walk (CTRW), 4, 7, 39, 52, 78, 80, 94, 98, 129, 131, 132, 186–188, 193, 194, 203, 205, 223, 225, 230, 273, 310, 333, 353, 364, 384, 433
    equation, 435
    simulation, 101, 102, 203
    subdiffusion, 8, 22
convolution, 210, 216, 218
    semigroup, 218
Coulomb potential, 1D, 422
counting process, 353, 359, 364
coupled Langevin equations, 15
Cox–Weiss solution formula, 230
critical frequency, 299
current density, 400

data, 381
Debye, 222
degradation, 79, 98
detailed balance, 188, 190
dielectric function, 221
dielectric relaxation, 221
differential growth equation, 388
diffusion, 225, 332
    Bochner–Levy, 223
    fractional, 225
    Montroll–Weiss, 223
Diffusion and Transport in Washboard Potentials, 316
diffusion comb model, 434
diffusion equation, 130
diffusion on fractal supports, 24
Dirac comb, 422
discretization, 215, 217
dissipative Hamiltonian dynamics, 446
dissipative system, 223
distributed-order derivatives, 109
Donsker-Varadhan, 94
double $\delta$-potential barrier, 422
dynamic susceptibility, 63, 69, 71–73
dynamical rule, 212
dynamical system, 211
    fractional, 209
dynamics, 209, 210, 384
    fractional, 209, 222
dynamics in the field of 1D lattice, 422

effective diffusivity, 138
effective relaxation time, 60, 64, 67, 72
eigenfunction, 399
eigenvalue, 399
empirical allometry, 378
empirical allometry relation, 379
energy of stationary states, 419
energy spectra, 421
ensemble average, 20, 24, 186
ergodic subdiffusion, 21
ergodicity, 5, 22, 302
ergodicity breaking, 10, 185–188, 199, 202–205
escape rate, 58, 60, 63, 65, 66, 73
even states, 400
excess wing, 221
exponential dependence, 144

Feller–Takayasu diamond, 237
Feynman 3D quantum kernel, 412
Feynman path integral, 394
Feynman quantum mechanical kernel, 408

Feynman–Kac equation, 185, 187, 192, 204
Feynman–Kac formula, 173
finite square potential well, 422
finite temperature Casimir energy, 500
first arrival, 19
first order moment, 138
first passage, 17, 18
first passage leapover, 19
first passage problems, 83, 87
  escape problem, 87, 89, 97
first passage time, 37, 197, 205
Fisher–Kolmogorov, 145, 155
Fisher–Kolmogorov equation, 129
Fisher–Kolmogorov–Petrovskii–Piskunov equation, 173
flow, 212
fluctuation dissipation relation, 312
fluid mechanics, 136
Fokker–Planck equation, 52, 53, 55, 56, 58, 63, 65, 66, 73, 225, 383
Fokker–Planck operator, 187, 192, 195
Fourier transforms, 231, 233, 236, 238, 243, 246, 384
Fourier–Fick's prescription, 130
Fourier–Laplace transform, 383
Fox $H$-function, 6, 84, 405, 407, 411, 412, 423, 425
Fox $H$-function representation, 241
fractal, 170, 484
  dimension, 379, 395
  scaling, 389
fractal time, 267
fractional Lévy $\alpha$-stable motion, 332
fractional Bohr atom, 415
fractional Bohr radius, 416
fractional boundary conditions, 494
fractional Brownian motion, 24, 296, 310, 333, 485
fractional calculus, 429
fractional canonical classical mechanical action, 404
fractional Caputo operator, 9
fractional constitutive equations, 4
fractional density matrix, 423
fractional derivative, 96, 220
  Caputo, 111, 134, 230, 241, 243, 245, 310, 431, 435
  Grüwald-Letnikov, 81
  integral representation, 220
  $\lambda$-truncated, 137
  Riemann–Liouville, 81, 193, 245, 473, 474
  Riesz, 81
  Riesz–Feller, 230, 241, 242
  substantial, 187, 204
  truncated, 155
  vector, 275
  Weyl, 431, 432
fractional differential equation, 221, 396, 413, 424, 425
fractional diffusion, 4, 51, 53, 66, 73, 223, 379
  operator, 383
  process, 229, 246, 252–254, 260
  tempered, 155
fractional diffusion equation, 8, 52–54, 62, 67, 80, 107, 110, 112, 124, 131, 229, 230, 246, 382, 384, 385
  tempered, 282
fractional directional derivatives, 278
fractional dissipative systems, 225
fractional drift
  equation, 244, 251
  process, 251
fractional dynamical semigroup, 458
fractional dynamical system, 209
fractional dynamics, 26, 209, 222
  theorem, 217
fractional eigenvalue problem, 413
fractional Euclidean action, 422
fractional Feynman–Kac equation, 185, 187, 188, 192, 193, 197, 200, 203, 204
  a general functional, 193
  backward equation, 194–196, 204
  Lévy flights, 193

fractional Fick's law, 268, 272
fractional Fokker–Planck equation (FFPE), 186, 192, 193, 199, 309, 334, 346, 430, 435
fractional Fokker–Planck–Smoluchowski equation, 10
fractional generalization of the Bloch equation, 424
fractional globally Hamiltonian systems, 476
fractional Hamiltonian, 424
fractional Hamiltonian dynamics, 462
fractional Hamiltonian operator, 397, 398, 418
fractional Hamiltonian system, 475
fractional Heisenberg equation, 456, 463, 465
fractional hydrogen-like atom, 415
fractional integral
    Riemann–Liouville, 245, 474
    Weyl, 440
fractional kicked rotor, 438, 440
fractional Klein–Gordon equation, 485
fractional Klein–Gordon field, 483, 486–490, 494, 502
fractional Klein–Gordon operator, 487
fractional Kramers equation, 169
fractional Lévy stable motion, 333
fractional Langevin equation, 24, 293, 311, 333, 346
fractional Laplacian, 39
fractional locally Hamiltonian systems, 476
fractional Markovian dynamics of quantum states, 461
fractional master equation, 225
fractional Neumann boundary condition, 486, 488, 489, 502
fractional Neumann condition, 490, 497
fractional order moments, 5
Fractional oscillator, 417
fractional particle on a circle, 439
fractional Poisson process, 353, 354, 359, 368
fractional power of quantum Markovian superoperator, 456
fractional power of superoperators, 452
fractional probability current density vector, 401
fractional quantum dynamics, 395, 429
fractional quantum Markovian equation in the interaction representation, 457
fractional quantum mechanical kernel, 405
    1D free particle, 408
fractional quantum mechanics, 393, 395, 396, 426, 485
fractional reaction-dispersion equation, 279
fractional relaxation, 4
fractional Schrödinger equation, 193, 396, 421, 426, 485
    1D, 397
    3D, 397
fractional statistical mechanics, 395, 422, 425, 426
fractional subdiffusion equation, 79
fractional time derivative, 430
fractional time evolution, 218
fractional time Schrödinger equation (FTSE), 430, 432
fractional von Neumann equation, 462
free particle kernel, 405
frequency of radiation, 416
friction kernel, 291
front acceleration, 150
front propagation, 129
front speed, 148
functionals, 185, 186
    anomalous, 187, 204
    Brownian, 187, 204

Galilei invariance, 12
geminate recombination, 97
generalized central limit theorem, 26, 37

generalized diffusion coefficient, 80
generalized Einstein relation, 309, 313
generalized Einstein–Stokes relation, 11
Generalized Langevin Equation, 311
generalized Mittag-Leffler function, 315
generalized Poisson law, 360
generator, 220, 221
 infinitesimal, 220, 221
glass, 220, 223
glass transition, 223
glassy relaxation, 221
GLE dynamics in periodic potentials, 318
glycerol, 223
Green's function, 137, 143, 433, 437, 439, 443
group, 212, 218
 one-parameter, 212

Hamiltonian function, 394
Hamiltonian operator, 394
Havriliak–Negami, 222
Heaviside step function, 97, 99
heavy tail, 271
Heisenberg equation, 456, 477
Helmholtz conditions, 476
Hermitian operator, 398
histogram, 381
Hookean force, 12
host network, 386, 388
Hurst index, 266
hypothesis, 380, 382

identification and validation tools, 335
inertia, 169
infinite divisibility, 238
infinite potential well, 413, 414
infinitesimal generator, 220, 221
integral, 64, 67
integral relaxation, 60, 72
integral relaxation time, 54, 58, 60, 64, 72
integral, effective and the longest relaxation times, 51, 67
intermediate asymptotic, 129, 151
internal time, 15
interspecies allometry relation, 377
intraspecies allometry, 378
invariant measure, 214
 ergodic, 214
invasive species, 279
inverse $\beta$-stable subordinator, 334
inverse Mellin transform, 406, 407
inverse power-law
 Nutting, 6
 probability densities, 376
inverse temperature, 422
inversion operator, 400
inversion transformation, 400
irreversibility, 218
irreversibility problem
 normal, 219
 reversed, 219
Itô stochastic differential equation, 34, 37
iterated maps, 9

jump length, 7

Kac's theorem, 215
Kato's formula, 453
kernel, 403, 404
Klein–Kramers equation, 41, 47
Kramers problem, 41, 42
KWW, 222

Lévy flights, 36, 39–42, 48, 131
 exponentially tempered, 155
Lévy motion, 37, 48, 271
Lévy noise, 48
Lévy process, 165, 353, 354, 356
 $\alpha$-stable, 135
Lévy density, 136
Lévy index, 395, 396
Lévy motion, 395, 396
Lévy path integral, 395, 426
Lévy paths, 395

Lévy stable law, 35, 36
Lévy stochastic processes, 129
Lévy–Khintchine representation, 135
laboratory time, 15
lag time, 20
Lamperti PDF, 196, 204
Langevin equation, 36, 47, 180, 286
Laplace integral, 246
Laplace transform, 231, 233, 235, 384
Laplacian, 397
leading edge approximation, 146
linear potential field, 421
local limit theorem, 217
long-range dependence, 333
long-tailed waiting time distribution, 78, 99
longest relaxation time, 51, 57, 59, 64, 67, 72, 73
Lévy flights, 5, 8, 15, 18

M-Wright function, 235, 236, 244, 250, 252
Markov kernel, 365, 366
Markov process, 34
Markovian, 5
Markovian embedding, 321
Markovian Lévy flights, 45
Markovian/non-Markovian process, 230, 252, 253, 310
Martin–Siggia–Rose action, 176
massive fractional Klein–Gordon field, 497
massless fractional Klein–Gordon field, 493
master equation, 161, 163
maximal displacement, 198, 204
mean first passage time, 38, 39, 41, 42, 46
mean-squared displacement, 290, 295, 331
measure, 214, 215, 217
  invariant, 214, 215
  preserving system, 214
  preserving transformation, 214, 215, 217
    induced, 215
    Kakutani's, 215

Mellin transform, 406
Mellin–Barnes integral representation, 244
memory effects, 89, 292
metabolic rate, 377
method of images, 17
Mittag-Leffler binomial, 221
Mittag-Leffler distribution, 236, 252
Mittag-Leffler function, 5, 44, 46, 48, 54, 62, 63, 65, 86, 88, 133, 202, 221, 231–235, 244, 250, 294, 314, 433
Mittag-Leffler process, 244, 249, 252
mixed state, 220
momentum operator, 402
momentum representation, 410, 412
Montroll–Weiss diffusion, 223
Montroll–Weiss (MW) equation, 132
morphogen gradient, 80, 90, 98
multifractional Brownian motion, 502
multifractional Klein–Gordon field, 502
multifractional Ornstein–Uhlenbeck process, 502
multifractional Riesz–Bessel field, 502

natural science, 376
network, 377, 386
Newton, 211
Niels Bohr postulates, 415
non-ergodic subdiffusion, 21
non-Markovian Lévy flights, 43–46, 48
nonlinear anomalous mobility, 318
normal diffusion, 51, 53–55, 57, 58, 62–73

observable, 211–213
  time evolution of, 212
occupation time, 187, 195, 204
odd states, 400
one-sided Lévy distribution, 309
operational time, 273
operational time (vs. physical time), 230, 246–248, 251, 253–255, 260
order statistics, 87
origin, 379, 380, 382
Ornstein–Uhlenbeck process, 167, 177
overdamped Langevin equation, 299

Pareto distribution, 382
parity, 400
parity conservation law, 399
particle tracking, 280
partition function, 423, 424
Pascal principle, 93
path integral, 395, 403
phase space, 214, 215
phase space path integral, 404
plasma physics, 136
Poincaré's recurrence theorem, 215
Poisson process, 353, 354, 357, 359, 363, 364
pollution, 271
potential barrier, $\delta$, 422
potential well, $\delta$, 420
power-law indices, 387
power-law jumps, 274
power-law waiting times, 274
probability, 379, 383
probability distribution
    $\alpha$-stable, 395
probability measure, 214, 215
    induced, 215
    space, 214
propagator, 88
propylene carbonate, 222
$p$-variation, 341, 347

quantum
    comb model, 435
    dissipation, 441
    like ratchet, 441
    localization, 441
    map, 441
quantum mechanical fractional oscillator, 418
quantum mechanical kernel, 412
quantum mechanical oscillator, 419
quantum mechanical probability density, 401
quantum mechanics, 394, 395
quantum operation, 451
quantum Riesz fractional derivative, 396–399, 412
quantum-mechanical amplitude, 403
quark confinement, 417
quarkonium, 417, 418
quenched energy landscapes, 10

random, 383
random allometry coefficients, 380
random clock, 308
random fractal, 266, 270
random scatterers, 169
random walk, 26, 223, 225, 252, 254, 354, 365, 366
    directing, 230, 255
    leading, 230, 254, 255
    parent, 254, 255
    subordinated, 254, 255
reaction-anomalous-diffusion, 145
reaction-diffusion equation, 78, 172
reaction-diffusion systems, 132
reaction-subdiffusion equation, 90, 94, 96, 98
reactivity, 80, 94
    constant, 99
    exponential, 101
    piecewise constant, 99
    reactivity field, 90, 97
recurrence, 215, 216
recurrence time, 215, 216
recurrent character, 215
relaxation, 222
    Cole–Davidson, 222
    Debye, 222
    dielectric, 222
    Havriliak–Negami, 222
    Kohlrausch–Williams–Watts (KWW), 222
relaxation function, 54, 56, 57, 59, 65, 72
renewal process, 7, 179
renormalized topologically generated mass, 491, 492
repulsive, 497
repulsive Casimir force, 494
rescaling constants, 217
Riemann–Liouville derivative, 110
Riemann–Liouville fractional derivative, 43, 135, 293, 431
Riemann–Liouville fractional operator, 9

Riemann–Liouville operator, 310
Riesz fractional derivative, 111
Riesz representation theorem, 214
Riesz–Weyl derivative, 37
Riesz–Weyl space fractional operator, 9
Rydberg constant, 416

sample $p$-variation, 335, 341
sample MSD, 335, 339
sample paths, 254, 256–260
scale invariant solution, 410
scale transformation, 409
scale transformation group, 409
scaling, 378, 382, 384
scatter distribution, 24
Schrödinger equation, 396
second Einstein relation (linear response), 11, 22
self-adjoint operator, 399
self-interacting fractional Klein–Gordon field, 490
self-similar, 333
self-similar process, 36
self-similar scaling, 388
self-similarity, 236, 238, 252
semiclassical approximation, 418
semigroup, 213, 218
semi-Markov, 354
semi-Markov kernel, 365
semi-Markov process, 365, 366
semi-Markovian, 7
separation of variables, 87
shot noise representation, 280
single particle tracking, 20
single trajectory time averages, 20
Smoluchowski, 85
Smoluchowski–Fokker–Planck equation, 37, 41, 46–48
space-fractional diffusion equation, 269
space-time fractional diffusion equation, 273
Sparre Andersen scaling, 40, 45
Sparre Andersen universality, 19
spatial inversion, 400
stable
  density, 237, 239, 244, 246, 248, 253
  process, 248, 249, 251
state, 213
  mixed, 216, 220
  pure, 213, 214
statistical fractals, 379
statistical validation, 343
sticking phenomena, 10
stochastic, 382
Stone's theorem, 433
stretched exponential, 44, 87, 91, 129, 144
strong anticipation, 376, 379, 386, 389
subdiffusion, 5, 43, 44, 129, 296, 332
sub-diffusive, 155
sub-Ohmic, or fracton thermal bath, 312
subnetwork, 387, 388
subordination, 14, 17, 43, 47, 48, 166, 218, 334
  integral, 252, 253
  parametric, 252–254
subvelocity, 317
superdiffusion, 5, 129, 296, 332
super-diffusive, 155
super-diffusive propagation of fronts, 146
superslow diffusion, 109, 115
superstatistics, 174
survival probability, 17, 38, 39, 79, 80, 82–85, 87, 89, 90, 92, 93
  asymptotic, 92
susceptibility, 221, 223
  dielectric, 221, 223
  frequency dependent, 221, 223
symmetrized fractional derivative, 400
symmetry breaking, 492
system
  conservative, 214, 223
  dissipative, 223
  dynamical, 211

target problem, 80–82, 85, 97
Tauberian theorem, 86, 271
tempered Lévy processes, 136
tempered Lévy flights, 129
tempered stable, 281
time, 60, 72
  absolute, 211
  arrow, 213
  flow, 211
  recurrence, 215, 216
  waiting, 224
time averaged mean squared displacement, 20
time evolution, 212–214, 218
  adjoint, 214
  composite, 221
  fractional, 216, 218, 220
  induced, 215–217
  macroscopic, 216
  microscopic, 214, 216, 220
time flow, 211
time instant, 211
  initial, 212
time series, 20
time translation, 213
time-average, 185, 186
time-averaged position, 199, 200, 202, 204
time-independent fractional Schrödinger equation, 402, 413
tissue patterning, 98, 102
topological symmetry breaking, 490
translation, 210, 211, 213, 214, 216, 218
  left, 213
  right, 213, 218
trapping problem, 79
truncated Lévy flights, 129

universal asymptotical scaling, 315
universal fluctuations of anomalous mobility, 315
universality class of subdiffusion, 319
universality class of the CTRW-based subdiffusion, 327

variability, 381
variance, 138
velocity, 314
velocity correlation function, 290, 295
velocity operator, 401
velocity–velocity correlation function, 25
virial theorem, 415
viscoelastic force, 310
viscoelastic materials, 4
viscoelasticity, 310

waiting time, 7, 224
wave function, 396, 398, 402, 405, 410, 412, 413
weak ergodicity breaking, 315
Weyl quantization, 470, 471
Wiener process, 34, 36
Wigner function, 443
Wright function, 234, 235, 253

zero temperature Casimir energy, 501
zeta function regularization method, 489
zeta regularization, 497